Universitext

Springer
Berlin
Heidelberg
New York
Barcelona
Hong Kong
London
Milan
Paris
Tokyo

Egon Börger
Erich Grädel
Yuri Gurevich

The Classical Decision Problem

Springer

Egon Börger
Università di Pisa
Dipartimento di Informatica
Corso Italia, 40
56125 Pisa, Italy
e-mail: boerger@di.unipi.it

Erich Grädel
RWTH Aachen
Lehrgebiet Math. Grundlagen
der Informatik
52056 Aachen, Germany
e-mail:
graedel@informatik.rwth-aachen.de

Yuri Gurevich
Foundations of Software Engineering
Microsoft Research
One Microsoft Way
Redmond, WA 98052, USA
e-mail: gurevich@microsoft.com

Library of Congress Cataloging-in Publication Data applied for

Die Deutsche Bibliothek - CIP-Einheitsaufnahme
Boerger, Egon:
The classical decision problem / Egon Boerger ; Erich Grädel ; Yuri Gurevich. - 2. printing of the 1. ed.. - Berlin ; Heidelberg ; New York ; Barcelona ; Hong Kong ; London ; Milan ; Paris ; Tokyo : Springer, 2001
(Universitext)
ISBN 3-540-42324-9

Second Printing 2001 of the First Edition 1997, which was originally published in the series *Perspectives in Mathematical Logic*, ISBN 3-540-57073-X

Mathematics Subject Classification (2000):
0302, 03B10, 03B25, 03C13, 03D15, 03D35, 68Q25

ISBN 3-540-42324-9 Springer-Verlag Berlin Heidelberg New York

Springer-Verlag Berlin Heidelberg New York
a member of BertelsmannSpringer Science+Business Media GmbH

http://www.springer.de

Typesetting: Camera-ready copy produced by the authors using a Springer $\TeX$ macro package
Cover design: *design & production* GmbH, Heidelberg

Printed on acid-free paper SPIN: 10844969 46/3142LK - 5 4 3 2 1 0

Preface

This book is addressed to all those – logicians, computer scientists, mathematicians, philosophers of science as well as the students in all these disciplines – who may be interested in the development and current status of one of the major themes of mathematical logic in the twentieth century, namely the classical decision problem known also as Hilbert's *Entscheidungsproblem*. The text provides a comprehensive modern treatment of the subject, including complexity theoretic analysis.

We have made an effort to combine the features of a research monograph and a textbook. Only the basic knowledge of the language of first-order logic is required for understanding of the main parts of the book, and we use standard terminology. The chapters are written in such a way that various combinations of them can be used for introductory or advanced courses on undecidability, decidability and complexity of logical decision problems. This explains a few intended redundancies and repetitions in some of the chapters. The annotated bibliography, the historical remarks at the end of the chapters and the index allow the reader to use the text also for quick reference purposes.

The book is the result of an effort which went over a decade. Many people helped us in various ways: with English, with pictures and latex, with comments and information. It is a great pleasure to thank David Basin, Bertil Brandin, Martin Davis, Anatoli Degtyarev, Igor Durdanovic, Dieter Ebbinghaus, Ron Fagin, Christian Fermüller, Phokion Kolaitis, Alex Leitsch, Janos Makowsky, Karl Meinke, Jim Huggins, Silvia Mazzanti, Vladimir Orevkov, Martin Otto, Eric Rosen, Rosario Salomone, Wolfgang Thomas, Jurek Tyszkiewicz, Moshe Vardi, Stan Wainer and Suzanne Zeitman. This list is incomplete and we apologize to those whose names have been inadvertently omitted. We are specially thankful to Saharon Shelah for his help with the Shelah case and to Cyril Allauzen and Bruno Durand for providing an appendix with a new, simplified proof for the unsolvability of the domino problem. Also, we use this opportunity to thank Springer Verlag, the Omega group, the Heidelberg Academy of Sciences and in particular Gert Müller for the patience and belief in our longstanding promise to write this book.

August 1996

Egon Börger
Erich Grädel
Yuri Gurevich

Table of Contents

Part II. Decidable Classes and Their Complexity

1. Introduction: The Classical Decision Problem

1.1 The Original Problem

The original *classical decision problem* can be stated in several equivalent ways.

- The *satisfiability problem* (or the *consistency problem*) for first-order logic: given a first-order formula, decide if it is consistent.
- The *validity problem* for first-order logic: given a first-order formula, decide if it is valid.
- The *provability problem* for a sound and complete formal proof system for first-order logic: given a first-order formula, decide if it is provable in the system.

Recall that a formula is *satisfiable* (or *consistent*) if it has a model. It is *valid* (or *logically true*) if it holds in all models where it is defined. A proof system is *sound* if every provable formula is valid; it is *complete* if every valid formula is provable.

It was Hilbert who drew attention of mathematicians to the classical decision problem and made it into a central problem of mathematical logic. He called it *das Entscheidungsproblem*, literally "the decision problem". In the beginning of this century, he was developing the formalist programme for the foundations of mathematics (see [263, 264, 525]) and thus was interested in axiomatizing various branches of mathematics by means of finitely many first-order axioms. In principle, such an axiomatization reduces proving a mathematical statement to performing a mechanical derivation in a fixed formal logical system, see below. Obviously, the *Entscheidungsproblem* is very important in this context:

> *... stellt sich ... die Frage der Widerspruchsfreiheit als ein Problem der reinen Prädikaten-Logik dar ... Eine solche Frage ... fällt unter das "Entscheidungsproblem".*[1] [267, page 8]

Hilbert and Ackermann formulated a sound formal proof system for first-order logic and conjectured that the system is complete [266]. Later Gödel

[1] ... the question of consistency presents itself as a problem of the pure predicate logic ... Such a question ... falls under the "Entscheidungsproblem".

proved the completeness [184]. The proof is found in standard logic textbooks, e.g. [57, 142, 146, 307, 471]. For our purposes, the details of a formal system are of no importance. We will simply assume that some sound and complete formal proof system for first-order logic has been fixed. Notice that there is a mechanical procedure that derives all valid first-order formulae in some order.

To explain how proving a mathematical statement reduces to performing a mechanical derivation, assume that T is a finitely axiomatizable mathematical theory. Without loss of generality, the axioms have no free individual variables (that is, are sentences); indeed, if an axiom has free individual variables, replace it with its universal closure. Let α be the conjunction of the axioms, β another first-order sentence (a mathematical claim in the terminology of Hilbert), and γ the implication $\alpha \rightarrow \beta$. Then β is a theorem of T if and only if γ is valid if and only if γ is provable in the fixed formal proof system. Thus the mathematical question whether β is a theorem of T reduces to the logical question whether γ is valid which, in its turn, reduces to the question whether the mechanical procedure mentioned above derives γ.

Many important mathematical problems reduce to logic this way [266, 267]. Let us add another example.

Example. Reduction of the Riemann Hypothesis to the validity problem for some first-order sentence γ. Recall that a Diophantine equation is an equation $P(x_1, \ldots, x_k) = 0$ where P is a polynomial with integers coefficients and the variables x_i range over integers. In [98], the authors exhibit a Diophantine equation E that is solvable if and only if the Riemann Hypothesis fails. It suffices to find a finitely axiomatizable theory T and a sentence β such that β is provable in T if and only if E is solvable; the desired γ is then the implication $\alpha \rightarrow \beta$ where α is the conjunction of (the universal closures of) the axioms of T.

Recall that the standard arithmetic $\mathcal{A}$ is the set of natural numbers with distinguished element 0, the successor function, addition, multiplication and the order relation $\leq$. Let L be the first-order language of $\mathcal{A}$. Robinson's system Q is a finitely axiomatizable theory in L such that an arbitrary existential L-sentence φ is provable in Q if and only if it holds in $\mathcal{A}$ [307]. (A similar theory is called N in [471].)

Choose T to be Q. It suffices to construct an existential L-sentence β in such a way that E is solvable if and only if β holds in $\mathcal{A}$.

In fact, an arbitrary Diophantine equation D can be expressed by an existential formula β_D in such a way. Since a disjunction of existential sentences is equivalent to an existential sentence, it suffices to check that an existential L-sentence can express the given equation $P(x_1, \ldots, x_k) = 0$ together with an atomic constraint $x_i \geq 0$ or $x_i \leq 0$ for every variable x_i. But this is obvious. For example, an equation $x^3 - y^5 + 1 = 0$ with constraints $x \leq 0, y \leq 0$ is equivalent to an equation $(-x)^3 - (-y)^5 + 1 = 0$ with constraints $x \geq 0, y \geq 0$

which is equivalent to an equation $y^5 + 1 = x^3$ with constraints $x \geq 0, y \geq 0$ which is obviously expressible by an existential L-sentence.

The classical decision problem is called the main problem of mathematical logic by Hilbert and Ackermann:

> *Das Entscheidungsproblem ist gelöst, wenn man ein Verfahren kennt, das bei einem vorgelegten logischen Ausdruck durch endlich viele Operationen die Entscheidung über die Allgemeingültigkeit bzw. Erfüllbarkeit erlaubt. (...) Das Entscheidungsproblem muss als das Hauptproblem der mathematischen Logik bezeichnet werden.*[2] [266, pp 73ff]

Hilbert and Ackermann were not alone in their evaluation of the importance of the classical decision problem. Their attitude has been shared by other leading logicians of the time. Bernays and Schönfinkel wrote:

> *Das zentrale Problem der mathematischen Logik, welches auch mit den Fragen der Axiomatik im engsten Zusammenhang steht, ist das* Entscheidungsproblem.[3] [35].

Herbrand's paper [253] starts with:

> *We could consider the fundamental problem of mathematics to be the following. Problem A: What is the necessary and sufficient condition for a theorem to be true in a given theory having only a finite number of hypotheses?*

The paper ends with:

> *The solution of this problem would yield a general method in mathematics and would enable mathematical logic to play with respect to classical mathematics the role that analytic geometry plays with respect to ordinary geometry.*

In [254], Herbrand adds:

> *In a sense it* [the classical decision problem – BGG] *is the most general problem of mathematics.*

Ramsey wrote that his paper was

> *concerned with a special case of one of the leading problems in mathematical logic, the problem of finding a regular procedure to determine the truth or falsity of any given logical formula.* [435, p. 264]

[2] The Entscheidungsproblem is solved when we know a procedure that allows for any given logical expression to decide by finitely many operations its validity or satisfiability. (...) The Entscheidungsproblem must be considered the main problem of mathematical logic.

[3] The cental problem of mathematical logic, which is also most closely related to the questions of axiomatics, is the *Entscheidungsproblem*.

The roots of the classical decision problem can be traced while back. Philosophers were interested in a general problem-solving method. The medieval thinker Raimundus Lullus called such a method *ars magna*. Leibniz was the first to realize that a comprehensive and precise symbolic language (*characteristica universalis*) is a prerequisite for any general problem solving method. He thought about a calculus (*calculus ratiocinator*) to resolve mechanically questions formulated in the universal language. A universal symbolic language, restricted to mathematics, had to wait until 1879 when Frege published [171]; the language allowed Russel and Whitehead [446] to embed virtually the whole body of then known mathematics into a formal framework.[4] Leibniz distinguished between two different versions of *ars magna*. The first version, *ars inveniendi* , finds all true scientific statements. The other, *ars iudicandi* , allows one to decide whether any given scientific statement is true or not [255].

In the framework of first-order logic, an *ars inveniendi* exists: the collection of valid first-order formulae is recursively enumerable, hence there is an algorithm that lists all valid formulae. The classical decision problem can be viewed as the *ars iudicandi* problem in the first-order framework. It can be sharpened to a yes/no question: Does there exist an algorithm that decides the validity of any given first-order formula? Some logicians felt sceptical about ever finding such an algorithm. It wasn't clear, however, whether the scepticism could be justified by a theorem. John von Neumann wrote:

> *Es scheint also, daß es keinen Weg gibt, um das allgemeine Entscheidungskriterium dafür, ob eine gegebene Normalformel a beweisbar ist, aufzufinden. (Nachweisen können wir freilich gegenwärtig nichts. Es ist auch gar kein Anhaltspunkt dafür vorhanden, wie ein solcher Unentscheidbarkeitsbeweis zu führen wäre.) (...) Und die Unentscheidbarkeit ist sogar die Conditio sine qua non dafür, daß es überhaupt einen Sinn habe, mit den heutigen heuristischen Methoden Mathematik zu treiben. An dem Tage, an dem die Unentscheidbarkeit aufhörte, würde auch die Mathematik im heutigen Sinne aufhören zu existieren; an ihre Stelle würde eine absolut mechanische Vorschrift treten, mit deren Hilfe jedermann von jeder gegebenen Aussage entscheiden könnte, ob diese bewiesen werden kann oder nicht.*
> *Wir müssen uns also auf den Standpunkt stellen: Es ist allgemein unentscheidbar, ob eine gegebene Normalformel beweisbar ist oder nicht. Das einzige, was wir tun können, ist, (...), beliebig viele beweisbare Normalformeln aufzustellen. (...) Auf diese Art können wir von vielen Normalformeln feststellen, daß sie beweisbar sind. Aber auf diesem Weg kann uns niemals die Feststellung gelingen, daß eine Normalformel nicht beweisbar ist.* [5] [525, pp 11–12]

[4] See the forthcoming book by M. Davis [96] in this connection.

[5] It appears thus that there is no way of finding the general criterion for deciding whether or not a well-formed formula a is provable. (We cannot, however, at

Gödel's Incompleteness Theorem [185] was a breakthrough in logic. Can one use a similar method to prove the nonexistence of a decision algorithm for the classical decision problem? In an appendix to his paper "The fundamental problem of mathematical logic" Herbrand wrote:

> *Note finally that, although at present it seems unlikely that the decision problem can be solved, it has not yet been proved that it is impossible to do so.* [254]

Herbrand, Gödel and Kleene developed a very general notion of recursive functions [307]. In 1936, Church put forward a bold thesis: Every computable function from natural numbers to natural numbers is recursive in the sense of Herbrand-Gödel-Kleene. He showed that no recursive function could decide the validity of first-order sentences and concluded that that there was no decision algorithm for the classical decision problem [80].

Independently, Alan Turing introduced computing devices which are called now Turing machines. He put forward a similar thesis: a function from strings to strings is computable if and only if it is computable by a Turing machine [513]. He showed that no Turing machine could decide the validity of first-order sentences and also concluded that there is no decision algorithm for the classical decision problem. The equivalence of Church's and Turing's theses was quickly established. The Church-Turing thesis was largely accepted and thus it was accepted that the yes/no version of the classical decision problem was solved negatively by Church and Turing.

1.2 The Transformation of the Classical Decision Problem

By the time of Church's and Turing's theses, the area of the classical decision problem had already a rich and fruitful history. Numerous fragments of first-order logic were proved decidable for validity and numerous fragments were shown to be as hard as the whole problem. What does it mean that a fragment F is as hard for validity as the whole problem? This means that there exists

---

the moment demonstrate this. Indeed, we have no clue as to how such a proof of undecidability would go.) (...) The undecidability is even the *conditio sine qua non* for the contemporary practice of mathematics, using as it does heuristic methods, to make any sense. The very day on which the undecidability would cease to exist, so would mathematics as we now understand it; it would be replaced by an absolutely mechanical prescription, by means of which anyone could decide the provability or unprovability of any given sentence.

Thus we have to take the position; it is generally undecidable, whether a given well-formed formula is provable or not. The only thing we can do is (...) to construct an arbitrary number of provable formulae. In this way, we can establish for many well-formed formulae that they are provable. But in this way we never succeed to establish that a well-formed formula is not provable.

an algorithm A that transforms an arbitrary formula φ into a formula in F in such a way that $A(\varphi)$ is valid if and only if φ is so; such a fragment is called a *reduction class* for validity. Actually, it had been more common to speak about satisfiability and finite satisfiability, that is satisfiability in a finite structure. Reduction classes for satisfiability (respectively finite satisfiability) are defined similarly.

To convey a feeling of the field, let us quote some early results on fragments of pure first-order predicate logic (first-order logic without function symbols or equality). But first let us recall that a *prenex* formula is a formula with all its quantifiers up front. View a string in the four-letter alphabet $\{\forall, \exists, \forall^*, \exists^*\}$ as a regular expression denoting a collection of strings in the two-letter alphabet $\{\forall, \exists\}$. For example, $\forall^3\exists^*$ denotes the collection of strings of the form $\forall^3\exists^j$ where j is an arbitrary natural number, and $\exists^*\forall^2\exists^*$ denotes the collection of strings of the form $\exists^i\forall^2\exists^j$ where i and j are arbitrary natural numbers.

In 1915, Löwenheim [365] gave a decision procedure for the satisfiability of predicate formulae with only unary predicates. He proved also that formulae with binary predicates form a reduction class for satisfiability. In 1931, Herbrand [254] sharpened the latter result showing that just three binary predicates suffice. In 1936, Kalmár [295] showed that one binary predicate suffices.

In 1920, Skolem [477] showed that $\forall^*\exists^*$ sentences form a reduction class for satisfiability. In 1928, Bernays and Schönfinkel [35] gave a decision procedure for the satisfiability of $\exists^*\forall^*$ sentences. In 1928, Ackermann [16] gave a decision procedure for the satisfiability of $\exists^*\forall\exists^*$ sentences. Gödel [186], Kalmár [293] and Schütte [457], separately in 1932, 1933 and 1934 respectively, discovered decision procedures for the satisfiability of pure $\exists^*\forall^2\exists^*$ sentences. In another paper, Gödel proved that every satisfiable $\exists^*\forall^2\exists^*$ sentence has a finite model and that $\forall^3\exists^*$ sentences form a reduction class for satisfiability [187]. (See [234] for a popular introduction to the classical decision problem.)

The reaction of the logicians to the discoveries of Church and Turing was that the classical decision problem was wider than the yes/no version of it. Here is one of the earliest reactions:

> *Solche Reduktionen des Entscheidungsproblems werden hoffentlich vorteilhaft sein für systematische Untersuchungen über die Zählausdrücke, z.B. wenn man versuchen will eine Übersicht zu bekommen, für welche Klassen von solchen man das Entscheidungsproblem wirklich lösen kann. Bekanntlich hat A. Church bewiesen, dass eine allgemeine Lösung dieses Problems nicht möglich ist.*[6] [482]

[6] Such reductions [a reference to the reductions proposed by Skolem in the paper cited — BGG] will hopefully be advantageous for systematic investigations of first-order formulae, for example if one would like to try to arrive at a complete picture, for which classes of such formulae one can really solve the Entschei-

The logicians started to think about the classical decision problem as a classification problem.

- Which fragments are decidable for satisfiability and which are undecidable?
- Which fragments are decidable for finite satisfiability and which are undecidable?
- Which fragments have the finite model property and which contain axioms of infinity (that is satisfiable formulae without finite models)?

For a long time the classical decision problem remained a central problem of mathematical logic. With the development of computational complexity theory, the problem has been refined. If a fragment of first-order logic is decidable for satisfiability, then indeed there is an absolutely mechanical procedure, that is an algorithm, for deciding the satisfiability or unsatisfiability of any given sentence. But what is the computational complexity of determining satisfiability? Similarly, if a given fragment is decidable for finite satisfiability, what is the computational complexity of determining finite satisfiability?

Of course, the unrestricted classifiability problem is hopeless. There are just too many fragments. Some of them are of no interest to anybody. Some of them involve particular branches of mathematics. Consider for example the satisfiability problem for sentences $\alpha \wedge \beta$ where α is (the universal closure of) the conjunction of the axioms of fields and β is an arbitrary formula in the vocabulary of fields; this problem rightfully belongs to field theory rather than logic.

Eventually, the classical decision problem became to mean the restriction of the classification problem described above to traditional fragments. This description is admittedly not precise but it gives a good guidance which we will follow. One can argue that the complexity issue does not really belong to the traditional classical decision problem. This is true too, but it is impossible to ignore the complexity issue these days, in particular because of the relevance of the logical decision procedures to theorem proving and model checking methods. We will try to cover the known complexity results.

As we have mentioned above, for a long time the classical decision problem remained a central problem of mathematical logic. The literature on the subject is huge and contains a great wealth of material. The classical decision problem served as a laboratory for various logic methods[7] and especially reduction methods. The classification results have been used not only in logic but also in theoretical computer science. In particular, they have been used as a guide to the study of zero-one laws for fragments of second-order logic. Classical techniques inspired some proofs on the zero-one laws and some of classical techniques have been further extended. See [235, 313, 314, 315, 414, 415] in this connection.

dungsproblem. As it is known, A. Church has proved that a general solution of this problem is not possible.

[7] By the way, Ramsey proved his famous combinatorial lemma in a paper on the classical decision problem [435].

There is a number of books devoted to the classical decision problem. In the 1950s, Ackermann gave a comprehensive treatment of the solvable cases known at the time [18], and Surányi gave a complementary comprehensive treatment of reduction classes known at the time [498]. The book [133] of Dreben and Goldfarb illustrates the potential of the so-called Herbrand expansion technique in establishing solvability. The complementary book [351] of Lewis covers many reduction results on classical fragments of pure predicate logic. Together the two books give a systematic treatment of decision problems for predicate logic without functions or equality.

Nevertheless, much of the wealth has never appeared in a book form. Moreover, by now, the work on the classical decision problem is by and large completed (though some open problems remain of course) and most of the major classifications have not been ever covered in book form. That is exactly what we intend to do in this book.

1.3 What Is and What Isn't in this Book

We give most attention to the most traditional fragments of first-order logic, namely, to collections of prenex formulae given by restrictions on the quantifier prefix and/or vocabulary. (Recall that there is a simple algorithm for transforming an arbitrary first-order formula to an equivalent one in the prenex form.)

Strings in the two-letter alphabet $\{\forall, \exists\}$ will be called *prefixes*. A *prefix set* is a set of prefixes. An *arity sequence* is a function p from the set of positive integers to the set of non-negative integers augmented with the first infinite ordinal ω.

Definition 1.3.1 (Prefix-Vocabulary Classes). For any prefix set Π and any arity sequences p and f, $[\Pi, p, f]$ (respectively, $[\Pi, p, f]_=$) is the collection of all prenex formulae φ of first-order logic without equality (respectively with equality) such that

- the prefix of φ belongs to Π,
- the number of n-ary predicate symbols in φ is $\leq p(n)$, and
- the number of n-ary function symbols in φ is $\leq f(n)$.
- φ has no nullary predicate symbols with the exception of the logic constants *true* and *false*, no nullary function symbols and no free variables.

Let us explain the last clause. We will speak about logic without equality but the same applies to logic with equality. It is easy to see that the status (decidable or undecidable) of the (finite) satisfiability problem for a prefix-vocabulary class does not change if nullary predicate symbols are allowed. Now let us consider the rôle of nullary function symbols, that is individual constants. Let $C = [\Pi, p, f]$ and C' be the version of C when one is allowed to use say 7 individual constants. It is easy to see that the status of the (finite) satisfiability problem for C' is that of the (finite) satisfiability problem for $[\Pi', p, f]$ where $\Pi' = \{\exists^7\pi : \pi \in \Pi\}$. Instead of individual constants, we could speak about free individual variables. Thus allowing individual constants or free individual variables does not give us more classes either.

The definition of prefix-vocabulary classes above seems to be excessively general. Call a prefix set *closed* if it contains all substrings (even not contiguous substrings) of its prefixes. Clearly, one can restrict attention to closed prefix sets. Further, call a prefix set Π *standard* if either it is the set of all prefixes or else it can be given by a string w in the four-letter alphabet $\{\forall, \exists, \forall^*, \exists^*\}$. In the first case Π is denoted *all*. Thus, every standard prefix set has a succinct notation. Furthermore, we can require without loss of generality that w is *reduced* in the following sense: $\forall^*$ cannot have $\forall$ as a neighbor, and similarly $\exists^*$ cannot have $\exists$ as a neighbor. For example, a string $\forall^*\forall\exists\exists^*$ reduces to $\forall^*\exists^*$; clearly the two strings define the same prefix set.

Call an arity sequence p *standard* if it satisfies the following condition: $p(n) = \omega$ whenever the sum $p(n) + p(n+1) + \cdots$ is infinite. Every standard sequence can be given a succinct notation. The standard arity sequence that assigns ω to each n will be denoted *all*. Any other standard sequence p has a tail of zeroes, $0 = p(m) = p(m+1) = \cdots$, and will be denoted by the sequence $(p(1), p(2), \ldots, p(m-1))$. In case $m = 1$, for readability, we denote p with (0) rather than $()$. Similar notation can be used for non-standard sequences with a tail of zeroes. Notice that every arity sequence reduces (in a sense made more precise in Sect. 2.3) to a standard arity sequence. For example, $[all, (0, \omega), (0)] \subseteq [all, (\omega, \omega), (0)]$ and every sentence $\varphi \in [all, (\omega, \omega), (0)]$ can be easily rewritten as an equivalent sentence in $[all, (0, \omega), (0)]$: just replace formulae $R(x)$ with formulae $R'(x, x)$ where R' is a binary predicate symbol that does not occur in φ.

Definition 1.3.2. A prefix-vocabulary class $[\Pi, p, f]$ or $[\Pi, p, f]_=$ is *standard* if Π, p and f are standard.

The classification problem for the prefix-vocabulary fragments admits a complete solution in a form of a finite table. In particular, there are only finitely many minimal undecidable fragments with closed prefix sets, and all these minimal fragments are standard. This follows from the Classifiability Theorem of Gurevich proved in Sect. 2.3. Accordingly, in the main body of the book, the prefix-vocabulary classes of interest will be almost exclusively standard classes. The Classifiability Theorem has provided guidance for research and it provides guidance for this book.

Let us review briefly the contents of the book. The main part of Chapter 2 is devoted to the reduction theory which we explain from scratch and develop to a certain depth. The reduction theory helps us to give simpler proofs and proper lower complexity bounds. The rest of Chapter 2 is devoted to the Classifiability Theorem.

In Chapters 3 and 4, we give a complete treatment of the undecidable prefix-vocabulary fragments of first-order logic (with or without function symbols, with or without equality). In Chapter 5, we present various other undecidable fragments mainly defined in terms of additional restrictions on the propositional structure of the formulae; we study in particular Krom and Horn formulae which have played an important rôle in the theory of logic programming.

In Chapters 6 and 7, we treat the decidable prefix-vocabulary fragments of first-order logic (with or without function symbols, with or without equality). Together with the results of Chapters 3 and 4 this gives a complete classification of the decidable and undecidable prefix-vocabulary classes. Tables 1.1 and 1.2 summarize the decidability/undecidability results on prefix-vocabulary fragments.

Undecidable Cases

A: Pure predicate logic (without functions, without =)

(1)	$[\forall\exists\forall, (\omega, 1), (0)]$	(Kahr 1962)
(2)	$[\forall^3\exists, (\omega, 1), (0)]$	(Surányi 1959)
(3)	$[\forall^*\exists, (0, 1), (0)]$	(Kalmár-Surányi 1950)
(4)	$[\forall\exists\forall^*, (0, 1), (0)]$	(Denton 1963)
(5)	$[\forall\exists\forall\exists^*, (0, 1), (0)]$	(Gurevich 1966)
(6)	$[\forall^3\exists^*, (0, 1), (0)]$	(Kalmár-Surányi 1947)
(7)	$[\forall\exists^*\forall, (0, 1), (0)]$	(Kostyrko-Genenz 1964)
(8)	$[\exists^*\forall\exists\forall, (0, 1), (0)]$	(Surányi 1959)
(9)	$[\exists^*\forall^3\exists, (0, 1), (0)]$	(Surányi 1959)

B: Classes with functions or equality

(10)	$[\forall, (0), (2)]_=$	(Gurevich 1976)
(11)	$[\forall, (0), (0, 1)]_=$	(Gurevich 1976)
(12)	$[\forall^2, (0, 1), (1)]$	(Gurevich 1969)
(13)	$[\forall^2, (1), (0, 1)]$	(Gurevich 1969)
(14)	$[\forall^2\exists, (\omega, 1), (0)]_=$	(Goldfarb 1984)
(15)	$[\exists^*\forall^2\exists, (0, 1), (0)]_=$	(Goldfarb 1984)
(16)	$[\forall^2\exists^*, (0, 1), (0)]_=$	(Goldfarb 1984)

Table 1.1. Minimal Undecidable Standard Classes

Decidable Cases

A: Classes with the finite model property

(1)	$[\exists^*\forall^*, all, (0)]_=$	(Ramsey 1930)
(2)	$[\exists^*\forall^2\exists^*, all, (0)]$	(Gödel 1932, Kalmár 1933, Schütte 1934)
(3)	$[all, (\omega), (\omega)]$	(Löb 1967, Gurevich 1969)
(4)	$[\exists^*\forall\exists^*, all, all]$	(Maslov-Orevkov 1972, Gurevich 1973)
(5)	$[\exists^*, all, all]_=$	(Gurevich 1976)

B: Classes with infinity axioms

(6)	$[all, (\omega), (1)]_=$	(Rabin 1969)
(7)	$[\exists^*\forall\exists^*, all, (1)]_=$	(Shelah 1977)

Table 1.2. Maximal Decidable Standard Classes

We give also a fairly complete complexity analysis of the decidable cases. One open problem is to find the exact complexities of the satisfiability and finite satisfiability problems for the Shelah class. For most of the maximal decidable standard fragments, the satisfiability problem has a very high computational complexity, typically deterministic or nondeterministic exponential time, the complexity is even non-elementary in the case of the Rabin class. At the end of Chapter 6 we also present a classification of the standard classes that have the finite model property and of those having infinity axioms. The decidability results in Chapter 7 rely (in our exposition) on a reduction to S2S, the monadic second-order theory of the infinite binary tree. The decidability of S2S, proved by Rabin [430], is one of the most important and difficult decidability theorems for mathematical theories. We give a complete proof of this result in Sect. 7.1. In Chapter 8 we present some other decidable cases of the decision problem. In addition, the book contains a quite extensive annotated bibliography and an appendix, written by Cyril Allauzen and Bruno Durand, containing a new simplified proof for the unsolvability of the unconstrained domino problem which is used at many places in this book.

Some classifications appear for the first time in a book: For example, the classifications of prefix-vocabulary fragments in the cases of logic with equality, functions or both. All complexity results appear for the first time in a book. There are many new proofs, e.g. those (assisted by Shelah) related to the Shelah class. There are also many new results.

On the other hand, there are many closely related topics that we do not cover in this book. We are concerned here with fragments of first-order logic and do not deal with decision problems for second-order logic, higher-order logic, intuitionistic logic (see [385, 412]), linear logic (see the forthcoming book [362]) or any other logic. We do not deal with decision problems for mathematical theories formalized in first-order or any other logic; in this connection see [89, 97, 148, 166, 231, 432, 506].

Furthermore, even though the classical decision problem is more or less finished in its most classical form, there are various other natural versions and extensions of it that we do not deal with here systematically. For example, we do not deal with classifications based on the resolution calculus; in this connection see [163, 340]. But we do discuss various extensions of the classical decision problem and various open problems on our way. Let us mention some extensions and open problems here.

Extend the classifiability theorem in various directions. This is very important; without a proper direction, it is hard even to remember a myriad of specific results.

Extend the prefix-vocabulary classification to important undecidable mathematical theories; see [229] in this connection. Find the computational complexity of decidable prefix-vocabulary classes of important mathematical theories (see [201, 206]); in many cases even the computational complexity of the theory itself is unknown. It would also be interesting to extend the classification to different logics.

We were interested whether a given fragment contains a formula without finite models. Does a given fragment contain a formula without recursive models? This direction is still covered by the Classifiability Theorem; in particular there are finitely many minimal prefix-vocabulary classes with formulae without recursive models and each of them is standard. Instead of recursivity, one can speak about other kinds of descriptive or computational complexity. Similarly, does a given fragment contain an axiom of an essentially undecidable theory? Since the fragment may be not closed under conjunction, it is meaningful to ask if the fragment includes a finite set of sentences that form an axiomatization of an essentially undecidable theory. Also, one may restrict attention to infinite models of certain complexity: primitive recursive models, recursive models, models of such and such Turing degrees, Borel models, etc.

In cases of fragments of reasonably low complexity bound, develop practical solutions of the decision problem. This problem is well recognized as a major bottleneck for *e.g.* model checking [70], an important current method for computer verification of hardware and software correctness claims.

One extension of the classical decision problem is related to the strictness of reductions. If one cares only about satisfiability, it suffices to require that a reduction transforms a given formula α into a formula α' which is satisfiable if and only if α is so. We usually care about satisfiability and finite satisfiability and thus consider so called *conservative reductions* when it is required that (i) α' is satisfiable if and only if α is satisfiable, and (ii) α' is finitely satisfiable if and only if α is finitely satisfiable. One may be interested in even stricter reductions. For example, one may require that α and α' have the same spectra or – more generally – that there is a simple connection between the spectra. (On several occasions, Surányi insisted that there should exist a general method that transforms a given model of α' to a model of α.) On the other hand, one may consider not only recursive but also arithmetical, Borel, etc. transformations.

There are many more specific problems. One is to examine Boolean combinations of prefix-vocabulary classes; see Section 5.4 in this connection.

The book is addressed to a wide audience and not only to professional logicians. There are scattered remarks and exercises addressing more special audiences (logicians, people familiar with logic programming, etc.) but the main body requires only the familiarity with basic notions of mathematical logic. (This does not mean of course that all parts are easy to read; some proofs are quite involved even after much simplification). Finally, let us note that sometimes we will omit the adjective "first-order"; formulae, languages and theories are by default first-order in this book.

Part I

Undecidable Classes

2. Reductions

This chapter starts from scratch, presupposing only basic notions from predicate logic and complexity theory. We introduce a simple but general form of the basic reduction technique which will be used throughout the book. We show characteristic examples for the variety of applications of the reduction technique to decision problems in predicate logic and complexity theory, thus providing the ground and motivation for the systematic treatment of those questions in the following chapters. In doing this we also fix the terminology which will be used throughout the book. The reader who has already some knowledge about logical decision problems may go directly to subsequent chapters and come back here only as needed. We start with undecidability results. First we prove the theorem of Church and Turing on the algorithmic unsolvability of Hilbert's *Entscheidungsproblem* and Trakhtenbrot's variant of that theorem for finite satisfiability. We prove both theorems by a reduction from the halting problem for Turing machines.

We define the fundamental notion and exhibit simple examples of first-order *reduction classes* for satisfiability and finite satisfiability. By showing that a class K of formulae is a reduction class for satisfiability, we really prove that the satisfiability problem for K is complete for the co-r.e. sets under recursive reductions. Similarly, if K is a reduction class for finite satisfiability, then the finite satisfiability problem for K is complete for the r.e. sets. Reduction classes will concern us for much of the book; their introduction here serves also the purpose to illustrate the crucial use of canonical models (often also called *Herbrand* models) first brought into this area by Büchi. In this introductory chapter we mainly rely upon refinements of Turing's method to express machines by logical formulae. These formulae essentially define by logical means what in computer science is called the *semantics of machine programs*. This allows us to give a uniform treatment of classical questions about *simple* formulae – namely those which define the semantics of programs – which in a natural way yields *undecidability results* related to finitely axiomatizable first-order theories, to models of satisfiable formulae, to interpolation procedures, and to explicit definitions (even over finite domains) of implicitly defined predicates.

We exhibit analogous impossibility or completeness results in the realm of the decidable. They can be obtained by applying the reduction technique to space or time bounded machines. The descriptions of such machines typically need only limited logical expressive means. Our first example is the analogue of Hilbert's *Entscheidungsproblem* for *propositional logic* which turns out to be the fundamental problem in complexity theory, namely the problem whether P = NP. The theorem of Cook and Levin that the satisfiability problem for propositional logic is NP-complete [91, 343] is proved by applying the reduction technique to propositional descriptions of non-deterministic Turing machine computations with polynomial time bounds.

Descriptions of polynomial space bounded Turing machine computations by formulae of quantified propositional logic yields the result of Stockmeyer that the decision problem for quantified Boolean formulae is PSPACE-complete.

Another interesting example arises from the *spectrum problem.* The spectrum of a sentence is the class of cardinalities of its finite models. The spectrum problems is to characterize the class of spectra of first-order sentences, and more generally, of the class of spectra of (the strong) logic of order n for arbitrary n. The reduction technique yields a characterization of NP in terms of generalized spectra (Fagin's Theorem); more generally it yields a characterization of the spectra of strong n-th order sentences as those sets which are accepted by non-deterministic Turing machines in n-fold exponential time. Fagin's Theorem has inspired similar logical characterizations of other complexity classes; we will explain a number of such results in Sect. 2.2.3.

We proceed to illustrate the effect of some natural syntactic restrictions on the complexity of logical decision problems; in particular we consider prenex formulae with restrictions on the prefix, on the arity and the number of predicate or function symbols, and on the propositional structure of first-order formulae. We use the reduction technique to exhibit two formulae classes whose satisfiability problems are complete for PSPACE and EXPTIME respectively, i.e. complete for the sets accepted by deterministic Turing machines with polynomial space or exponential time bounds respectively.

We conclude the chapter with Gurevich's Classifiability Theorem that justifies abstractly the study of decision problems for prenex formulas with restrictions on the prefix, on the number and the arities of predicate symbols, and on the number and the arities function symbols.

2.1 Undecidability and Conservative Reduction

2.1.1 The Church-Turing Theorem and Reduction Classes

It is a well known feature of first-order logic that it is rich enough to express large parts of mathematics, and in particular a great variety of problems that occur naturally in dealing with algorithms and computers. This includes many problems which are known to be algorithmically unsolvable. Effective reductions of such undecidable problems to the decision problem of first-order logic are the common feature of the many different proofs of the unsolvability of Hilbert's *Entscheidungsproblem.*

The essential ingredients of such an undecidability proof are as follows. Given an instance p of a known undecidable problem P, one constructs a *reduction formula* α_p and proves that p has a positive solution iff α_p is logically valid (or satisfiable, etc.). Thus, the undecidability of the given decision problem P yields the undecidability of the validity (or satisfiability, etc.) problem for the class of all reduction formulae α_p.

We follow here the approach of Turing [513] who proved the undecidability of the *Entscheidungsproblem* by reducing to it an unsolvable problem for Turing machine programs. We provide economical descriptions of non-deterministic Turing and register machines variants of which will be used throughout the book. This will give us also an occasion to present a simple but fundamental observation, made by Büchi [64], how canonical models (also called Herbrand models) can be used to relate in a simple and transparent way combinatorial problems to logical decision problems.

Economical Description of Turing Machines. In order to show the uniformity of the procedure and for future use, we start by formulating a general reduction scheme for non-deterministic Turing machine programs. The specialization of this scheme for the halting problem of deterministic Turing machines will then yield a proof for the following fundamental result.

Theorem 2.1.1 (Church, Turing). *The* Entscheidungsproblem *is algorithmically unsolvable.*

Later other specializations of the same scheme will also be given. For a description of Turing machine computations, configurations (instantaneous descriptions) C are expressed by logical statements $\underline{C}$; each single computation step by which the given machine program M produces a successive configuration C' from a given C, is described as logical derivation of $\underline{C'}$ from $\underline{C}$ using a program formula STEP_M:

$$\underline{C} \wedge \mathrm{STEP}_M \vdash \underline{C'}.$$

Initial formulae START_C and *end formulae* $\mathrm{END}_{C'}$ will specify the initial and final configurations C, C'. (We will often omit C, C' when they are clear from the context.) Thus the problem of whether M, started in configuration C, will eventually reach C', will be reflected directly as a logical decision problem, namely whether the formula $\mathrm{END}_{C'}$ is a logical consequence of START_C and STEP_M.

Let M be a (possibly non-deterministic) program for a one-tape Turing machine (which we assume without loss of generality to be one-way infinite and with cells numbered in the natural way by 0,1,2,...).

We describe STEP_M as a scheme in the *basic formulae* $H(t,x)$, $T_j(t,x)$, $I_i(t)$, $S(t,t')$. These basic formulae are left unspecified here; for the time being the reader might think of them as atomic formulae. Depending on the case where we want to apply the scheme, these basic formulae will be interpreted as first order atomic formulae, as propositional variables, as atomic formulae of order n, etc. The basic formulae come with parameters t (for time), x (for tape cells, usually identified with their order numbers), j (for letters, usually identified with number indices), i (for states of the machine, usually identified with numbers) and the following *intended interpretation* over the natural numbers:

- $H(t, x)$ iff the head position at time t is cell x;
- $T_j(t, x)$ iff the tape at time t contains letter j in cell x;
- $I_i(t)$ iff at time t the machine is in state i;
- $S(t, t')$ iff the successor of t is t'.

For a configuration C_t of M at time t let $\underline{C_t}$ be a formula that describes C_t in terms of appropriate basic formulae.

The reader might think for example of the conjunction of basic formulae for all parameters needed to describe C_t. This can be a finite or an infinite conjunction, depending on whether computations with finite or potentially infinite time or space resources (finitely or infinitely many t, t', x) are considered; in other cases the parameters t, t', x might be quantified individual variables; we will see these and other examples below. The program formula scheme STEP_M defined below satisfies the following

Lemma 2.1.2 (Simulation Lemma). *Let C_0 be an arbitrary initial configuration of M and $\mathfrak{A}$ a model of $\mathrm{STEP}_M \wedge \underline{C_0}$. Let t be arbitrary. Assume that there is an M-computation of length t which starts at C_0 and that there are corresponding S-chains of length t (of time and cell parameters) in $\mathfrak{A}$. Then there is at least one M-configuration C_t, reachable by M in t steps from C_0 along the given S-chains, for which $\mathfrak{A}$ is a model of $\underline{C_t}$.*

Definition of STEP_M. Let (the program of) M consist of instructions (i, j, k, p, m) with the following meaning: being in internal state i and reading letter j, print letter k, move the reading head one cell to the right, left or not at all (for $p = 1, -1, 0$ respectively) and go to state m. Without loss of generality we assume that, for each pair (i, j), all instructions in M that begin with (i, j) have the same value of p. STEP_M is defined as the conjunction of the following formulae for all parameters i, m (for states), j, k (for letters) and all t, t', x, x', y concerned according to the underlying time and space-bounds:

- right movement instructions:

$$I_i(t) \wedge H(t, x) \wedge T_j(t, x) \wedge S(t, t') \wedge S(x, x') \rightarrow \bigvee_{(i,j,k,+1,m)\in M} I_m(t') \wedge H(t', x') \wedge T_k(t', x) \tag{2.1}$$

- instructions without movement: the same with $H(t', x)$ in the conclusion

– left movement instructions:

$$I_i(t) \wedge H(t,x') \wedge T_j(t,x') \wedge S(t,t') \wedge S(x,x') \rightarrow \bigvee_{(i,j,k,-1,m)\in M} I_m(t') \wedge H(t',x) \wedge T_k(t',x') \qquad (2.2)$$

– no change outside the head cell:

$$H(t,x) \wedge y \neq x \wedge T_j(t,y) \wedge S(t,t') \rightarrow T_j(t',y) \qquad (2.3)$$

Note that this conjunction is finite or infinite, depending on whether the time and space resources are finite or infinite. Usually we treat halting states i of M as states with idle instructions $(i,j,j,0,i)$ thus allowing infinite configuration sequences which become constant when a halting state is encountered; however, sometimes it is convenient to consider halting states as states without any instructions $(i,\ldots)$ whatsoever.

Exercise 2.1.3. Prove the Simulation Lemma by induction on t.

Remark. Note that the Simulation Lemma is really a scheme that receives a concrete meaning once a particular logic is specified. For the reader who knows other proofs in the literature, our formulae might look frugal: they describe only what is needed for a machine transition to be reflected in a model of the formula.

There are no conditions that, at any moment, the machine is in only one internal state with only one reading head and only one symbol written in a given cell. The preceding exercise should help the reader to convince himself/herself that these frugal formulae are indeed sufficient for our purposes.

Since the reduction formulae are composed from loosely specified basic formulae, they can be used for different logics, corresponding to different definitions of the basic formulae and different compositions of the latter into the reduction formulae: as (quantified or unquantified) propositional formulae, as first- or higher-order formulae, etc. Using the Simulation Lemma, we now prove the Church-Turing Theorem by specifying the basic formulae as atomic first-order formulae and by composing them using quantifiers over time and space parameters. It suffices to specify the preceding scheme for the following variant of the *halting problem* for deterministic Turing machines with say 1 as a unique halting state:

H_1 = {M: Starting in state 0 with the head at the left end of the empty tape, M halts in state 1.}

It is well known that the set H_1 is recursively enumerable but not recursive. For such machines we can indeed prove the following equivalence.

Reduction Property: Let ρ_M be the universal closure of

$$\text{STEP}_M \wedge \text{START} \wedge \text{NONSTOP} \wedge \text{ORDER}$$

with START, NONSTOP and ORDER as defined below. Then M, started in state 0 with head positioned at the left end of the empty tape, eventually halts in state 1 if and only if ρ_M is contradictory.

Since the formulae ρ_M can be effectively constructed from the programs M, the undecidability of the halting problem implies the undecidability of the non-satisfiability problem (and therefore of the satisfiability problem) for the class of all reduction formulae ρ_M. This implies the undecidability of the validity problem for formulae $\neg\rho_m$ and therefore *a fortiori* for the class of all first-order formulae. This proves the Church-Turing Theorem.

Let ORDER be a formula that axiomatizes a total order K with successor relation S, intended to be interpreted over the natural numbers in the usual way. ORDER is the conjunction of:

$$\begin{aligned}
&\forall x \neg Kxx\\
&\forall x \forall y \forall z (Kxy \wedge Kyz \rightarrow Kxz)\\
&\forall x \forall y (Kxy \vee x = y \vee Kyx)\\
&\forall x \forall y (Sxy \leftrightarrow (Kxy \wedge \neg\exists z (Kxz \wedge Kzy))).
\end{aligned}$$

Let the order for this proof also satisfy $\forall x \exists y Sxy$, whereby it becomes infinite. Let 0 be an individual constant, intended to be interpreted as number 0. Let I_i be monadic and H, T_j, S binary predicate symbols, let t, x be individual variables. The variables will be universally quantified; intuitively they range over the natural numbers. Define the initial formula by

$$\text{START} := I_0(0) \wedge H(0,0) \wedge \forall x T_0(0,x)$$

expressing this way that the machine starts at internal state 0 with the head at the left end of the empty tape.

Define the end formula by

$$\text{NONSTOP} := \forall t \neg I_1(t)$$

expressing that at no time will the halting state 1 be reached.

It is now routine to check that the indicated intended interpretation satisfies ρ_M if M, started at state 0 with head positioned at the left end of the empty tape, does not halt at state 1. If ρ_M has a model then, by the definition of ORDER and the assumption that 1 is the only halting state, we can conclude from the Simulation Lemma that M does not halt. Note that the additional assumption for S made in this proof guarantees the existence of sufficiently long (indeed infinite) S-chains in models of the reduction formulae.

We will use variants of this basic reduction scheme throughout the book. The reduction property shows that unsolvability of the decision problem is obtained not only for the class of all formulae, but for the subclass of all reduction formulae ρ_M. Here is a different example, motivated by semigroup theory.

Exercise 2.1.4. [371] A *Thue process* often also called Thue system is a semigroup T with a finite number of generators g_i ("letters" representable as individual constants) and a finite number of identities $V_j = W_j$ ("defining relations ") imposed on concatenations of generators ("words"). Reduce the word problem for Thue systems – the question whether $V = W$ holds for a given Thue system T and given words V, W of T – to the satisfiability problem for first order logic with equality as the only predicate symbol and with terms built up from individual constants and one binary function symbol. By the unsolvability of the word problem for Thue systems, this gives another proof for the Theorem of Church and Turing.

Exercise 2.1.5. Derive from the previous exercise the undecidability of the first order theory of semigroups. This is sharpened in [372] to the $\forall\exists$-positive theory of a free semigroup.

Reduction Classes. This brings us to a fundamental feature of such reduction proofs, which will be the object of extensive study in this book, namely that the combinatorial structure of the reduced decision problem is reflected in the syntactical structure of the class of reduction formulae. To obtain undecidable classes of formulae of simple syntactical structure, we will have to look for sophisticated logical encodings of undecidable decision problems with "simple" combinatorial structure. The ultimate goal is to classify undecidable classes of "minimal" syntactic structure, i.e. such that further restriction yields a decidable class of formulae. In our proofs we will pay particular attention to make it transparent how reduction formulae encode the "data structure" – the basic objects and the basic transformations – of the reduced computational problem. In doing this we will also obtain natural and explicit ways to "look at" or "interpret" logical formulae of a great variety of classes in a procedural manner as (describing) "algorithms" or "programs"; thus the reduction formulae will represent a logical operational definition of the semantics of those programs. This relates the study of logical decision problems to the study of expressiveness and complexity of program classes in logic programming languages (one representative of which is Prolog where the procedural interpretation of Horn formulae has received some practical significance through efficient implementations).

The above proof shows more than the mere undecidability of the class of reduction formulae. As a matter of fact, the halting problem for Turing machines used above is complete for the recursively enumerable sets. This means that one can find for each r.e. set $Y \subseteq \mathbb{N}$ a recursive function f

which associates to each number n a Turing machine program $f(n)$ such that the following equivalence holds: $n \in Y$ iff $f(n)$, started in state 0 with head positioned at the left end of the empty tape, eventually halts in state 1. Since the reduction procedure associating $\neg\rho_M$ to M is also a recursive function and since, by Gödel's Completeness Theorem, the valid first-order formulae form a recursively enumerable set, it follows that the validity problem for the formulae $\neg\rho_{f(n)}$ is not only recursively unsolvable but in fact complete for the r.e. sets.

In the context of the *Entscheidungsproblem*, this completeness phenomenon has played an eminent rôle. The *Entscheidungsproblem* (with respect to validity) can be reduced to the validity problem for the class of formulae $\neg\rho_M$ by a recursive function f which associates with each first order formula β a formula $f(\beta)$ of form $\neg\rho_{M(\beta)}$ in such a way that β is valid iff $f(\beta)$ is valid. The class X of our reduction formulae $\neg\rho_M$ incorporates therefore the whole *Entscheidungsproblem* in the sense that the validity problem for first-order logic can be effectively (many-one) reduced to the validity problem for X. Such a class of formulae is called a *reduction class* with respect to validity.

The notion of reduction class was a central notion for the attempts to settle the *Entscheidungsproblem*: the search for algorithms to solve the decision problem for large classes of formulae was complemented by investigations to reduce the *Entscheidungsproblem* to the decision problem for small reduction classes, with the hope that the two efforts would match. From the unsolvability of the *Entscheidungsproblem* it obviously follows that each reduction class has an unsolvable validity problem (which is in fact hard for the r.e. sets). After the negative solution of the *Entscheidungsproblem*, the interest in reduction classes has been mainly motivated by a classification problem: are there natural classifications of classes of first order formulae into decidable and undecidable ones and which of the latter are of maximal computational complexity (i.e. reduction classes)? Can the borderline between decidability and undecidability be drawn sharply and on the basis of which logical criteria? This classification problem will occupy us for much of the book, and its methodological status will be analysed at the end of this chapter.

Actually is has been more common to speak abut satisfiability rather than validity. We therefore fix here the notion of a reduction class in terms of satisfiability.

Definition 2.1.6. A class X of formulae is called a *reduction class* (for satisfiability) if there exists a recursive function f which maps each first order formula ψ to a formula $f(\psi) \in X$ such that ψ is satisfiable if and only if $f(\psi)$ is satisfiable.

Exercise 2.1.7. Show that the class of all formulae of the pure predicate calculus is a reduction class. Hint: Eliminate function symbols and equality.

Canonical (Herbrand) Models. In the above proof of the Church-Turing Theorem, the basic data of Turing machine computations – letters, states and tape positions – are logically represented using predicate symbols T_j, I_i, H, and numbers. To assure the successor structure of the natural numbers (through the individual constant 0 and the successor relation S) it was sufficient to add an order axiom. In Exercise 2.1.4 on Thue systems the basic data are words which can be represented as terms built up from finitely many 0-ary function symbols (individual constants which stand for the letters of the underlying alphabet) and a binary function symbol (which is interpreted as the concatenation of words). To enforce the semigroup structure it is sufficient to impose certain identities on these terms and the associativity of the binary operation. In many reduction proofs in the literature such axiomatizations of the intended domain of objects become very complicated and constitute much of the difficulty of the whole reduction procedure. It was Büchi who in 1962 made the simple but crucial observation that a well known theorem of Skolem in many cases guarantees a simple way to logically "implement" a given data structure by a corresponding domain of terms, without any explicit axiomatization of this object domain. Since we want to use this fundamental method freely in the sequel, we are going to explain it here. The method allows us for arbitrary first order formulae ψ without equality to restrict attention, after Skolemization, to so-called canonical interpretations in the search for models satisfying ψ. Canonical interpretations of ψ are interpretations over the domain of terms that are built using only the function symbols appearing in ψ and where terms are interpreted by themselves. We start therefore from the Skolem Normal Form Theorem for first order formulae.

Theorem 2.1.8 (Skolemization). *With every first-order formula ψ one can effectively associate a universal formula $\varphi = \forall x_1 \cdots \forall x_n \eta$ (where η is quantifier-free) such that:*

- *ψ and φ have the same free individual variables and the same predicate symbols.*
- *The function symbols of φ are those of ψ augmented by some new function symbols.*
- *ψ and φ are satisfiable over the same domains.*
- *$\varphi \models \psi$.*

The universal formula φ is called a *Skolem normal form* or also *functional form* of ψ. We view individual constants as nullary function symbols.

Proof. The proof consists in iterated application of the following elimination of an existential quantifier, applied from left to right to the prenex normal form of the originally given formula:

Lemma 2.1.9. *Let ψ be a first order formula of form $\forall x_1 \cdots \forall x_n \exists y \alpha$. Choose a new n-ary function symbol f and let $\varphi = \forall x_1 \cdots \forall x_n \alpha[y/fx_1 \cdots x_n]$*

be the result of deleting $\exists y$ from ψ and of replacing y in α by $fx_1 \cdots x_n$. Then ψ and φ are satisfiable over the same domains and $\varphi \models \psi$.

On the basis of the Axiom of Choice the proof of this lemma is obvious. □

The universal formula produced by this method is unique up to inessential renaming of symbols; it may be called *the* Skolem normal form of the original formula.

We can restrict attention to closed formulae, i.e. sentences, because a formula ψ with free variables $x_1, \ldots, x_n$ is satisfiable over the same domains as its existential closure $\exists x_1 \cdots \exists x_n \psi$.

Definition 2.1.10 (Canonical Domain). Let ψ be a first order sentence and φ its Skolem normal form with constants $c_1, \ldots, c_n$ and function symbols $f_1, \ldots, f_m$ of positive arity. (If no constants occur in ψ, i.e. if $n = 0$, then a new constant c is added.) The set of terms built up from these constants and function symbols is the *canonical domain* or the *Herbrand universe* of ψ.

Here are some examples of canonical domains for quantifier and function free formulae β.

- The set of "natural numbers" $0, S0, SS0, \ldots$ is the canonical domain of prenex sentences $\exists u \forall x \exists v \forall y \beta$.
- The set of words over alphabet $\{a_1, \ldots, a_n\}$ represents the canonical domain of prenex sentences $\forall x \exists v_1 \cdots \exists v_n \beta$. For example, a word $a_1 a_2 a_1$ corresponds to a term $f_1 f_2 f_1 c$.
- The canonical domain of prenex sentences $\forall x \forall y \exists v \beta$ is the set of terms built from individual constant c by means of a binary operation $*$. Notice that the operation is not supposed to be associative and thus one cannot ignore parentheses. Example: $((c * (c * c)) * c)$. Notice further that any expression can be reconstructed from the pattern of parentheses. Thus the domain can be represented by purely parenthetical expressions

Remark. The condition that in building the canonical domain of a given formula ψ, we start from a new constant in case the Skolem normal form of ψ has no 0-ary function symbol, assures the non-emptiness of these domains. Note that if the Skolem normal form contains at least one function symbol of positive arity, then the canonical domain is countably infinite.

Definition 2.1.11 (Canonical Model). A structure is a *canonical structure* or *Herbrand structure* of a given sentence ψ if its universe is the canonical domain of ψ and each function symbol f of the Skolem normal form of ψ is interpreted as the term building operator $\overline{f}$, i.e., by $\overline{f}(t_1, \ldots, t_m) = ft_1 \cdots t_m$. A formula ψ is called *canonically satisfiable* if it is satisfiable by a canonical structure. Such a structure will be called a *canonical model* of ψ.

We are now ready for a succinct formulation of the above mentioned theorem of Skolem which allows us to restrict attention to canonical models without loss of generality. The universal closure of a formula φ with the free variables $x_1, \ldots, x_n$ is the sentence $\forall x_1 \cdots \forall x_n \varphi$.

Theorem 2.1.12 (Skolem). *The universal closure of a quantifier free formula without equality is satisfiable iff it is canonically satisfiable.*

Proof. From a given model $\mathfrak{A}$ for $\forall x_1 \cdots \forall x_n \varphi$ one obtains a canonical model $\mathfrak{A}^*$ by interpreting the relation symbols P of ψ by

$$P^* := \{(t_1, \ldots, t_n) : \mathfrak{A} \models P t_1 \cdots t_n\}.$$

□

Exercise 2.1.13. [64] Show that in the reduction formulae ρ_M in the proof of the Church-Turing Theorem, the successor relation S and the clauses of ORDER in which it appears can be dispensed with if 0 is canonically interpreted as number 0 and $'$ as $+1$.

Economical Description of Register Machines. We illustrate the use of Skolem's Theorem for a simple reduction of the halting problem for register machines. We follow an idea from [2, 39] for eliminating also the explicit mentioning of the time (or length) of computations used in reduction formulae for Turing machines. In this way we obtain a further refinement – sharpening the propositional structure of the quantifier free part to so-called Krom and Horn formulae – and a simplification of the proof of the reduction property.

The following definitions go back to the American mathematicians A. Horn and M. Krom.

Definition 2.1.14 (Krom and Horn Formulae). We call a disjunction of atomic and negated atomic formulae a *clause*. A *Horn clause* is a clause with at most one non-negated atom. Alternatively, a Horn clause can be written as an implication $\alpha \to \beta$ where α is a conjunction of non-negated atoms and β is either an atom or the logical constant *false*. A *Horn formula* is a first-order formula in prenex normal form whose quantifier-free part is a conjunction of Horn clauses. We write HORN for the class of all Horn formulae.

A *Krom clause* is a is a clause with at most two constituents. A *Krom formula* is a first-order formula in prenex normal form whose quantifier-free part is a conjunction of Krom clauses and KROM denotes the class of all Krom formulae.

A two-register machine M is similar to a Turing machine but instead of a tape it has two registers. Each register contains a natural number. In one step, the machine can add 1 to the content of one of the registers or test whether the content of the given register is zero and if not then subtract 1.

In addition, M goes to a new state depending on which registers are empty (that is contain zero). The basic data structure of register machines is the set of natural numbers with zero and the successor function. It can be realized as the canonical domain of a sentence $\exists u \forall x \exists v \forall y_1 \cdots \forall y_n \beta$ where β is quantifier free and does not contain function symbols. This means that the variables $x, y_1, \ldots, y_n$ can be used directly as ranging over register contents and that a register machine configuration C can be completely described by a single atomic formula $\underline{C}$ that involves all the register variables. This allows us to express the M-reachability problem $C \Rightarrow_M D$ (which means that M, started in C, reaches D) directly as a logical derivability problem $\mathrm{STEP}_M, \underline{C} \vdash \underline{D}$, avoiding the need to introduce a further variable to describe the time (or length) of M's computations. Starting from machines with two registers, which are well known to have an r.e. complete halting problem (see [393, 469]), we can restrict ourselves to using only two universal quantifiers. We thus obtain:

Theorem 2.1.15 (Aanderaa, Börger). *One can effectively associate with every deterministic 2-register machine program M a prenex Krom and Horn sentence $\psi_M = \exists u \forall x \exists v \forall y \alpha$ of the pure predicate calculus with Skolem normal form $\forall x \forall y \mathrm{STEP}_M$, and encode M-configurations C by atomic formulae $\underline{C}$ so that:*

Reduction Property: *For all M-configurations C, D we have that $C \Rightarrow_M D$ if and only if the formula*

$$\forall x \forall y \mathrm{STEP}_M \wedge \underline{C} \rightarrow \underline{D}$$

is logically valid.

Further, ψ_M contains only 2-place predicate symbols.

Proof. Let M be an arbitrary 2-register machine program with instructions I_i for $i = 0, \ldots, r$. The effect of program M on arbitrary M-configurations (i, m, n) – with state i and register contents m,n in the first and second registers respectively – is defined locally through the action performed by instruction I_i on configurations (i, m, n). Correspondingly we will define the *program formula* STEP_M as a conjunction of all formulae ε_i each of which expresses the meaning of the corresponding M-instruction I_i. The reduction property tells us that the effect of I_i that we have to formalize is restricted to a reachability question for configurations – namely whether starting from C and applying M-instructions we can reach D. We will therefore represent configurations (i, m, n) by $\underline{(i, m, n)} := K_i mn$, with binary predicate symbols K_i and where, thanks to Skolem's Theorem, the register contents m, n can be identified with the corresponding logical terms (built up from 0 applying the successor function $'$).

Addition instructions $I_i = (i, r, j)$: at state i, add 1 to the content of register r and then to go to state j. In case $r = 1$, they are formalised by implications $\varepsilon_i := K_ixy \rightarrow K_jx'y$, so that if a configuration (i, m, n) is reached then also the configuration $(j, m+1, n)$ is reached as well. The case of $r = 2$ is similar.

Subtraction instructions $I_i = (i, r, j, k)$: at state i, test whether the content of the register r is 0; if yes then go to state j, and if not then subtract 1 from (the content of) register r and go to state k. In case $r = 1$, they are formalised by the conjunction ε_i of the following two implications, reflecting the two possible test result: $K_ix'y \rightarrow K_kxy$ and $K_i0y \rightarrow K_j0y$. The case of $r = 2$ is similar.

This ends the definition of the ε_i and therefore of STEP_M and ψ_M. From the preceding explanations it is easy to prove the reduction property. Indeed, assume that $\forall x \forall y \mathrm{STEP}_M \wedge \underline{C} \rightarrow \underline{D}$ is logically valid. The intended interpretation satisfies the premise $\forall x \forall y \mathrm{STEP}_M \wedge \underline{C}$ and therefore satisfies the conclusion $\underline{D}$. This implies $C \Rightarrow_M D$. Conversely, assume $C \Rightarrow_M D$. For each canonical model $\mathfrak{A}$ of $\forall x \forall y \mathrm{STEP}_M \wedge \underline{C}$, show the following *simulation property*: for each t and each M-configuration E which is reached by M in t steps starting from C, we have that $\mathfrak{A} \models \underline{E}$. This is proved by a straightforward induction on t, using the truth of $\underline{C}$ in $\mathfrak{A}$ in the induction base case and the truth of the appropriate ε_i for the induction step. From $C \Rightarrow_M D$, it then follows that $\mathfrak{A} \models \underline{D}$. □

Corollary 2.1.16. *The class* $\exists\forall\exists\forall \cap$ KROM $\cap$ HORN *with only binary predicate symbols is a reduction class.*

Proof. Since a formula ψ is not logically valid iff $\neg\psi$ is satisfiable, Theorem 2.1.15 implies: $C \not\Rightarrow_M D$ iff $\forall x \forall y \mathrm{STEP}_M \wedge \underline{C} \wedge \neg \underline{D}$ is satisfiable. The same holds for the non-Skolemized prenex normal form $\beta_{M,C,D}$ which is in the class $\exists\forall\exists\forall \cap$ KROM $\cap$ HORN with only binary predicate symbols. We now specialize M to a program that enumerates all logically valid formulae in the following sense: ψ is satisfiable iff M, started with configuration $C(\psi) = (0, m(\psi), 0)$ — with an appropriate 2-register machine encoding $m(\psi)$ of ψ — does not halt in state 1 with empty registers. Specializing therefore C to $C(\psi)$ and D to $(1, 0, 0)$ yields the desired reduction: ψ is satisfiable iff $f(\psi)$ is satisfiable, where $f(\psi) = \beta_{M,C(\psi),D}$. □

Exercise 2.1.17. Prove the analogous versions of Theorem 2.1.15 and Corollary 2.1.16 for the prefix classes $\forall\exists\exists\forall$ and $\forall\exists \wedge \forall\exists\forall$. Hint: Use terms x^+, with another "successor" function $+$, as 0.

Exercise 2.1.18. Prove the analogous versions of Theorem 2.1.15 and Corollary 2.1.16 for the prefix $\exists\forall \wedge \forall\exists\forall$.

Exercise 2.1.19. (see [6]) Prove that the class of all relational Krom and Horn sentences with equality, with prefix of form $\forall\exists\forall\exists$, and only binary predicate symbols is a reduction class. Hint: Modify the formulae from Corollary 2.1.16 using an axiomatization of 0 by the formula

$$\forall x \exists v \forall y \exists u (Nu \wedge (Nx \rightarrow x = u) \wedge (Ny \rightarrow y = u)).$$

The rest of Sect. 2.1.1 is for the advanced reader and can be skipped.

Exercise 2.1.20 (Advanced). (see [227, 49]) Prove the reduction class property for the class of all universal Horn sentences with only one variable and a quantifier free conjunction built from equalities and inequalities in which only one-place function symbols occur. Hint: Use the encoding by terms

$$\underline{(i,p,q)} := (k_i r_1^p r_2^q n x$$

with function symbols k_i for M-states i and function symbols r_1, r_2 for the registers. Require that $r_1 r_2 x = r_2 r_1 x$. The idea behind n is to create zero out of x. Recall that we should be able to test whether a given register is empty. To this end, introduce special unary function symbols n_1, n_2 and require the following where j and k are distinct members of $\{1, 2\}$:

$$n_k n x = n x \wedge n_k r_k x \neq r_k x \wedge (n_k x = x \leftrightarrow n_k r_j x = r_j x)$$

so that $n_k x = x$ iff $x = r_j^p n y$ for some p and y.

Remark. The formulae STEP_M constructed in the preceding proof for 2-register machine programs M can be seen as a logical definition for the semantics of M. Since STEP_M is a Horn formula one can look at it also as a (pure) Prolog program. Each step of the register machine program corresponds to the use — in the simulating logical deduction — of an implication in the reduction formula and therefore to a resolution step of the corresponding Prolog program. The reader is invited to verify that the equivalence proof of the reduction property shows

$$C \Rightarrow_M^1 D \text{ iff } \mathrm{STEP}_M, \underline{C} \vdash_{res}^1 \underline{D},$$

where $\Rightarrow_M^1$ and $\vdash_{res}^1$ denote reachability in one step by M-computation and by the resolution calculus, respectively (see [56]). This close correspondence allows us to transfer many complexity-theoretic properties from machine decision problems to logical decision problems and vice versa. We will see various applications of this correspondence later in this chapter, but let us cite here already a few simple examples.

Example 2.1.21. The class of Horn formulae (or pure Prolog programs) with only binary relations, number terms $0, x, x', y$, and procedure bodies of length ≤ 1 is computation universal in the sense that, under appropriate presentation of inputs and outputs, every partial recursive function is computable by a program in that class. Similar universality results for pure Prolog programs have been rediscovered and proved by other methods in the 1970s, see [20, 268, 279, 460, 503]. See also Exercises 2.1.22 and 2.1.42.

Exercise 2.1.22 (Advanced). Find a single computation-universal Horn formula (pure Prolog program). Hint: Start from a computation-universal register machine program or interpret the state index in the reduction formulae as third argument of a ternary predicate symbol.

Example 2.1.23. The above reduction procedure provides a simple but nevertheless general scheme to derive undecidability proofs for run-time properties of Prolog programs from undecidable register machine properties. The proof exploits the step-by-step correspondence between computations of register machine programs M and their translation into the particular Prolog programs STEP_M. See [55] for details.

Example 2.1.24. The halting problem is recursively isomorphic to its implementation as a logical derivability (and by the Completeness Theorem therefore also semantical entailment) problem: For every set X of formulae, let $Ded(X) := \{\psi \in X : \vdash \psi\}$ and

$$\begin{aligned} H_D(M) &:= \{C : C \Rightarrow_M D\} \\ F_D(M) &:= \{\forall x \forall y \text{STEP}_M \wedge \underline{C} \rightarrow \underline{D} : C \text{ is an } M\text{-configuration}\} \end{aligned}$$

Then $H_D(M) \equiv Ded(F_D(M))$ where $\equiv$ indicates the existence of a recursive bijection. This recursive bijection allows one to identify the complexity of various decision problems for recursive classes of formulae when the complexity of the corresponding decision problems for the corresponding recursively enumerable sets of integers is known. (We tacitly assume a standard arithmetical representation of recursive classes of formulae.)

Here are some examples which are interesting in the context of logical decision problems:

- The emptiness problem $\{F : Ded(F) = \emptyset\}$ ("F contains no logically valid formula") for recursive classes F of formulae is Π_1-complete. Indeed this is true for the emptiness problem of r.e. sets W_x which can be identified with the sets $H_D(x)$, where D is the canonical stop configuration with empty registers.
- The totality resp. infinity problem $\{F : F \subseteq Ded(F)\}$ ("F contains only logically valid formula") resp. $\{F : Ded(F) \text{ is infinite}\}$ ("F contains infinitely many logically valid formulae") for recursive classes F of formulae is Π_2-complete because so are the corresponding problems for r.e. sets.
- The decision problem $\{F : Ded(F) \text{ is recursive}\}$ and the reduction class problem $\{F : F \text{ is a reduction class with respect to validity}\}$ are Σ_3-complete, because so are the problems of recursivity and of Σ_1-completeness of r.e. sets respectively. Note that as we have seen above, F is a reduction class with respect to validity iff $Ded(F)$ is Σ_1-complete. (See [44].)

The recursive bijection between register machine halting problems and their implementation as logical decision problems also carries over degree-theoretic representation theorems from the former to the latter, see [58].

Using the Friedberg-Muchnik Theorem (see e.g. [472, p. 57]), one gets the positive answer to the question stated as an open problem in [498, p. 177] whether there are undecidable classes of formulae which are not reduction classes. In this connection, see also [243]. The theorem in [58] shows that classes whose decision problem has degree complexity between 0 and $0'$ are as (un-)natural as those degrees. This answers the corresponding question in [532, p. 54].

Example 2.1.25. In Exercise2.1.20, we had to consider only inputs of form $\underline{C}$, i.e. consisting only of an atomic formula. Therefore the step-by-step correspondence holds even for the so called (positive) unit resolution, an important refinement of general resolution where one of the two premises of rule application has to be a (non negated) literal. In [252] it has been shown that (positive) unit resolution provides a complete calculus for (definite) Horn clauses, which for propositional Horn formulae yields a polynomial time satisfiability test and a polynomial time interpolation procedure (see [94]). See also [279] for further results derived from the step-by-step correspondence between machine computation and logical resolution.

Remark. The Krom form of the program formulae STEP_M could be achieved because register machine configurations (i, m, n) have been completely encoded by a single atomic formula, namely $K_i mn$. This has allowed us to express the semantics of M-instructions I_i by implications in ε_i which have one premise (for an arbitrary given configuration $K_i mn$) and one conclusion (for the immediately succeeding configuration). This yields the step-by-step correspondence between the register machine program computations and the corresponding logical (resolution) deductions. The Horn form of the reduction formulae STEP_M could be achieved because the given programs M were deterministic; in case of non-deterministic programs M the conclusions of the implications in ε_i would become (positive) disjunctions, with one (non negated) disjunct for each of the choices of I_i.

Remark. The program formulae STEP_M can also be easily interpreted as various classical computation formalisms; examples are semi-Thue systems, Markov algorithms, Thue systems, Post canonical (normal) calculi, Post correspondence systems, partial implication propositional calculi, etc.. Also for these interpretations one can establish a step-by-step correlation between the computation steps of M and those of STEP_M. Again interesting complexity theoretic properties are carried over from the interpreted to the interpreting formalism. Thus the formulae STEP_M appear as logical form of conceptually different but semantically equivalent computational interpretations which have the same or closely related complexity theoretic properties (see [50, 52, 58, 60, 85]).

2.1.2 Trakhtenbrot's Theorem and Conservative Reductions

In many cases, the decidability of the satisfiability problem for a formula class has been proved by showing that the given class has the *finite model property*: Every satisfiable formula in the class also has a finite model. Due to the facts that *(i)* up to isomorphism, the finite structures of a given finite vocabulary are recursively enumerable, and *(ii)* the property that a given finite structure is a model of a given first-order sentence is decidable (in fact of very low complexity), it follows that the satisfiability problem of every formula class with the finite model property is recursive.

It is easy to exhibit satisfiable formulae without finite models. An example is the following formula $\prec_K$ (in the class $[\forall\exists\forall] \cap \text{KROM} \cap \text{HORN}$) which axiomatizes a relation K with the intended canonical interpretation "Kxu iff x is smaller than u" over the natural numbers:

$$\prec_K := \forall x \exists u \forall y (\neg Kxx \wedge Kxu \wedge (Kyx \rightarrow Kyu)).$$

Exercise 2.1.26. Prove that $\prec_K$ has only infinite models.

From the unsolvability of Hilbert's *Entscheidungsproblem* we can conclude more than the existence of such *infinity axioms* (formulae that are satisfiable but without finite models). There cannot be an effective reduction of satisfiability to finite satisfiability and the class of infinity axioms is not recursive.

Definition 2.1.27. Let X be a class of formulae. We write

- *Sat*(X) for the the set of $\psi \in X$ that are satisfiable;
- *Val*(X) for the set of logically valid $\psi \in X$;
- *Fin-sat*(X) for the set of $\psi \in X$ that have a finite model;
- *Inf-axioms*(X) for *Sat*(X) – *Fin-sat*(X), the *infinity axioms* of X;
- *Non-sat*(X) for the set of unsatisfiable $\psi \in X$.

If X is the set FO of all first-order formulae we usually adopt the simplified notation *Sat*, *Val*, *Fin-sat* etc., instead of *Sat*(FO), *Val*(FO), *Fin-sat*(FO) etc.

To establish the desired strengthening we will use the following concept from recursion theory.

Definition 2.1.28. Two disjoint sets X, Y are called *recursively inseparable* if there is no recursive set R such that $X \subseteq R$ and $R \cap Y = \varnothing$.

Exercise 2.1.29. Show that the classes *Inf-axioms* and *Non-sat* are recursively inseparable.

The natural question remains whether the finite satisfiability problem is recursively solvable or not. A slight change of the reduction formulae constructed for the Church-Turing Theorem gives Trakhtenbrot's negative answer to the problem.

Theorem 2.1.30 (Trakhtenbrot). *The sets of all finitely satisfiable, all unsatisfiable and all only infinitely satisfiable formulae Fin-sat, Non-sat, and Inf-axioms are pairwise recursively inseparable.*

Proof. After the preceding exercise it remains to prove the recursive inseparability of *Non-sat*, *Fin-sat* and of *Inf-axioms*, *Fin-sat*. The proof idea is to apply the reduction procedure in the Theorem of Church and Turing to recursively inseparable halting problems. In particular, this means to enrich the reduction formulae appropriately. Note that the sets *Fin-sat* and *Non-sat* and their union – the complement of *Inf-axioms* – are recursively enumerable.

Here are halting problems, which are known to be recursively enumerable and pairwise recursively inseparable (see e.g.[57, 445]). Let $i = 1, 2$.

$$\begin{aligned} H_i &= \{M : (0,0)00\ldots \Rightarrow_M (i,0)00\ldots, \text{ M a TM program}\} \\ H &= \{M : (0,0)00\ldots \not\Rightarrow_M \textit{Halt}, \text{ M a TM program }\} \end{aligned}$$

Without loss of generality we restrict attention to programs which have only the two stop states 1 and 2, where *Halt* stands for an arbitrary configuration with halting state i ($i = 1, 2$) and $(m, 0)00\ldots$ denotes the configuration with state m and reading head at the left end of the empty one-way infinite tape.

The reduction property is refined as follows to formulae $\rho_{M,\prec}$ which are obtained from ρ_M in the Church-Turing Theorem by replacing ORDER with $\prec_{K,M}$ — a "relativization" of the above defined relation $\prec_K$ to visited tape cells, see below. For the sake of simplicity we also remove the successor relation S by interpreting $0, x'$ canonically.

Reduction Property: For all Turing machine programs M with stop states $1, 2$ the following hold:
(i) $M \in H_1$ iff $\rho_{M,\prec}$ is contradictory
(ii) $M \in H_2$ iff $\rho_{M,\prec}$ is finitely satisfiable.

The reduction property transfers the recursive inseparability from H_2, H_1 and H, H_2 to *Fin-sat*, *Non-sat* and *Fin-sat*, *Inf-axioms*: for if there were a recursive R with *Fin-sat* $\subseteq R$ and *Non-sat* $\cap R = \varnothing$ (resp. *Inf-axioms* $\cap R = \varnothing$), then the recursive set $\{M : \rho_{M,\prec} \in R\}$ would separate the sets H_2, H_1 resp. H_2, H.

It remains therefore to show the refined reduction property. Intuitively speaking, the formula $\rho_{M,\prec}$ guarantees the initial configuration $(0,0)00\ldots$, takes the closure with respect to M-computation steps and explicitly excludes final configurations with stop state 1. It does not say anything about the new halting state 2 and therefore will allow us to build finite models for computations of M which terminate in state 2.

Let us disregard for a moment the conjunct $\prec_{K,M}$ of $\rho_{M,\prec}$. The first claim is the same as the reduction property in the Church-Turing Theorem. Also for the direction from left to right in the second claim we can proceed as before: if $M \in H_2$, then the intended interpretation yields a model. This model can be made into a finite model by restricting the domain of elements to $\{0, \ldots, k+1\}$ — where k is the length of the given M-computation terminating in state 2 — and by defining $(k+1)'$ (the "successor" of $k+1$) as $k+1$. (Note that in k steps a Turing machine program, started in cell 0, can visit no cell to the right of cell k.)

For the reverse direction in the second claim – which was used for the proof of the recursive inseparability of *Inf-axioms* and *Fin-sat* – we do the following.

- We sharpen the stop condition for the considered Turing machine programs M in order to ensure that the sequence of M-configurations, computed starting from the initial configuration $(0,0)00\ldots$, becomes cyclic only when M halts. (Given our normalization assumption on stop states, halting of M means that the configuration sequence becomes constant when reaching a stop state.) This implies that any configuration sequence which does not become cyclic will be infinite.
- We restrict the class of possible models of the logical description of such computations in order to reflect that if a computation does not become cyclic, then also all its logical models have to be infinite.

The sharpened stop condition can be assumed for M without loss of generality. (For example one can include a "step counter" which will be erased only just before halting.) It can be reflected in the logical models of Turing machine computations by restricting the existential claim in the infinity axiom $\prec_K$ (see the above exercise) to numbers of cells visited during the computation of M. This will ensure what we need for the direction from right to left in the second claim of the reduction property. Namely, for each of the objects possibly occurring during a Turing machine computation, each model must have a distinct representative. There must be infinitely many if the computation does not become cyclic (and therefore visits infinitely many cells); finitely many will suffice if the computation comes to a halt.

Here is the M-restricted version $\prec_{K,M}$ of the infinity axiom $\prec_K$ which we add as conjunct in $\rho_{M,\prec}$:

$$\forall x \exists u \forall y (\neg Kxx \wedge (Hyx \rightarrow Kxu) \wedge (Kyx \rightarrow Kyu)).$$

In the models constructed so far we can interpret K as the usual smaller relation and thereby also satisfy the conjunct $\prec_{K,M}$. In addition we can prove now that $\rho_{M,\prec}$ has only infinite models if $M \notin H_2$. In fact if M does not halt, started in $(0,0)00\ldots$, then in the course of that (acyclic) computation each natural number n must occur at least once as number of a visited cell; in each model of $\rho_{M,\prec}$ their representatives $\underline{n}$ are linearly ordered by the relation K, so that the model must be infinite. □

Exercise 2.1.31. [64] Show that $\rho_{M,\prec}$ can be chosen as formula of the form $\exists x Zx \wedge \forall x \exists u \forall y \beta$ where Z is a monadic predicate symbol and β a formula which contains no other variables than x, u, y and in which only the three binary predicate symbols H, T_0, K and finitely many monadic predicate symbols occur. Hint: Use the universality of Turing machines with binary alphabet $\{0,1\}$, interpret Zx as $x = 0$ and incorporate START into β as $Zy \to \mathrm{START}[0/y]$.

Remark. For use in the next chapter observe also that due to the subformula $Zxx \to Hxx$, Büchi's reduction formulae also satisfy $\exists y Hyy$.

Exercise 2.1.32. Prove Trakhtenbrot's Theorem using register machine formulae $\psi_{M,\prec}$ with Skolem normal form

$$\forall x \forall y (\mathrm{STEP}_M \wedge \underline{(0,0,0)} \wedge \neg \underline{(1,0,0)} \wedge \prec_{K,M}) ,$$

where

$$\prec_{K,M} := \neg Kxx \wedge (Kyx \to Kyu) \wedge \bigwedge_i ((K_i xy \to Kxu) \wedge (K_i yx \to Kxu)).$$

Exercise 2.1.33. [299] Prove that there is an effective reduction of the decision problem for validity to the decision problem for finite satisfiability. Hint: Use Herbrand's Theorem; see also the proof in [498, Theorem XXI].

For the proof of Trakhtenbrot's Theorem we have strengthened the reduction procedure in such a way that it works simultaneously for two halting problems. One can apply this refinement to an enriched version of the "logic machine" M introduced to establish the reduction class property in Corollary 2.1.16. This will bring us to an interesting sharpening of the reduction class property. Recall that $C(\psi)$ is the starting configuration $(0, m(\psi), 0)$ of M, with an appropriate encoding $m(\psi)$ of ψ, and consider the following two special halting problems of such a machine M:

$$H'_{M,i} = \{\psi : C(\psi) \Rightarrow_M (i,0,0),\ M \text{ a 2-RM program } \}$$

where halting in state 1 with empty registers is equivalent to the non satisfiability of ψ (validity of $\neg\psi$) and halting in state 2 with empty registers is equivalent to the finite satisfiability of ψ. We add to the reduction formulae $\beta_{M,C(\psi),(1,0,0)}$ in the proof of Corollary 2.1.16 the conjunct $\prec_{K,M}$ constructed in the exercise to Trakhtenbrot's Theorem. Thereby we obtain refined reduction formulae $g(\psi) = \beta_{M,\prec,C(\psi),(1,0,0)}$ with Skolem normal form

$$\forall x \forall y (\mathrm{STEP}_M \wedge \underline{(0, m(\psi), 0)} \wedge \neg \underline{(1,0,0)} \wedge \prec_{K,M})$$

which reduce simultaneously satisfiability and finite satisfiability of arbitrary first order formulae ψ.

Exercise 2.1.34. Prove the preceding claim.

This exercise motivates the following definition.

Definition 2.1.35 (Conservative Reduction). Let X and Y be recursive classes of formulae. A *conservative reduction* from X to Y is a recursive function $g : X \to Y$ such that for all $\psi \in X$:

(i) ψ is satisfiable if and only if $g(\psi)$ is satisfiable;
(ii) ψ is finitely satisfiable if and only if $g(\psi)$ is finitely satisfiable.

A formula class X is a *conservative reduction class* if there exists a conservative reduction from the set of all first-order formulae to X.

Exercise 2.1.36. Let X be a conservative reduction class. Prove that the sets *Non-sat*(X), *Fin-sat*(X), and *Inf-axioms*(X) are pairwise recursively inseparable.

Exercise 2.1.37. Show that the classes $[\exists\forall\exists\forall] \cap$ KROM $\cap$ HORN and $[\forall\exists\exists\forall] \cap$ KROM $\cap$ HORN with only binary predicate symbols are conservative reduction classes.

The following theorem tells us that to establish the conservative reduction class property for a class of formulae it is sufficient to many-one reduce to it the unsatisfiable and the finitely satisfiable formulae (*semi-conservative* reduction). In terms of our reduction properties this means that we can prove the simulation property without having to care about finiteness of the given model and restrict attention to the intended model for the finite case. This simplifies sometimes the reduction proofs.

Definition 2.1.38. A *semi-conservative reduction* from X to Y is a recursive function $f : X \to Y$ such that

(i) $f($*Non-sat*$(X)) \subseteq$ *Non-sat*(Y)
(ii) $f($*Fin-sat*$(X)) \subseteq$ *Fin-sat*(Y)

Theorem 2.1.39 (Gurevich). *If there exists a semi-conservative reduction from the set of all first-order formulae to a recursive class $X \subseteq$ FO, then X is a conservative reduction class.*

Proof. The proof uses a recursion theoretic argument which can be reconstructed from [485, Chap. V]. It uses the following variant of the inseparability concept for pairs of sets. A pair (A, B) of disjoint sets is called *effectively inseparable* if there exists a recursive function f such that for each pair (i, j) of Gödel numbers of disjoint r.e. supersets W_i of A and W_j of B it holds that $f(i, j) \notin W_i \cup W_j$. The recursion theoretic argument says that for every pair of disjoint r.e. sets P_1, P_2 and every every pair of recursively enumerable, effectively inseparable sets R_1, R_2, there exists a recursive function g such that $P_1 = g^{-1}(R_1)$ and $P_2 = g^{-1}(R_2)$. We only have to apply this argument to the disjoint r.e. sets *Non-sat*, *Fin-sat* and the effectively inseparable r.e. sets *Non-sat*(X), *Fin-sat*(X). The sets *Non-sat*(X) and *Fin-sat*(X) are effectively inseparable since two effectively inseparable sets, namely *Non-sat* and *Fin-sat*, can be recursively embedded in them. (Note that *Non-sat* and *Fin-sat* are effectively inseparable by the fact that the two effectively inseparable sets H_1 and H_2 can be recursively embedded into them as shown by the above reduction property). □

Exercise 2.1.40. Show that each semi-conservative reduction preserves recursive inseparability of contradictory and finitely satisfiable formulae, i.e. formally: Let $f : X \to Y$ be a semi-conservative reduction from X to Y. Then the following two properties hold:

1. If *Non-sat*(X) and *Fin-sat*(X) are recursively inseparable, then so are *Non-sat*(Y) and *Fin-sat*(Y).
2. There exists a recursive function $g : X \to Y$ such that for each $\psi \in X$ it holds that ψ is (finitely) satisfiable if and only if $g(\psi)$ is (finitely) satisfiable.

2.1.3 Inseparability and Model Complexity

In the preceding sections we have related the halting problems of machine programs to the unsatisfiability and finite satisfiability problems of logical formulae. We show in this section how by a slightly different use of that reduction the complexity of machine halting problems can also be easily related to the complexity of program formulae models or of the logical theories constituted by those formulae. This gives a computational insight into some classical, seemingly unrelated questions. We have only to push a bit further the close relation we have discovered between programs — computational mechanisms to derive, step by step, successive configurations from given ones — and their logical descriptions as formulae on the basis of which, step by step, all logical (configuration representing) conclusions are drawn from given (configuration encoding) assumptions.

In the Aanderaa-Börger Theorem, we can read the reduction property as saying that the formula $\forall x \forall y \text{STEP}_M$, starting from the assumption $\neg\underline{(1,0,0)}$, "disproves" or "refutes" (the logical description of) all configurations which are rejected by M, i.e. which lead to the rejecting halting configuration $(1,0,0)$. Formally:

$$C \Rightarrow_M (1,0,0) \quad \text{iff} \quad \forall x \forall y \text{STEP}_M \wedge \neg\underline{(1,0,0)} \to \neg\underline{C} \text{ is logically valid.}$$

Thereby it is to be expected that this formula determines a logical theory the complexity of which is related to that of the halting problem

$$H_{M,1} := \{m : (0,m,0) \Rightarrow_M (1,0,0)\}.$$

Similarly we can expect that the complexity of models of this formula is related to the complexity of $H_{M,1}$. Indeed we show that such formulae are without recursive models and determine a non-recursive theory if M has recursively inseparable halting problems $H_{M,i} (i = 1, 2)$. It suffices to guarantee that all M-rejected configurations are disproved, and further, that all configurations C are "proved" which are M-accepted i.e. which lead to the accepting halting configuration $(2,0,0)$. This can be guaranteed by describing "inverse" M-computations from $(2,0,0)$ to C. We introduce the new "starting" configuration by adding $\underline{(2,0,0)}$ and take the closure with respect to the inverses of all STEP_M-implications. The resulting formula γ_M determines a logical theory and has models whose complexity cannot be recursive, if the halting problems $H_{M,i}$, $i = 1,2$ are recursively inseparable as we are going to show in more detail now.

Technically it suffices to sharpen the reduction property in the Aanderaa-Börger Theorem as follows.

Theorem 2.1.41 (Aanderaa, Börger (Second Version)). *Let τ_M be the result of replacing $\rightarrow$ by $\leftrightarrow$ in the program formula STEP_M for 2-register machine programs M. Let γ_M be $\forall x \forall y \tau_M \wedge \neg\underline{(1,0,0)} \wedge \underline{(2,0,0)}$. Then the following holds for all M-configurations C:*

(i) M rejects C iff γ_M disproves C, i.e.: $C \Rightarrow_M (1,0,0)$ iff $\gamma_M \vdash \neg\underline{C}$
(ii) M accepts C iff γ_M proves C, i.e.: $C \Rightarrow_M (2,0,0)$ iff $\gamma_M \vdash \underline{C}$

The claim holds in particular for the calculus of unit resolution.

Proof. If $C \Rightarrow_M (1,0,0)$ then the assertion follows from the reduction property in the first version of the Aanderaa-Börger Theorem because τ_M logically implies STEP_M.

If $C \not\Rightarrow_M (1,0,0)$ then the formula $\gamma_M \wedge \underline{C}$ has a canonical model in the interpretation

$$K_j = \{(m,n) : (j,m,n) \not\Rightarrow_M (1,0,0)\}.$$

If $C \Rightarrow_M (2,0,0)$, let $C_0, \ldots, C_k$ be the configuration sequence determined by M and starting configuration $C_0 = C$, breaking off with $C_k = (2,0,0)$. We paraphrase the proof of the reduction property for STEP_M in the first version of the theorem. Starting with $\underline{C_{k-0}} := K_2 00$, the inverse of each step of the given M-computation is simulated by a logical deduction (say unit resolution) step through which (the resolvent) $\underline{C_{k-t-1}}$ arises from $\underline{C_{k-t}}$ using the inverse of the corresponding implication in the associated ε_i. (Formally this is an induction on $t \leq k$). Thus $\underline{C}$ is logically deducible from γ_M (by unit resolution).

If $\gamma_M \vdash \underline{C}$, then each model of γ_M also satisfies $\underline{C}$. Clearly

$$K_j(m,n) \text{ iff } (j,m,n) \Rightarrow_M (2,0,0)$$

yields a canonical model of γ_M. Therefore $\underline{C}$ also holds in this model. This means that $C \Rightarrow_M (2,0,0)$. □

Exercise 2.1.42. Show that each k-ary computable function f can be computed by a binary pure Prolog program (a set of Krom-Horn clauses) C_f with number terms $x, y_i, 0, x'$ $(0 \leq i \leq k+1)$ such that for all m, n we have that: $f(m) = n$ iff $C_f \models Fmn$ iff $C_f \vdash_{unit-res} Fmn$. Hint: describe the r.e. graph of f as acceptance set of an appropriate register machine program.

Definition 2.1.43. A theory T is *essentially undecidable* if it is consistent and every consistent extension of it is undecidable. For a theory T the set of *T-refutable sentences* is $\neg T := \{\psi : T \models \neg\psi\}$.

Note that the formula γ_M in Theorem 2.1.41 is a KROM$\cap$HORN formula without equality.

Corollary 2.1.44. *If the halting problems $H_{M,i} (i = 1, 2)$ of M are recursively inseparable, then*

(i) γ_M is satisfiable but without recursive models
(ii) the first order theory $T_M := \{\psi : \gamma_M \models \psi\}$ axiomatized by the binary pure Prolog program γ_M and the set $\neg T_M := \{\psi : \gamma_M \models \neg\psi\}$ of sentences refuted by γ_M are recursively inseparable. In particular T_M is essentially undecidable.

Proof. γ_M is satisfied by the canonical interpretation: $\underline{C}$ is true iff $C \not\Rightarrow (1, 0, 0)$. Let $T = T_M$. The consistency of T follows from the satisfiability of γ_M. Each set R that separates T_M from $\neg T_M$ yields, by the reduction properties of Theorem 2.1.41, the separating set $\{n : \underline{(0, n, 0)} \in R\}$ of $H_{M,1}$ and $H_{M,2}$. Similarly for each consistent extension T' of T, the "refutation set"

$$R(T') := \{n : T' \models \neg\underline{(0, n, 0)}\}$$

separates $H_{M,1}$ and $H_{M,2}$. For the same reason, in each model $\mathfrak{A}$ of γ_M, the "refutation set"

$$R(\mathfrak{A}) := \{n : \mathfrak{A} \models \neg\underline{(0, n, 0)}\}$$

separates $H_{M,1}$ and $H_{M,2}$. Therefore there are no such recursive $R, T', \mathfrak{A}$ if $H_{M,1}$ and $H_{M,2}$ are recursively inseparable. □

Löwenheim [365] was the first to discover infinity axioms in first-order logic. Hilbert and Bernays [267] conjectured that there exist sentences in pure predicate logic that are satisfiable but without recursive models. The conjecture has been proved by Kreisel [326] and Mostowski [397] through first-order axiomatizations of a certain system of set theory going back to von Neumann, Bernays and Gödel. The proof was simplified by Rabin [428], and Mostowski [398] gives another proof which rests on a first-order description of Post canonical calculi. Vaught [520] strengthens the result by showing that besides monadic predicates only one binary predicate is needed for such formulae. The impossibility of recursive models for our (Krom) formula γ_M given above is lost when we pass to its non Skolemized prenex form in pure predicate logic. Indeed, each satisfiable Krom formula in pure predicate logic admits a recursive model (see [11, 147]). But one can easily transform γ_M into a non-Krom variant γ'_M of pure predicate logic which has no recursive models; it suffices to replace the Skolem functions occuring in γ_M by corresponding axiomatizations as we are going to prove now.

Let γ'_M be the conjunction of the following formulae, resulting from γ_M by elimination of the Skolem functions $0,'$.

1. $\forall x \exists v Sxv$ (existence of successors) with a new binary predicate symbol S.
2. $\exists u \forall y \tau_{M,0}$ where $\tau_{M,0}$ is the conjunction of all conjuncts of γ_M in which 0 occurs and with 0 replaced by u.
3. $\forall x \forall v \forall y (Sxv \rightarrow \tau_{M,'})$, where $\tau_{M,'}$ is the conjunction of all conjuncts of γ_M in which x' occurs and with x' replaced by v. (Note that we really mean the formula obtained by further reduction of the quantifier free part to conjunctive normal form. This formula is a Horn formula.)

Corollary 2.1.45. *If the halting problems $H_{M,i}(i = 1,2)$ of M are recursively inseparable, then the variant γ'_M of γ_M defined above is a satisfiable Horn formula of pure predicate logic without recursive models.*

Proof. γ'_M is satisfiable because γ_M is; obviously S is interpreted by the graph of the successor function and 0 by an appropriate u. Let $\mathfrak{A}$ be any model of γ'_M. Choose an element u_0 which satisfies $\exists u \forall y \tau_{M,0}$ in $\mathfrak{A}$; using the subformula $\forall x \exists v Sxv$ enumerate a subset $\omega = \{u_0, u_1, \ldots\}$ such that $\mathfrak{A} \models Su_n u_{n+1}$ for all n. Therefore the restriction of $\mathfrak{A}$ to ω yields a model for γ_M by interpreting 0 as u_0 and u'_n as u_{n+1}. If $\mathfrak{A}$ were recursive, then the enumeration procedure and therefore the restricted model for γ_M would be recursive, in contradiction to Theorem 2.1.41. □

Exercise 2.1.46. The following variant γ''_M of γ'_M – a Krom formula with equality but without function symbols – is satisfiable but without recursive models, if the halting problems $H_{M,i}(i = 1,2)$ of M are recursively inseparable:

$$\exists u \forall y \tau_{M,0} \wedge \forall x \exists v \forall y (Sxv \wedge \tau_{M,'}) \wedge \forall x \forall y \forall z \exists w (Sxy \rightarrow y = w) \wedge (Sxz \rightarrow z = w).$$

Hint: Because of the functionality of S, γ''_M logically implies γ'_M.

Exercise 2.1.47. The following formula δ_M is satisfiable but without $\Pi_1 \cup \Sigma_1$ models, if the halting problems $H_{M,i}(i = 1,2)$ of M are recursively inseparable:

$$\delta_M := \exists x_0 \cdots \exists x_r \exists y_0 \cdots \exists y_r \gamma'''_M \wedge \forall x \forall y \bigwedge_{i \leq r} (Px_i xy \leftrightarrow \neg Py_i xy).$$

Here, P is a ternary predicate symbol, γ'''_M arises from γ'_M by replacing all atomic formulae $K_i st$ by $Px_i st$ and Sst by $Px_r st$ where r is the number of states of M. Hint: Via parametrisation, P represents each K_i and S as well as their negations. Therefore a model for δ_M with $\Pi_1 \cup \Sigma_1$-predicate P would yield (by the negation theorem of recursion theory) a recursive interpretation of the K_i and S which satisfies γ'_M.

Remark. The lower bound for the complexity of models of δ_M cannot be improved because each consistent first order theory with a theorem predicate in Δ_2 and Δ_1-language has a model in the Boolean closure of Σ_1 (see [72]).

Exercise 2.1.48. Let E_n be the n-th class of the Grzegorczyk hierarchy of primitive recursive functions. From $(E_{n+1} - E_n)$-separable sets, construct E_{n+1}-formulae which have E_{n+1}-models but no E_n-model.

From the existence of finitely axiomatizable theories with recursively inseparable provability and refutability predicates, proved in Corollary 2.1.44, one can infer an algorithmic limitation for interpolation in first order logic.

Definition 2.1.49. Let $\psi \to \varphi$ be a logically valid implication of first-order formulae. An *interpolant* for $\psi \to \varphi$ is a formula θ containing only relation and function symbols that appear in both ψ and φ, such that $\psi \to \theta$ and $\theta \to \varphi$ are logically valid.

Craig's Interpolation Theorem (see e.g. [76, pp. 87–89] or [270, p. 301]) states that every logically valid implication of first-order formulae has an interpolant. Further, if neither ψ nor φ contains equality then $\psi \to \varphi$ has an interpolant in which the equality sign does not occur. (However, if only one of the formula ψ, φ is equality-free, then an interpolant without equality need not exist.)

Theorem 2.1.50 (Kreisel). *There is no recursive interpolation function Int which for every valid implication $\psi \to \varphi$ of first-order formulae without equality computes an interpolant Int(ψ, φ) without equality.*

Proof. We fix a function $\varphi \mapsto \varphi'$ where φ' is obtained from φ by renaming all function and predicate symbols. We show that for any total interpolation function *Int* and every consistent first order theory T without equality axiomatized by a single sentence ψ, the set of sentences

$$R := \{\alpha : \mathit{Int}(\psi \wedge \alpha, \psi \to \alpha') = \mathit{true}\}$$

separates the theorems of T from the set of T-refutable sentences.

Indeed, suppose that $\psi \models \alpha$ or $\psi \models \neg\alpha$. Then the implication $(\psi \wedge \alpha) \to (\psi' \to \alpha')$ is valid. Therefore

$$\beta := \mathit{Int}(\psi \wedge \alpha, \psi' \to \alpha')$$

is an interpolant for $\psi \wedge \alpha$ and $\psi' \to \alpha'$. Since due to renaming, the latter formulae have no function or predicate symbol in common, and since the formulae do not contain equality, β must be one of the truth constants *true* or *false*. If $\beta = true$ then, by the interpolation property, $\models \psi' \rightarrow \alpha'$ and therefore $\psi \models \alpha$. If $\beta = false$ then, by the interpolation property $\models \neg(\psi \wedge \alpha)$, hence $\psi \models \neg\alpha$. Therefore $\alpha \in R$ if $\psi \models \alpha$, and $\alpha \notin R$ if $\psi \models \neg\alpha$.

This set R would be recursive if *Int* were. Since we have seen above that there are finitely axiomatized theories of this form with recursively inseparable refutability and provability sets, there can be no recursive interpolation function. □

Remark. Maehara and Tait have given a proof of the interpolation theorem from which one can extract an effective procedure to compute an interpolant, given a *derivation* for $\models \psi \rightarrow \varphi$ (see [157]).

Remark. In the preceding logical definitions of program semantics we have deliberately produced "abstract" formulae which do not uniquely determine their models, but only assure a formalization of that part of information in the latter which is needed to reconstruct the computational problem. This economy in logical specification results, as we have seen, in simple proofs and smooth complexity correlations between machine computations and the simulating logical deductions. It allows us also easy adaptability of the reduction formulae to various interpretations. This will become even clearer from the applications in the next section. In the context of Craig's Interpolation Theorem and the related Definability Theorem for first order logic due to Beth, one can give an interesting example where the opposite attitude pays off, providing a proof for a theorem by Gurevich [230] showing a fundamental complexity theoretic limitation of interpolation and of explication of implicitly defined predicates over finite structures. The computationally needed specification is refined by additional auxiliary conditions on the machine environment which determine the computational model uniquely; this provides a short implicit definition of complete (terminating but possibly long) machine computations, each explicit equivalent or interpolant of which over finite structures is very long (at least of the length of the described computation). For details see [57, pages 455-461].

2.2 Logic and Complexity

In the preceding section we applied the method of first order program description to express halting problems without any restriction on the length or the memory requirement of the described computations. Thereby the degree of undecidability of these unrestricted halting problems was carried over to the decision problem of the associated classes of formulae, of their theories and models. In this section we show how to use the logical specification of computations to formalize time-restricted or space-restricted halting problems in syntactically restricted classes of formulae with solvable decision problem.

We thus establish a close relationship between logical expressibility and computational complexity. (In this section we suppose that the reader is familiar with the basic notions of complexity theory, see e.g. [29, 57, 416, 530].)

2.2.1 Propositional Satisfiability

The propositional logic analogue of the undecidability of the decision problem for first order logic is the NP-completeness of SAT, i.e. the satisfiability problem for propositional logic. This result follows from a simple propositional logic specification of the program formula scheme STEP_M defined in the preceding section.

Theorem 2.2.1 (Cook, Levin). *The satisfiability problem for formulae of propositional logic in conjunctive normal form is* NP*-complete.*

Proof. It is easy to see that this problem is in NP: For an arbitrary input formula α in conjunctive normal form guess nondeterministically an assignment ε of truth values to the variables occurring in α. Then compute the truth value $\varepsilon(\alpha)$ of α under this assignment, and accept iff $\varepsilon(\alpha) = 1$. Obviously this can be done in polynomial time in the length of α.

For the hardness part, let L be any problem in NP, accepted by a nondeterministic Turing machine program in time s. We want to reduce this problem by a deterministic polynomial-time algorithm to the satisfiability problem for propositional formulae. For this purpose we compute propositional reduction formulae

$$\gamma_{M,n,s} := \mathrm{STEP} \wedge \mathrm{START}_{s,n} \wedge \mathrm{ACCEPT}_{s,1}.$$

These formulae are in conjunctive normal form and, for "input" variables $x_0, \ldots, x_{n-1}$, the following condition holds.

Reduction Property: M accepts input $w_0 \cdots w_{n-1} \in \{0,1\}^n$ in s steps iff the formula $\gamma_{M,n,s}(x_0/w_0, \ldots, x_{n-1}/w_{n-1})$ is satisfiable.

This will establish the completeness part of the theorem.

As STEP we specify the scheme STEP_M, defined in the preceding section for arbitrary Turing machine programs, as follows. We restrict the parameters (for time and therefore also for space) by $t, t', x, x', y \leq s$ and read the basic formulae $H(t,x), T_j(t,x), I_i(t)$ as pairwise distinct propositional variables. The conjuncts $S(u,v)$ – meaning $v = u + 1$ – and $y \neq x$ are read as external conditions on the admissible parameters and are not part of the formula itself. The initial configuration – namely $(0, w_0)w_1 \cdots w_{n-1}2 \cdots 2$ (where 2 stands for the blank tape symbol) with reading head position 0 in state 0 – is described by the formula $\mathrm{START}_{s,n}$ defined as

$$I_0(0) \wedge H(0,0) \wedge \bigwedge_{0 \leq j < n} (T_1(0,j) \leftrightarrow x_j) \wedge (T_0(0,j) \leftrightarrow \neg x_j) \wedge \bigwedge_{n \leq j \leq s} T_2(0,j)$$

(Note that we want the input formula to depend only on the length n of input and not on the input $w_0 \cdots w_n$) itself; otherwise one could take simply the conjunction of all $T_1(0,j)$ for $w_j = 1$ and all $T_0(0,j)$ for $w_j = 0$ instead of the equivalences.) The acceptance condition of reaching (no state different from) state 1 at the final moment s is expressed by

$$\mathrm{ACCEPT}_{s,1} := \bigwedge_{i \neq 1} \neg I_i(s)$$

It is easy to see that $\gamma_{M,n,s}$ can be written in conjunctive normal form. The reduction property holds by the Simulation Lemma and the intended interpretation given in the preceding section. Note the reason for the seemingly roundabout way to express in $\mathrm{ACCEPT}_{s,1}$ state 1 as the accepting state. Namely, the Simulation Lemma assures for non-deterministic programs M that for each time t at least one of the states i which is reachable by M at time t is reflected in models of the program formula by the truth of $I_i(t)$; but the Simulation Lemma does not guarantee this for a particular i.

□

Exercise 2.2.2. Prove the NP-completeness of 3-SAT, i.e. the satisfiability problem for propositional formulae

$$\psi := \bigwedge_i (Y_{i,1} \vee Y_{i,2} \vee Y_{i,3})$$

in conjunctive normal form with (at most) three literals in each clause.

Exercise 2.2.3. ([91], see also [284]) Prove that the satisfiability problem for propositional Krom formulae (2-SAT) is in P. Hint: Use the resolution calculus, the fact that the resolvent of two Krom formulae is a Krom formula and the fact that over n propositional variables there are at most $(2n)^2$ Krom clauses.

In fact, more sophisticated arguments show that 2-SAT is complete for nondeterministic logarithmic space.

Exercise 2.2.4. Devise a polynomial-time satisfiability test for propositional Horn formulae. For an interesting strengthening of such an algorithm to show the completeness of unit resolution for first order Horn formulae and the polynomial-time decidability of propositional unit resolution, see [252] and [283].

Remark. For deterministic programs M which compute an n-ary Boolean function f, our reduction formulae

$$\gamma_{M,n,s} := \mathrm{STEP} \wedge \mathrm{START}_{s,n} \wedge \mathrm{ACCEPT}_{s,1}$$

(for appropriate s) define f via

$$f(w_0, \dots, w_{n-1}) = 1 \text{ iff } \gamma_{M,n,s}(x_0/w_0, \dots, x_{n-1}/w_{n-1}) \text{ is satisfiable}$$

for each input $w_0, \dots, w_{n-1} \in \{0,1\}^n$. In the light of the preceding exercise it is interesting to note that these formulae are Horn except for the formalization of input 0 in $\text{START}_{s,n}$; furthermore, modulo trivial transformations each instance $\gamma_{M,n,s}(x_0/w_0, \dots, x_{n-1}/w_{n-1})$ is Horn. This has led to define the *Horn complexity* of Boolean functions f, which measures the minimal length of definitions of f which are Horn except for their "input variables" x_j. Horn complexity is strongly related to the P = NP problem; as a matter of fact Horn complexity is polynomially equivalent to Turing complexity and network complexity of Boolean functions, see [4, 5]. For another use of the technique of economical description of Turing machines by propositional formulae see [23].

A slightly more sophisticated propositional specification of the program formula scheme STEP_M using formulae of *quantified* propositional logic allows a description of arbitrary Turing machine computations with polynomial space bound. This establishes the PSPACE-completeness of QBF, the decision problem for quantified propositional logic.

Theorem 2.2.5 (Stockmeyer). *The decision problem for quantified propositional logic is* PSPACE*-complete.*

Proof. We first show that the problem is PSPACE-hard.

Let A be in PSPACE and M be a deterministic Turing machine program that decides A with space $p(n)$. Since the number of different M-configurations that use up to s tape cells is bounded by 2^{cs} for some constant c, it follows that M decides A in time $2^{cp(n)}$.

We associate, by a polynomial-time procedure, with each input word w for M a quantified propositional formula ACCEPTED_w which is valid if and only if w is accepted by M.

As in previous proofs, such a formula will consist of a start, a program steps and a stop part. We adopt the same propositional description $\underline{C}$ of machine configurations C (restricted to tapes of length $p(n)$) as for the Cook-Levin Theorem, but the variables $I_i(t), H(t,x), T_j(t,x)$ are grouped into disjoint sets U^t depending on their time parameter t. We have to avoid the exponential growth of the reduction formulae that would result if we used all the $2^{cp(n)}$ variable classes (one per time parameter) that occur in the *explicit* formalization of all M-computation steps. This problem can be solved by exploiting an important technique of *reusing variables* along the following lines. One has to describe by a short formula $\text{REACH}_{t+1}(X,Y)$ (using variables from two sets X, Y) that the configuration coded by Y can be reached from the one coded by X by an M-computation of length $\leq 2^{t+1}$. For this purpose the given computation is split into two computations of length at most 2^t. This allows us to define $\text{REACH}_{t+1}(X,Y)$ inductively using only one occurrence of $\text{REACH}_t(U,V)$; this single occurrence is imposed for two possible substitutions: $U = X, V = Z$ for the first half and $U = Z, V = Y$ for the second half, where Z is a new variable set representing an intermediate configuration (see below for the precise definition of $\mathrm{REACH}_{t+1}(X,Y)$).

For given space bound $s = p(n)$ let therefore $U, V, W, X, Y, \ldots$ be pairwise disjoint sets of propositional variables $I_i, H(x), T_j(x)$ as in the Cook-Levin Theorem.

For every M-configuration C (with space bound s) we can write the propositional formula $\underline{C}(X)$ saying that X encodes C, in the sense that the unique assignment $X \to \{0,1\}$ making $\underline{C}(X)$ true is the one corresponding to the indented meaning of the variables $I_i, H(x), T_j(x)$ in X with respect to configuration C.

By the same techniques as in previous proof we then construct the following propositional formulae:

- The initial configuration $C(w)$ with input word w is formalized by the initial formula
$$\mathrm{START}_w(X) := \underline{C(w)}(X).$$
- $\mathrm{CONF}(X)$ expresses that X indeed describes an M-configuration (at least one state, one head position and in each tape cell one letter).
- $\mathrm{ACC}(X)$ says that X represents an accepting configuration, i.e. that M is in an accepting state.
- $\mathrm{NEXT}(X,Y)$ expresses that Y represents the successor configuration of X.
- $\mathrm{EQ}(X,Y)$ expresses that X is equal to Y. Formally $\mathrm{EQ}(X,Y)$ is the conjunction of the equivalences of corresponding variables in the sets X and Y.

Note that so far we didn't need any quantifiers. We now construct, by induction on t, formulae $\mathrm{REACH}_t(X,Y)$ with the following

Reduction Property: Let C, D be M-configurations with space bound $p(n)$. An assignment of X and Y makes $\mathrm{REACH}_t(X,Y)$ true if and only if X and Y encode, via this assignment, two M-configurations C and D such that M reaches D from C in at most 2^t steps.

For the base of the induction, note that $\mathrm{REACH}_0(X,Y)$ has to express that Y either represents the same configuration as X or its immediate successor. Thus, we put

$$\mathrm{REACH}_0(X,Y) := \mathrm{CONF}(X) \wedge \mathrm{CONF}(Y) \wedge [\mathrm{EQ}(X,Y) \vee \mathrm{NEXT}(X,Y)].$$

Inductive step: $\mathrm{REACH}_{t+1}(X,Y)$ has to express that M reaches in at most 2^{t+1} steps the configuration coded by Y from the configuration coded by X. This can be obtained by stating that M can go from X to some intermediate configuration Z in at most 2^t steps and then from Z to Y in at most 2^t steps. However, the straightforward idea to use

$$\exists Z(\mathrm{CONF}(Z) \wedge \mathrm{REACH}_t(X,Z) \wedge \mathrm{REACH}_t(Z,Y))$$

results in a formula of exponential length. But with the idea sketched above, we can instead use the equivalent formula

$$\begin{aligned}\mathrm{REACH}_{t+1}(X,Y) \quad := \quad & \exists Z(\mathrm{CONF}(Z) \wedge \forall U \forall V((\mathrm{EQ}(U,X) \wedge \mathrm{EQ}(V,Z)) \vee \\ & (\mathrm{EQ}(U,Z) \wedge \mathrm{EQ}(V,Y)) \rightarrow \mathrm{REACH}_t(U,V)).\end{aligned}$$

Note that REACH_{t+1} has length $|\mathrm{REACH}_t(U,V)| + q(n)$ for some polynomial q (roughly $p(n)^2 \log p(n)$). Thus $\mathrm{REACH}_t(X,Y)$ has length $O(tq(n))$ and can be constructed in polynomial time.

The theorem now follows by defining the formula

$$\mathrm{ACCEPTED}_w := \exists X \exists Y(\mathrm{START}_w(X) \wedge \mathrm{REACH}_{cp(n)}(X,Y) \wedge \mathrm{ACC}(Y)).$$

Obviously the formula $\mathrm{ACCEPTED}_w$ can be constructed in polynomial time from w.

Exercise 2.2.6. Prove the reduction property.

It remains to show that there is an algorithm that uses only polynomial space to test the validity of quantified propositional formulae. Since the validity of $\exists x\alpha$ is equivalent to the validity of $\alpha(x/0) \vee \alpha(x/1)$, it is possible to split the test of validity of $\exists x\alpha$ into two subtests checking truth of $\alpha(x/0)$ and $\alpha(x/1)$ under a given truth assignment to propositional variables. Since these two tests are independent of each other, they can be computed using the same space.

Exercise 2.2.7. Fill in the details for the above outlined algorithm. Hint: see the proof of Savitch's theorem in [449].

□

2.2.2 The Spectrum Problem and Fagin's Theorem

Trakhtenbrot's Theorem suggests a special case of the general problem of classifying the non-isomorphic models of logical theories, namely to characterise the classes of cardinalities of finite models of first-order sentences. This problem has been formulated as the *spectrum problem* [455], and has attracted considerable interest in complexity theory. Fagin's generalization of spectra for existential second-order sentences [152] has revealed a fundamental relation between complexity theory and finite model theory and constitutes a landmark for the development of the latter. In this subsection we will exhibit a characterization of spectra and generalized spectra in terms of computational complexity. It is obtained by an appropriate interpretation of the reduction scheme for the description of Turing machine by first-order formulae introduced above.

It is reasonable in this context to restrict attention to finite models. Indeed, by the Löwenheim-Skolem Theorem, every first-order sentence has either no infinite models (in which case it also cannot have models of arbitrarily large finite cardinality), or has infinite models of arbitrary cardinality. Furthermore we will consider only formulae with equality because equality-free formulae are closed under domain extension of models. Last but not least we will consider only relational formulae. (Note that n-ary function symbols can always be represented through their $n+1$-ary graph predicate.)

These considerations have led to the following definition.

Definition 2.2.8 (Hasenjäger, Markwald, Scholz). The *spectrum* of a formula ψ is defined as the set of cardinalities of the finite models of ψ, i.e.

$$\text{spectrum}(\psi) := \{k \in \mathbb{N} : \psi \text{ has a model with } k \text{ elements.}\}$$

Exercise 2.2.9. The spectra of first-order sentences with empty vocabulary (i.e. with $=$ as the only predicate symbol) are the finite and the co-finite sets. Hint: If a closed equality formula ψ has an infinite model, then $\neg\psi$ has no infinite models so that spectrum($\neg\psi$) is finite.

Exercise 2.2.10. The spectra of first-order existential or universal sentences are, respectively, the final segments $\{m : m \geq n\}$ and the initial segments $\{m : m \leq n\}$ of $\mathbb{N}$. It was proved by Ramsey [435] that the spectra of prenex sentences with prefix of form $\exists \cdots \exists \forall \cdots \forall$ are finite or co-finite. To establish this result Ramsey proved his famous combinatorial theorem.

Fagin's Theorem. We now present Fagin's characterization of the complexity class NP in terms of generalized spectra. The spectrum of a first-order sentence φ of relational vocabulary $\tau = \{R_1, \ldots, R_t\}$ can be viewed as the set of finite models of the existential second-order sentence

$$(\exists R_1) \cdots (\exists R_t)\varphi.$$

Since all relation symbols are quantified, this is a formula over the empty vocabulary, i.e. its models are just sets. Thus there is a one to one correspondence between the spectra of first-order sentences and the classes of non-isomorphic finite models of existential second-order sentences *over the empty vocabulary.* By allowing different vocabularies for existential second-order sentences, this naturally leads to the notion of a generalized spectrum [152].

Definition 2.2.11. A *generalized spectrum* is the class of finite models of a sentence in existential second-order logic.

For instance, the class of bipartite graphs is a generalized spectrum. It is defined by the sentence

$$(\exists R)\forall x \forall y(Exy \rightarrow (Rx \leftrightarrow \neg Ry)).$$

Exercise 2.2.12. Prove that the class of Hamiltonian graphs, the class of k-colourable graphs (for any fixed k) and the class of graphs that admit a perfect matching are generalized spectra. (A perfect matching in a graph $G = (V, E)$ is a set $M \subseteq E$ of edges such that every node belongs to precisely one edge of M.)

To formulate Fagin's Theorem precisely we have to describe how structures are encoded by strings. At least implicitly, such an encoding requires that we select an ordered representation of the structure.

Definition 2.2.13. The class $\mathcal{O}(\tau)$ of *ordered* τ-structures is the class of all structures $(\mathfrak{A}, <)$ where $\mathfrak{A}$ is a finite τ-structure and $<$ is a linear order on A (the universe of $\mathfrak{A}$).

For any ordered structure $(\mathfrak{A}, <)$ of cardinality n and any k we can identify the Cartesian product A^k with the set $\{0, \dots, n^k - 1\}$, by associating each k-tuple with its rank in the lexicographical ordering induced by $<$ on A^k. Ordered structures can be encoded as binary strings in many natural ways. The particular choice of an encoding is not important. We only need the following conditions to be satisfied.

Definition 2.2.14. An encoding $code : \mathcal{O}(\tau) \to \Gamma^*$ (over any finite alphabet Γ) is *good* if it identifies isomorphic structures, its values are polynomially bounded, if it is first-order definable and if it allows to compute efficiently the values of atomic statements. Formally this means that the following conditions are satisfied:

(i) $code(\mathfrak{A}, <) = code(\mathfrak{B}, <)$ if and only if $(\mathfrak{A}, <) \cong (\mathfrak{B}, <)$.
(ii) $|code(\mathfrak{A}, <)| \leq p(|A|)$ for some polynomial p.
(iii) For all $k \in \mathbb{N}$ and all symbols $\sigma \in \Gamma$ there exists a first-order formula $\beta_\sigma(x_1, \dots, x_k)$ of vocabulary $\tau \cup \{<\}$ such that for all $(\mathfrak{A}, <) \in \mathcal{O}(\tau)$ and all $\bar{a} \in A^k$ the following equivalence holds:

$$(\mathfrak{A}, <) \models \beta_\sigma[\bar{a}] \text{ iff the } \bar{a}\text{-th symbol of } code(\mathfrak{A}, <) \text{ is } \sigma.$$

(iv) There exists a algorithm which, given $code(\mathfrak{A}, <)$, a relation symbols R of τ and (a representation of) a tuple $\bar{a}$, decides whether $\mathfrak{A} \models R\bar{a}$, and whose space complexity is logarithmically bounded (with respect to the cardinality of A).

A convenient encoding is given as follows. Let $<$ be a linear order on A and let $\mathfrak{A} = (A, R_1, \dots, R_t)$ be a τ-structure of cardinality n. Let ℓ be the maximal arity of $R_1, \dots, R_t$. With each relation R of arity j we associate the string $\chi(R) = w_0 \cdots w_{n^j-1} 0^{n^\ell - n^j} \in \{0,1\}^{n^\ell}$ where $w_i = 1$ if the i-th tuple of A^j belongs to R, and $w_i = 0$ otherwise. Now, set $code(\mathfrak{A}, <) = 1^n 0^{n^\ell - n} \chi(R_1) \cdots \chi(R_t)$.

Exercise 2.2.15. Prove that this encoding is good. In fact this encoding lends itself to a very simple logical description in the following sense: If besides (or instead of) the linear ordering $<$, the corresponding successor relation S, and the constants $0, e$ for the first and last element with respect to $<$, are available, then the encoding is definable by *quantifier-free* formulae $\beta_\sigma(\bar{x})$.

For the rest of this section we fix a good encoding function and understand structures to be represented by their encodings. So when we say that a Turing machine decides a property of τ-structures we actually mean that M decides the set of encodings of the structures with that property. Similarly we identify a class K of τ-structures with the set

$$code(K) := \{code(\mathfrak{A}, <) : \mathfrak{A} \in K \text{ and } < \text{ is a linear order on } A\}.$$

Note that encoding a structure involves selecting an ordering on the universe. In general, different orderings will produce different encodings. However, we want to consider properties of structures, not of their encodings, so we consider only classes K that are closed under isomorphisms, i.e. $\mathfrak{A} \cong \mathfrak{B}$ and $\mathfrak{A} \in K$ imply $\mathfrak{B} \in K$.

It thus makes sense to ask whether such a K belongs to a complexity class, like P or NP. On the other side, any language $L \subseteq \Gamma^*$ can also be considered as a class of structures over the vocabulary $\{<\} \cup \{P_a : a \in \Gamma\}$. Indeed, a word $w = w_0 \dots w_{m-1} \in \Gamma^*$ is described by the structure $\mathfrak{B}(w)$ with universe $\{0, \dots, m-1\}$ with the usual interpretation of $<$, and with $P_a = \{i : w_i = a\}$.

Once we have a representation of structures as strings, i.e. as inputs for Turing machines, we can investigate the computational complexity of properties of finite structures. In particular, we can ask how difficult it is to decide the class of models of a logical sentence. For first-order sentences this is very simple.

Proposition 2.2.16. *Let ψ be a first-order sentence. Then*

$$\{\mathfrak{A} : \mathfrak{A} \text{ finite}, \mathfrak{A} \models \psi\} \in \text{LOGSPACE}.$$

Proof. Let ψ be in prenex normal form, say $\psi = Q_1x_1 \cdots Q_rx_r\varphi$. To check whether a given structure $\mathfrak{A}$ of cardinality n is a model of ψ, a Turing machine needs $r \log n$ bits of workspace to cycle systematically through all (representations of) r-tuples $a_1, \dots, a_r$ and to check whether $\mathfrak{A} \models \varphi[a_1, \dots, a_r]$. □

Exercise 2.2.17. Show that the set of finite models of a first-order sentence ψ can be decided by Boolean circuits of constant depth and polynomial size, i.e. $\{\mathfrak{A} : \mathfrak{A} \models \psi\} \in \text{AC}^0$.

Theorem 2.2.18 (Fagin). *Let K be an isomorphism-closed class of finite structures of some fixed non-empty finite vocabulary. Then K is in* NP *if and only if K is a generalized spectrum.*

Proof. First we show how to decide a generalized spectrum. Let $\psi := \exists P_1 \cdots \exists P_r \varphi$ be an existential second-order sentence. We describe a non-deterministic polynomial-time algorithm which, given an encoding $code(\mathfrak{A}, <)$ of a structure $\mathfrak{A}$, decides whether $\mathfrak{A} \models \psi$. First, M guesses relations $P_1, \dots, P_r$ on A. The relation P_i is determined by a binary string of length n^{s_i} where s_i is the arity of P_i and $n = |A|$. Then M decides whether $(\mathfrak{A}, P_1, \dots, P_r) \models \varphi$. Since φ is first-order, by the preceding proposition this can be done with logarithmic space and thus in polynomial time.

So the computation of M consists of guessing a polynomial number of bits, followed by a deterministic polynomial-time computation. Obviously, M decides the class of finite models of ψ.

Conversely, let K be an isomorphism-closed class of τ-structures and M be a non-deterministic one-tape Turing machine which, on input $code(\mathfrak{A}, <)$, decides in polynomial time whether $\mathfrak{A}$ belongs to K. We assume that M has a initial state 0, a unique rejecting state 2, and that all computations of M on input $code(\mathfrak{A}, <)$ accept or reject after at most $n^k - 1$ steps (where n is the cardinality of $\mathfrak{A}$). We describe the computation of M by an adaptation of the reduction scheme for the economical description of Turing machines (see pp. 19–21).

Suppose first that, on every input structure $\mathfrak{A}$, we have an ordering $<$, the corresponding successor relation S and constants $0, e$ for the first and last element of A with respect to $<$. To represent the n^k time and space parameters of the computation we identify numbers up to $n^k - 1$ with tuples in A^k. Note that the corresponding successor relation on k-tuples is definable from $S, 0, e$ by a quantifier-free formula. Indeed

$$S(\bar{x}, \bar{y}) \equiv \bigvee_{i \leq k} \Big(\bigwedge_{j<i} (x_j = e \wedge y_j = 0) \wedge S(x_i, y_i) \wedge \bigwedge_{j>i} x_j = y_j \Big).$$

The predicates H and T_σ are interpreted as $2k$-ary relations and I_i as k-ary relations over A. More precisely,

$$\begin{aligned} H &:= \{(\bar{t}, \bar{a}) \in A^k \times A^k : \text{at time } \bar{t}, \text{ the head of } M \text{ is on position } \bar{a}\} \\ T_\sigma &:= \{(\bar{t}, \bar{a}) \in A^k \times A^k : \text{at time } \bar{t}, \text{ cell } \bar{a} \text{ contains the symbol } \sigma\} \\ I_i &:= \{\bar{t} \in A^k : \text{at time } \bar{t}, M \text{ is in state } i\}. \end{aligned}$$

For any fixed input length n, any M-configuration C and any time $t < n^k$ we obtain $\underline{C_t}$ by taking the conjunction over the atomic statements $H(\bar{t}, \bar{a})$, $T_\sigma(\bar{t}, \bar{a})$ and $I_i(\bar{t})$ that hold for C at time $\bar{t}$.

The begin of the computation is described by a formula START which expresses that the configuration of M at time $t = 0$ is $C_0(\mathfrak{A}, <)$, the input configuration with input $code(\mathfrak{A}, <)$. Recall that a good encoding is represented by first order formulae $\beta_\sigma(\bar{x})$ (condition *(iii)* of the definition of good encodings). We set

$$\text{START} := H(\bar{0},\bar{0}) \wedge I_0(\bar{0}) \wedge \bigwedge_\sigma(\beta_\sigma(\bar{x}) \rightarrow T_\sigma(\bar{0},\bar{x})).$$

Further, let STEP_M be the conjunction over the formulae (2.1)–(2.3) of the reduction scheme (see page 20), with arguments $x, t, \ldots$ replaced by k-tuples $\bar{x} = x_1, \ldots, x_k, \bar{t} = t_1, \ldots, t_k, \ldots$ of variables. Finally let $\text{ACCEPT} := \neg I_2(\bar{t})$ where 2 is the unique rejecting state of M.

The formula ψ_M is defined as the universal closure of $\text{START} \wedge \text{STEP}_M \wedge \text{ACCEPT}$. If M accepts $code(\mathfrak{A},<)$ then there clearly exists an expansion $\mathfrak{B}$ of $\mathfrak{A}$ such that $\mathfrak{B} \models \psi_M$. Conversely, let $\mathfrak{B} = (\mathfrak{A}, <, S, 0, e, H, \bar{T}, \bar{I})$ an expansion of a τ-structure $\mathfrak{A}$ by a linear order $<$, the corresponding $S, 0, e$, and appropriate $H, \bar{T}, \bar{I}$ such that $\mathfrak{B} \models \psi_M$. It follows by induction on t that there exists a computation $C_0(\mathfrak{A},<) = C_0, C_1, \ldots, C_t, \ldots$ of length n^k such that $\mathfrak{B} \models \underline{C_t}$ for all t. Since $\psi_M \models \neg\underline{C_t}$ for every rejecting configuration C and every t, it follows that M accepts the input $code(A,<)$. Thus

$$(\mathfrak{A}, <, S, 0, e) \models (\exists H)(\exists \bar{T})(\exists \bar{I})\psi_M \quad \text{iff} \quad M \text{ accepts } code(\mathfrak{A},<).$$

On $\mathfrak{A}$, an ordering need not be present, but in existential second-order logic we can introduce one by quantifying over a binary relation $<$ and using a first-order axiom α saying that $<$ is a linear order. Of course, $S, 0, e$ are first-order definable from $<$ and thus need not be explicitly quantified over. We obtain that

$$\mathfrak{A} \in K \quad \text{iff} \quad \mathfrak{A} \models (\exists <)(\exists H)(\exists \bar{T})(\exists \bar{I})(\alpha \wedge \psi_M).$$

This proves that K is a generalized spectrum. □

Remark. Properties expressible in existential-second order logic are sometimes called Σ^1_1-properties. Hence Fagin's Theorem is often stated in the more succinct form: $\text{NP} = \Sigma^1_1$.

Exercise 2.2.19. Prove that every set in NP can be defined by a Σ^1_1-sentence whose first-order part has an $\forall^*\exists^*$-prefix. Furthermore, prove that this cannot be reduced to $\forall^*$.

Exercise 2.2.20. Derive the Cook-Levin Theorem (the NP-completeness of SAT) from Fagin's Theorem.

There are several interesting consequences of Fagin's Theorem. Asser [26] has formulated an outstanding open problem, namely whether the complement of every spectrum is again a spectrum. For generalized spectra this problem turns out to be equivalent to the problem whether NP = Co-NP.

Corollary 2.2.21. *The class of generalized spectra is closed under complementation if and only if* NP = Co-NP.

Corollary 2.2.22. *Let K be an isomorphism-closed class of finite structures of some fixed non-empty vocabulary τ. Then* code(K) *is in the polynomial-time hierarchy* PH *if and only if there exists a second-order sentence α such that K is the class of finite models of α.*

Fagin's characterization of generalized spectra implies also a characterization of spectra of first-order formulae which had been proved in a different way in [285].

Corollary 2.2.23 (Jones, Selman). *A set of natural numbers is a first-order spectrum if and only if it is element of* NEXPTIME.

Corollary 2.2.24 (Fagin). *Every set of natural numbers in* EXPTIME *is a categorical spectrum. (A spectrum is* categorical *if and only if it is the spectrum of a sentence that has, up to isomorphism, at most one model in any finite cardinality.)*

Exercise 2.2.25 (Advanced). Prove these two corollaries.

The Spectrum Theorem. The notion of spectra is not restricted to first-order logic. We present in this paragraph a solution for the spectrum problem for logics of arbitrary finite order.

We first discuss the notion of n-th order predicate logic. In fact there exist two such notions, which we refer to as *weak* and *strong* n-th order logic. Note that a first-order formula can contain second-order objects (predicates and function symbols) but quantification is allowed only over first-order objects. A formula is *weak second order* if it is second order (i.e. quantification over predicates and functions is allowed) but does not contain predicates (or functions) of predicates. A formula of *strong second order* can contain predicates of predicates (which cannot be quantified however). This distinction extends to predicate logic of order n: In weak n-th order logic we have only predicates and function up to n-th order whereas in strong n-th order logic we may have predicates and functions of order $n+1$ but quantification is restricted to objects of order n.

In the literature on the spectrum problem, n-th order logic usually meant strong n-th order logic. However, in most other branches of logic, notably model theory and finite model theory, the understanding of n-th order logic is weak n-th order logic (in particular for second-order logic).

The spectrum problem for n-th order logic is the problem of characterizing $\mathrm{SPECTRA}_n$, the class of spectra of formulae of strong n-th order predicate logic. The Spectrum Hierarchy Theorem gives a solution in terms of computational complexity. It was proved piecemeal in [32, 79, 285, 443]. We obtain it from an appropriate specification of the schema for economical description of Turing machine programs. The reader who is not interested in its general type theoretical form may consider m to be simply 1. We write $\exp_m(n)$ for the m-fold iterated exponential function, i.e. $\exp_0(n) = n$ and $\exp_{m+1}(n) = 2^{\exp_m(n)}$.

Theorem 2.2.26 (Spectrum Hierarchy Theorem). *For each $m \geq 1$, the class* SPECTRA$_m$ *coincides with the class of sets S of positive integers which are accepted by non-deterministic Turing machines in m-fold exponential time, i.e.*

$$\text{SPECTRA}_m = \{S \subseteq \mathbb{N} : S \in \text{NTIME}(\exp_m(O(n)))\}.$$

Proof. We first associate with each non-deterministic one-tape Turing machine M a reduction formula of order $m \geq 1$ whose spectrum is the set of (positive) numbers which is accepted by M in m-fold exponential time. To simplify the formulae describing the input, we give the construction in terms of *unary* input 1^n; in doing so we use the fact that the time bound $\exp_m(n^c)$, for some arbitrarily chosen but fixed constant c and unary input 1^n, corresponds to an $m+1$-fold exponential bound in the length of the binary representation of n. It is therefore sufficient to construct, for each nondeterministic Turing machine M and each c, a reduction formula

$$\text{STEP}_{M,c} \wedge \text{START}_c \wedge \text{STOP}_{M,c} \wedge \text{ORDER}_K(Z, S)$$

of order $m+1$ such that the following holds.

Reduction Property: For each number $n \geq 2$, M accepts input 1^n in $\exp_m(n^c)$ steps if and only if

$$\text{STEP}_{M,c} \wedge \text{START}_c \wedge \text{STOP}_{M,c} \wedge \text{ORDER}_K(Z, S)$$

is satisfiable over the universe $n = \{0, \ldots, n-1\}$.

The problem consists in describing computations of length $\exp_m(n^c)$ over a domain of n elements. The problem is similar to the one in the proof for Fagin's Theorem. Here we have to assure that possible models of our program formulae contain sufficiently many "objects" to encode the time and space parameters; in addition we need a successor relation among these objects. A simple solution is offered by the possibility to speak, in formulae of order $m+1$, about objects obtained by m iterations of the power set construction. We start with the c-fold Cartesian product of the set n. Formally, denote by $*$ the *type* (of the elements) of the domain, and define

$$\tau_0 := \underbrace{(* \ldots *)}_{c\text{-times}}, \quad \tau_{i+1} := (\tau_i).$$

τ_0 is the type of (the elements of) the c-fold Cartesian product of the domain of type $*$, τ_{i+1} the type of (the elements of) the power set of a set of (elements of) type τ_i. Over finite domain (say $n = \{0, 1, \ldots, n-1\}$) of n elements of type $*$ there are $\exp_m(n^c)$ elements of type τ_m These elements can be totally ordered by a formula $\text{ORDER}_K(Z, S)$, of order $m+1$, which formalises an order K with successor relation S and zero-predicate Z (i.e. satisfying $\exists z Zz \wedge \forall x(Zx \leftrightarrow \neg\exists y Kyx)$).

This allows us to specify our program formula schema to a program formula $\mathrm{STEP}_{M,c}$ of order $m+1$ as follows. In STEP_M all parameters t, t', x, x', y are treated as variables (for elements) of type τ_m and are universally quantified. Consequently the predicate symbols H, T_j, K, S and I_i, Z are of type $(\tau_m \tau_m)$ and (τ_m) respectively and therefore of order $m+1$. Their intended interpretation remains unchanged except for expressing the time and tape cell parameters by the order number $|x|$ of elements x in the linear ordering K of objects of type τ_m.

For the stop formula $\mathrm{STOP}_{M,c}$ we can use the same formula as in the proof of the Cook-Levin Theorem. We express the last configuration of the (non-deterministic) computation by referencing the last element in the K-ordering:

$$\mathrm{STOP}_{M,c} := \forall t((\neg \exists t' \; Ktt') \rightarrow \bigwedge_{i \neq 1} \neg I_i t)$$

What remains is to formalize the beginning of the computation on some a priori unknown input 1^n. We use an *embedding formula* to formalize an order-preserving embedding of the given domain of individuals into the initial segment of the K-ordering of elements of type τ_m. The domain of individuals is some $n = \{0, 1, \ldots, n-1\}$ and is ordered using a first-order formula $\mathrm{ORDER}_{K'}(Z', S')$. The graph F of an embedding function can then be formalized as conjunction of the following formulae:

$\forall u \exists x Fux$ (existence)

$\forall u \forall v \forall x \forall y (Fux \wedge Fvy \rightarrow (S'uv \leftrightarrow Sxy))$ (preservation of order)

$\forall u \forall x (Fux \wedge Z'u \rightarrow Zx)$ (preservation of minima)

$\mathrm{ORDER}_{K'}(Z', S')$

Note that order-preservation ensures that F is the graph of an injective function that maps n onto an initial K-segment.

We are now in a position to define the start formula START_c as conjunction of the embedding formula and the following formulae:

$\forall t(Zt \rightarrow I_0 t \wedge Htt)$ (at time 0: state 0, head position 0

$\forall t(Zt \rightarrow \forall x(\exists u Fux \rightarrow T_1 tx))$ (input 1 in domain of embedding)

$\forall t(Zt \rightarrow \forall x((\neg \exists u Fux) \rightarrow T_0 tx))$ (blank input 0 outside domain of F)

This concludes the definition of the reduction formula. It remains to prove that the reduction property is satisfied. If M accepts input n in at most $\exp_m(n^c)$ steps, then the configuration sequence of length $\exp_m(n^c)$ becomes constant at the first occurrence of the accepting state 1. Therefore the above indicated intended interpretation over n yields a model of

$$\mathrm{STEP}_{M,c} \wedge \mathrm{START}_c \wedge \mathrm{STOP}_{M,c} \wedge \mathrm{ORDER}_c.$$

Conversely the Simulation Lemma from Sect. 2.1.1 holds for an arbitrary model of this formula with universe n if in the configuration formulae $\underline{C_t}$ of C_t the time and space parameters t, x are replaced by their order numbers $|t|, |x|$ in the given model. The truth, in that model, of the stop formula then assures that at the final configuration, M can be only in its accepting state. □

Exercise 2.2.27. Prove that the satisfiability over n of formulae of order m is decidable by non-deterministic Turing machines in m-fold exponential time. Hint: Use the fact that for appropriate k, for each variable of any type τ occuring in the formula, there are at most $\exp_m(n^k)$ elements of type τ over a domain of n elements of type $*$. (See [443] for an explicit exponential arithmetic description of the finite satisfiability problem of m-th order formulae.)

Exercise 2.2.28. Deduce Fagin's Theorem from the Spectrum Theorem.

Note that, by the proof of the Spectrum Hierarchy Theorem, a set $S \subseteq \mathbb{N}$ is a first-order spectrum if and only if $\{1^n : n \in S\} \in \mathrm{NP}$.

2.2.3 Capturing Complexity Classes

Fagin's Theorem suggests the question whether there are similar logical descriptions of other important complexity classes, in particular for P. This question has led to the development of an important branch of finite model theory, called *descriptive complexity theory*. While computational complexity theory investigates the amount of machine resources (e.g. time or space) necessary to solve a given problem, descriptive complexity asks for the 'logical resources': in which logic is the given problem definable.

It has turned out that there are intimate connections between computational and descriptive complexity as long as one considers structures where a linear order is explicitly given (or definable in an appropriate way). On ordered structures, model-theoretic characterizations are known for most of the major complexity classes, and it is fair to say that logic plays a similar rôle as any abstract machine model.

However, if no linear order is available, then the relationship between complexity and definability is much more complicated. It is not known whether *any* model-theoretic characterization of polynomial-time complexity is possible on arbitrary finite structures. The reduction method, as used in the proof of Fagin's Theorem and Theorems 2.2.36 and 2.2.45 below, is not applicable without a linear order on the input structure. The problem whether there exists a 'logic for polynomial-time' is explained and discussed in [100, 232, 413].

In this section, we explain model-theoretic characterizations of P and NLOGSPACE on ordered structures. We show that variants of the reduction scheme used in the proof Fagin's Theorem yield descriptions of these complexity classes in terms of fragments of second-order logic (or, equivalently, in terms of generalized Horn and Krom spectra). We also show that the more common characterizations in terms of fixed-point logic and transitive closure logic can be easily derived from these results.

Definition 2.2.29. Let L be logic, $\mathcal{S}$ be a class of finite structures and $\mathcal{C}$ a complexity class. We say that L *captures* $\mathcal{C}$ *on* $\mathcal{S}$ if, for every isomorphism-closed class $K \subseteq \mathcal{S}$ of fixed finite vocabulary, the following are equivalent.

(i) There exists a sentence $\psi \in L$ such that $K = \{\mathfrak{A} \in \mathcal{S} : \mathfrak{A} \models \psi\}$.
(ii) The problem whether a given structure $\mathfrak{A} \in \mathcal{S}$ belongs to K is in the complexity class $\mathcal{C}$.

We simply say that L captures $\mathcal{C}$ in case that L captures $\mathcal{C}$ on the class of all finite structures.

Thus, Fagin's Theorem says that existential second order Σ_1^1 captures NP. Further, it is an almost immediate consequence of Fagin's Theorem that the polynomial-time hierarchy is captured by second-order logic (see [416, 491]).

Second-Order Horn Logic and Polynomial Time..

Definition 2.2.30. *Second-order Horn logic*, denoted, SO-HORN, is the set of second-order formulae of the form

$$(Q_1 R_1) \cdots (Q_m R_m) \forall y_1 \cdots \forall y_s \bigwedge_{i=1}^{t} C_i$$

where $Q_i \in \{\exists, \forall\}$, the R_i are relation symbols and each of the clauses C_i is a disjunction of atoms and negated atoms with at most one positive occurrence of a predicate R_j. Occurrences of equalities and inequalities, of predicates not belonging to $R_1, \ldots, R_m$, and negative occurrences of $R_1, \ldots, R_m$ are not restricted. Thus the quantifier-free part of the formulae in SO-HORN are Horn formulae with respect to the quantified predicates but not necessarily with respect to the 'input predicates' from the underlying vocabulary. Σ_1^1-HORN denotes the existential fragment of SO-HORN, i.e. the formulae where all second order quantifiers are existential.

It is convenient to write the clauses in 'logic programming notation'

$$H \leftarrow B_1 \wedge \cdots \wedge B_k.$$

The conjunction $B_1 \wedge \cdots \wedge B_k$ is the *body* of the clause; H is is either an atom $R_j \bar{u}$ or the symbol $\square$ indicating a contradiction and is called the *head* of the clause. (In this notation the predicates $R_1, \ldots, R_r$ always appear unnegated.)

Example. The problem GEN is a well-known P-complete problem [215, 283]. It may be presented as the set of structures (A, S, f, a) in the vocabulary of one unary predicate S, one binary function f and a constant a, such that a is contained in the closure of S under f. Clearly, the complement of GEN is also P-complete. It is defined by the following sentence of Σ_1^1-HORN:

$$(\exists R)\forall y \forall z\Big((Ry \leftarrow Sy) \land (Rfyz \leftarrow Ry \land Rz) \land (\Box \leftarrow Ra)\Big).$$

Example. The circuit value problem CVP is also P-complete [215], even when restricted to circuits with fan-in two over NAND-gates (also called Sheffer's stroke gates). Such a circuit can be considered as a structure (V, E, I^+, I^-, out) where (V, E) is a directed acyclic graph, I^+ and I^- are monadic and a is a constant; Exy means that node x is one of the two input nodes for y, I^+ and I^- contain the inputs node with value 1 and 0, respectively, and out stands for the output node.

We will take for granted that E is a connected, acyclic graph with fan-in two, sources $I^+ \cup I^-$ and sink out. Then the formula $(\exists T)(\exists F)\forall x \forall y \forall z \varphi$ where φ is the conjunction of the clauses

$$\begin{aligned}
Tx &\leftarrow I^+x \\
Fx &\leftarrow I^-x \\
Ty &\leftarrow Fx \land Exy \\
Fz &\leftarrow Tx \land Exz \land Ty \land Eyz \land y \neq z \\
\Box &\leftarrow Tx \land Fx \\
Tx &\leftarrow x = out
\end{aligned}$$

states that the circuit (V, E, I^+, I^-, out) evaluates to 1.

Exercise 2.2.31. To justify the definition of SO-HORN, show that the admission of quantifiers over functions or of first order prefixes of a more general form, would make the restriction to Horn clauses pointless. Any such extension of SO-HORN has the full power of second order logic.

Theorem 2.2.32. *Every sentence $\psi \in$ SO-HORN is equivalent to some sentence $\psi' \in \Sigma_1^1$-HORN.*

Proof. It suffices to prove the Theorem for formulae of the form

$$\psi := (\forall P)(\exists R_1) \cdots (\exists R_m)\forall \bar{z}\varphi$$

where φ is a conjunction of Horn clauses. Indeed, an arbitrary formula in SO-HORN may then be brought to existential form by successively removing the innermost universal second order quantifier. We first prove the following

Claim. *A formula $(\exists \bar{R})\forall \bar{z}\varphi(P, \bar{R}) \in \Sigma_1^1$-HORN is true for all predicates P (on a given structure $\mathfrak{A}$) if it holds for those predicates P that are false at at most one point.*

Let k be the arity of P. For every k-tuple $\bar{a}$, let $P^{\bar{a}} = A^k - \{\bar{a}\}$, i.e. the predicate that is false at $\bar{a}$ and true at all other points. By assumption there exist predicates $\bar{R}^{\bar{a}}$ such that

$$(\mathfrak{A}, P^{\bar{a}}, \bar{R}^{\bar{a}}) \models \forall \bar{z}\varphi.$$

For every predicate $P \neq A^k$ let

$$R_i := \bigcap_{\bar{a} \notin P} R_i^{\bar{a}}.$$

We claim that $(\mathfrak{A}, P, \bar{R}) \models \forall \bar{z} \varphi$.

Suppose that this is false; then there exists a relation $P \neq A^k$, a clause C of φ and an assignment $\pi : \{z_1 \ldots, z_s\} \to A$ such that $(\mathfrak{A}, P, \bar{R}) \models \neg C\pi$. We show that then there exists a tuple $\bar{a}$ such that also $(\mathfrak{A}, P^{\bar{a}}, \bar{R}^{\bar{a}}) \models \neg C\pi$.

If the head of $C\pi$ is $P\bar{u}$ then take $\bar{a} = \bar{u} \notin P$. If the head of $C\pi$ is $R_i\bar{u}$, then choose some $\bar{a} \notin P$ such that $\bar{u} \notin R_i^{\bar{a}}$; such a $\bar{a}$ must exist because $\bar{u} \notin R_i$. Otherwise (if the head is empty or an atom $Q\bar{u}$ where Q belongs to the vocabulary of $\mathfrak{A}$), take an arbitrary $\bar{a} \notin P$. The head of $C\pi$ is clearly false in $(\mathfrak{A}, P^{\bar{a}}, \bar{R}^{\bar{a}})$. The atom $P\bar{a}$ does not occur in the body of $C\pi$, because $\bar{a} \notin P$ and all atoms in the body of $C\pi$ are true in $(\mathfrak{A}, P, \bar{R})$; all other atoms of the form $P\bar{v}$ that might occur in the body of the clause remain true also for $P^{\bar{a}}$. Moreover every atom $R_i\bar{v}$ in the body remains true if R_i is replaced by $R_i^{\bar{a}}$ (because $R_i \subseteq R_i^{\bar{a}}$). This implies that the clause $(\mathfrak{A}, P^{\bar{a}}, \bar{R}^{\bar{a}}) \models \neg C\pi$ and thus

$$(\mathfrak{A}, P^{\bar{a}}, \bar{R}^{\bar{a}}) \models \neg \forall \bar{z} \varphi$$

which contradicts our assumption.

Thus the claim is established. This implies that the original formula ψ is equivalent to the conjunction

$$(\exists \bar{R}) \forall \bar{z} \varphi_0 \wedge \forall \bar{y} (\exists \bar{R}) \forall \bar{z} \varphi_1$$

where φ_1 (resp. φ_0) are obtained from φ by replacing every atom $P\bar{u}$ by $\bar{u} \neq \bar{y}$ (which is true iff $\bar{u} \in P^{\bar{y}}$), resp. by $(\bar{u} = \bar{u})$ (which is always true). It is easy to transform that conjunction into an equivalent formula in Σ_1^1-HORN. □

Theorem 2.2.33. *Let $\psi \in$ SO-HORN. Then the set of finite models of ψ is in* P.

Proof. By the previous theorem we can restrict attention to sentences $\psi = (\exists R_1) \cdots (\exists R_m) \forall \bar{z} \bigwedge_i C_i$ in Σ_1^1-HORN. Given any finite structure $\mathfrak{A}$ of appropriate vocabulary, we reduce the problem whether $\mathfrak{A} \models \psi$ to the satisfiability problem for a propositional Horn formula in the following way.

Replace the universal quantifiers $\forall z_i$ by conjunctions over the elements $z_i \in A$ and omit the quantifier prefix. Then substitute the relation symbols that belong to the vocabulary of $\mathfrak{A}$, including equalities and inequalities, by their truth values in $\mathfrak{A}$. If there is any clause that is already made false by this partial interpretation (i.e. the head is false and all atoms in the body are true) then reject ψ. Otherwise, omit all clauses that are already made true (i.e. the head is true or an atom in the body is false) and delete the already interpreted atoms from the remaining clauses. Consider the atoms $R_i\bar{u}$ as propositional variables. The resulting formula is a propositional Horn formula whose length is polynomially bounded in the cardinality of $\mathfrak{A}$ and which is satisfiable if and only if $\mathfrak{A} \models \psi$. It is known that the satisfiability problem for propositional Horn formulae can be solved in polynomial time (see e.g. [416, p. 79]). □

Recall that the class $\mathcal{O}(\tau)$ of *ordered* τ-structures is the class of all $(\mathfrak{A}, <)$ where $\mathfrak{A}$ is a finite τ-structure and $<$ a linear order $<$ on A. For describing polynomial-time complexity by SO-HORN the presence of a linear ordering does not suffice. We need to have a successor relation given explicitly.

Definition 2.2.34. A *successor structure* is a finite structure $(\mathfrak{A}, <, S, 0, e)$ where $<$ is a linear order on the universe of $\mathfrak{A}$, S is the corresponding successor relation and $0, e$ are the first resp. last elements of the universe with respect to $<$.

Theorem 2.2.35. *Every property of successor structures that is decidable in polynomial time is definable in* Σ_1^1-HORN.

Proof. This follows by an analysis of our proof for Fagin's Theorem. Indeed, if the Turing machine M happens to be deterministic then the sentence

$$(\exists H)(\exists \bar{T})(\exists \bar{I})\psi_M$$

constructed in that proof can easily be transformed to an equivalent sentence in Σ_1^1-HORN. To see this, recall that ψ_M is the universal closure of START $\wedge$ STEP$_M$ $\wedge$ ACCEPT. Here START is a conjunction of atomic statements and of implications $\beta_\sigma(\bar{x}) \rightarrow T_\sigma(\bar{0}, \bar{x})$. We already observed in Exercise 2.2.15 that, in the presence of $S, 0$ and e, we can assume that the $\beta_\sigma(\bar{x})$ are quantifier-free. Take the disjunctive normal form $\bigvee_i \beta_{\sigma,i}(\bar{x})$ of $\beta_\sigma(\bar{x})$ and replace the implication $\beta_\sigma(\bar{x}) \rightarrow T_\sigma(\bar{0}, \bar{x})$ by the conjunction of the clauses $\beta_{\sigma,i}(\bar{x}) \rightarrow T_\sigma(\bar{0}, \bar{x})$. Since the quantified predicates $H, \bar{T}, \bar{I}$ do not occur in $\beta_\sigma(\bar{x})$ these are clauses of the required form.

The formula STEP$_M$ is the conjunction of the formulae (2.1)–(2.3) of the reduction scheme (from page 20). If M is deterministic then these are indeed Horn clauses. (Note that the successor relation on k-tuples is definable from the basic successor relation via Horn clauses.) Finally ACCEPT $= \neg I_2(\bar{t})$ is also a Horn clause. □

We thus have established a logical characterization of polynomial time.

Theorem 2.2.36 (Grädel). SO-HORN *and* Σ_1^1-HORN *capture polynomial time on successor structures.*

Exercise 2.2.37. Prove that contrary to the case of Fagin's Theorem the assumption that a successor relation is explicitly given cannot be eliminated, since linear orderings and successor relations cannot be axiomatized by Horn formulae. In fact, even an explicitly given order $<$ does not suffice. Hint: Sentences in SO-HORN are preserved under substructures: If $\mathfrak{A} \models \psi$ and $\mathfrak{B} \subseteq \mathfrak{A}$ then also $\mathfrak{B} \models \psi$.

Fixed-Point Logics. There are a number of other model-theoretic characterizations of polynomial-time on ordered structures. The most important and well-known are those in terms of fixed-point logics [276, 517].

It is well-known that the expressive power of first-order logic is limited by the lack of a mechanism for unbounded iteration or recursion. The most notable example of a query that is not first-order expressible is the transitive closure (TC) of a relation. This has motivated the study of more powerful languages that add recursion in one way or another to first-order logic. Fixed-point logics are one possibility to achieve this.

Let τ be a vocabulary, P an k-ary predicate not in τ and $\psi(\bar{x})$ a formula of vocabulary $\tau \cup \{P\}$ with only positive occurrences of P, and let $\bar{x}$ be a k-tuple of variables. Then ψ defines for every τ-structure $\mathfrak{A}$ an operator $\psi^{\mathfrak{A}} : \mathcal{P}(A^k) \to \mathcal{P}(A^k)$ on the class of k-ary relations on A, namely

$$\psi^{\mathfrak{A}} : P \mapsto \{\bar{a} : (\mathfrak{A}, P) \models \psi[\bar{a}]\}.$$

Since P occurs only positively in ψ, this operator is monotone, i.e. $Q \subseteq P$ implies $\psi^{\mathfrak{A}}(Q) \subseteq \psi_{\mathfrak{A}}(P)$. Therefore this operator has a *least fixed point* which may be constructed inductively. Set $P^0 := \varnothing$, $P^{j+1} := \psi^{\mathfrak{A}}(P^j)$ and $P^\omega := \bigcup_{j<\omega} P^j$. For finite $\mathfrak{A}$ this process will reach the least fixed point P^ω in a polynomial number of steps.

Definition 2.2.38. The *least fixed-point logic* (FO + LFP) is defined by adding to the syntax of first order logic the *least fixed point formation rule:* If $\psi(\bar{x})$ is a formula of vocabulary $\tau \cup \{P\}$ with the properties stated above and $\bar{u}$ is a k-tuple of terms, then

$$[\mathrm{LFP}_{P,\bar{x}}\ \psi](\bar{u})$$

is a formula of vocabulary τ (to be interpreted as $P^\omega(\bar{u})$).

Example. Here is a fixed-point formula that defines the reflexive and transitive closure of the binary predicate E:

$$\mathrm{TC}(u,v) := [\mathrm{LFP}_{T,x,y}\ (x = y) \lor (\exists z)(Exz \land Tzy)](u,v).$$

Exercise 2.2.39. Prove that the problem GEN and the circuit value problem (see the examples for SO-HORN) are expressible in (FO + LFP).

Proposition 2.2.40. *Let ψ be a sentence in* (FO + LFP). *It is decidable in polynomial time whether a given finite structure $\mathfrak{A}$ is a model of ψ.*

This is obvious, given that the least fixed point of $\psi^{\mathfrak{A}}$ is reached after a polynomial number of iterations and that first-order operations are polynomial-time computable.

Theorem 2.2.41. *Every formula $\psi \in$* SO-HORN *is equivalent to some formula $\psi^* \in$* (FO + LFP).

Proof. By Theorem 2.2.32 we can assume that $\psi = (\exists R_1)\cdots(\exists R_m)\varphi \in \Sigma^1_1$-HORN. By combining the predicates $R_1, \ldots, R_m$ to a single predicate R of larger arity and by renaming variables it is easy to transform ψ into an equivalent formula

$$\psi' := (\exists R)\forall\bar{x}\forall\bar{y} \bigwedge_i C_i \wedge \bigwedge_j D_j$$

where C_i are clauses of the form $R\bar{x} \leftarrow \alpha_i(\bar{x}, \bar{y})$ (with exactly the same head $R\bar{x}$ for every i) and D_j are clauses of the form $\Box \leftarrow \beta_j(\bar{x}, \bar{y})$. Let R^ω be the least-fixed point (or equivalently, the minimal model) of the clauses C_i on $\mathfrak{A}$. R^ω is defined by the fixed-point formula

$$\alpha^\omega(\bar{z}) := [\mathrm{LFP}_{R,\bar{x}} \bigvee_i \exists\bar{y}\alpha_i(\bar{x}, \bar{y})](\bar{z}).$$

Let $\beta := \exists\bar{x}\exists\bar{y} \bigvee_j \beta_j(\bar{x}, \bar{y})$. Obviously,

$$\mathfrak{A} \models \psi \quad \Longleftrightarrow \quad (\mathfrak{A}, R^\omega) \models \forall\bar{x}\forall\bar{y} \bigwedge_i C_i \wedge \bigwedge_j D_j \quad \Longleftrightarrow \quad (\mathfrak{A}, R^\omega) \models \neg\beta.$$

Thus ψ is equivalent to the formula $\psi^* := \neg\beta[R\bar{z}/\alpha^\omega(\bar{z})]$ obtained from $\neg\beta$ by substituting all occurrences of atoms $R\bar{z}$ by $\alpha^\omega(\bar{z})$. Clearly this formula is in (FO + LFP). □

Theorem 2.2.42 (Immerman, Vardi). *On ordered structures, the least fixed point logic* (FO + LFP) *captures polynomial time.*

This is an immediate consequence of Theorem 2.2.36 and Theorem 2.2.41, together with the obvious fact that in (FO + LFP) the successor relation is definable from the linear ordering.

There are many other variants of fixed-point logic. We refer to [141] for more results on this subject.

Second-Order Krom Logic.

Definition 2.2.43. *Second order Krom logic*, denoted SO-KROM, is the set of second-order formulae

$$(Q_1R_1)\cdots(Q_mR_m)\forall y_1\cdots\forall y_s \bigwedge_{i=1}^{t} C_i$$

where every clause C_i is a disjunction of atoms and negated atoms with at most two occurrences of predicates $R_1, \ldots, R_m$. Such formulae are Krom (i.e. in 2-CNF) with respect to the quantified predicates. Σ^1_1-KROM is the existential fragment of SO-KROM. The intersection of Σ^1_1-HORN and Σ^1_1-KROM is denoted by Σ^1_1-KROM-HORN.

Example. The graph accessibility problem ("Is there a path in the graph (V, E) from a to b?") is complete for NLOGSPACE via first order translations. Its complement is expressible by a formula from Σ_1^1-KROM-HORN:

$$(\exists T)\forall x \forall y \forall z \Big(Txx \land (Txz \leftarrow Txy \land Eyz) \land (\Box \leftarrow Tab)\Big).$$

As in the case of SO-HORN it is also known that every sentence of SO-KROM is equivalent to a sentence of Σ_1^1-KROM (see [207]).

Proposition 2.2.44. *Let* $\psi \in$ SO-KROM. *Then the set of finite models of* ψ *is in* NLOGSPACE.

The proof is analogous to the proof of Theorem 2.2.33. It uses the fact that the satisfiability problem for propositional Krom formulae is in NLOGSPACE.

On successor structures, SO-KROM captures NLOGSPACE. We indicate the general idea of the proof. Suppose that M is an $O(\log n)$-space bounded nondeterministic Turing machine with an input tape, carrying a representation $code(\mathfrak{A}, <)$ of an input structure, and one or more separate work-tapes. A *reduced configuration* of M reflects the control state of M, the content of the work tapes and the positions of the heads on the input tape and the work tapes. Thus a configuration is specified by a reduced configuration together with the input. Given that reduced configurations of M on input $code(\mathfrak{A}, <)$ have logarithmic length with respect to $|A|$, we can represent them by tuples $\bar{c} = c_1, \ldots, c_r \in A^r$ for fixed r. The *initial* reduced configuration on any input $code(\mathfrak{A}, <)$ is represented by the tuple $\bar{0}$. Assume that M has a single accepting state, say state 1, and let the first component of a reduced configuration describe the state; then the condition that $\bar{y}$ represents an *accepting* reduces configuration is expressed by $\text{ACCEPT}(\bar{y}) := (y_1 = 1)$. Further, is not difficult (although a bit lengthy) to write down a quantifier-free formula $\text{NEXT}(\bar{x}, \bar{y})$ such that, for every successor structure $(\mathfrak{A}, S, 0, e)$ and every tuple $\bar{c}$ representing a reduced configuration,

$$(\mathfrak{A}, S, 0, e) \models \text{NEXT}[\bar{c}, \bar{d}]$$

if and only if $\bar{d}$ represents a reduced successor configuration of $\bar{c}$ on input $(\mathfrak{A}, <)$. Taking the disjunctive normal form $\text{NEXT}(\bar{x}, \bar{y}) = \bigvee_i \text{NEXT}_i(\bar{x}, \bar{y})$ we can express that M does *not* accept the input $code(\mathfrak{A}, <)$ by the the sentence

$$\begin{aligned}\psi_M \quad := \quad & (\exists R)\forall \bar{x} \forall \bar{y} \big(R\bar{0} \land \bigwedge_i (R\bar{y} \leftarrow R\bar{x} \land \text{NEXT}_i(\bar{x}, \bar{y})) \\ & \land (\Box \leftarrow R\bar{y} \land \text{ACCEPT}(\bar{y})\big).\end{aligned}$$

This proves that, on successor structures, the complement of every problem in NLOGSPACE is definable in SO-KROM. Since NLOGSPACE is closed under complement, and since the formula ψ_M is in fact in Σ_1^1-KROM-HORN we have proved the following result.

Theorem 2.2.45 (Grädel). *On successor structures, the logics* SO-KROM, Σ^1_1-KROM *and* Σ^1_1-KROM-HORN *capture* NLOGSPACE.

Remark. The characterizations of P and NLOGSPACE by second-order Horn and Krom logics can also be reformulated in terms of generalized spectra. The notion of a generalized spectrum can be appropriately modified to the notions of a *generalized Horn spectrum* and a *generalized Krom spectrum.* Let a *model class* be any isomorphism-closed class of structures of some fixed finite signature. Fagin's Theorem and Grädel's Theorems 2.2.36 and 2.2.45 can the be summarized as follows.

- A model class of finite structures is NP iff it is a generalized spectrum.
- A model class of successor structures is in P iff it is a generalized Horn spectrum.
- A model class of successor structures is in NLOGSPACE iff it is a generalized Krom spectrum.

Transitive Closure Logic. A similar technique yields Immerman's earlier characterization of NLOGSPACE in terms of transitive closure logic [277].

Definition 2.2.46. *Transitive closure logic*, denoted (FO + TC), is obtained by augmenting the syntax of first order logic by the following rule for building formulae:

Let $\varphi(\bar{x},\bar{y})$ be a formula with variables $\bar{x} = x_1,\dots,x_k$ and $\bar{y} = y_1,\dots,y_k$, and let $\bar{u}$ and $\bar{v}$ be two k-tuples of terms. Then

$$[\mathrm{TC}_{\bar{x},\bar{y}}\ \varphi(\bar{x},\bar{y})](\bar{u},\bar{v})$$

is a formula, which says that the pair $(\bar{u},\bar{v})$ is contained in the reflexive, transitive closure of the binary relation on k-tuples that is defined by φ.

Of course, it is understood that φ can contain other free variables than $\bar{x}$ and $\bar{y}$; these will be free also in the new formula. Moreover, transitive closure logic is closed under the usual first order operations. We thus can build Boolean combinations of TC-formulae, we can nest TC-operators etc.

Exercise 2.2.47. Show that for every $\psi \in$ (FO + TC), the set of finite models of ψ is decidable in NLOGSPACE.

The same idea as in the proof of Theorem 2.2.45 shows that, on ordered structures, (FO + TC) captures NLOGSPACE. The condition that an $O(\log n)$ space bounded Turing machine M accepts *code*$(\mathfrak{A},<)$ is expressed by the formula

$$\exists\bar{z}\big(\mathrm{ACCEPT}(\bar{z}) \wedge [\mathrm{TC}_{\bar{x},\bar{y}}\ \mathrm{NEXT}(\bar{x},\bar{y})](\bar{0},\bar{z})\big).$$

Theorem 2.2.48 (Immerman). *On ordered structures,* (FO + TC) *captures* NLOGSPACE.

Exercise 2.2.49. Prove that, on arbitrary finite structures, SO-KROM is strictly weaker than (FO + TC).

Remark. Another characterization of NLOGSPACE, in terms of *narrow Henkin quantifiers,* is due to Blass and Gurevich [36].

2.2.4 A Decidable Prefix-Vocabulary Class

As the last example in this section we introduce a theme that will be treated in more detail in subsequent chapters. We derive two results on the complexity of the decision problem of fragments of the so-called *Bernays-Schönfinkel class.* This class is denoted by $[\exists^*\forall^*]$ and consists of all prenex sentences of the pure predicate calculus (without function symbols and equality) with prefix of the form

$$\exists u_1 \cdots \exists u_m \forall x_1 \cdots \forall x_n.$$

It was one of the first formula classes for which the satisfiability problem was shown to be recursive [35]; much later it turned out to be one of the two maximal decidable prefix classes. This eminent rôle of the Bernays-Schönfinkel class is confirmed by the fact that the restrictions of the class to Krom formulae or to Horn formulae yields classes whose decision problem is complete for outstanding complexity classes. We use here the reduction method to establish the hardness results; the decision procedures establishing the corresponding upper complexity bounds will be given in Chap. 8.

Theorem 2.2.50. *The satisfiability problem for the Bernays-Schönfinkel class, restricted to Krom sentences, is* PSPACE*-hard. This is even true for the subclass* $[\exists^2\forall^*] \cap \mathrm{KROM} \cap \mathrm{HORN}$.

Proof. We paraphrase the proof of Theorem 2.1.15 of Aanderaa and Börger. Here we reduce the problem $C_0(w) \Rightarrow_{M,p(|w|)} C_{acc}$ of acceptance for deterministic Turing machine programs M with polynomial space bound $p(|w|)$, for input words w. We reduce it to the problem to logically derive $\underline{C_{acc}}$ from the Krom and Horn formula $\forall x_1 \ldots \forall x_m \mathrm{STEP}_{M,w} \wedge \underline{C_0(w)}$. Since this formula will be element of the class in question and of length bounded by $p(|w|)$ (and since PSPACE is closed under complementation), this suffices to establish the claim.

To simplify notation we imagine M-configurations with tape length m given in the form $C = j_1 \ldots j_{l-1}(i,j)j_{l+1} \ldots j_m$, where j_λ, j represent letters of the alphabet of M, i an internal state of M and the occurrence of the pair (i,j) the reading head position. The logical encoding $\underline{C}$ is defined as the atomic formula $Cj_1 \ldots j_{l-1}(i,j)j_{l+1} \ldots j_m$ where $j_\lambda, (i,j)$ are read as individual constants and C as m-ary predicate symbol. Given input word $w_1 \ldots w_n$ and $m := p(|w|)$, the intended interpretation of Ct is that $C_0(w) \Rightarrow_{M,m} t$, where $C_0(w)$ denotes the initial configuration $(0, w_1)w_2 \ldots w_n 0 \ldots 0$.

The program formula $\mathrm{STEP}_{M,w}$ is therefore defined as conjunction of the implications $Cx(i,j)y \to Cx(i',j')y$ (for print instructions (i,j,j',i') of M) and $Cx(i,j)j'y \to Cxj(i',j')y$ (for right movement instructions (i,j,r,i') of M) and analogous implications for left move instructions. Here $x = x_1 \dots x_{l-1}$ and $y = x_{l+1} \dots x_m$ or $y = x_{l+2} \dots x_m$ for all $l \leq m$ (for print instructions) or $l < m$ (for right movement instructions).

As initial and final formulae we obviously set START $:= \underline{C_0(w)}$ and END $:= \underline{C_{acc}}$. Without loss of generality $C_{acc} = (1,0)0 \cdots 0$, i.e. M accepts in state 1 with reading head at the left end of the empty tape.

Exercise 2.2.51. Verify that M accepts w iff the formula

$$\forall x_1 \cdots \forall x_m \mathrm{STEP}_{M,w} \wedge \underline{C_0(w)} \wedge \underline{C_{acc}}$$

is contradictory.

To obtain the reduction with only two existential quantifiers it suffices to carry out the Turing machine tape encoding over the two element alphabet {0,1}. □

Remark. Using the determinacy of M, the preceding construction can easily be modified to make the reduction formulae *determinate* in the sense of Prolog (see [421]).

Theorem 2.2.52. *The satisfiability problem for the restriction of the Bernays-Schönfinkel class to Horn formulae is* EXPTIME-*hard.*

Proof. We reduce the acceptance problem "M accepts input w in $< 2^{c|w|}$ steps" for deterministic Turing machine programs M. We reduce the problem to the satisfiability problem of a Bernays-Schönfinkel Horn formula

$$\overline{\forall}\mathrm{STEP}_M \wedge \mathrm{START}_w \wedge \mathrm{END} \wedge \eta$$

Here START_w, END describe the initial and final accepting configuration and STEP_M the program M, using an auxiliary formula η defined below. $\overline{\forall}$ denotes the universal closure.

Consider M-computations of length $\leq \ell := 2^{c|w|}$, started with input word w. To simplify the notation we represent such a computation as an $\ell \times \ell$ matrix μ where in the t-th row the entry $\mu(x,t)$ is the letter j in non-reading head positions x at time t or the pair (i,j) of state i of M and letter j in the reading head position x of M at time t. For a short description of such computations we divide the part of the tape that has been used into segments of length $n := c|w|$. We encode time and tape parameters $t, u < 2^n$ in a segment by binary sequences $\overline{t}, \overline{u}$ of length n. The logical representation $\underline{C}$ of configurations C is defined as conjunction of atomic formulae $T(\overline{x},\overline{u},\overline{t})$ with the following intended interpretation (with respect to the given computation):

$$T(\overline{x},\overline{u},\overline{t}) \text{ is true iff } \overline{x} \text{ is the } \overline{u}\text{-th tape segment at time } \overline{t}.$$

In these atomic formulae we treat letters j, state-letter pairs (i, j) and binary digits as individual constants.

For the acceptance condition, namely that M at time $2^n - 1$ will be in state 1, we assume without loss of generality that at time $2^n - 1$ the reading head position will in any case be at the left end of the empty tape. Since the time $2^n - 1$ moment is encoded in binary by $\overline{1}$, the acceptance condition can be expressed by :

$$\text{END} := \bigwedge_{i \neq 1} \neg T((i, 0)0 \ldots 0, \overline{0}, \overline{1})$$

Initial configurations have the form $(0, w_0)w_1 \ldots w_{m-1}\overline{0}\overline{0} \ldots \overline{0}$ with input word $w = w_0 \ldots w_{m-1}$ followed by a sequence $\overline{0}$ of blanks in the 0-th tape segment, with initial M-state 0, reading head position 0 and sequences $\overline{0}$ of blanks in all other segments to the right. These configurations can be represented by the initial formula

$$\text{START}_w := T((0, w_0)w_1 \ldots w_{m-1}\overline{0}, \overline{0}, \overline{0}) \wedge \forall \overline{u} T_r(\overline{0}, \overline{u}, \overline{0})$$

where the auxiliary predicate T_r formalizes a 1-segment shift of T to the right. This means that the intended interpretation of $T_r(\overline{x}, \overline{u}, \overline{t})$ is $T(\overline{x}, \overline{u + 1}, \overline{t})$.

In the program formula we will use also auxiliary predicates T_l for a 1-segment shift of T to the left ("left neighbour segment") and T_+ for a shift to the next moment, i.e. with the intended interpretation:

$$\begin{array}{lll} T_l(\overline{x}, \overline{u}, \overline{t}) & \text{iff} & T(\overline{x}, \overline{u - 1}, \overline{t}) \\ T_+(\overline{x}, \overline{u}, \overline{t}) & \text{iff} & T(\overline{x}, \overline{u}, \overline{t + 1}) \end{array}$$

Let $\overline{M}$ be the function determined by M that assigns to each triple $(\overline{x}, \overline{y}, \overline{z})$ of neighbouring tape segments at time t the successor tape segment $M(\overline{x}, \overline{y}, \overline{z})$ at time $t+1$ of segment $\overline{y}$. Then the program formula STEP_M can be defined as conjunction of all the relevant implications

$$T_l(\overline{x}, \overline{u}, \overline{t}) \wedge T(\overline{y}, \overline{u}, \overline{t}) \wedge T_r(\overline{z}, \overline{u}, \overline{t}) \to T_+(\overline{M}(\overline{x}, \overline{y}, \overline{z}), \overline{u}, \overline{t}).$$

Special conjuncts correspond to transitions in the first and the last segment.

Exercise 2.2.53. Spell out this program formula in full.

What remains to define is the auxiliary formula η that provides the intended interpretation of the auxiliary predicates T_r, T_l, T_+. They are intended to describe neighbouring segments or segments at the next time moment. Consider for instance T_l. We have to ensure that in our models $T(\overline{x}, \overline{y}, \overline{z})$ implies $T_l(\overline{x}, \overline{y + 1}, \overline{z})$ and vice versa. Therefore essentially we have to formalize binary addition and subtraction of 1 on subsequences $\overline{y}$ (representing numbers up to $2^n - 1$) of sequences $\overline{x}, \overline{y}, \overline{z}$. This is easily achieved as follows. First we shift the sequence $\overline{y}$ stepwise from right to left, simultaneously changing 1 to 0 (reflecting 1+1=0) until the first appearance of 0; then this 0 is changed to 1 (reflecting 0+1=1); then the remaining digits are shifted unchanged to the left.

To formalize this idea let A and P be new predicate symbols to represent the two phases of "adding" and "permuting"; let c be a new individual constant to mark the beginning of the "counter" $\overline{y}$; let $\overline{x}, \overline{y}, \overline{z}$ be sequences of (universally quantified) variables of length n. The conjunction of the following formulae assures the desired transition from $T(\overline{x}, \overline{y}, \overline{z})$ to $T_l(\overline{x}, \overline{y+1}, \overline{z})$ where $\overline{y}$ represents a number less than $2^n - 1$.

Start: Mark the left end of the counter and start the addition:

$$T(\overline{x}, \overline{y}, \overline{z}) \rightarrow A(\overline{x}, c, \overline{y}, \overline{z}).$$

Addition: Add 1 and shift 0 to the left (formalizing 1+1=0) until 0 is encountered; shift this 0 to the left as 1 (formalizing 0+1=1) and start then the permutation to the left:

$$(A(\overline{x}, \overline{y}, 1, \overline{z}) \rightarrow A(\overline{x}, 0, \overline{y}, \overline{z})) \wedge (A(\overline{x}, \overline{y}, 0, \overline{z}) \rightarrow P(\overline{x}, 1, \overline{y}, \overline{z})).$$

Permutation: Permute the remaining counter digits unchanged (for $i = 1, 2$):

$$P(\overline{x}, \overline{y}, i, \overline{z}) \rightarrow P(\overline{x}, i, \overline{y}, \overline{z}).$$

Stop: Stop the subprocess when the left end of the counter is reached:

$$P(\overline{x}, \overline{y}, c, \overline{z}) \rightarrow T_l(\overline{x}, \overline{y}, \overline{z}).$$

Similar conjuncts have to be added to η for going from $T_l(\overline{x}, \overline{y+1}, \overline{z})$ to $T(\overline{x}, \overline{y}, \overline{z})$, and analogously for T_r and T_+.

Exercise 2.2.54. Show that these formulae describe the acceptance problem of M as stated above.

□

Remark. It is interesting that Theorems 2.2.50 and 2.2.52 were first proved in the context of dependency theory for relational databases. In fact Chandra, Lewis and Makowsky [74] proved that the *inference problem for full implicational dependencies* is EXPTIME-complete; in the same paper they showed that the restriction to dependencies with only one atom in the antecedent is PSPACE-complete.

A full implicational dependency (FID) is in fact just a universal relational Horn sentence. The inference problem is the question whether a given finite conjunction of FID's implies another given FID. Clearly this just the (un)satisfiability problem for a $\exists^*\forall^*$-Horn sentence (in fact a conjunction of a universal and an existential Horn sentence). If the body of each FID is atomic then the inference problem corresponds to an $\forall^*\exists^*$-sentence in Krom-Horn form. We refer to [14] for background on database theory.

These results were later reproved, without the connection to dependency theory in [106, 421].

Exercise 2.2.55. Show that the satisfiability problem for Bernays-Schönfinkel formulae is hard for nondeterministic exponential time. Hint: Replace in the preceding proof the conclusions $T_+(\overline{M}(\overline{x},\overline{y},\overline{z}),\overline{u},\overline{t})$ by disjunctions $\bigvee_{\overline{v}\in\overline{M}(\overline{x},\overline{y},\overline{z})} T_+(\overline{v},\overline{u},\overline{t})$. For a different proof, see Sect. 6.2 of this book

2.3 The Classifiability Problem

2.3.1 The Problem

We have seen that Hilbert's *Entscheidungsproblem* was solved negatively, whereas many special cases of it, that is restrictions of the *Entscheidungsproblem* to particular classes of formulae, were solved positively. This leads in a natural way to the investigation of the exact boundary between decidable and undecidable classes of formulae. The *Entscheidungsproblem* turns into a meta problem to classify classes into decidable and undecidable.

However, there are continuum many classes of formulae. In the context of algorithmic solutions of decision problems, it is reasonable to restrict attention to constructively presented classes. Historically most attention has been given to classes of prenex formulae given by syntactical restrictions on the vocabulary, the quantifier prefix, the form of the quantifier free part. Notice that the classification problem arises not only in the connection with recursive decidability and undecidability. It also arises in the realm of decidable classes, for example, in terms of decidability within given resource bounds (easy versus hard). There are many other variations of the classifiability problem. Pure predicate logic can be extended with function variables, equality and maybe additional predicate or function constants satisfying some axioms; see other examples at the end of this section.

The theorem of Gurevich [222] that we present in this section, guarantees a complete and satisfactory solution of various classification problems for prefix-vocabulary classes of prenex formulas. The idea is that these classes form a well partially ordered set (we will recall the definition of that important notion) where the collection of positive classes (that is classes with a positive solution) is closed downward. It follows that a finite collection of minimal negative classes characterizes *all* negative classes: each negative class dominates at least one of the minimal negative classes. Moreover, the minimal negative classes are *standard* in a sense which will be made precise below and thus have succinct names. The number of maximal standard positive classes is finite as well (but not every standard positive class is dominated by a maximal standard positive class). Therefore, in the main body of the book, the prefix-vocabulary classes of interest will be almost exclusively standard classes.

Gurevich's Classifiability Theorem can be further extended; one such extension is found in Sect. 5.4. It is the main organizing principle of this book.

2.3.2 Well Partially Ordered Sets

In this section we recall the basics of the theory of well partially ordered sets (wpo sets) including the Finite Sequence Theorem (also known as Higman's Lemma [260]). A history of the subject and further references can be found in [335]. Our proof of the Finite Sequence Theorem is a version of Nash Williams' proof [403].

Definition 2.3.1. A *quasi order* (or a *pre-order*) is a reflexive and transitive binary relation. A *quasi ordered set* (in short a *qoset*) is a set with a quasi order. A *partial order* is an antisymmetric quasi order, in other words a quasi order where $x \leq y \leq x$ implies $x = y$. A partially ordered set (in short a *poset*) is a set with a partial order.

Definition 2.3.2 (Wpo and Wqo Sets). A qoset is a *well quasi ordered set* (in short a *wqo set*) if, for every infinite sequence $a_1, a_2, \ldots$, there exist $i < j$ such that $a_i \leq a_j$. A *well partially ordered set* (in short, a *wpo set*) is a poset that is wqo.

Call elements x and y of a qoset $(A, \leq)$ *equivalent*, write $x \equiv y$, if $x \leq y$ and $y \leq x$. Order the equivalence classes in the natural way: $[x] \leq [y]$ if $x \leq y$. (Notice that the choice of representatives is immaterial: if $x \leq y$, $x' \in [x]$ and $y' \in [y]$ then $x' \leq x \leq y \leq y'$.) The result is a poset called the *quotient poset* of $(A, \leq)$.

Exercise 2.3.3. A qoset $(A, \leq)$ is wqo if and only if its quotient poset is wpo. Further suppose that B is a subset of A that contains exactly one element from every equivalence class. Then $(A, \leq)$ is wqo if and only if the substructure $(B, \leq)$ is a wpo set.

Elements x, y of a qoset are *incomparable* if neither $x \leq y$ nor $y \leq x$. A set of pairwise incomparable elements is an *antichain*. $x < y$ will mean that $x \leq y$ and $x \not\equiv y$. Of course, $y \geq x$ means $x \leq y$, and $y > x$ means $x < y$. A sequence $a_1 \leq a_2 \leq \cdots$ is *increasing* or *weakly increasing*, and a sequence $b_1 < b_2 < \cdots$ is *strictly increasing*. Similarly, a sequence $a_1 \geq a_2 \geq \cdots$ is *decreasing* or *weakly decreasing*, and a sequence $b_1 > b_2 > \cdots$ is *strictly decreasing*. (We speak about sequences which are either finite or of type ω, like natural numbers.)

Exercise 2.3.4. Let $(A, \leq)$ be a qoset. The following statements are equivalent.

1. The qoset is wqo.
2. Every infinite sequence has an infinite weakly increasing subsequence.
3. All strictly decreasing sequences are finite, and all antichains are finite.
4. Every nonempty subset X has at most one minimal element and there is only a finite number of inequivalent minimal elements of X.

Hint: Prove $1 \to 2 \to 3 \to 4 \to 1$.

Even though qosets occur naturally in many applications, it is often more convenient to deal with posets.

Exercise 2.3.5. Let $(A, \leq)$ be a poset. The following statements are equivalent.

1. The poset is wqo.
2. Every nonempty subset contains at least one and at most finitely many minimal elements.

A subset X of a qoset $(A, \leq)$ is *upward closed* if $y \geq x \in X$ implies $y \in X$. The *upward closure* of X is the set $\{y : (\exists x \in X)(y \geq x)\}$. The notions *downward closed* and *downward closure* are defined similarly.

Exercise 2.3.6. Suppose that $(A, \leq)$ is a wpo set, $X \subseteq A$, M the set of the minimal elements of X, and $\overline{M}$ is the upward closure of M. Then $X \subseteq \overline{M}$. If X is upward closed then $X = \overline{M}$.

Exercise 2.3.7. Suppose that $(B, \leq)$ is a substructure of a qoset $(A, \leq)$. If $(A, \leq)$ is wqo, then so is $(B, \leq)$. If $(B, \leq)$ is wqo and $A - B$ is finite, then $(A, \leq)$ is wqo.

Recall that a homomorphism from a qoset $(A_1, \leq_1)$ to a qoset $(A_2, \leq_2)$ is a mapping $f : A_1 \to A_2$ such that $x \leq_1 y$ implies $fx \leq_2 fy$. If, in addition, f maps A_1 onto A_2, then $(A_2, \leq_2)$ is a homomorphic image of $(A_1, \leq_1)$.

Lemma 2.3.8. *If a qoset (respectively poset) $(A_2, \leq_2)$ is a homomorphic image of a wqo set $(A_1, \leq_1)$ then $(A_2, \leq_2)$ is wqo (respectively wpo).*

Proof. Let f be a homomorphism from $(A_1, \leq_1)$ onto $(A_2, \leq_2)$ and consider any infinite sequence $fx_1, fx_2, \ldots$. Since $(A_1, \leq_1)$ is wqo, there exist $i < j$ such that $x_i \leq_1 x_2$. But then $fx_1 \leq_1 fx_2$. □

Recall that the direct product of posets $(A, \leq)$ and $(B, \leq)$ is the direct product $A \times B$ of the universes ordered as follows: $(a_1, b_1) \leq (a_2, b_2)$ if and only if $a_1 \leq a_2$ and $b_1 \leq b_2$. The direct product of several posets is defined similarly. (It would be more correct to use different names for the two given orders. It may happen a priori that $x \leq y$ with respect to one order but not the other. Alternatively, one can assume without loss of generality that A and B are disjoint.) The following lemma follows from the Finite Sequence Theorem (and the fact that the wqo property is preserved by substructures) but we will prove it anyway, as a warm-up.

Lemma 2.3.9. *Direct products of finitely many wqo sets are wqo.*

Proof. It suffices to prove the lemma for the direct product of two qosets $(A, \leq), (B, \leq)$ with disjoint universes. Let S be be an arbitrary sequence of pairs $(a_1, b_1), (a_2, b_2), \ldots$. Since $(A, \leq)$ is wqo, the sequence $a_1, a_2, \ldots$ has an infinite weakly increasing subsequence $a_{f1} \leq a_{f2} \leq \cdots$. Since $(B, \leq)$ is wqo, the sequence $b_{f1}, b_{f2}, \ldots$ has an infinite weakly increasing subsequence $b_{g1} \leq b_{g2} \leq \cdots$. Thus S has an infinite weakly increasing sequence $(a_{g1}, b_{g1}) \leq (a_{g2}, b_{g2} \leq \cdots$. □

Any set A can be viewed as an alphabet. Then finite sequences of elements of A are *words* over A. Instead of $x = (a_1, \ldots, a_m)$, we write $x = a_1 \ldots a_m$; the length m will be denoted $|x|$. Let A^* be the collection of all words over A. A quasi order $\leq$ on A can be extended to the following *embedding order* on finite sequences: $a_1 \ldots a_m \leq b_1 \ldots b_n$ if there exists a monotone one-to-one map f from $[1 \ldots m]$ into $[1 \ldots n]$ such that $a_i \leq b_{fi}$ for all $i = 1, \ldots, m$.

Exercise 2.3.10. If $(A, \leq)$ is a poset then so is $(A^*, \leq)$.

Theorem 2.3.11 (Finite Sequence Theorem). *If $(A, \leq)$ be a wqo set then so is $(A^*, \leq)$.*

Proof. Let x, y range over A^*. Call an infinite sequence $x_1, x_2, \ldots$ *bad* if there exist no $i < j$ such that $x_i \leq x_j$. By contradiction, suppose that there are bad sequences in $(A^*, \leq)$. Choose

- x_1 to be a shortest word such that some infinite bad sequence starts with x_1,
- x_2 to be a shortest word such that some infinite bad sequence starts with x_1, x_2,
- x_3 to be a shortest word such that some infinite bad sequence starts with x_1, x_2, x_3, and so on.

This gives a particular infinite bad sequence $x_1, x_2, \ldots$. Let a_n be the first letter of x_n, and y_n be the rest of x_n, so that $x_n = a_n y_n$. Since $(A, \leq)$ is wqo, the sequence $a_1, a_2, \ldots$ has an infinite weakly increasing subsequence $a_{f1}, a_{f2}, \ldots$. Let $m = f1$. It is easy to see that the sequence

$$x_1, \ldots, x_{m-1}, y_{f1}, y_{f2}, \ldots$$

is bad. But this contradicts the choice of x_m. □

An ordinary finite alphabet A can be seen as wpo alphabet where the letters are pairwise incomparable. The order $x \leq y$ on words is the relation x is a (not necessarily contiguous) subword of y.

Corollary 2.3.12. *The set of all words over a finite alphabet together with the subword relation is a wpo set.*

We we will need one additional lemma from [222] about pwo sets.

Lemma 2.3.13. *Let $(A, \leq)$ be a poset where every finite subset X has a least upper bound* $\sup(X)$. *Suppose that $(A, \leq)$ has a wpo substructure $(B, \leq)$ such that, for every $x \in A$, there is a finite $Y \subseteq B$ such that $x = \sup(Y)$ in $(A, \leq)$. Then $(A, \leq)$ is wpo.*

Proof. By the Finite Sequence Theorem (Theorem 2.3.11), the set B^* of finite sequences of elements of B together with the embedding order $\leq^*$ is wqo. Since $(B, \leq)$ is a poset, $(B^*, \leq^*)$ is wpo. For every finite seqence $x = (b_1, \ldots, b_n)$ let $f(x) = \sup\{b_1, \ldots, b_n\}$ in $(A, \leq)$. It is easy to see that f is a homomorphism from $(B^*, \leq^*)$ onto $(A, \leq)$. By Lemma 2.3.8, $(A, \leq)$ is wpo. □

2.3.3 The Well Quasi Ordering of Prefix Sets

In this section we prove that a natural quasi ordering of sets of quantifier prefixes is a well quasi ordering. Further, we define standard prefix sets with succinct names and prove that every prefix set is equivalent to a finite union of standard sets.

Strings in the alphabet $\{\forall, \exists\}$ will be called *prefixes.* According to Corollary 2.3.12, the subword relation on prefixes is a well partial order. This order will be denoted by $\leq$.

A *prefix set* is simply a set of prefixes. Call a prefix set *closed* if it is downward closed. In other words, a prefix set Π is closed if and only if $x \leq y \in \Pi$ implies $x \in \Pi$. The closure of a prefix set Π will be denoted $\overline{\Pi}$.

Definition 2.3.14. A prefix set Π_1 *dominates* a prefix set Π_2, symbolically $\Pi_1 \geq \Pi_2$, if $\overline{\Pi_2} \subseteq \overline{\Pi_1}$.

It is this domination ordering that will be proved being well quasi ordering. It suffices to prove that the inclusion order of closed prefix classes is a well partial order. In general, we will be primarily interested in closed prefix classes.

Some prefix set can be given by regular expressions. Let λ be the regular expression for the empty set. Since we are interested primarily in closed prefix sets, we change slightly the semantics of regular expressions over the prefix alphabet $\{\forall, \exists\}$ so that each of them denotes a closed prefix set. Namely, the regular expression $\forall$ will denote the set $\{\lambda, \forall\}$ (rather than $\{\forall\}$); similarly, the regular expression $\exists$ will denote the set $\{\lambda, \exists\}$. The operations of concatenation, union and iteration (the Kleene star operation) have their usual meaning. If e is a regular expression, let (e) be the prefix class denoted by e.

Example $(\forall\exists\forall) = \{\lambda, \forall, \exists, \forall\forall, \forall\exists, \exists\forall, \forall\exists\forall\}$, and $(\forall^*\exists^2) = \{A^i\exists^j : 0 \leq i, 0 \leq j \leq 2\}$.

The alphabet $\{\forall, \exists, \forall^*, \exists^*\}$ will be called the *extended prefix alphabet.* Strings over the extended prefix alphabet will be called *generalized prefixes.*

Definition 2.3.15. A prefix set is *standard* if either it is the set of all prefixes, or else it can be given by a generalized prefix.

The standard prefix sets are partially ordered by inclusion.

Lemma 2.3.16. *The poset of standard prefix sets is wpo.*

Proof. By Corollary 2.3.12, the collection of generalized prefixes with the subword relation is a wpo set. The poset of proper standard prefix sets is a homomorphic image of this wpo set (under the homomorphism $e \mapsto (e)$). By Exercise 2.3.7, the poset of standard prefix sets is wpo. □

We would like to give canonic names to standard prefix sets. The class of all prefixes will be denoted *all*.

Definition 2.3.17. A generalized prefix is *reduced* if (i) the only possible neighbours of a letter $\forall^*$ are letters $\exists$ and $\exists^*$, and (ii) the only possible neighbours of a letter $\exists^*$ are letters $\forall$ and $\forall^*$.

Lemma 2.3.18. *Every proper standard prefix set Π has a unique presentation by means of a reduced generalized prefix.*

Proof. First we prove the existence of the desired presentation. Since Π is proper, it is given by some generalized prefix. Let w be a shortest generalized prefix such that $(w) = \Pi$. If w equals $u\forall\forall^*v$ or $u\forall^*\forall v$ or $u\forall^*\forall^*v$ then $(u\forall^*v) = \Pi$ which contradicts the choice of w. Similarly $\exists$ and $\exists^*$ cannot be neighbours in w.

Second, we prove the uniqueness of the desired presentation. It suffices to prove that different reduced generalized prefixes denote different prefix sets.

Every reduced generalized prefix w is a concatenation of strings $u_1 \dots u_m$ where each u_i is a letter $\forall^*$, a letter $\exists^*$, a maximal contiguous string of $\forall$'s, or a maximal contiguous string of $\exists$'s. The strings u_i will be called *blocks*. The number m will be called the *block length* of w.

Assume that $w \neq \lambda$. Then every prefix in the standard prefix set w has at most $m-1$ quantifier alternations and there exists a prefix with exactly $m-1$ quantifier alternations in (w). For example, if $m = 3$ and starts with an $\forall$ or $\forall^*$ then (w) contains the string $\forall\exists\forall$ and every string in (w) has the form $\forall^i\exists^j\forall^k$.

Now let $w_1 \neq w_2$ be reduced generalized strings of block length m_1 and m_2 respectively. Without loss of generality, $m_1 \geq m_2 > 0$. If $m_1 > m_2$ then (w_1) contains a string with $m_1 - 1$ quantifier alternations but no such string belongs to (w_2). Thus $m_1 = m_2$. Let $m = m_1$. Let $u_1, \dots, u_m$ be the blocks of w_1, and $v_1, \dots, v_m$ be the blocks of w_2.

We illustrate the rest of the proof on the case $m = 3$. Without loss of generality, w_1 starts with an $\forall$ or $\forall^*$. If w_2 starts with an $\exists$ or $\exists^*$ then the string $\forall\exists\forall$ belongs to $(w_1) - (w_2)$. Thus w_2 also starts with an $\forall$ or $\forall^*$. If one of the blocks u_1, v_1 is $\forall^*$ and the other is $\forall^n$, then one of the sets contains the prefix $\forall^{n+1}\exists\forall$ and the other class doesn't.

If one of the blocks u_1, v_1 is $\forall^*$ and the other is $\forall^n$, then one of the sets contains the prefix $\forall^{n+1}\exists\forall$ and the other class doesn't. If one of the blocks u_2, v_2 is $\forall^*$ and the other is $\forall^n$, then one of the sets contains the prefix $\forall\exists^{n+1}\forall$ and the other doesn't. If one of the blocks u_3, v_3 is $\forall^*$ and the other is $\forall^n$, then one of the sets contains the prefix $\forall\exists\forall^{n+1}$ and the other doesn't. These three statement remain true if $\forall*$ is replaced with any integer $k > n$. It follows that $u_1 = v_1$, $u_2 = v_2$ and $u_3 = v_3$. □

Lemma 2.3.19. *The union of an increasing sequence of standard prefix sets is standard.*

Proof. We use the terminology of the previous proof. Consider an increasing sequence

$$(w_1) \subseteq (w_2) \subseteq (w_3) \subseteq \cdots$$

where each w_i is a reduced generalized prefix of block length m_i. Let $\Pi = \bigcup_i (w_i)$. If the sequence of numbers m_i is unbounded then Π is the set of all prefixes. Thus we may suppose that the sequence of numbers m_i is bounded. Without loss of generality, these numbers are all equal to some number m, so that each w_i splits into m blocks $u_{i1} \dots u_{im}$.

For each positive integer $j \leq m$, there are only two cases: either each u_{ij} is universal (that is of the form $\forall^n$ or $\forall^*$) or each u_{ij} is existential (that is of the form $\exists^n$ or $\exists^*$). In the first case, let $q = \forall$ (and $q^* = \forall^*$); in the second case, let $q = \exists$ (and $q^* = \exists^*$). We define a new block u_j. If any $u_{ij} = q^*$, set $u_j = q^*$. Otherwise, let k_{ij} be the length of u_{ij}. If the sequence of numbers k_{ij} is unbounded, set $u_j = q^*$. Otherwise let $k_j = \sup_i(k_{ij})$ and set $u_j = q^{k_j}$. It is easy to see that $(u_1 \dots u_m) = \Pi$. □

Definition 2.3.20. Let Π be a closed prefix set. A *component* of Π is a maximal standard subset of Π.

Lemma 2.3.21. *Let Π be a closed prefix set. The number of components of Π is finite and Π is the union of its components.*

Proof. The components of Π are incomparable and thus form an antichain. But the poset of standard sets is wpo (Lemma 2.3.16); hence the antichain is finite (Exercise 2.3.5). Since Π is closed, every prefix in Π belongs to a standard subset of Π and therefore to a maximal standard subset of Π. □

Theorem 2.3.22. *Closed prefix classes with the inclusion relation form a wpo set. Arbitrary prefix classes with the domination relation form a wqo set.*

Proof. To prove the first statements, use Lemma 2.3.13. The desired wpo substructure is formed by standard classes. To prove the second statement, use Exercise 2.3.3. □

2.3.4 The Well Quasi Ordering of Arity Sequences

We study sequences that arise naturally in the study of fragments of first-order logic given by restrictions of the following sort: use at most p_n predicate symbols of arity n, or use at most f_n function symbols of arity n. We prove that these sequences together with an appropriate quasi ordering form a wqo set. Further, we introduce standard arity sequences with succinct names and prove that every arity sequence is equivalent to a unique standard sequence.

Definition 2.3.23. An *arity sequence* is a function from positive integers to the set of natural numbers extended with the first infinite ordinal ω.

We think about an arity sequence p as a sequence $(p_1, p_2, \ldots)$ where $p_i = p(i)$. A tail of zeros may be omitted. This allows us to speak about finite arity sequences; for example, $(\omega, 1)$ is a finite arity sequence. Accordingly, the sequence of zeroes will be called the *empty* sequence. For readability, the empty sequence will be denoted (0) rather than ().

Replacing a predicate (respectively function) symbol by a predicate (respectively function) symbol of higher arity can only increase expressibility. This justifies the following definition where p and q are arity sequences.

Definition 2.3.24. An arity sequence p *dominates* an arity sequence q, symbolically $p \geq q$, if $\sum_{i \leq j} p_j \geq \sum_{i \leq j} q_j$ for all i.

Clearly the domination order is a quasi order. We will prove that the qoset of arity sequences if wqo. The sequence

$$\hat{p} = (\sum_{1 \leq j} p_j, \sum_{2 \leq j} p_j, \ldots)$$

will be called the *associate sequence* of p. Thus, arity sequences are equivalent (in the qoset of arity sequences) if and only if they have the same associate sequence.

Definition 2.3.25. An arity sequence p is *standard* if it satisfies the following condition: for every i, if $\sum_{i \leq j} p_j$ is infinite then $p_i = \omega$.

The standard sequence $(\omega, \omega, \ldots)$ will be denoted *all*.

Lemma 2.3.26. *In the qoset of arity sequences, every sequence is equivalent to a unique standard sequence.*

Proof. It is easy to see that all arity sequences without tails of zeroes are equivalent to *all* which is the only standard sequence without a tail of zeros.

Suppose that p is a finite sequence $p_1, \ldots, p_m$. If $p_1 + \cdots + p_m$ is finite then p is standard and inequivalent to any other arity sequence. Otherwise let i be the least index such that $p_i + \cdots + p_m$ is infinite. It is easy to see that the sequence $(\omega, \ldots, \omega, p_{i+1}, \ldots, p_m)$ is the only standard sequence equivalent to p. □

Theorem 2.3.27. *The qoset of arity sequences is wqo.*

Proof. First we prove that the subqoset of finite sequences is wqo. Ordinals $\leq \omega$ (that is finite ordinals and ω) form a well ordered set. By the Finite Sequence Theorem (Theorem 2.3.11), finite sequences of such ordinals with the embedding relation form a wpo set. By Lemma 2.3.8, it suffices to prove that $(p_1, \ldots, p_m) \leq (q_1, \ldots, q_n)$ in the qoset of finite arity sequences if there exists a monotone one-to-one function $f : [1..m] \to [1..n]$ such that every $p_i \leq q_{fi}$. Since f is monotone and one-to-one, we have that, for every i,

$$\sum_{i \leq j} p_j \leq \sum_{i \leq j} q_{fj} \leq \sum_{i \leq j} q_j .$$

It follows that the poset of finite standard sequences if wpo. By Exercise 2.3.7, the poset of all standard sequences is wpo. By Exercise 2.3.3, the qoset of arity sequences if wqp. □

2.3.5 The Classifiability of Prefix-Vocabulary Sets

In this subsection, we prove the Classifiability Theorem. It shows the importance of so-called standard prefix-vocabulary classes. In the main body of the book, the prefix-vocabulary classes of interest will be almost exclusively standard classes.

Definition 2.3.28 (Prefix-Vocabulary Classes). Let Π be a prefix set, and p, f arity functions. The *prefix-vocabulary class* $[\Pi, p, f]$ (respectively $[\Pi, p, f]_=$) is the collection of prenex sentences φ of first-order logic without equality (respectively, first-order logic with equality) such that

- the quantifier prefix of φ belongs to Π,
- for all $i > 0$, φ has at most p_i predicate symbols of arity i and at most f_i function symbols.
- φ has no nullary predicate symbols with the exception of logic constants *true* and *false*, and no individual constants.

In case $f = (0)$, we may write $[\Pi, p]$ (respectively $[\Pi, p]_=$) instead of $[\Pi, p, (0)]$ (respectively $[\Pi, p, (0)]_=$).

Of course, we could allow nullary predicate variables, individual constants and free variables. The reason for their exclusion has been explained in Sect. 1.3. The class $[\lambda, (0), (0)]$ will be called *trivial*; it contains only *true* and *false*. Any $[\Pi, (0), f]$, $[\lambda, p, q]$ or $[\lambda, p, q]_=$ is trivial.

The class $[\Pi, p, f]_=$ can be called the class $[\Pi, p, f]$ of logic with equality. In the rest of the book, we use the notation $[\Pi, p, f]$ and $[\Pi, p, f]_=$ but here it will be convenient to fix a logic $\mathcal{L}$ with or without equality and deal with classes $[\Pi, p, f]$ appropriate to $\mathcal{L}$. We say that p (respectively f) is the *predicate arity sequence* (respectively *function arity sequence*) of $[\Pi, p, f]$.

Exercise 2.3.29. For every nontrivial prefix-vocabulary class, the set of prefixes, the predicate arity sequence and the function arity sequence are defined uniquely.

For future convenience, we will use only the name $[\lambda, (0), (0)]$ for the trivial class, so that every prefix-vocabulary class has a unique name. This simplifies some of the following statements.

Exercise 2.3.30. Prove that for any $[\Pi, p, f]$, the following statements are equivalent

(i) $[\Pi, p, f]$ has only finitely many sentences up to logical equivalence,
(ii) Π is finite, p is finite and f is empty.

Definition 2.3.31. $[\Pi, p, f]$ *dominates* $[\Pi', p', f']$, symbolically $[\Pi, p, f] \geq [\Pi', p', f']$, if Π dominates Π', p dominates p' and f dominates f'.

Exercise 2.3.32. If $[\Pi', p', f'] \leq [\Pi, p, f]$ then $[\Pi', p', f']$ conservatively reduces to $[\Pi, p, f]$.

Lemma 2.3.33. *The domination ordering of the prefix-vocabulary classes is a well quasi order.*

Proof. By Theorem 2.3.22, the domination ordering of prefix sets is wqo. By Theorem 2.3.27, the domination ordering of arity sequences is wqo. It remains to use the collection of wqo sets is closed under finite direct products (Lemma 2.3.9) and homomorphisms (Lemma 2.3.8). □

Definition 2.3.34. A class $[\Pi, p, f]$ is *standard* if Π, p and f are standard. It is *closed* if Π is closed and p, f are standard.

Notice that the trivial class is standard and closed.

Exercise 2.3.35.
- A closed class K_1 dominates a class K_2 if and only $K_2 \subseteq K_1$.
- Every prefix-vocabulary class $[\Pi, p, f]$ is equivalent to (that is, dominates and is dominated by) the closed prefix-vocabulary class $[\overline{\Pi}, p, f]$.
- Equivalent closed classes are equal.
- The closed classes together with the inclusion relation form a wpo set.

Lemma 2.3.36. *Every closed prefix-vocabulary class $[\Pi, p, f]$ is a finite union of standard classes with the same predicate arity sequence and the same function arity sequence.*

Proof. By Lemma 2.3.21, Π is a finite union of standard components $\Pi_1, \ldots, \Pi_m$. But then $[\Pi, p, f] = \bigcup_i [\Pi_i, p, f]$. □

Consider a collection $\mathcal{D}$ of prefix-vocabulary classes that is *downward closed*, so that $K_1 \leq K_2 \in \mathcal{D}$ implies $K_1 \in \mathcal{D}$. The complement $\mathcal{U}$ of $\mathcal{D}$ is *upward closed*) so that $K_1 \geq K_2 \in \mathcal{U}$ implies $K_1 \in \mathcal{U}$. Notice that both $\mathcal{D}$ and $\mathcal{U}$ are closed under the equivalence relation. Let $\mathcal{M}$ be the collection of minimal closed members of $\mathcal{U}$ and $\overline{\mathcal{M}}$ be the upward closure of $\mathcal{M}$.

Lemma 2.3.37. *$\mathcal{M}$ is finite and $\mathcal{U} = \overline{\mathcal{M}}$.*

Proof. Since closed prefix-vocabulary classes form a wpo, $\mathcal{M}$ is finite. It is obvious that $\mathcal{U} = \overline{\mathcal{M}}$. □

Assume additionally that $\mathcal{D}$ is closed under finite unions.

Lemma 2.3.38. *Every member of $\mathcal{M}$ is standard.*

Proof. Given a member K of $\mathcal{M}$, use Lemma 2.3.36 and present K is the of some different standard classes $K_1, \ldots, K_m$. By contradiction suppose that $m > 1$. Then each K_1 is strictly dominated by K. But K is a minimal member of $\mathcal{U}$. It follows that each $K_i \in \mathcal{D}$. But then $K \in \mathcal{D}$ which is impossible. □

Lemma 2.3.39. *There exist only a finitely many maximal standard members of $\mathcal{D}$.*

Proof. Use the fact that closed classes form a wpo set. □

Notice, however, that not every standard member of $\mathcal{D}$ is dominated by some maximal standard member. For example, if $\mathcal{D}$ consists of the classes with finite prefix set (including the trivial class), then it has no maximal members whatsoever. However, in cases of interest to us (e.g. when $\mathcal{D}$ consists of classes with solvable satisfiability problem), maximal standard members of $\mathcal{D}$ exist and play an important rôle.

We formulate some of our conclusions as the Classifiability Theorem for prefix-vocabulary classes.

Theorem 2.3.40 (Classifiability Theorem). *Let $\mathcal{D}$ be a downward closed collection of prefix-vocabulary classes closed under finite unions. Further, let $\mathcal{U}$ be the complement of $\mathcal{D}$, and $\mathcal{M}$ the collection of minimal closed classes in $\mathcal{U}$. Then $\mathcal{M}$ is finite, every member of $\mathcal{M}$ is standard, and $\mathcal{U}$ is the upward closure of $\mathcal{M}$. In addition, the number of maximal standard members of $\mathcal{D}$ is finite.*

The predicate symbols of $\mathcal{L}$ are in fact predicate variables. It is easy to see that the classifiability theorem remains true if $\mathcal{L}$ is extended with predicate constants satisfying certain axioms. (The only change is that the classes $[\Pi, (0), f]$ are not necessarily trivial.) Similarly one can extend $\mathcal{L}$ with function constants or with predicate and function constants. For example, one may suppose that $\mathcal{L}$ contains a binary operation satisfying the axioms of group theory.

There are many interesting collections $\mathcal{D}$ satisfying the conditions of the Classifiability Theorem applies. We give some examples where $\mathcal{L}$ may contain predicate and/or function constants

1. Classes with decidable satisfiability problem. More generally, classes where the satisfiability problem is decidable within a given complexity class like polynomial space, exponential time, etc.
2. The analogue of 1 for finite satisfiability.
3. The analogues of 1 (respectively 2) for Krom and/or Horn formulae.
4. Classes where the following task, depending on a given sentence φ, can be performed within a given complexity bound: If φ has a finite model, produce such a model; otherwise output NONE.
5. Classes with the finite model property.
6. Classes with the zero-one law. Roughly speaking, a formula class K satisfies the zero-one law if, as n grows to infinity, the fraction μ_n of n-element models satisfying φ tends to either zero or one, for every sentence $\varphi \in K$ (see [88] for a survey on zero-one laws). Instead of the zero-one law, one can speak about the limit law (μ_n has a limit), the slow oscillation law ($\lim_{n\to\infty} \mu_{n+1} - \mu_n = 0$), etc.
7. Classes K such that the class K' of second-order sentences $\{(\exists \bar{P})\varphi : \varphi \in K\}$ satisfies the zero-one law (or the limit law, etc.) [235, 314, 315].

The reader can easily find additional examples. The Classifiability Theorem can be extended in various directions. One example of such extension is found in Sect. 5.4.

In some cases, the collection $\mathcal{D}$ is closed not only under finite unions but also under arbitrary unions. For example, any union of classes with the finite model property has the finite model property. In such cases, the finite collection of maximal standard classes of $\mathcal{D}$ gives another finite presentation of $\mathcal{D}$.

Corollary 2.3.41. *Let $\mathcal{D}$ be a downward closed collection of prefix-vocabulary classes that is closed under arbitrary unions. Then a standard class K belongs to $\mathcal{D}$ if and only if it is a part of a maximal standard class in D.*

2.4 Historical Remarks

Hilbert's *Entscheidungsproblem* has been answered negatively via different methods by Church [80] and Turing [513]. Subsequently many different proofs for this result were given, usually by reducing an unsolvable combinatorial decision problem to the *Entscheidungsproblem.* An exception is Kalmárs' proof in [301].

Although reductions of the Entscheidungsproblem have been formulated explicitly already by Löwenheim [365] and Skolem [477] and numerous other papers cited in [267, I§4], the notion of a *reduction class* seems to appear in print for the first time in [82, §47]. Three years later Surányi's book [498] gives a comprehensive treatment of the reduction classes known at that time (see a detailed list in the annotated bibliography). The best finite prefixes obtained are $\forall^3\exists$, $\exists\forall\exists\forall^2$ and $\forall\exists\exists\forall^2$ (or $\exists\forall\exists^3\forall$ and $\forall\exists^4\forall$ if equality is present); among the minimal prefix-vocabulary classes with infinite prefix only $[\exists^*\forall^3\exists, (0,1)]$, $[\forall^3\exists^*, (0,1)]$, $[\forall^*\exists, (0,1)]$ are established as reduction classes.

Büchi's paper [64] was a breakthrough: He combined Turing's proof method with the use of Skolem's theorems on Skolem normal form and canonical models. This allowed him to establish the conservative reduction class property for $[\exists \wedge \forall\exists\forall, (\omega, 3)]$ by an elementary proof and prepared the ground for getting to the minimal reduction class $[\forall\exists\forall, (\omega, 1)]$ (see the next chapter of this book). Aanderaa [2] and Börger [39] independently refined the Turing-Büchi method further, they showed how the explicit reference to the time component can be avoided if properties of computations are formalized where the time needed to reach that property is irrelevant. This allowed them to impose further conditions on the propositional structure of reduction formulae and, in particular, to refine the conservative Büchi prefix class $[\exists\forall\exists\forall]$ to Krom and Horn formulae (for further details on Krom and Horn formulae see the Chap. 5 in this book).

The Aanderaa-Börger method was developed further in [45, 48] and was shown to be related to the study of the degree complexity of combinatorial decision problems in [46, 58, 59, 60]. In [53] the method is interpreted as defining the semantics of programs by logical formulae and the economical description of Turing machines appearing in Sect. 2.1.1 is defined and used to provide uniform simple proofs for standard completeness results of logical and combinatorial decision problems (see also [53, 55, 61] from where the proofs in this chapter are taken). This includes also the Cook-Levin Theorem which appears in [91, 343]. Stockmeyer's Theorem appears in [493], and also [490] from where we have adapted the proof. The historical development of the theorems of Skolem (and Löwenheim) which are used throughout in this book is well explained in [82, §45 and §49].

Trakhtenbrot's Theorem appears in [509, 510]. It was found independently by Craig [92]. Our proof is an adaption of the proof in [40] to the economical description of Turing machines in [53]. Kalmár [299] proves Trakhtenbrot's Theorem by an effective reduction of the validity problem to the finite satisfiability problem. Gurevich's theorem on semi-conservative reductions appears in [227] and our proof is taken from there.

Trakhtenbrot's Theorem triggered the definition of the notion of a spectrum and the formulation of the Spectrum Problem in [455]. For the history of this problem see the first section of [53]; since Fagin's work (see also [151]–[155]) this problem and its generalizations have played a crucial rôle in finite model theory. Our proof of the Spectrum Hierarchy Theorem is from [53].

Descriptive complexity theory, i.e. the design and study of logical languages that capture complexity classes was motivated by Fagin's Theorem and explicitly proposed as a research program by Immerman [276, 277]. But there were earlier relevant results, most notably the characterization of the regular languages by means of monadic second-order logic [62, 511]. Immerman and Vardi [276, 517] proved that, on ordered structures, the problems solvable in polynomial time are exactly those definable in least fixed-point logic. Immerman systematically studied the problem of capturing complexity classes by logical languages and came up with logical characterizations for most popular complexity classes. For instance, logarithmic space complexity classes are captured by various forms of transitive closure logics [277]. The characterizations of polynomial time and nondeterministic logarithmic space complexity in terms of Horn and Krom fragments of second-order logic are due to Grädel [207]. We refer to the survey papers [232, 278] and to the monograph [141] for more information on descriptive complexity theory.

The study of complexity results for decidable prefix classes of first-order logic originates in [352, 175]. The lower bounds of Sect. 2.2.4 appeared first (in disguised form) in [74] and later more explicitly in [106, 421]. Our proofs are adaptions of the proofs appearing there. See also [61] and the results in Chap. 5–8 of this book.

The classifiability problem was addressed after Church and Turing had proved the unsolvability of the *Entscheidungsproblem*. But even before that, syntactical classifications, especially by the prefix structures and vocabulary, had been used to put some order into the myriad of individual results related to the classical decision problems. After completing the solution of the prefix-vocabulary problem for pure predicate logic and coming up the final classification in [219], Gurevich analysed the underlying reasons for the possibility of a finite solution. That analysis lead him to the Classifiability Theorem [222]; in the same paper he gave the final classification of the prefix-vocabulary problem for pure logic with functions. Gurevich did not know the theory of well partially ordered sets; he rediscovered the notion and developed a portion of the theory necessary for the Classifiability Theorem.

3. Undecidable Standard Classes for Pure Predicate Logic

This chapter is devoted to the classification of prefix-vocabulary classes in pure predicate logic, i.e. first-order logic without equality or function symbols, with respect to the question whether the satisfiability problem for these classes is decidable. *A posteriori,* the same classification is obtained if satisfiability is restricted to finite satisfiability.

Gurevich's Classifiability Theorem (Theorem 2.3.40) tells us that there is a finite number of minimal undecidable prefix-vocabulary classes of the form $[\Pi, (p_1, p_2, \ldots)]$ where Π is an extended prefix (a word over $\forall, \exists, \forall^*, \exists^*$) and each p_i is natural numbers or ω. We prove in this chapter results which establish what these minimal undecidable classes are; we further show that all of them are indeed conservative reduction classes.

This is summed up by the following main theorem.

Theorem 3.0.1 (Reduction Classes for Pure Predicate Logic). *A prefix-vocabulary class $[\Pi, (p_1, p_2, \ldots)]$ (without function symbols or equality) is undecidable (and indeed is a conservative reduction class) if it contains at least one of the following nine classes:*

Classes with finite prefix:
- $[\forall\exists\forall, (\omega, 1)]$ *(Kahr 1962)*
- $[\forall^3\exists, (\omega, 1)]$ *(Surányi 1959)*

Classes with $\forall^*$ in the prefix:
- $[\forall^*\exists, (0, 1)]$ *(Kalmár-Surányi 1950)*
- $[\forall\exists\forall^*, (0, 1)]$ *(Denton 1963)*

Classes with $\exists^*$ in the prefix:
- $[\forall\exists\forall\exists^*, (0, 1)]$ *(Gurevich 1966)*
- $[\forall^3\exists^*, (0, 1)]$ *(Kalmár-Surányi 1947)*
- $[\forall\exists^*\forall, (0, 1)]$ *(Kostyrko-Genenz 1964)*
- $[\exists^*\forall\exists\forall, (0, 1)]$ *(Surányi 1959)*
- $[\exists^*\forall^3\exists, (0, 1)]$ *(Surányi 1959)*

The decidability results for the classes $[all, (\omega)]$, $[\exists^*\forall^*, all]$ and $[\exists^*\forall^2\exists^*, all]$ proved in Chap. 6 will complete the classification by providing the "only if" part. Indeed it is easy to verify that each standard class $[\Pi, p]$ is either essentially finite and thus decidable for trivial reasons, or is contained in one of the three decidable classes just mentioned, or is a reduction class containing at least one of the classes mentioned in Theorem 3.0.1. (A class $[\Pi, p]$ is essentially finite if Π defines a finite set of prefixes and p a (up to renaming) finite vocabulary.)

Further, note that the classification problem for *prefix classes* (in pure predicate logic) is completely solved by the undecidable classes $[\forall\exists\forall]$ and $[\forall^3\exists]$ (containing the classes of Kahr and Surányi), and the decidable classes $[\exists^*\forall^*]$ of Bernays-Schönfinkel and $[\exists^*\forall^2\exists^*]$ of Gödel-Kalmár-Schütte.

We will show in this chapter that the nine classes listed in Theorem 3.0.1 are indeed conservative reduction classes. This obviously implies the theorem. The difficult cases are the Kahr class and the Gurevich class for each of which we reserve a separate section. It is easy to give conservative reductions of Kahr's class to the other minimal prefix-vocabulary classes that do not contain $\exists^*$ in the prefix (see Sect. 3.1.4). Analogously Gurevich's class can be reduced to the other classes which are minimal among the prefix-vocabulary classes that contain $\exists^*$ in the prefix (see Sect 3.3.6). We devote Sect. 3.2 to the Kalmár-Surányi class $[\forall^3\exists^*, (0,1)]$ in order to present in an explicit manner the method of *existential interpretation.* This method provides conservative reductions between theories under strict control of the prefix structure and underlies the proof for the Gurevich class. The following figure surveys the reductions of this chapter.

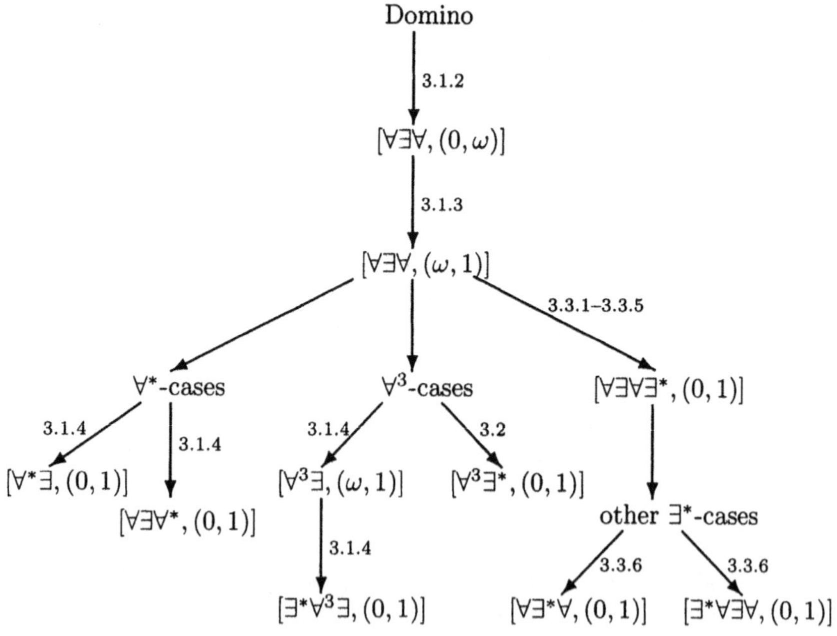

Figure 3.1. Reductions for standard classes

3.1 The Kahr Class

This section deals with the minimal conservative reduction classes among the prefix-vocabulary classes that contain only a bounded number of existential quantifiers. The main and difficult case here is the Kahr class $[\forall\exists\forall, (\omega, 1)]$.

Theorem 3.1.1 (Kahr). $[\forall\exists\forall, (\omega, 1)]$ *is a conservative reduction class.*

To prove this result we proceed as follows. In Sect. 3.1.1 we introduce *domino problems*, an important class of undecidable combinatorial problems that have played an important rôle for proving undecidability and lower complexity results in various branches of mathematical logic. In Sect. 3.1.2 we prove the Kahr-Moore-Wang class $[\forall\exists\forall, (0, \omega)]$ to be a conservative reduction class by formalizing an appropriate domino problem. In Sect. 3.1.3 we show how one can interpret formulae of the Kahr-Moore-Wang class in coloured graphs; then we bring these graphs into a normal form which lends itself to a description by a formula in Kahr's class. This yields a conservative reduction of the Kahr-Moore-Wang class to Kahr's class.

Sect. 3.1.4 is devoted to the remaining minimal classes that do not contain $\exists^*$ in the prefix. First we show that the only other minimal reduction class with finite prefix, namely Surányi's reduction class $[\forall^3\exists, (\omega, 1)]$, can easily be obtained by a conservative reduction from Kahr's class. (By the way, there is a trivial reduction of the latter class to the minimal Surányi class $[\exists^*\forall^3\exists, (0, 1)]$ that contains $\exists^*$ in the prefix.) Then we prove also that the classes with $\forall^*$ in the prefix, namely the Kalmár-Surányi class $[\forall^*\exists, (0, 1)]$ and the Denton class $[\forall\exists\forall^*, (0, 1)]$, are conservative reduction classes by reducing Kahr's class to them.

3.1.1 Domino Problems

Domino problems are a very simple and general form of combinatorial problems. They were introduced by Wang [531, 532] as a tool for proving the unsolvability of the $\forall\exists\forall$-prefix class in the pure predicate calculus. In the last thirty years they have been used to establish many undecidability results and lower complexity bounds for various systems of propositional logic, for subclasses of first order logic and for decision problems in mathematical theories (see e.g. [78, 201, 203, 196, 206, 227, 245, 246, 288, 351, 355, 448]).

The original, 'unconstrained' version of a domino problem is given by a finite set of dominoes or tiles, each of them an oriented unit square with coloured edges; the question is whether it is possible to cover the first quadrant in the Cartesian plane by copies of these tiles, without holes and overlaps, such that adjacent dominoes have matching colours on their common edge. The set of tiles is finite, but there are infinitely many copies of each tile available; rotation of the tiles is not allowed. Variants of this problem require that certain places (e.g. the origin, the bottom row or the diagonal) are tiled by specific tiles.

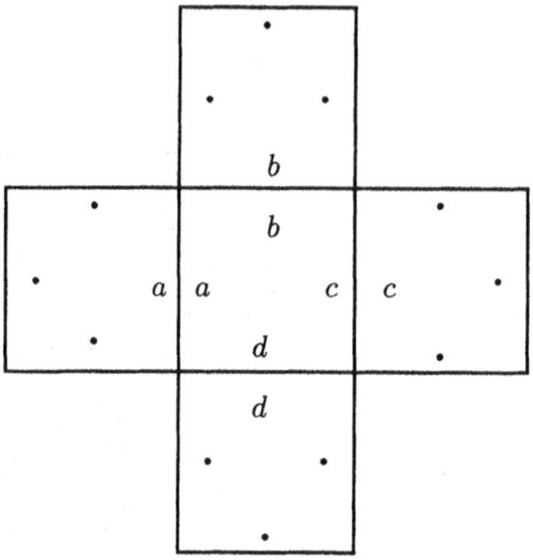

Figure 3.2. Domino adjacency condition

All these problems are undecidable; more precisely, they are complete for the co-r.e sets (Π_1^0-complete). Domino problems with higher degrees of unsolvability were defined by requiring that some dominoes appear infinitely often in the tiling; these *recurring domino problems* are Σ_1^1-complete, i.e. they sit in the first level of the analytical hierarchy [245, 246]. In most cases, the undecidability is established by a straightforward encoding of an appropriate halting problem for Turing machines: successive rows of the tiling represent successive configurations of the Turing machine. An exception is the unconstrained domino problem in its original form; it is more difficult to handle because the constraints on the tiling of certain places are necessary to encode the beginning of the computation. The unconstrained domino problem was proved to be undecidable by Berger [33]; an essential part of his proof was the construction of a set of tiles that admits only non-periodic tilings. A simpler proof is due to R. Robinson [440].

If the space to be tiled is not an infinite portion of the plane, but a finite square or rectangle we obtain *bounded domino problems.* Variants of these are complete in various important complexity classes such as NP or PSPACE, and again these results are proved by straightforward encodings of Turing machine computations. Roughly, the dimensions of the tiled rectangle correspond to the time and space restrictions of the Turing machine. Recently domino problems have been generalized to *domino games* which capture the behaviour of alternating procedures [78, 198, 203]. Domino games are two person games; the problem whether the first player has a winning strategy in m moves corresponds to acceptance of an alternating Turing machine within m alternations, and again, the size of the board is related to time and space.

Thus, domino problems are very flexible and capture the essential properties of computations; on the other hand they have a very simple geometrical and combinatorial structure, and their formulation is independent of the details of a particular machine model. Therefore reductions from domino problems tend to be simpler than direct encodings of computations. This makes them a powerful tool for proving the undecidability of 'simple' formula classes in first order logic. The most famous of these results is the unsolvability of the $\forall\exists\forall$-prefix class in the pure predicate calculus, established by Kahr, Moore and Wang [288]; another example is the class $[\forall, (0), (2)]_=$ proved to be unsolvable by Gurevich [228]. Domino problems have turned out to be very useful also in complexity theory. Lewis and Papadimitriou [355] and also Savelsbergh and van Emde Boas [448] have argued that BOUNDED TILING has some advantages over SAT (the satisfiability problem for propositional formulae) as a 'master' NP-complete problem. Harel and Chlebus have established lower bounds for various propositional logics by reductions from domino problems [245, 78] and Grädel has shown that domino problems yield also good lower complexity bounds for simple formula classes in mathematical theories [200, 206].

Definition 3.1.2. A *domino system* $\mathcal{D}$ is a triple (D, H, V) where D is a finite set of dominoes and H, $V \subseteq D \times D$ are two binary relations. Let S be any of the spaces $\mathbb{Z} \times \mathbb{Z}$, $\mathbb{N} \times \mathbb{N}$ or $\mathbb{Z}_t \times \mathbb{Z}_t$. We say that $\mathcal{D}$ *tiles* S if there exists a *tiling* $\tau : S \to D$ such that for all $(x, y) \in S$:

(i) If $\tau(x, y) = d$ and $\tau(x + 1, y) = d'$ then $(d, d') \in H$;
(ii) if $\tau(x, y) = d$ and $\tau(x, y + 1) = d'$ then $(d, d') \in V$.

This definition of a domino system is equivalent to the more intuitive description by unit squares with coloured edges. In fact, H (resp. V) just contain those pairs (d, d') of dominoes for which the right (upper) colour of d is equal to the left (lower) colour of d'. Conversely, given $\mathcal{D} = (D, H, V)$ as above, take a unit square tile for each triple (d, d', d'') with $(d, d') \in H$ and $(d, d'') \in V$ and colour its left and lower edge with d, its right edge with d' and its upper edge with d''.

Proposition 3.1.3. *A domino system $\mathcal{D}$ admits a tiling of $\mathbb{Z} \times \mathbb{Z}$ if and only if it admits a tiling of $\mathbb{N} \times \mathbb{N}$.*

Proof. It is clear that a tiling of $\mathbb{Z} \times \mathbb{Z}$ also gives a tiling of $\mathbb{N} \times \mathbb{N}$. The converse is a nice application of König's Lemma . Suppose that τ is a tiling of $\mathbb{N} \times \mathbb{N}$ by $\mathcal{D}$. There exists at least one domino d such that for all n there exist $i, j > n$ with $\tau(i, j) = d$. Fix such a d. Further, for every $k \in \mathbb{N}$, let S_k be the square $\{-k, \dots, -1, 0, 1, \dots, k\} \times \{-k, \dots, -1, 0, 1, \dots, k\}$.

We define a finitely branching tree whose nodes are the correct tilings τ_k of S_k by $\mathcal{D}$ such that $\tau_k(0, 0) = d$. The root is the unique such tiling of S_0 and the children of a tiling τ_k are the possible extensions to tilings τ_{k+1} of S_{k+1}. This tree contains paths of any finite length. By König's Lemma it also contains an infinite path from the root, which means that $\mathcal{D}$ admits a tiling of $\mathbb{Z} \times \mathbb{Z}$. □

Exercise 3.1.4. [531, 532] Show that the origin constrained domino problem – where a given 'origin domino' has to be placed in position $(0,0)$ – is algorithmically undecidable. Hint: Simulate the proof of the Church-Turing Theorem given in Chap. 2 by reducing Turing machine computations to tilings of $\mathbb{N}\times\mathbb{N}$. Note that for this proof, the origin domino is necessary to make sure that the first row of the tiling encodes the initial configuration of the given Turing machine (see [351, Chapter I B.2]).

In the context of conservative reductions we are also interested in periodic solutions of domino problems.

Definition 3.1.5. A domino system $\mathcal{D}$ is said to admit a *periodic tiling* of the space S if there is a tiling τ of S by $\mathcal{D}$ which has a horizontal and a vertical period $h, v > 0$ respectively. This means that for all points $(x,y) \in S$ we have that

$$\tau(x,y) = \tau(x+h,y) = \tau(x,y+v).$$

Exercise 3.1.6. Prove that this periodicity condition is equivalent to the seemingly more general one where the two translations that leave the tiling invariant are not necessarily parallel to the coordinate axis. More precisely: Show that tiling τ is periodic in the sense of Definition 3.1.5 if and only if there exist linearly independent vectors (a,b) and $(c,d) \in \mathbb{N}\times\mathbb{N}$ such that

$$\tau(x,y) = \tau(x+a,y+b) = \tau(x+c,y+d)$$

for all points $(x,y) \in S$.

Berger [33] has shown that the domino problem for $\mathbb{Z}\times\mathbb{Z}$ and $\mathbb{N}\times\mathbb{N}$ is undecidable. Gurevich and Koryakov [237] strengthened Berger's undecidability result to an inseparability result.

Theorem 3.1.7 (Berger, Gurevich-Koryakov). *The set of domino systems that admit, respectively, no tiling and a periodic tiling of $\mathbb{Z}\times\mathbb{Z}$ or $\mathbb{N}\times\mathbb{N}$ are recursively inseparable.*

A new proof of this theorem (by C. Allauzen and B. Durand) is given in Appendix A of this book. In fact the proof shows that the recursively inseparable halting problems H_i ("M on empty input eventually halts in state i") and H ("the computation of M on empty input is infinite and never cycles"), that we already used in the proof of Trakhtenbrot's Theorem, can be reduced to the domino problem as follows.

There exists a recursive function that associates with every Turing machine M a domino system $\mathcal{D}$ satisfying the following

Reduction Property:

(i) If $M \in H_1$ then $\mathcal{D}$ admits a periodic tiling of the plane.
(ii) If $M \in H_2$ then $\mathcal{D}$ does not tile the plane
(iii) If $M \in H$ then $\mathcal{D}$ tiles the plane, but only aperiodically.

Together with Theorem 2.1.39 on semi-conservative reductions this gives us a very convenient and powerful method to prove that formula classes are conservative.

Corollary 3.1.8. (Semi-Conservative Reduction from the Domino Problem) *A formula class X is a conservative reduction class if there exists a recursive function that associates with every domino system $\mathcal{D}$ a formula $\psi_{\mathcal{D}} \in X$ such that*

(i) If $\mathcal{D}$ admits a periodic tiling of $\mathbb{N} \times \mathbb{N}$, then $\psi_{\mathcal{D}}$ has a finite model.
(ii) If $\mathcal{D}$ does not tile $\mathbb{N} \times \mathbb{N}$, then $\psi_{\mathcal{D}}$ is not satisfiable.

Proof. First, we recall the following fact from the proof of Trakhtenbrot's Theorem. Since *Fin-sat* and *Non-sat* are recursively enumerable we can effectively associate with every first order formula φ a Turing machine M (with two halting states and whose infinite computations never cycle) such that $M \in H_1$ if φ has a finite model and $M \in H_2$ if φ is logically invalid. Second, the proof of the Theorem of Berger and Gurevich-Koryakov, presented in Appendix A, gives us a reduction from H_1, H_2 to domino systems that, respectively, admit a periodic tiling and no tiling of $\mathbb{N} \times \mathbb{N}$. Third, we have the given reduction from these domino problems to *Fin-sat*(X) and *Non-sat*(X).

The composition of these three reductions is a semi-conservative reduction from FO to X. Hence X is conservative. □

3.1.2 Formalization of Domino Problems by $[\forall\exists\forall, (0, \omega)]$-Formulae

In this section we prove the following theorem.

Theorem 3.1.9 (Kahr-Moore-Wang, Gurevich-Koryakov). *The class $[\forall\exists\forall, (0, \omega)]$ is a conservative reduction class.*

Proof. By Corollary 3.1.8 it suffices to reduce domino problems to formulae of the Kahr-Moore-Wang class such that domino systems that admit periodic tilings are represented by finitely satisfiable formulae, and domino systems without a tiling of $\mathbb{N} \times \mathbb{N}$ by unsatisfiable formulae. Hence, we effectively construct for each domino system $\mathcal{D}$ a formula $\psi_D \in [\forall\exists\forall, (0, \omega)]$ such that

(i) If $\mathcal{D}$ does not tile $\mathbb{N} \times \mathbb{N}$, then $\psi_{\mathcal{D}}$ is not satisfiable,
(ii) If $\mathcal{D}$ admits a periodic tiling of $\mathbb{N} \times \mathbb{N}$, then $\psi_{\mathcal{D}}$ has a finite model.

We define the reduction formula $\psi_{\mathcal{D}}$ for a given domino problem $\mathcal{D} = (D, H, V)$ by giving its Skolem normal form $\forall x \forall y \varphi(x, x', y)$. The vocabulary of $\psi_{\mathcal{D}}$ consists of binary predicate symbols P_d, one for each domino $d \in D$. The intended models describe tilings. Suppose, for instance, that τ is a tiling of $\mathbb{N} \times \mathbb{N}$ by $\mathcal{D}$. Then $\psi_{\mathcal{D}}$ has a model with universe $\mathbb{N}$ and relations

$$P_d = \{(i, j) : \tau(i, j) = d\}.$$

The quantifier-free part $\varphi(x, x', y)$ is the conjunction of the following two formulae. The first conjunct expresses that at most one domino is placed at each point (x, y):

$$\bigwedge_{d \neq d'} \neg(P_d xy \wedge P_{d'} xy).$$

The second conjunct of $\varphi(x, x', y)$ expresses that the horizontal and vertical adjacency conditions are satisfied:

$$\bigvee_{(d,d') \in H} (P_d xy \wedge P_{d'} x'y) \wedge \bigvee_{(d,d') \in V} (P_d yx \wedge P_{d'} yx').$$

It remains to prove properties *(i)* and *(ii)*.

For *(i)* let $\mathfrak{A}$ be a Skolem model of $\psi_{\mathcal{D}}$. Since $\psi_{\mathcal{D}}$ is an $\forall\exists\forall$-formula, its Skolem models have universe $\mathbb{N}$. We define a tiling τ of $\mathbb{N} \times \mathbb{N}$ by setting

$$\tau(i, j) = d \text{ iff } \mathfrak{A} \models P_d ij.$$

This produces a correct tiling. Indeed, $\psi_{\mathcal{D}}$ ensures that τ is well-defined and that the horizontal and vertical adjacency conditions are satisfied. To see *(ii)* let τ be a periodic tiling of $\mathbb{N} \times \mathbb{N}$ by $\mathcal{D}$ with horizontal period h and vertical period v. Let t be the least common multiple of h and v, so that for all natural numbers i, j we have that

$$\tau(i, j) = \tau(i + t, j) = \tau(i, j + t).$$

This allows us to restrict the intended interpretation of P_d to $\mathbb{Z}_t$ (with the usual successor function modulo t). Obviously $(\mathbb{Z}_t, (P_d)_{d \in D})$ is a finite model of $\psi_{\mathcal{D}}$. □

Exercise 3.1.10. Show the conservative reduction class property for the subclass of formulae $\forall x \exists u \forall y \varphi$ in $[\forall\exists\forall, (0, \omega)]$ where only atomic subformulae of form Pxy, Puy, Pyx appear (see [288] and [319]). Hint: Replace occurrences of $P_d yu$ in $\psi_{\mathcal{D}}$ by $Q_d uy$ with new predicate symbols Q_d axiomatized by $Q_d xy \leftrightarrow P_d yx$ (see [288, 319]).

Exercise 3.1.11. Show that the Gödel reduction class $[\forall^3\exists, (0, \omega)]$ is conservative [187]. Hint: Reduce the Kahr-Moore-Wang class $[\forall\exists\forall, (0, \omega)]$ to the Gödel reduction class. First replace $\psi := \forall x \exists u \forall y \alpha \in [\forall\exists\forall, (0, \omega)]$ by

$$\varphi := \forall x \forall y \forall z \exists v (Fxv \wedge (Fxy \rightarrow y = v)) \wedge (Fxz \rightarrow \alpha[u/z]),$$

where the new predicate symbol F represents the graph of the Skolem function of ψ. Then replace in φ the equality $y = v$ by Gyv – for a new predicate symbol G – and add axioms that force G to be a congruence relation, i.e., an equivalence relation satisfying for each predicate symbol R in α the condition $Gxy \rightarrow ((Rxz \leftrightarrow Ryz) \wedge (Rzx \leftrightarrow Rzy))$.

Formalization of Turing Machines by $[\forall\exists\forall, (0, \omega)]$-Formulae. In order to correct an error which for a long time has passed unobserved in the literature we sketch here in the form of an extended exercise a proof variant for the reduction class property of the Kahr-Moore-Wang class. It directly formalizes Turing machine computations and thereby does not rely on the inseparability result for domino problems, but the reduction is not conservative.

Exercise 3.1.12. [441] Show without using domino problems that the Kahr-Moore-Wang class is a reduction class. Formalize directly the halting problem for Turing machines M, say over alphabet $\{0, 1\}$, through formulae Red(M), such that M, started in the initial configuration $C_0 = (0, 0, 0)$ with state 0 and reading head in cell 0 of the empty tape, eventually halts if and only if Red(M) is satisfiable.

Sketch of a solution. Since in t steps M can inspect and change at most the cells $\{0, 1, \ldots, t-1\}$, one can denote the configuration C_t, reached by M after t steps, by $C_t = (i_t, h_t, a_{t,0} \ldots a_{t,t})$ with internal state i_t, reading head position h_t and content $a_{t,j}$ of the j-th tape cell. The computation $C_0, \ldots, C_t$ can thus be encoded by a sequence of intervals R_i of length 2^i in the i-th diagonal $D_i = \{(i + x, x) \mid x \in \mathbb{N}\}$, as pictorially represented in Fig. 3.3.

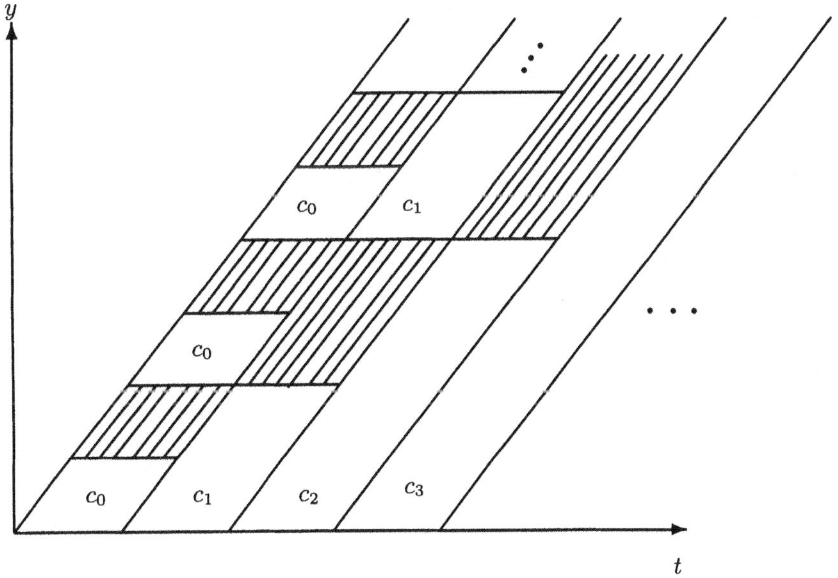

Figure 3.3. TM-encodings in diagonal

Formally such a *pan pipe*, starting at (x_0, x_0), is defined using a pattern of alternating red–black interval colourings (see below) such that R_i contains only red points and B_i only black points, i.e. formally

$$(1) \qquad R_i = \{(i + x_0 + j, x_0 + j) \mid 0 \leq j < 2^i\} \subseteq D_i,$$
$$(2) \qquad B_i = \{(i + x_0 + j, x_0 + j) \mid 2^i \leq j < 2^{i+1}\} \subseteq D_i.$$

For $x_0 > 0$ it is also assumed that the point $(i + x_0 - 1, x_0 - 1)$ is black. Denote $i + x_0$ by basis(i, x) for elements $(i + x, x) \in R_i$.

Using the economical description of Turing machines (see Chapter 2) the intended interpretation of the predicate symbols H and T_k (for reading head position and content k of the tape cells) can therefore be relativized as follows:

- $H(t + x, x)$ iff the head position at time t is cell $x -$ basis(t, x)
- $T_k(t + x, x)$ iff the tape at time t contains letter k in cell $x -$ basis(t, x)

The intended interpretation of the instruction predicates I_i is relativized by:

- $I_i(t + x, x)$ iff at time t the machine is in state i (i.e. $i_t = i$).

Define

$$\text{Red}(M) := \forall x \forall y(\ \text{STEP}_M \wedge \text{START} \wedge \text{NON–STOP} \wedge \text{SEGM}).$$

The initial M-configuration is described by

$$\text{START} := I_0(x, x) \wedge (R(x, x) \rightarrow T_0(x, x) \wedge H(x, x))$$

expressing that on the main diagonal the initial state 0 holds and each read point represents the initial reading head position 0 with cell content 0. The condition that M does never halt in state 1 is formalized by

$$\text{NON–STOP} := \neg S_1(x, y).$$

The formula STEP_M describes the effect of each M-instruction for the transition from C_t, supposed to be encoded in a red D_t–interval of length t, to C_{t+1}, which has to be encoded in the neighbouring red D_{t+1}-interval of length $t + 1$.

The *tape cell inscription* at moment $t + 1$ is formalized by three groups of formulae in STEP_M. For the tape cell after the execution of a printing instruction (i, k, l) ("in state i print letter k and go to state l") we have two types of conjuncts. One is for the reading head position (where the new tape symbol will be k) and one for the non–working cells (where the tape symbol, say j, does not change, for any symbol j in the alphabet of M):

$$I_i xy \wedge Hxy \wedge Rx'y \rightarrow T_k x'y,$$
$$\bigwedge_j \neg Hxy \wedge Rx'y \wedge T_j xy \rightarrow T_j x'y.$$

For non–printing instructions $(i, \ldots)$ of M the following conjuncts (for each letter j of M) assure that also the content of the working cell, say j, is preserved:

$$I_i xy \wedge Hxy \wedge Rx'y \wedge T_j xy \to T_j x'y.$$

The last conjunct requires that black points are never reading head positions and carry the black symbol 0:

$$Bxy \to \neg Hxy \wedge T_0 xy.$$

Note that this blank symbol will be transferred to a neighbouring read point in the next diagonal by the preceding tape inscription conjuncts.

The *next state*, at moment $t+1$, is formalized by again three groups of conjuncts in STEP_M. The next state for any instruction (i, op, j) of M with non–test operation op is formalized by

$$I_i xy \to I_j x'y.$$

The effect of a test instruction (i, j, k, l) ("in state i, if the reading head scans letter j then go to instruction k, else go to instruction l") is formalized by the following two conjuncts:

$$I_i xy \wedge Hxy \wedge T_j xy \to I_k x'y,$$

$$I_i xy \wedge Hxy \wedge \neg T_j xy \to I_l x'y.$$

The third group of conjuncts expresses that all the points of a diagonal encode the same state:

$$\bigwedge_i I_i xy \leftrightarrow I_i x'y'.$$

The *working cell* does not change for print-instructions or test-instructions. This is formalized by the corresponding conjuncts

$$I_i xy \wedge Rx'y \to (Hxy \leftrightarrow Hx'y).$$

The new working cell after execution of a left movement instruction (i, left, j) is illustrated by Fig. 3.4.

This is formalized by the following conjunct:

$$\bigwedge_{(i,\mathit{left},j)\in M} I_i xy' \wedge Rx'y' \to (Hxy' \leftrightarrow Hxy).$$

The effect of right move instructions (i, right, j) is illustrated by Fig. 3.5 and formalized by the conjunct

$$\bigwedge_{(i,\mathit{right},j)\in M} I_i xy \wedge Rx'y \to (Hxy \leftrightarrow Hx''y').$$

The segmentation formula SEGM has to guarantee that in models of $\text{Red}(M)$ there is enough red space in D_t to describe t computation steps of M. What is needed is resumed in the following

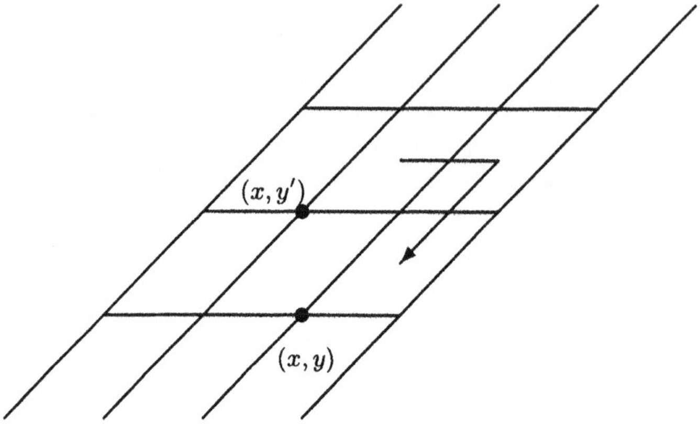

Figure 3.4. Encoding of a left move

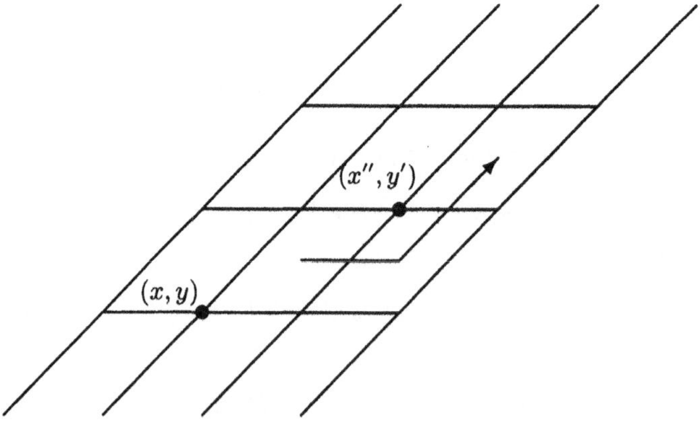

Figure 3.5. Encoding of a right move

pan pipe property: For each t there is an initial point $(t+p, p)$ (with $0 < p$) of a red interval of length 2^t in D_t to the left of which, up to the main diagonal, lie initial points $(i+p, p)$ of red D_i–intervals of length 2^i for all $0 \leq i < t$.

By a red D_t-interval of length $q+1$ we mean a sequence

$$(t+p, p), \ldots (t+p+j, p+j), \ldots, (t+p+q, p+q)$$

of points in R such that (a) $(t+p-1, p-1) \in B$ or $p = 0$ and (b) $(t+p+q+1, p+q+1) \in B$.

The fact that the points on or under the main diagonal and no others will be coloured is formalized by the following conjunct in SEGM for a greater–than relation G, with the intended interpretation that $G(p,q)$ is true iff $p \geq q$:

$$Gxx \wedge (Gxy \rightarrow Gx'y) \wedge (Gyx' \rightarrow Gyx) \wedge \neg Gxx'.$$

The following two SEGM-conjuncts ensure that every point below the main diagonal has exactly one of the colours "red" or "black" and that no other point is coloured:

$$Gxy \rightarrow (Rxy \vee Bxy) \wedge \neg(Rxy \wedge Bxy)$$
$$\neg Gxy \rightarrow \neg Rxy \wedge \neg Bxy.$$

The intended segmentation on the main diagonal into intervals of length one is ensured by the SEGM-conjunct

$$Rxx \leftrightarrow Bx'x'.$$

Note that due to the restriction to an $\forall\exists\forall$-prefix one cannot determine whether the point $(0,0)$ and therefore the points $(p,0)$ on the x-axis are coloured black or red. (For the intended interpretation we assume without loss of generality that each point on the x-axis is red.)

We have to assure the continuation of the segmentation from one diagonal to the next diagonal on the right. For this purpose we require that exactly each second *colour change point* (p,q) in D_t – i.e. with successor point $(p+1, q+1)$ of a different colour than(p,q) – passes on that colour change property to its right neighbour in D_{t+1}. This implies the doubling of the length of red and black intervals in D_{t+1}. The following SEGM–conjunct expresses that black and only black colour change points are passed to the right; colour change points are formalized by the predicate C:

$$Cxy \leftrightarrow (Rxy \wedge Bx'y') \vee (Bxy \wedge Rx'y'),$$
$$Gxy \rightarrow ((Cxy \wedge Bxy) \leftrightarrow Cx'y).$$

This concludes the definition of Red(M) except for the elementary transformations to bring all the conjuncts into the required Skolem normal form.

Exercise 3.1.13. Show by a counterexample that the reduction of the preceding exercise is not conservative. (This shows that the conservativity claim for a similar reduction presented in [532] is wrong.)

3.1.3 Graph Interpretation of $[\forall\exists\forall, (0, \omega)]$-Formulae

In this section we show how one can transform every model of a formula in the Kahr-Moore-Wang class into a coloured graph; we then bring these graphs into a normal form – namely with orthogonal edges between neighbouring points – that lends itself to a description by a formula in Kahr's class. This yields a conservative reduction of the Kahr-Moore-Wang class to Kahr's class. Together with the reduction in the preceding subsection this establishes Kahr's Theorem.

Theorem 3.1.14. *There is a conservative reduction of the Kahr-Moore-Wang class* $[\forall\exists\forall, (0, \omega)]$ *to Kahr's class* $[\forall\exists\forall, (\omega, 1)]$.

Proof. The idea of the proof can be described as follows. Let $\psi := \forall x \forall y \alpha$ be the Skolem normal form of a formula in $[\forall\exists\forall, (0, \omega)]$ with binary predicate symbols $R_1, \ldots, R_n$. Let R be a new binary predicate symbol and $\mathfrak{A} = (\mathbb{N}, R_1, \ldots, R_n)$ be a model of ψ with universe $\mathbb{N}$. Every point (p, q) with sequence $R_1pq, \ldots, R_npq$ of truth values is represented by an elementary square with global coordinates (p, q) containing n distinguished points $\bar{i}$ with local coordinates (ρ_i, σ_i); the local coordinates are required to satisfy $\rho_i, \sigma_i < r$ where r has to be chosen sufficiently large (see below), in particular $r > n$. The truth of R_ipq in $\mathfrak{A}$ is reflected by the truth of R at the point $(pr + \rho_i, qr + \sigma_i)$, i.e., the point $\bar{i}$ with local coordinates (ρ_i, σ_i) in the elementary square with global coordinates (p, q). Such a point where R is true will be called a coloured point. This encoding is pictorially represented in Fig. 3.6.

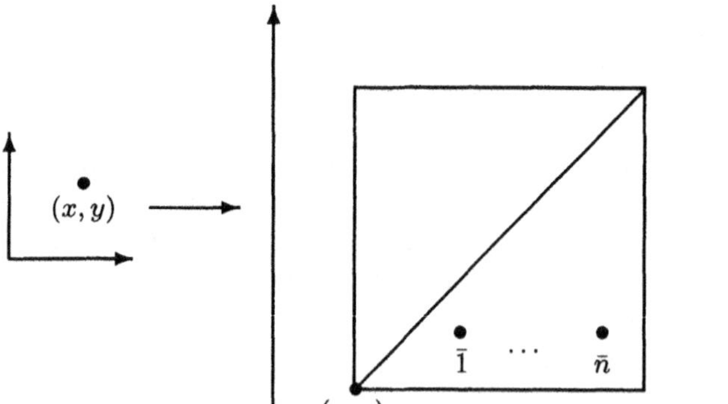

Figure 3.6. Encoding of $R_i(x, y)$ by $R(xr + \rho_i, yr + \sigma_i)$

Implications $R_i(a, b) \to R_j(c, d)$ in ψ are thus translated into implications of the form

$$R(ar + \rho_i, br + \sigma_i) \to R(cr + \rho_j, dr + \sigma_j).$$

These implications form an expression $G(\psi)$ and are interpreted as arrows in a graph that propagate the property of being coloured. More precisely such an implication is viewed as expressing that for all elementary squares with global coordinates $(a,b),(c,d)$ respectively there is an arrow from the point with local coordinates (ρ_i,σ_i) in the elementary square with global coordinates (a,b) to the point with local coordinates (ρ_j,σ_j) in the square (c,d). Let $\mathcal{G}(\psi)$ be the class of graphs that satisfy $G(\psi)$.

The problem consists in formalizing these implications for R by using only one occurrence of the Skolem function $'$, namely in x'.

We use a refinement of the grid to transform the arrows into paths of arrows between neighbouring or mirror image points. A geometrical argument will suffice to provide enough space for constructing these paths in such a way that they do not interfere with each other. The graph of these paths is then easily formalized by a reduction formula in $[\forall\exists\forall,(\omega,1)]$ which is (finitely) satisfiable if and only if the graph formula $G(\psi)$ has a (finite) graph model in $\mathcal{G}(\psi)$.

As a preparatory step we will transform ψ into a normal form amenable to the above outlined geometrical interpretation.

Thus the proof splits into three steps.

Step 1: We transform the given formula $\psi \in [\forall\exists\forall(0,\omega)]$ with Skolem normal form $\forall x\forall y\alpha$ into a formula φ with Skolem normal form $\forall x\forall y\beta$ such that ψ is (finitely) satisfiable if and only if φ is (finitely) satisfiable, each predicate symbol occurs in φ at most three times and β is a conjunction of formulae of the following forms:

(1) atom R_ixy
(2) internal $R_ixy \to R_jxy$
(3) external (3*a*) $R_ixy \to R_jx'y$ (3*b*) $R_ix'y \to R_jxy$
(4) negation (4*a*) $R_ixy \to \neg R_jxy$ (4*b*) $\neg R_ixy \to R_jxy$
(5) mirror $R_ixy \to R_jyx$
(6) conjunction $R_ixy \wedge R_jxy \to R_kxy$.

Step 2. Each formula ψ obtained through Step 1 is transformed into an expression $G(\psi)$ defining a class $\mathcal{G}(\psi)$ of coloured graphs with vertices from $\mathbb{N}\times\mathbb{N}$ such that ψ is (finitely) satisfiable if and only if $G(\psi)$ has a (finite) model in $\mathcal{G}(\psi)$.

Step 3. The graph description obtained in Step 2 is refined to an expression $G'(\psi)$ which has a (finite) model if and only if $G(\psi)$ does. In this refinement the arrows connect only neighbouring or mirror image points, i.e. go only from (a,b) to $(a+1,b)$, $(a,b+1)$ or to (b,a). $G'(\psi)$ is translated into a formula $Red(G'(\psi)) \in [\forall\exists\forall,(\omega,1)]$ such that $G'(\psi)$ has a (finite) model if and only if $Red(G'(\psi))$ does.

Steps 1–3 imply that ψ is (finitely) satisfiable if and only if $Red(G'(\varphi))$ is.

Step 1. Let ψ be an arbitrary formula in $[\forall\exists\forall,(0,\omega)]$ with Skolem normal form $\forall x\forall y\alpha$. We can assume without loss of generality that α contains only atomic subformulae of form $Pxy, Px'y, Pyx$. (See Exercise 3.1.10 and the Exercise 3.1.15 below.) Without loss of generality we also assume that α is built up from these atomic formulae using only negation and conjunction. We transform ψ in three steps.

Step 1.1. The goal of this step is to make sure that atomic subformulae Ryx or $Rx'y$ of α appear in the reduced formula only in equivalences of form $Pyx \leftrightarrow Qxy$ or $Px'y \leftrightarrow Qxy$, respectively.

It suffices to replace in α each atomic subformula $P_ix'y$ and P_jyx by Q_ixy and R_jxy respectively where Q_i, R_j are new predicate symbols. Let the result be α'. Add to α' the conjunction of all equivalences $P_ix'y \leftrightarrow Q_ixy$ and $P_jyx \leftrightarrow R_jxy$. The result is a formula $\psi_1 := \forall x \forall y \alpha' \wedge \alpha''$ which is clearly (finitely) satisfiable if and only if $\forall x \forall y \alpha$ is.

Exercise 3.1.15. Show how one could use the same technique to replace atomic formulae of form $Pxx, Pyy, Pxx', Px'x$ in such a way that those diagonal expressions appear only in equivalences of form $Pxx \leftrightarrow Qxx$.

Step 1.2. The goal of this step is to eliminate the propositional structure of α' in ψ_1 and to express it using only implications of forms (1)–(6).

It suffices to replace, by induction on the subformulae of α', each formula $\neg Pxy$ in α' by Rxy, where R is a new relation symbol, and to add the conjunct $\neg Pxy \leftrightarrow Rxy$. The same is done for $Pxy \wedge Qxy$ adding the conjunct $Rxy \leftrightarrow (Pxy \wedge Qxy)$. (Clearly one has also to rename the variables x, y in implications and to replace implications $Rxy \rightarrow (Pxy \wedge Qxy)$ by the two implications $Rxy \rightarrow Pxy$ and $Rxy \rightarrow Qxy$.) This results in a formula ψ_2 of the required form which clearly is (finitely) satisfiable if and only if ψ_1 is.

Step 1.3. The goal of this step is to reduce to three the number of occurrences of predicate symbols in ψ_2.

For each P which has $m > 3$ occurrences in ψ_2 choose new predicate symbols $P_1, \ldots, P_m$, replace the i-th occurrence of P by P_i (for $1 \leq i \leq m$) and add the following implications:

$$Pxy \rightarrow P_1xy$$

$$\bigwedge_{1 \leq i < m} P_ixy \rightarrow P_{i+1}xy$$

$$P_mxy \rightarrow Pxy.$$

The result is a formula $\varphi := \forall x \forall y \beta$ of the required form which is (finitely) satisfiable if and only if ψ_2 and therefore ψ is.

Step 2. We construct for given ψ obtained in Step 1 an expression $G(\psi)$ that describes a class $\mathcal{G}(\psi)$ of coloured graphs with nodes in the Gaussian quadrant $\mathbb{N} \times \mathbb{N}$ such that ψ is (finitely) satisfiable if and only if $G(\psi)$ has a (finite) model in $\mathcal{G}(\psi)$.

Let $R_1, \ldots, R_n$ be the predicate symbols occuring in ψ. Let R be a new binary predicate symbol and choose a sufficiently large number r, in particular

$r > n$. (Further conditions on the size of r will be added below.) Choose for each $1 \leq i \leq n$ a pair (ρ_i, σ_i) of numbers $\rho_i, \sigma_i < r$ satisfying conditions (C1)–(C3) to be explained below. As explained above we will interpret (ρ_i, σ_i) as local coordinates of the point $(pr+\rho_i, qr+\sigma_i) \in \mathbb{N}\times\mathbb{N}$ in the elementary square with global coordinates $(p, q) \in \mathbb{N}\times\mathbb{N}$. For a given model $\mathfrak{A} = (\mathbb{N}, R_1, \ldots, R_n)$ of ψ, the intended interpretation of R – which reflects the encoding of the truth R_ipq by colouring the point $(pr + \rho_i, qr + \sigma_i)$ – is

$$R := \{(pr + \rho_i, qr + \sigma_i) : \mathfrak{A} \models R_ipq\}.$$

$\mathcal{G}(\psi)$ is now constructed by encoding each of the implications (2)–(6) in ψ by a corresponding subformula of $G(\psi)$ which describes a pattern of arrows in the intended graph interpretation of $\mathcal{G}(\psi)$.

Each conjunct $R_ixy \to R_jxy$ of type (2) in ψ is encoded as a pattern of internal arrows leading from any point $(pr + \rho_i, qr + \sigma_i)$ to the point $(pr + \rho_j, qr + \sigma_j)$ in the same elementary square (see Fig. 3.7), for all $p, q \in \mathbb{N} \times \mathbb{N}$; let us stress the fact that the pattern of arrows is the same in all elementary squares (with the possible exception of boundary squares, see below). The conjunct $R(pr+\rho_i, qr+\sigma_i) \to R(pr+\rho_j, qr+\sigma_j)$ in $G(\psi)$ which translates (2) is intended to mean that if an instance of the source point is coloured, then also the corresponding instance of the target point is.

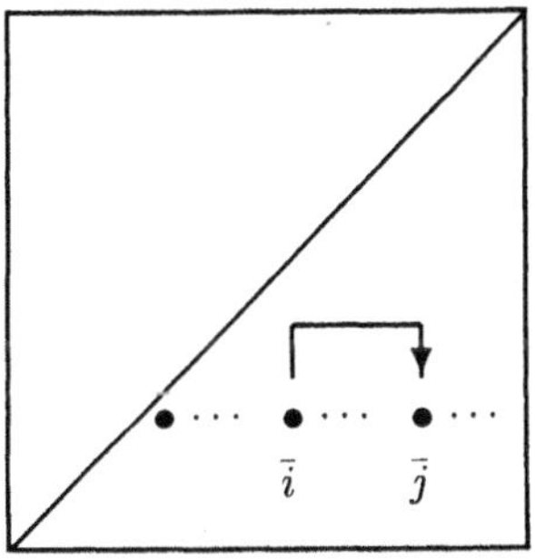

Figure 3.7. Encoding of $R_ixy \to R_jxy$

Conjuncts of type (3) are translated in the same way. Here the arrows are external, they go from points in an elementary square to points in the neighbouring square to the right (3a) (see Fig. 3.8) or to the left (3b).

Consider now conjuncts of type (4) containing a negation sign, say $R_ixy \to \neg R_jxy$. We choose for each such conjunct one neighbouring point of (ρ_j, σ_j), say $\neg(\rho_j, \sigma_j)$, which is not (yet) coloured, i.e. which during the construction of $G(\psi)$ and the corresponding graph interpretation has not (yet) been used as source or target of any outgoing or incoming arrow. There is such a neighbour because each predicate symbol occurs in ψ at most three times and thereby can produce at most tree outgoing or incoming arrows. In addition we also have to make sure that the neighbouring point, chosen in order to represent negation, is not itself a distinguished point k used for

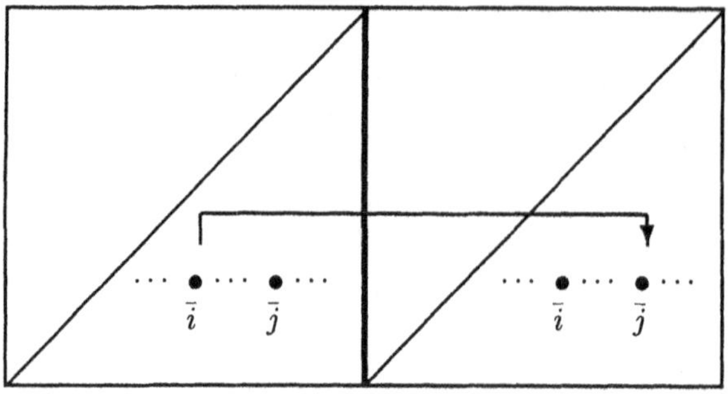

Figure 3.8. Encoding of $R_i xy \to R_j x'y$

encoding another relation symbol R_k. For later use we require even a little more, namely the following conditions:

$$(C1) \qquad \rho_i < \rho_j \wedge \sigma_i < \sigma_j, \text{ for all } i < j,$$
$$(C2) \qquad 11 \leq \rho_1, \sigma_1, \rho_{i+1} - \rho_i, \sigma_{i+1} - \sigma_i \wedge$$
$$\rho_n + 11, \sigma_n + 11 < r.$$

These conditions ensure that a) the encoding points increase with the index of the encoded relation symbol, and that b) for each encoding point, starting at any of its four neighbours, we find at least ten points in a row which are not themselves encoding points.

The conjunct $R_i xy \to \neg R_j xy$ in ψ is therefore encoded by arrows leading from each point with local coordinates ρ_i, σ_i to the point $\neg(\rho_j, \sigma_j)$ in the same elementary square. In $G(\psi)$ this is expressed by the conjunct

$$R(pr + \rho_i, qr + \sigma_i) \to R(pr + \neg(\rho_j, \sigma_j)_1, qr + \neg(\rho_j, \sigma_j)_2)$$

where $()_1, ()_2$ denote the first and second projection.

Conjuncts of type (4b) are formalized in a similar way. For each $\neg R_i xy \to R_j xy$ in ψ we draw an arrow from the point with local coordinates $\neg(\rho_i, \sigma_i)$ to the point with local coordinates (ρ_j, σ_j) in the same elementary square, expressed by the implication

$$R(pr + \neg(\rho_i, \sigma_i)_1, qr + \neg(\rho_i, \sigma_i)_2) \to R(pr + \rho_j, qr + \sigma_j).$$

As a consequence of this encoding of implications with negated literals we require that a graph, in order to be a model of $G(\psi)$, has to satisfy the following: for each occurrence of a literal $\neg R_j xy$ in ψ, any point with local coordinates (ρ_j, σ_j) is coloured if and only if the neighbouring point $\neg(\rho_j, \sigma_j)$ is not.

For conjuncts of type (6) in ψ we write the corresponding $G(\psi)$-conjunct

$$R(pr + \rho_i, qr + \sigma_i) \wedge R(pr + \rho_j, qr + \sigma_j) \to R(pr + \rho_k, qr + \sigma_k).$$

This expresses the following condition for graphs in $\mathcal{G}(\psi)$ to be models of $G(\psi)$ (see Fig. 3.9): if in an elementary square the two points with local coordinates $(\rho_i, \sigma_i), (\rho_j, \sigma_j)$ are coloured, then so is the point with local coordinates (ρ_k, σ_k) in the same elementary square.

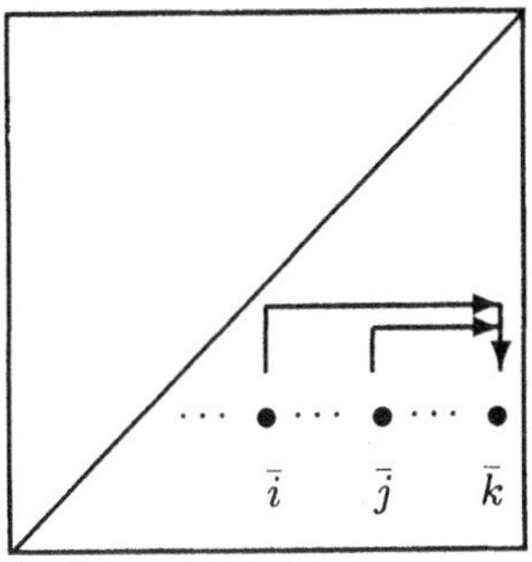

Figure 3.9. Encoding of $R_i xy \wedge R_j xy \rightarrow R_k xy$

For conjuncts $R_i xy \rightarrow R_j yx$ of type (5) in ψ we write the corresponding $G(\psi)$-conjunct

$$R(pr + \rho_i, qr + \sigma_i) \rightarrow R(qr + \rho_j, pr + \sigma_j).$$

This expresses the following condition for graphs in $\mathcal{G}(\psi)$ to be models of $G(\psi)$: if in an elementary square with global coordinates (p, q) the point with local coordinates (ρ_i, σ_i) is coloured, then so is the point with local coordinates (ρ_j, σ_j) in the reflection of that elementary square in the main diagonal, i.e. the square with global coordinates (q, p).

This ends the construction of $G(\psi)$ and the explanation of the class $\mathcal{G}(\psi)$ of coloured graphs over $\mathbb{N} \times \mathbb{N}$ which support the intended interpretation of R. By a finite model in $\mathcal{G}(\psi)$ we mean a model where the domain is a subset $\mathbb{Z}_p \times \mathbb{Z}_p \subseteq \mathbb{N} \times \mathbb{N}$ with $p = mr$ for some m.

From the construction it is clear that ψ has a (finite) model if and only if $G(\psi)$ has a (finite) model in $\mathcal{G}(\psi)$.

Step 3. The goal of this step is to orthogonalize the arrows in $G(\psi)$ along the $\mathbb{N} \times \mathbb{N}$-grid. This means that the arrows are replaced by sequences of arrows linking points to either one of their neighbours or to their reflection in the main diagonal. The formula $G'(\psi)$ describing these normalized graphs is easily recognized to belong to $[\forall\exists\forall, (\omega, 1)]$ and to have a (finite) model if and only if $G(\psi)$ does.

We have to go again through the cases (2)–(6) of conjuncts in $G(\psi)$. We indicate for each case the arrow transformation and its formalization in $[\forall\exists\forall, (\omega, 1)]$. This formalization is based upon the following axiomatization $mod(P_0, \ldots, P_r)$ of the $mod(r)$–function to express local coordinates $0 \leq \rho < r$, where $P_0, \ldots, P_{r-1}$ are monadic predicate symbols:

$$\bigvee_{0 \leq \rho < r} P_\rho x \wedge \bigwedge_{0 \leq \rho \neq \sigma < r} \neg(P_\rho x \wedge P_\sigma x) \quad \text{(Partitioning)}$$

$$\wedge \bigwedge_{0\leq\rho<r} (P_\rho x \leftrightarrow P_{\rho+1}x') \wedge (P_{r-1}x \leftrightarrow P_0 x') \quad \text{(Periodicity)}.$$

It is easy to see how to orthogonalize internal arrows coming from $G(\psi)$-implications $R(xr+\rho_i, yr+\sigma_i) \rightarrow R(xr+\rho_j, yr+\sigma_j)$ of type (2). It suffices to construct an equivalent path of arrows which go only from points (p,q) to neighbouring points $(p\pm 1, q)$ or $(p, q\pm 1)$ without crossing any other arrow, see Fig. 3.10. Given condition (C1) and the fact that each predicate symbol can occur at most three times, it is possible to choose r sufficiently large in order to find enough space for such crossing free paths.

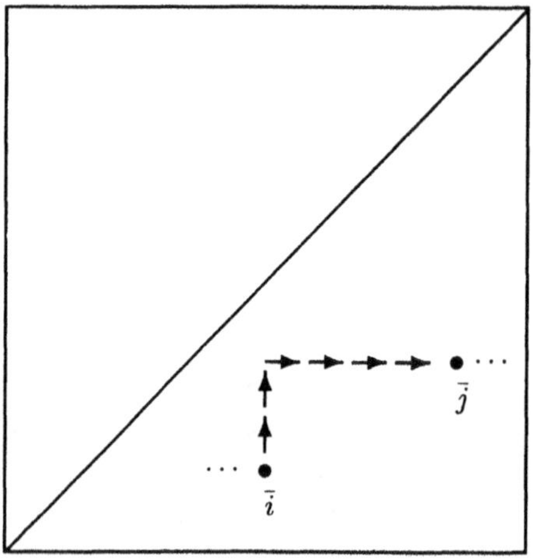

Figure 3.10. Orthogonalization of $i \rightarrow j$

Such a path can be formalized by a conjunction of $[\forall\exists\forall, (\omega, 1)]$ formulae of the following form:

$$P_{\rho_i} y \wedge P_{\sigma_i} x \wedge Ryx \rightarrow Ryx'$$

$$P_{\rho_i} y \wedge P_{\sigma_i+1} x \wedge Ryx \rightarrow Ryx'$$

$$\vdots$$

$$P_{\rho_j-1} x \wedge P_{\sigma_j} y \wedge Rxy \rightarrow Rx'y$$

This transformation and formalization clearly preserves models and finite models.

In the same way we can orthogonalize the arrows from or to negation points, i.e. coming from implications of type (4a,b); in the same way we formalize this in $[\forall\exists\forall, (\omega, 1)]$-conjuncts, see Fig. 3.11.

For external arrows coming from $G(\psi)$-implications $R(xr+\rho_i, yr+\sigma_i) \rightarrow R(x'r+\rho_j, yr+\sigma_j)$ of type (3) we do a similar transformation and formalization, as illustrated by Fig. 3.12.

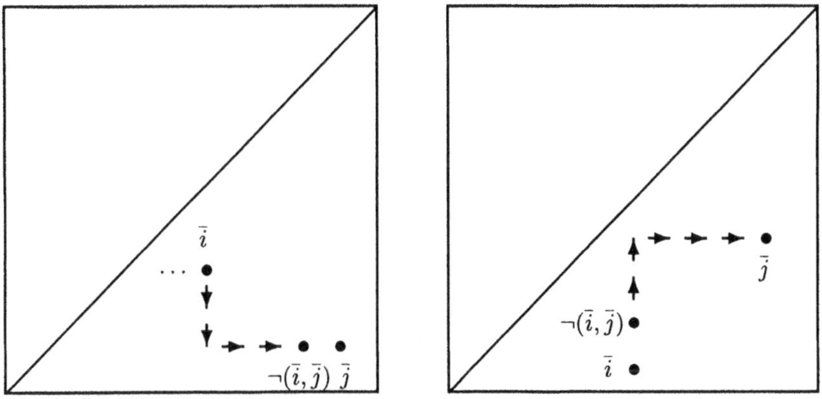

Figure 3.11. Orthogonalization of $R_ixy \to \neg R_jxy$ and $\neg R_ixy \to R_jxy$

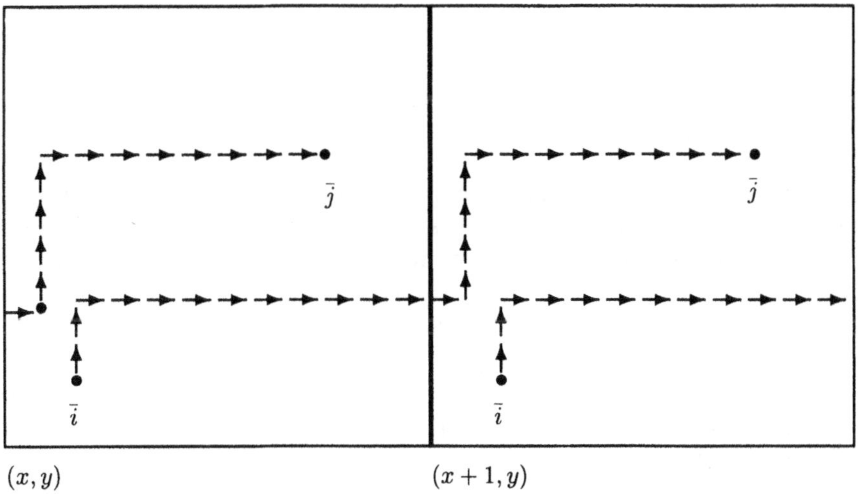

Figure 3.12. Orthogonalization of $R_ixy \to R_jx'y$

In order to prevent undesirable conflicts with paths to be constructed now for arrows coming from implications of type (5) or (6), we require the following additional property (C3) for the choice of local coordinates:

$$(C3) \quad 6 \leq \rho_i - \sigma_i \text{ for all } i \leq n.$$

This condition guarantees that the encoding points lie below the diagonal of the elementary squares (because of $\sigma_i < \rho_i$) and that their distance from that local diagonal is at least 6. This prevents also auxiliary points like $\neg(\rho_i, \sigma_i)$ to be above the local diagonals — similarly for auxiliary points for arrows from implications of type (5) and (6).

For each conjunct $R(xr + \rho_i, yr + \sigma_i) \rightarrow R(yr + \rho_j, xr + \sigma_j)$ of type (5) in $G(\psi)$ we choose a fresh point $jump(\rho_i, \sigma_i)$ at distance say two from $\bar{i}$ in the same elementary square, draw a new path from $\bar{i}$ to $jump(\rho_i, \sigma_i)$, draw an arrow from there to its mirror image over the main diagonal and continue with a fresh path from that point to the point $\bar{j}$ in the mirror image square. This is illustrated in Fig. 3.13. To avoid crossings we assume in particular that before this transformation step, for any elementary square the point $jump(\rho_i, \sigma_i)$ and its mirror image were both still free from incoming and outgoing arrows. (Note also that only for implications of this type we do use, for the first time, points above the diagonal.) This transformation can be formalized by a conjunction of formulae as above where the crucial jump conjunct is:

$$P_{j_1}x \wedge P_{j_2}y \wedge Rxy \rightarrow Ryx \quad \text{where } (jump(\rho_i, \sigma_i))_k = j_k.$$

Clearly this transformation preserves models and finite models.

For each conjunct $R(xr + \rho_i, yr + \sigma_i) \wedge R(xr + \rho_j, yr + \sigma_j) \rightarrow R(xr + \rho_k, yr + \sigma_k)$ of type (6) in $G(\psi)$ we construct two fresh paths as illustrated in Fig. 3.14. The construction is split into 6 parts.

First we choose an appropriate fresh point $conj(i, j, k)$. We choose this point in such a way that it has distance 5 from $\bar{k}$ and that neither $conj(i, j, k)$ nor (going in the direction of $\bar{k}$) its neighbour $conj(i, j, k)'$ nor the next point $conj(i, j, k)^*$ nor their mirror images over the main diagonal, have been coloured yet. Let $(\rho, \sigma), (\rho', \sigma'), (\rho^*, \sigma^*)$ be the local coordinates of $conj(i, j, k)$, $conj(i, j, k)'$, $conj(i, j, k)^*$ respectively. (In Fig. 3.14 we have chosen $\rho' = \rho + 1, \rho^* = \rho' + 1, \rho + 5 = \rho_k, \sigma = \sigma' = \sigma^* = \sigma_k$.)

Part 1. In every elementary square we draw a new path from the point $\bar{i}$ to the point $conj(i, j, k)$; this is formalized as before by conjuncts of $G'(\psi) \in [\forall\exists\forall, (\omega, 1)]$ of the following form :

$$P_{\rho_i}y \wedge P_{\sigma_i}x \wedge Ryx \rightarrow Ryx'$$

$$\vdots$$

$$P_{\rho}y \wedge P_{\sigma-1}x \wedge Ryx \rightarrow Ryx'$$

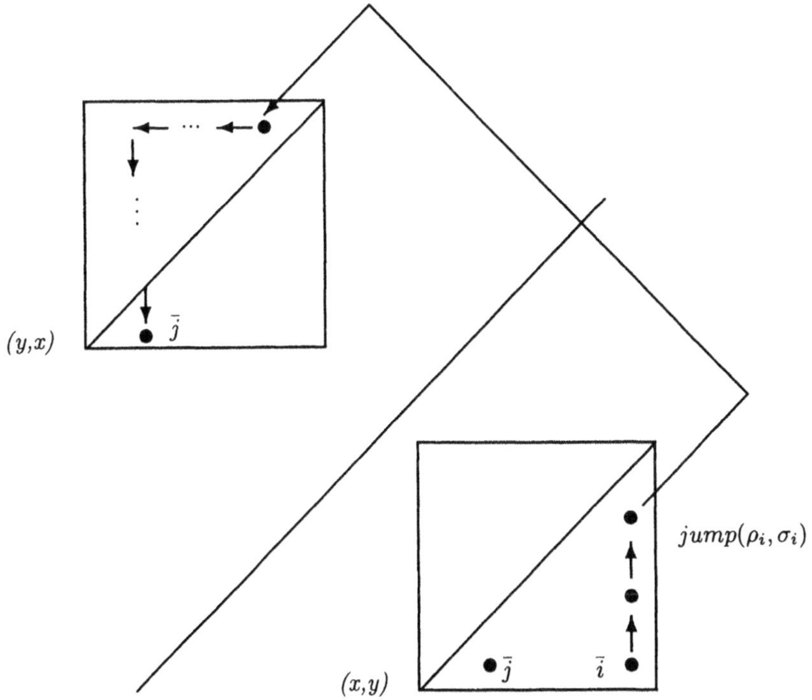

Figure 3.13. Orthogonalization of mirror arrows $R_i xy \to R_j yx$

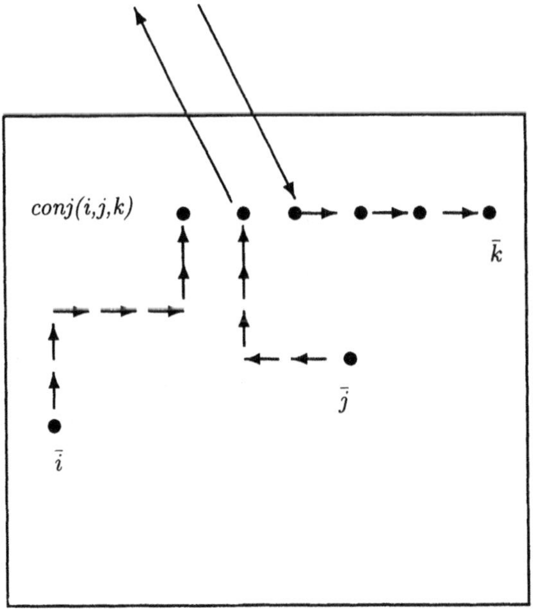

Figure 3.14. Orthogonalization of conjunction arrows $R_i xy \wedge R_j xy \to R_k yx$

Part 2. In the same way we construct and formalize for each elementary square a new path from the point $\bar{j}$ to the point $conj(i,j,k)'$.

What we want to describe now is the following formula:

$$Rxy \wedge Rx'y \wedge P_{\rho'}x' \wedge P_{\sigma'}y \rightarrow Rx''y.$$

We can then ensure by paths of the usual type (see Part 6 below) that also the nearby point $\bar{k}$ is coloured. We have to avoid however the nesting of the successor function in $Rx''y$. This is done by jumping twice over the main diagonal as described in Parts 3–5 of the construction.

Part 3. We draw in each elementary square an arrow from $conj(i,j,k)'$ to its mirror image over the main diagonal; the conjunctive rôle of this jump over the diagonal is formalized by the following subformula of $G'(\psi)$ expressing that if a point $(pr+\rho', qr+\sigma')$ as well as its (in our example, left) neighbouring point with local coordinates (ρ, σ) are both colored, then also the mirror image $(qr+\sigma', pr+\rho')$ over the main diagonal is colored:

$$P_{\rho'}x' \wedge P_{\sigma'}y \wedge Rxy \wedge Rx'y \rightarrow Ryx'.$$

Part 4. In each elementary square we construct a path from the node with local coordinates (σ', ρ') to the node with local coordinates (σ^*, ρ^*). This can be formalized as above by appropriate conjuncts in $G'(\psi)$.

Part 5. We draw arrows from points $(qr+\sigma^*, pr+\rho^*)$ to their mirror images $(pr+\rho^*, qr+\sigma^*)$ over the main diagonal. This is formalized by the following $G'(\psi)$-conjunct:

$$P_{\rho^*}x \wedge P_{\sigma^*}y \wedge Ryx \rightarrow Rxy.$$

Part 6. In each elementary square we construct a path from the node with local coordinates (ρ^*, σ^*) to the node $\bar{k}$. This is formalized by the following $G'(\psi)$-conjuncts:

$$P_{\rho^*}x \wedge P_{\sigma^*}y \wedge Rxy \rightarrow Rx'y,$$

$$P_{\rho_k - 1}x \wedge P_{\sigma_k}y \wedge Rxy \rightarrow Rx'y.$$

By this construction we have normalized the arrows in $\mathcal{G}(\psi)$ which come from conjunctive axioms of type (6) to equivalent paths which are built up from steps to immediate neighbours and from jumps to the mirror image over the main diagonal; these normalized paths lend themselves to formalization in $[\forall\exists\forall, (\omega, 1)]$.

□

3.1.4 The Remaining Cases Without $\exists^*$

In this section we first prove the conservative reduction class property for the second of the two minimal such classes with finite prefix, namely Surányi's reduction class $[\forall^3\exists, (\omega, 1)]$. Our proof provides a conservative reduction of Kahr's class to it. As an easy corollary we obtain the conservative reduction class property also for the Surányi class $[\exists^*\forall^3\exists, (0, 1)]$. We then show that there is also a simple conservative reduction of Kahr's class to the Kalmár-Surányi class $[\forall^*\exists, (0, 1)]$ and to the Denton class $[\forall\exists\forall^*, (0, 1)]$. Note that the last three cases establish undecidability for small classes of formulae of graph theory.

Surányi's Reduction Class $[\forall^3\exists, (\omega, 1)]$. In this paragraph we prove the following theorem.

Theorem 3.1.16 (Surányi). $[\forall^3\exists, (\omega, 1)]$ *is a conservative reduction class.*

Proof. We give a conservative reduction of Kahr's class to Surányi's class. Let $\forall x \forall y \alpha$ be the Skolem normal form of an arbitrary formula $\psi \in [\forall\exists\forall, (\omega, 1)]$ with monadic predicates $Q_1, \ldots, Q_n$ and one binary predicate R. For the first proof step we assume that the binary atomic formulae which occur in ψ are all among $Rxy, Ryx, Rx'y$.

The idea of the proof consists in coding each point a of a model $\mathfrak{A} = (A, Q_1^{\mathfrak{A}}, \ldots, Q_n^{\mathfrak{A}}, R^{\mathfrak{A}}) \models \psi$ by a triple (a_0, a_1, a_2) of points in a new structure $\mathfrak{B} = (B, P_0^{\mathfrak{B}}, P_1^{\mathfrak{B}}, P_2^{\mathfrak{B}}, Q_1^{\mathfrak{B}}, \ldots, Q_n^{\mathfrak{B}}, R^{\mathfrak{B}})$ where a_0 can be identified with a. The successor function $'$ in $\mathfrak{A}$ is naturally extended to the new points as indicated in Fig. 3.15; formally this is encoded into R by requiring the truth of Ra_0a_1, Ra_1a_2 and Ra_2a_0 in $\mathfrak{B}$ (see below axiom β_0).

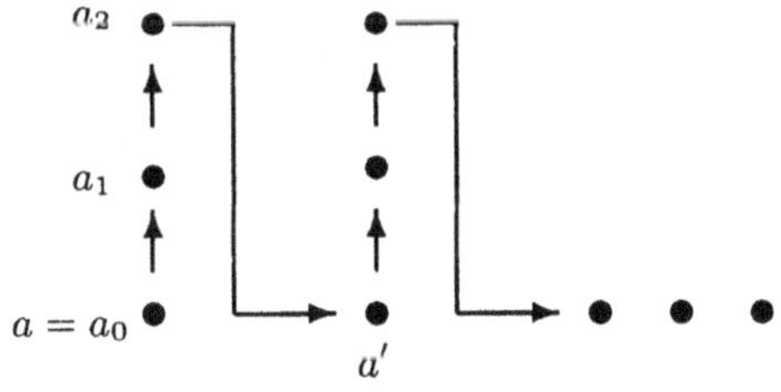

Figure 3.15. Tripling points of the given model

In this way a_1 becomes the successor of $a = a_0$. As a consequence, in the extended model the copy a_1 of a has to play the rôle of a' with respect to the encoding of the truth values $Q_1^{\mathfrak{A}}a', \ldots, Q_n^{\mathfrak{A}}a', R^{\mathfrak{A}}a'b$ for any b. In this respect also the second copy a_2 of a must be equivalent to a and therefore to a_1. This means that the predicates $Q_j^{\mathfrak{A}}$ are encoded into predicates $Q_j^{\mathfrak{B}}$ of the new model such that $Q_j^{\mathfrak{A}}a'$, $Q^{\mathfrak{B}}a_1$ and $Q^{\mathfrak{B}}a_2$ are equivalent for all $a \in A$, i.e. the following holds:

$$(*) \qquad \begin{array}{lll} a_1 \in Q_j^{\mathfrak{B}} & \text{iff} & a' \in Q_j^{\mathfrak{A}} \\ a_2 \in Q_j^{\mathfrak{B}} & \text{iff} & a' \in Q_j^{\mathfrak{A}} \\ a_0 \in Q_j^{\mathfrak{B}} & \text{iff} & a \in Q_j^{\mathfrak{A}}. \end{array}$$

This intended interpretation is easily axiomatized by (the Skolem form of) an $[\forall\exists, (\omega, 1)]$-formula β_0 as follows:

$$\beta_0 := Rxx' \wedge mod(P_0, P_1, P_2) \wedge (P_1x \vee P_2x \rightarrow \bigwedge_{1\leq j\leq n} (Q_jx \leftrightarrow Q_jx')),$$

Here $mod(P_0, P_1, P_2)$ describes a modulo-3 structure that encodes the successor relation described above among the copies a_0, a_1, a_2 of a (see Fig. 3.15), i.e. $mod(P_0, P_1, P_2)$ is the formula

$$\bigvee_{0\leq j\leq 2} P_jx \wedge \bigwedge_{0\leq i<j\leq 2} \neg(P_ix \wedge P_jx) \wedge \bigwedge_{j=i+1 \pmod 3} (P_ix \rightarrow P_jx').$$

In order to ensure the equivalence of a_1, a_2 with a' for the binary relation R the intended interpretation of R in $\mathfrak{B}$ should satisfy the equivalence of $R^{\mathfrak{A}}a'b$ with $R^{\mathfrak{B}}a_1b_0$ and $R^{\mathfrak{B}}b_0a_2$. Thus, we require that

$$(**) \qquad \begin{array}{lll} (a_1, b_0) \in R^{\mathfrak{B}} & \text{iff} & (a', b) \in R^{\mathfrak{A}} \\ (b_0, a_2) \in R^{\mathfrak{B}} & \text{iff} & (a', b) \in R^{\mathfrak{A}} \\ (a_0, b_0) \in R^{\mathfrak{B}} & \text{iff} & (a, b) \in R^{\mathfrak{A}}. \end{array}$$

In the presence of $mod(P_0, P_1, P_2)$ this intended interpretation is easily axiomatized by a $[\forall^3, (\omega, 1)]$-formula β_1 defined as conjunction of two formulae $\beta_{1,1}$ (describing the equivalence of the arrows 1 and 2 depicted in Fig. 3.16) and $\beta_{1,2}$ (describing the equivalence of the arrows 2 and 3 in Fig. 3.16):

$$\beta_{1,1} := P_1x \wedge P_2y \wedge Rxy \wedge P_0z \rightarrow (Rxz \leftrightarrow Rzy),$$

$$\beta_{1,2} := P_0z \wedge P_2y \wedge Ryx \wedge P_0x \rightarrow (Rzy \leftrightarrow Rxz).$$

We therefore encode models of ψ by models of a reduction formula φ where points a, b are represented by P_0-points and a' is represented as a P_1-point.

$$\varphi := \forall x\beta_0 \wedge \forall x\forall y\forall z(\beta_1 \wedge (P_0x \wedge P_0y \wedge P_1z \wedge Rxz \rightarrow \alpha[x'/z])).$$

Claim. *ψ is (finitely) satisfiable if and only if φ is (finitely) satisfiable.*

Exercise 3.1.17. Show that the claim implies the theorem under the assumption mentioned at the beginning of the proof.

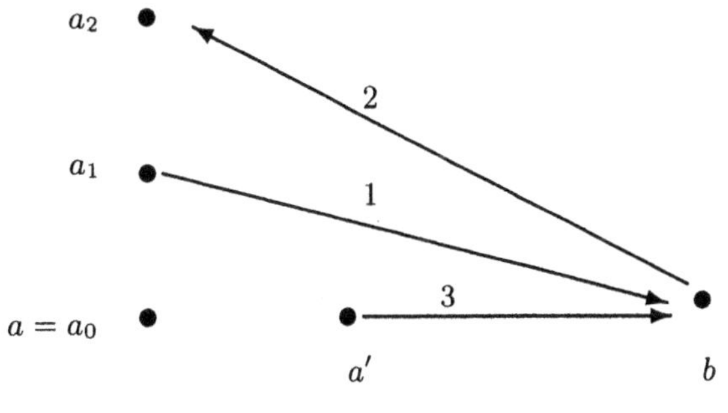

Figure 3.16. Equivalence of triple points

Proof of the claim. For a model $\mathfrak{A} = (A, Q_1^{\mathfrak{A}}, \ldots, Q_n^{\mathfrak{A}}, R^{\mathfrak{A}}) \models \psi$ we define $\mathfrak{B} = (B, P_0^{\mathfrak{B}}, P_1^{\mathfrak{B}}, P_2^{\mathfrak{B}}, Q_1^{\mathfrak{B}}, \ldots, Q_n^{\mathfrak{B}}, R^{\mathfrak{B}})$ with universe $B = A \times \{0,1,2\}$ by $P_j^{\mathfrak{B}} := A \times \{j\}$ and the intended interpretation given above (see $((*), (**))$ for $Q_j^{\mathfrak{B}}$ and $R^{\mathfrak{B}}$. It is easy to check that $\mathfrak{B}$ satisfies $\forall x \beta_0 \wedge \forall x \forall y \forall z \beta_1$. To show that $\mathfrak{B} \models \forall x \forall y \forall z P_0 x \wedge P_0 y \wedge P_1 z \wedge Rxz \to \alpha[x'/z]$ one proceeds by induction on the subformulae $\gamma(x, x', y)$ of α to verify that for all $a, b \in A$, $\mathfrak{A} \models \gamma[a, a', b]$ if and only if $\mathfrak{B} \models \gamma[(a,0),(a,1),(b,0)]$. By assumption on the atoms occuring in α, at the basis one has only three cases to consider: a) Rxy, b) Ryx, c) Rzy.

The first two cases are symmetric and follow by definition of $R(a,0)(b,0)$ in $\mathfrak{B}$. For Rzy we use that $\mathfrak{B} \models R(a,1)(b,0)$ iff $\mathfrak{A} \models Ra'b$. (Note that the equivalence would not hold for Ryz because Ryz is true in $\mathfrak{A}$ for successive points $(b,0),(b,1)$.)

For the converse let $\mathfrak{B}$ be a model of ψ^* with successor function $'$ satisfying $\forall x \beta_0$. Define $\mathfrak{A}$ as the restriction of $\mathfrak{B}$ to the elements satisfying $P_0 x$. Then for each $a \in B$ satisfying $P_0 a$ the formula $P_0 a'''$ holds by $mod(P_0, P_1, P_2)$. Furthermore since $\mathfrak{B} \models \forall x \beta_0 \wedge \forall x \forall y \beta_1$, it follows that $\mathfrak{B} \models Ra'''b \leftrightarrow Ra'b$ and $\mathfrak{B} \models Q_j a''' \leftrightarrow Q_j a'$ for all $a, b \in B$ satisfying $P_0 a \wedge P_1 b$; therefore $\alpha[a, a''', b]$ is equivalent in $\mathfrak{B}$ to $\alpha[a, a', b]$; since the latter is true in $\mathfrak{B}$, it follows that $\mathfrak{A} \models \forall x \forall y \alpha$. □

Exercise 3.1.18. Modify the construction in the proof to make it independent of the assumption that only binary subformulae $Rxy, Ryx, Rx'y$ occur.

The proof establishes the following stronger version of the theorem.

Corollary 3.1.19. *The class $[\forall\exists \wedge \forall^3, (\omega, 1)]$ of all formulae $\psi \wedge \varphi$ with only one binary predicate symbol and $\psi \in [\forall\exists, (\omega,1)]$ and $\varphi \in [\forall^3, (\omega, 1)]$ is a conservative reduction class.*

Corollary 3.1.20 (Surányi). *The class $[\exists^*\forall^3\exists, (0,1)]$ is a conservative reduction class.*

Proof. We give a reduction from Surányi's reduction class $[\forall^3\exists, (\omega,1)]$ to the class in question. Let $\psi := \forall x \forall y \forall z \exists u \alpha \in [\forall^3\exists, (\omega, 1)]$ be an arbitrary formula with monadic predicate symbols P_i $(1 \leq i \leq n)$ and one binary predicate symbol R. The reduction idea is to code each P_i into a branch $\{x : R(w_i, x)\}$ of R, where w_i is a *witness* for P_i, i.e. an element that is different from all the elements of the domain of a given model for ψ. As witnessing property we choose the property of being related to some special new element w_0 as illustrated in Fig. 3.17.

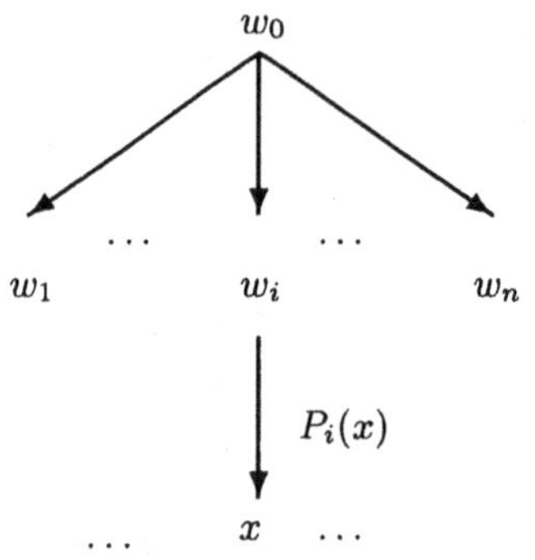

Figure 3.17. Witnessing monadic predicates

Formally we define therefore the reduction formula $\varphi \in [\exists^*\forall^3\exists, (0,1)]$ by

$$\varphi := \exists w_0 \cdots \exists w_n \forall x \forall y \forall z \exists u (\text{WITNESSES} \wedge \text{CLOSURE} \wedge \text{ENCODING}[\alpha])$$

where

$$\begin{aligned} \text{WITNESSES} &:= \bigwedge_{1 \leq i \leq n} R w_0 w_i \\ \text{CLOSURE} &:= \neg R w_0 u \\ \text{ENCODING}[\alpha] &:= \neg R w_0 x \wedge \neg R w_0 y \wedge \neg R w_0 z \rightarrow \alpha[P_i t / R w_i t]. \end{aligned}$$

Claim. *ψ is (finitely) satisfiable if and only if φ is (finitely) satisfiable.*

Exercise 3.1.21. Prove the claim.

□

Exercise 3.1.22. Prove that the Surányi class $[\exists^*\forall\exists\forall, (0,1)]$ is a conservative reduction class. Hint: Reduce the Kahr class. (A different proof which does not depend on the result of Kahr is given in the next section.)

Classes with $\forall^*$ in the Prefix. We prove here the following theorem.

Theorem 3.1.23. *The class $[\forall\exists \wedge \forall^*, (0,1)]$ of all formulae $\psi \wedge \varphi$ with only one binary predicate symbol and $\psi \in [\forall\exists, (0,1)]$ and $\varphi \in [\forall^*, (0,1)]$ is a conservative reduction class.*

By prenexing one obtains immediately the conservative reduction class property for the two minimal classes with $\forall^*$ in the prefix.

Corollary 3.1.24 (Kalmár, Surányi). *$[\forall^*\exists, (0,1)]$ is a conservative reduction class.*

Corollary 3.1.25 (Denton). *$[\forall\exists\forall^*, (0,1)]$ is a conservative reduction class.*

Proof. (of 3.1.23) We refine the encoding idea used already for the conservative reduction of Kahr's class to Surányi's reduction class. Let $\forall x \forall y \alpha$ be the Skolem normal form of an arbitrary formula $\psi \in [\forall\exists\forall, (\omega, 1)]$ with monadic predicate symbols $P_2, \ldots, P_n$ and binary predicate symbol R. Without loss of generality we assume $n \geq 3$.

The idea of the proof consists in coding each point a of a model $\mathfrak{A} = (A, P_2, \ldots, P_n, R) \models \psi$ by a tuple $(a_0, \ldots, a_n)$ of elements in a new structure $\mathfrak{B} = (B, Q)$ with one binary relation Q where a_0 can be identified with a. The successor function $'$ in $\mathfrak{A}$ is naturally extended to the new points as indicated in Fig. 3.18; formally this is encoded into Q by requiring that $\mathfrak{B} \models Q a_i a_j$ for $j = i + 1 \pmod{n+1}$.

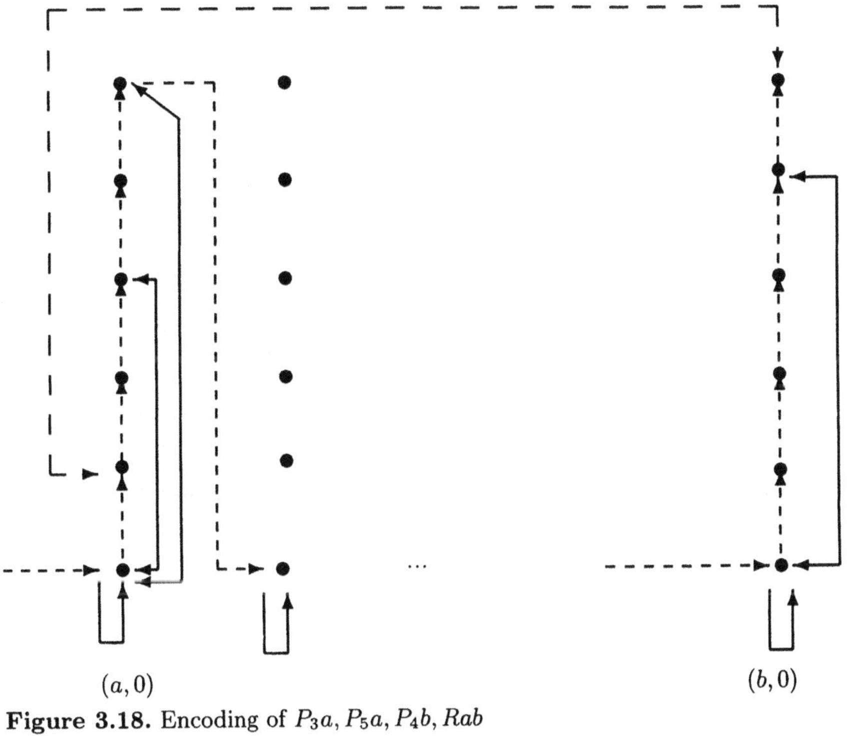

Figure 3.18. Encoding of $P_3 a, P_5 a, P_4 b, Rab$

Fig. 3.18 also shows the intended encoding of $P_i a$ and Rab which can be formally described as follows:

$$(*) \qquad \mathfrak{A} \models P_i a \iff \mathfrak{B} \models Qa_0a_i \wedge Qa_ia_0$$
$$\mathfrak{A} \models Rab \iff \mathfrak{B} \models Qa_1b_n \wedge Qb_na_1.$$

The given basis points $a = a_0$ can be characterized by reflexivity of Q on them, i.e. by requiring that

$$(**) \qquad \mathfrak{B} \models Qa_0a_0.$$

We express the successor relation by the formula

$$S(x,y) := Qxy \wedge \neg Qyx.$$

Further we express that $X = (x_0, \ldots, x_n)$ is an $(n+1)$-tuple of successive points, starting at basis x_0, (called a *chain*) by the formula

$$C(X) := Qx_0x_0 \wedge \bigwedge_{i<n} S(x_i, x_{i+1}).$$

To make this encoding idea work we use an $\forall\exists$-formula

$$\text{SUCCESSOR} := \forall x S(x, x')$$

to guarantee for each element the existence of a successor. To make the successor relation sufficiently unique we have to ensure that any successors y, z of x are equivalent for x with respect to Q (Q-indistinguishable). This is formalized as follows:

$$\text{UNIQUE} := S(x,y) \wedge S(x,z) \to ((Qyu \leftrightarrow Qzu) \wedge (Quy \leftrightarrow Quz)).$$

For every sequence of $n+1$ successive points we have also to guarantee that it contains exactly one basis point. This is formalized as follows:

$$\text{BASIS} := (\bigwedge_{i<n} S(x_i, x_{i+1})) \to (\bigvee_{i \le n} Qx_ix_i \wedge \bigwedge_{i<j\le n} \neg(Qx_ix_i \wedge Qx_jx_j)).$$

We can now define the reduction formula φ as the conjunction of SUCCESSOR and the universal closure of the formula

$$\eta \to \alpha[Rst/Qs_1t_n, P_it/Qt_0t_i]$$

where

$$\eta := \text{UNIQUE} \wedge \text{BASIS} \wedge C(X) \wedge C(Y) \wedge C(Z) \wedge S(x_n, y_0).$$

Claim. *φ is (finitely) satisfiable if and only if ψ is (finitely) satisfiable.*

Proof of the claim. If $\mathfrak{A} = (A, P_2, \ldots, P_n, R) \models \psi$, then we define $\mathfrak{B}$ over $B = A \times \{0, \ldots, n\}$ by the intended interpretation given above (see $(*)$, $(**)$ and Fig. 3.18). It is easy to check that $\mathfrak{B} \models \varphi$.

For the converse suppose that $\mathfrak{B} = (B, Q) \models \varphi$. Let $'$ be a function such that $\mathfrak{B} \models \forall x S(x, x')$. Over the restriction of B to the set A of all points a where Q is reflexive (i.e. $A = \{a \in B : \mathfrak{B} \models Qaa\}$) define

$$\begin{aligned} P_i &:= \{a \in A : \mathfrak{B} \models Qa_0a_i \wedge Qa_ia_0\} \\ R &:= \{(a,b) \in A \times A : \mathfrak{B} \models Qa_1b_n \wedge Qb_na_1\} \end{aligned}$$

where a_i denotes the result of i successive applications of the operation $'$ to a. Clearly $\mathfrak{A} = (A, P_2, \ldots, P_n, R) \models \alpha[a, a_{n+1}, b]$ for all $a, b \in A$ and therefore $\mathfrak{A} \models \psi$. □

3.2 Existential Interpretation for $[\forall^3\exists^*, (0,1)]$

In this section we prove that the Kalmár-Surányi class $[\forall^3\exists^*, (0,1)]$ is a conservative reduction class. We use the proof of this result to explain Gurevich's method of existential interpretation on a simple example before it is applied in a crucial way to the difficult Gurevich class in the next section.

Theorem 3.2.1 (Kalmár-Surányi). *The class $[\forall^3\exists^*, (0,1)]$ is a conservative reduction class.*

Proof. The proof splits into three steps.

Step 1. Reduction of Büchi's conservative reduction class $[\exists \wedge \forall\exists\forall, (\omega, 3)]$ to $[\forall^3\exists^2, (\omega, 4)]$ by axiomatizing the Skolem successor function through a new binary predicate representing its graph. (Starting from Kahr's class $[\forall\exists\forall, (\omega, 1)]$ would produce $[\forall^3\exists, (\omega, 2)]$.)

Step 2. Encoding of the finitely many monadic predicates P_i into branches $\{x : Q(w_i, x)\}$ of a new binary predicate symbol Q. The new elements w_i that "witness" the sets P_i are bound by existential quantifiers producing $[\forall^3\exists^*, (0,5)]$-formulae with equality. Here, equality is used to express the uniqueness of the witnesses.

Step 3. Encoding of the five predicates and equality resulting from Step 2 into one binary relation. It is here that for future use we make the technique of existential interpretation explicit in the form of several lemmata that are used to obtain the desired conservative reduction.

Step 1. In this step we give a conservative reduction of formulae $\psi \in [\exists \wedge \forall\exists\forall, (\omega, 3)]$ to formulae $\varphi \in [\forall^3\exists^2, (\omega, 4)]$ in such a way that every model of any of the reduction formulae satisfies $\exists v Hvv$ for one of the occuring binary predicate symbols.

In Exercise 2.1.31 to Trakhtenbrot's Theorem it has been shown that Turing machines can be encoded by Büchi-formulae

$$\psi := \exists v\alpha \wedge \forall x\exists u\forall y\beta \in [\exists \wedge \forall\exists\forall, (\omega, 3)]$$

in such a way that α contains the subformula Hvv. Let S be a new binary predicate symbol and define its rôle as the graph of the successor function by

$$\varphi := \exists v\alpha \wedge \forall x \exists u Sxu \wedge \forall x \forall u \forall y (Sxu \rightarrow \beta).$$

Then clearly ψ is (finitely) satisfiable if and only if φ is (finitely) satisfiable. Also each model of φ satisfies $\exists v Hvv$. By prenexing the formula φ one obtains the desired result of Step 1.

Step 2. In this step we give a conservative reduction from formulae $\psi \in [\forall^3\exists^2, (\omega, 4)]$ whose models satisfy $\exists v Hvv$ to formulae $\varphi \in [\forall^3\exists^*, (0,5)]_=$ (i.e. with equality).

Let $P_1, \ldots, P_m$ be the monadic and $Q_1, \ldots, Q_4$ the binary predicate symbols occuring in $\psi := \forall x \forall y \forall z \exists u \exists v \alpha$. Let Q be a new binary predicate symbol. The idea of the proof is to represent $P_i t$ in an appropriate branch of Q by $Qw_i t$ for a new element w_i (see the formula ENCODING[α] below). Thus the existence and the uniqueness of such "witnesses" w_i of P_i have to be ensured.

We generate the witnesses by imposing a directed chain $Qw_i w_{i+1} \wedge \neg Qw_{i+1} w_i$ of Q-reflexive points, starting from a distinguished point w_1 satisfying Hw_1w_1 (see Fig. 3.19). The uniqueness of the witnesses is formalized using equality. The domain where the given formula ψ is interpreted – namely on points that are not Q-reflexive – has to be closed; this is formulated by a closure axiom in φ.

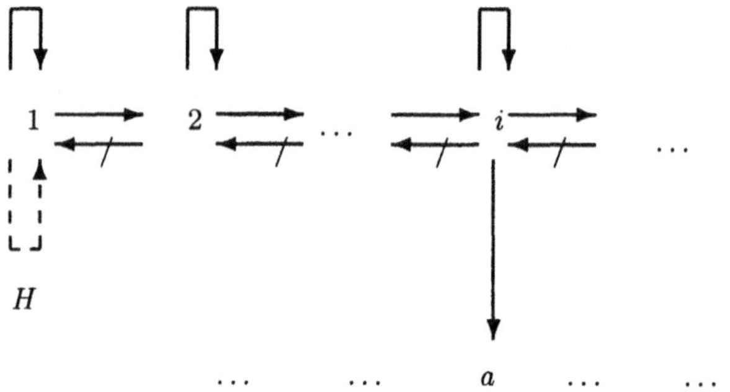

Figure 3.19. Encoding of $P_i a$

We define therefore:

$$\varphi := \forall x \forall y \forall z \exists u \exists v \exists w_1 \ldots \exists w_m (\text{WITNESSES} \wedge \text{CLOSURE} \wedge \text{ENCODING}[\alpha])$$

where

$$\begin{aligned} \text{WITNESSES} &:= \text{EXISTENCE} \wedge \text{UNIQUENESS} \\ \text{CLOSURE} &:= \neg Quu \wedge \neg Qvv \\ \text{ENCODING}[\alpha] &:= \neg Qxx \wedge \neg Qyy \wedge \neg Qzz \rightarrow \alpha[P_i t / Qw_i t] \end{aligned}$$

$$\begin{aligned} \text{EXISTENCE} &:= \text{FIRST}(w_1) \wedge \bigwedge\nolimits_{i<m} \text{NEXT}(w_i, w_{i+1}) \\ \text{FIRST}(x) &:= Qxx \wedge Hxx \\ \text{NEXT}(x,y) &:= Qxx \wedge Qyy \wedge Qxy \wedge \neg Qyx \end{aligned}$$

UNIQUENESS is the conjunction of the two formulae:

$$\text{FIRST}(x) \wedge \text{FIRST}(y) \to x = y$$

$$\text{NEXT}(x,y) \wedge \text{NEXT}(x,z) \to y = z.$$

It is now easy to verify that ψ is (finitely) satisfiable if and only if φ is (finitely) satisfiable.

Indeed if $\mathfrak{A} \models \psi$, set $\mathfrak{B} = (B, Q)$ where $B = A \cup \{w_1, \ldots, w_m\}$ for new elements w_i such that Q satisfies the formulae WITNESSES and CLOSURE (this means that $\mathfrak{B} \models \neg Qaa$ for each element a of the domain A of $\mathfrak{A}$) and such that $\mathfrak{B} \models Qw_ia$ if and only if $\mathfrak{A} \models P_ia$ for each $a \in A$. This results in a model $\mathfrak{B}$ that satisfies ENCODING$[\alpha]$ and therefore φ.

Conversely, assume that $\mathfrak{B} \models \varphi$. Restrict the domain of this model to $A := \{a \in B \mid \mathfrak{B} \models \neg Qaa\}$. Let w_i $(1 \le i \le m)$ be a Q-chain of elements such that

$$\mathfrak{B} \models \text{EXISTENCE}[w_1, \ldots, w_n].$$

Due to the CLOSURE axiom in φ the domain A is sufficiently closed. From $\mathfrak{B} \models$ ENCODING$[\alpha]$ we obtain therefore a model for ψ if we restrict $\mathfrak{B}$ to A and interpret P_i by $\{a \in A : \mathfrak{B} \models Qw_ia\}$.

For future use we will formulate below some of the characteristic features of the preceding reduction explicitly as lemmata.

Step 3. In this step we show how the five binary predicates and the equality which appear in the formulae resulting from step 2 can be encoded by one binary predicate. We first explain the proof strategy.

Proof strategy. Let $\mathfrak{A}$ be a model of $\psi \in [\forall^3\exists^*, (0,5)]_=$. Let $Q_1, \ldots, Q_5$ be the predicate symbols (different from equality) that occur in ψ and let R be a new binary predicate symbol. We create for each point $a_0 \in A$ a triple a_1, a_2, a_3 of new points which serve to encode the relations $Q_1, \ldots, Q_5$. Indeed each given relation Qx_0y_0 can be encoded as one of the six possible relations Rx_ix_j among the new points (where $i \neq j$, see Fig. 3.20). Formally, for each $(i,j) \in \{(1,2),(2,1),(1,3),(3,1),(2,3)\}$ let $Q_{i,j}$ be any of the predicates $Q_1, \ldots, Q_5$; let P_i $(0 \le i \le 3)$ be monadic predicates representing the domain of all elements of type x_i. Then $Q_{i,j}$ will be encoded by the formula

$$\text{CODE}(Q_{i,j}) := P_0x \wedge P_0y \wedge \exists x_i \exists y_j (P_ix_i \wedge P_jy_j \wedge Rxx_i \wedge Ryy_j \wedge Rx_iy_j).$$

The three new elements a_1, a_2, a_3 for each element a_0 of the given domain are provided by the following WITNESSES-axioms expressing that for $i = 1,2,3$ there exists for each P_0-element a (of the given universe A of $\mathfrak{A}$) exactly

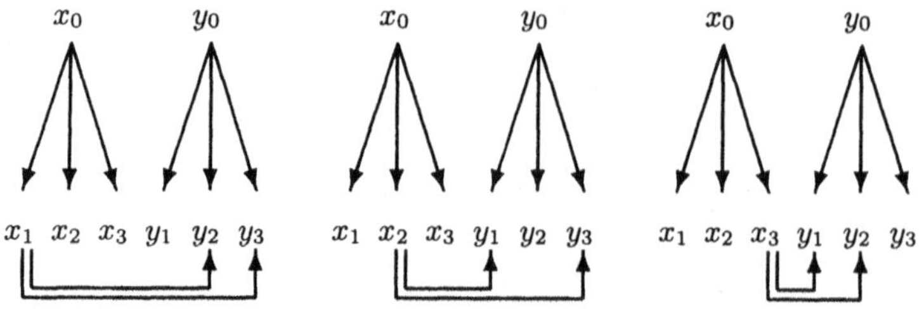

Figure 3.20. The 6 possible encodings of binary predicates

one P_i-element b (of the new domain where φ will be interpreted) such that $R[a, b]$ holds (see Fig. 3.21). Clearly it must also be ensured that there is at least one P_0-element (CLOSURE axiom). Formally:

$$\begin{aligned}
\text{WITNESSES} &:= \text{EXISTENCE} \wedge \text{UNIQUENESS} \wedge \text{CLOSURE},\\
\text{EXISTENCE} &:= \textstyle\bigwedge_{1\leq i\leq 3}(P_0x \rightarrow \exists w(P_iw \wedge Rxw)),\\
\text{UNIQUENESS} &:= P_0x \wedge P_iy \wedge Rxy \wedge P_iz \wedge Rxz \rightarrow y = z,\\
\text{CLOSURE} &:= \exists u P_0 u.
\end{aligned}$$

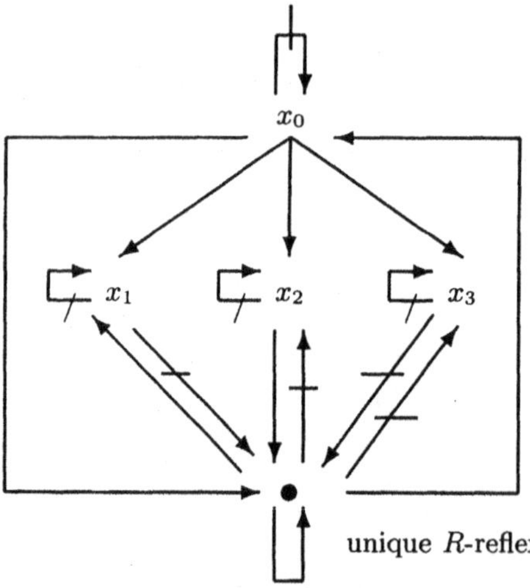

Figure 3.21. Domain structure with 3 new point copies

We have to define this partitioning into four predicates $P_0, \ldots, P_3$ using only the one binary predicate symbol R. The idea is to represent the truth values of $P_0, \ldots, P_3$ at a by the four possible patterns for the R-relation between a and a distinguished point, say u. We fix u as a unique R-reflexive point, i.e.

satisfying $R[u,u]$; then the sets $P_0, \ldots, P_3$ are defined by the formulae (see Fig. 3.21):

$$P_i(x) := \neg Rxx \wedge \exists u(\pm Rxu \wedge \pm Rux).$$

(We use $\pm\psi$ as abbreviation for any of ψ or $\neg\psi$.) To make this encoding work we have to ensure the existence of a unique R-reflexive point.

What follows is a stepwise introduction of the above outlined encoding features by $[\forall^3\exists^*, (0,1)]$-formulae. For future use we isolate the single reduction steps in the form of lemmata that are characteristic for the technique of existential interpretation.

Existential Interpretation. Gurevich's method of existential interpretations provides conservative reductions between theories which preserve certain modest prefix forms. The typical situation has been illustrated already through the conservative reductions of Kahr's class to various other prefix-vocabulary classes in the preceding section where the underlying theories T' and T are just logics (i.e. theories without non logical axioms): each T'-formula ψ is translated into a T-formula φ by replacing a certain atomic T'-subformula π by an interpreting T-formula π'. If both π' and $\neg\pi'$ are equivalent in T to existential formulae, then the translation of every $\forall^p\exists^*$-formula of T' is equivalent in T to a $\forall^p\exists^*$-formula and thus preserves the $\forall^p\exists^*$-prefix structure in reducing T' to T.

For a succinct formulation of the method of existential interpretation we introduce some terminology reflecting the fact that in dealing with general theories we are interested in their restriction to $\forall^p\exists^*$-formulae for some small number p.

By a *theory* T we understand in this book a first order theory, identified with the set of its theorems and usually represented by the set of its non logical axioms. We consider only theories with decidable language which is in the standard way identified with the vocabulary. (Finite) satisfiability of ψ *in* T means truth of ψ in some (finite) model of T, similarly for unsatisfiability in T. As usual a theory without non logical axioms is called a logic.

Definition 3.2.2. A theory T' is called a *p-extension* of a theory T if T and T' have the same vocabulary and T' is obtained from T by adding a finite number of axioms $\psi_1, \ldots, \psi_n$ such that the universal closure of each ψ_i is equivalent in T to a $\forall^p\exists^*$-sentence.

Example 1. Consider the logic T with vocabulary $Q_1, \ldots, Q_4, Q$ and equality as used in Step 2. It is 3-extended there to the theory T' obtained from T by adding the axiom $\exists w_1 \ldots \exists w_m$WITNESSES. The purpose of that axiom was to provide the witnesses w_i which assisted the elimination of monadic predicates symbols P_i from $\forall^3\exists^*$-formulae in T. Note that in Step 2 we have provided a semi-conservative reduction of the $\forall^3\exists^*$-fragment of T' to the $\forall^3\exists^*$-fragment of T.

Example 2. Let T_0 be the logic of a binary predicate R and T_1 be the 3-extension of T_0 by means of the following two axioms expressing the existence of an R-reflexive element with a "uniqueness" property.

$$\exists u Ruu,$$

$$Rxx \wedge Ryy \rightarrow (Rxz \rightarrow Ryz) \wedge (Rzx \rightarrow Rzy).$$

Clearly there is a semi-conservative reduction of the $\forall^3\exists^*$-fragment of T_1 to the $\forall^3\exists^*$-fragment of T_0.

(Semi-)conservative reductions that preserve the number of universal quantifiers will be used repeatedly and thus motivate a further definition.

Definition 3.2.3. A theory T' is called *p-reducible* to a theory T if there is a semi-conservative reduction of the $\forall^p\exists^*$-fragment of T' to the $\forall^p\exists^*$-fragment of T.

Lemma 3.2.4 (p-Extension Lemma.). *Show that each p-extension T' of a theory T is p-reducible to T.*

Exercise 3.2.5. Prove the p-Extension Lemma.

Example 3. Let T_2 be the 3-extension of T_1 of Example 2 by means of the axiom PARTITION which is the conjunction of the following four axioms explained in the proof strategy above.

$$\neg Rxx \wedge \exists u(\pm Rxu \wedge \pm Rux).$$

In the proof of Step 3 below we will use each of these axioms as the definition of a unary predicate $P_i (0 \leq i \leq 3)$ which can thus be considered as being introduced into T_2.

Note that each of the four axioms in Example 3, as well as their negation, is equivalent in T_1 to a purely existential formula (remember that $\exists u Ruu$ holds in every T-model). Since formulae with this property are useful for introducing predicates into $\forall^p\exists^*$-theories we give them a name.

Definition 3.2.6. A formula ψ is called a *neutral* formula of a theory T if both ψ and its negation are equivalent in T to existential formulae.

Lemma 3.2.7 (Neutral Definitions). *Let T be a theory, ψ be a formula in the vocabulary of T. Assume that the number m of free variables of ψ does not exceed p. If ψ is a neutral formula of T, then the theory T' obtained from T by introducing a new m-ary predicate P by means of ψ, i.e. by adding the axiom $Px_1 \cdots x_m \leftrightarrow \psi$, is p-reducible to T.*

Exercise 3.2.8. Prove the Lemma on Neutral Definitions .

In Example 2 above the "uniqueness" property expresses that two R-reflexive elements are indistinguishable by R. For (semi-)conservative reductions such an indistinguishability is often enough to encode equality. This motivates the following definition.

Definition 3.2.9. Elements a, b of a structure $\mathfrak{A}$ are called *indistinguishable* if for all first-order formulae $\psi(x)$ (in the vocabulary of $\mathfrak{A}$ and with exactly one free variable) it holds that $\mathfrak{A} \models \psi[a]$ if and only if $\mathfrak{A} \models \psi[b]$.

A structure $\mathfrak{A}$ is called *economical* if indistinguishable elements of $\mathfrak{A}$ are identical. The structure obtained from $\mathfrak{A}$ by identifying indistinguishable elements is called the *economical version* of $\mathfrak{A}$.

A formula $\varepsilon(x, y)$ *expresses indistinguishability* in a theory T if for every model $\mathfrak{A} \models T$ and for every pair a, b of elements it holds that a, b are indistinguishable in $\mathfrak{A}$ if and only if $\mathfrak{A} \models \varepsilon[a, b]$. (Of course ε is supposed to be in the vocabulary of T and to contain exactly two free individual variables.)

Lemma 3.2.10 (Equality as Neutral Indistinguishability). *Let T be a theory without equality and let ε be a neutral formula of T expressing indistinguishability in T. Assume that $p \geq 2$. Then the theory T' obtained from T by introducing equality by means of ε is p-reducible to T.*

Exercise 3.2.11. Prove Lemma 3.2.10.

Lemma 3.2.12 (Equality as Neutral Congruence). *Let T be a theory with finite vocabulary and without equality and let ε be a neutral formula of T with exactly two free variables. Assume that $p \geq \max(3, m)$ where m is the maximal arity of the predicate and function symbols in the vocabulary of T. Then the theory T' obtained from T by introducing equality by means of ε is p-reducible to T.*

Exercise 3.2.13. Prove Lemma 3.2.12. Hint: Let EQUIV(ε) be a neutral formula of T whose universal closure expresses that ε defines an equivalence relation. For each predicate symbol P of T let CONGRUENCE(P) be a neutral formula of T whose universal closure expresses that P does not distinguish between ε-equivalent elements. For each function symbol f of T let CONGRUENCE(f) be a neutral formula of T whose universal closure expresses that the values of f depend only upon the ε-equivalence classes of the arguments. Then use the p-Extension Lemma and Lemma 3.2.10 on equality as neutral indistinguishability.

Example 4. Let T_3 be the theory obtained from T_2 by introducing equality by means of the following formula:

$$(Rxx \wedge Ryy) \vee (\neg Rxy \wedge \bigvee_{0 \leq i \leq 3} (P_i(x) \wedge P_i(y))).$$

Therefore models of T_3 satisfy the following properties: a) there is exactly one R-reflexive element (see the T_1-axioms), b) R-irreflexive elements are partitioned by $P_0(x), \dots, P_3(x)$ into four classes (see the T_2-axioms), c) by the T_3-axiom, different elements in the same class are two-way connected by R. Since the additional T_3-axiom is neutral in T_2, it follows by Lemma 3.2.12 that T_3 is 3-reducible to T_2.

For conservative reductions of one theory to another one usually has to restrict the domain of elements where the encoding of the original formula is required to hold. An example is the restriction of the interpretation of $\alpha[P_i t/Qw_i t]$ in the formula ENCODING[α] of Step 2 above to Q-irreflexive points. Such a restriction is typically accompanied by closure axioms that make sure that there are enough elements in the restricted domain; an example can again be found in Step 2 above in the axiom CLOSURE. Since this situation is very frequent we give it a name. Due to the fact that in this chapter we only consider formulae without function symbols and equality the following definition is given for relational vocabularies only.

Definition 3.2.14. Let $\psi(x)$ be a first-order formula and $\mathfrak{A}$ be a structure with $\mathfrak{A} \models \exists x\psi(x)$. The *restriction* $\mathfrak{A}|\psi$ of $\mathfrak{A}$ by ψ is the substructure of $\mathfrak{A}$ with universe $\{a \mid \mathfrak{A} \models \psi[a]\}$. The restriction of a theory T by ψ is the theory $Th\{\mathfrak{A}|\psi : \mathfrak{A} \models T \wedge \exists x\psi(x)\}$, i.e. the set of sentences that hold in the ψ-restrictions of all T-models satisfying $\exists x\psi(x)$.

Lemma 3.2.15 (Neutral Model Restriction). *The restriction T' of a theory T by a neutral formula δ of T is p-reducible to T.*

Exercise 3.2.16. Prove Lemma 3.2.15. Hint: Construct for arbitrary T-formulae $\psi \in [\forall^p\exists^*]$ a suitable formula $\exists x\delta \wedge (\psi \mid \delta)$.

When interpreting a theory T' with equality in another theory T one can often show that T'-models are equivalent to quotient structures of T-models (with respect to some equivalence relation). In the context of conservative reductions this is made precise by the following definition of embeddings of theories.

Definition 3.2.17. Let T' be a theory of vocabulary σ whose sentences may contain equality; let ε be a neutral formula of a theory T (not necessarily with equality) whose vocabulary contains σ. T' is called *embedded* into T modulo ε if the following two conditions are satisfied:

- In each T-model $\mathfrak{A}$ the formula ε defines an equivalence relation on the domain of $\mathfrak{A}$; further ε-equivalent elements are indistinguishable in the σ-reduct $\mathfrak{A}|_\sigma$ of $\mathfrak{A}$ and the quotient structure $(\mathfrak{A}|_\sigma)/\varepsilon$ is a model of T'.
- For each finite T'-model $\mathfrak{B}$ there is a finite T-model $\mathfrak{A}$ whose quotient structure $(\mathfrak{A}|_\sigma)/\varepsilon$ is elementary equivalent to $\mathfrak{B}$, i.e. satisfies the same σ-sentences as $\mathfrak{B}$.

Lemma 3.2.18 (Neutral Embeddings). *If a theory T' with equality is embedded into a theory T modulo a neutral formula ε then T' is p-reducible to T.*

Exercise 3.2.19. Prove Lemma 3.2.18. Hint: Replace $s = t$ in given $\forall^p\exists^*$-formulae by $\varepsilon(s, t)$.

Concluding the proof of Step 3. Using the just formulated concepts of existential interpretation, the execution of the proof strategy for Step 3 takes the form of constructing successive 3-reductions starting from the logic T_0 of one binary predicate and reaching the result of Step 2, namely the logic T_7 of five binary predicates with equality. In order to let each single encoding idea stand out explicitly we will consider each step separately.

Step 3.1. (Introduction of a unique R-reflexive element.) Consider the logic T_1 of Example 2. By the p-Extension Lemma T_1 is 3-reducible to the logic T_0 of one binary predicate. Clearly every economical model of T_1 contains exactly one R-reflexive element.

Step 3.2. (Partitioning the R-irreflexive elements into four sets.) Consider the 3-extension T_2 of T_1 defined in Example 3. As already observed, the axioms of T_2 introduce four monadic predicates $P_i (0 \leq i \leq 3)$. Since these axioms are neutral T_1-formulae, Lemma 3.2.7 on neutral definitions implies that T_2 is 3-reducible to T_1.

Step 3.3. (Introduction of equality.) As shown in Example 4 the 3-extension T_3 of T_2 defined by introducing equality is 3-reducible to T_2.

Step 3.4. (Introducing witnesses for five binary predicates.) Let T_4 be the 3-extension of T_3 defined by means of the axioms WITNESSES defined in the proof strategy above. They state that there are P_0-elements and that for every P_0-element x and for $i = 1, 2, 3$ there is exactly one P_i-element y related to x by Rxy. By the p-Extension Lemma it follows that T_4 is 3-reducible to T_3.

Step 3.5. (Introducing five binary predicates.) Let T_5 be the theory obtained from T_4 by introducing binary predicates $Q_{i,j} \in \{Q_1, \dots, Q_5\}$ by means of the formulae CODE($Q_{i,j}$) defined in the proof strategy above. These formulae are neutral T_4-formulae. It follows from Lemma 3.2.7 on neutral definitions that T_5 is 3-reducible to T_4.

Step 3.6. (Restriction to P_0-elements.) Let T_6 be the restriction of T_5 by $P_0(x)$. Since $P_0(x)$ is a neutral T_5-formula, Lemma 3.2.15 on neutral model restrictions implies that T_6 is 3-reducible to T_5.

Step 3.7. (Embedding $[\forall^3\exists^*, (0,5)]$-formulae modulo equality.) Let T_7 be the logic with the vocabulary $\{Q_1, \dots, Q_5\}$ and equality. We use Lemma 3.2.18 on neutral embeddings to infer that T_7 is 3-reducible to T_6. Therefore it remains to show that T_7 is embedded into T_0 modulo equality.

Let $\mathfrak{A}$ be a model of T_7 with n elements. Using the intended interpretation explained in Steps 3.1–3.6 one can build a model $\mathfrak{B}$ of T_5 with $4n+1$ elements — adding the unique R-reflexive point and three fresh copies for each element of the given domain of $\mathfrak{A}$ — such that $\mathfrak{B}|P_0(x)$ restricted to the vocabulary of T_7 is isomorphic to $\mathfrak{A}$.

□

3.3 The Gurevich Class

This section deals with conservative reduction classes that are minimal among the prefix-vocabulary classes that contain $\exists^*$ in the prefix. The main and difficult case is the Gurevich class $[\forall\exists\forall\exists^*, (0,1)]$. We show it to be a conservative reduction class by a conservative reduction of the Kahr class $[\forall\exists\forall, (\omega, 1)]$ to it. We indicate in an exercise at the end of the section how the reduction can be made independent of the Kahr class by starting directly from the halting problem for Turing machines. Two other cases having $\exists^*$ in the prefix and subprefix $\forall\exists\forall$ easily follow from that result, namely the Surányi class $[\exists^*\forall\exists\forall, (0,1)]$ and the Kostyrko-Genenz class $[\forall\exists^*\forall, (0,1)]$. Note that all the results in this section establish undecidability for classes of formulae of graph theory.

3.3.1 The Proof Strategy

This section is entirely devoted to explaining the proof strategy for the following theorem.

Theorem 3.3.1 (Gurevich). *The class* $[\forall\exists\forall\exists^*, (0,1)]$ *is a conservative reduction class.*

The proof consists of a conservative reduction of the class $[\forall\exists\forall, (\omega, 1)]$. Having to deal with formulae whose prefix starts with $\forall x\exists z\forall y$, we will use without further mention the Skolem Normal Form Theorem and interpret (and write) z as successor x' of x. In this section K always denotes the unique binary predicate symbol.

The idea of the proof is to "encode" monadic atomic formulae, Px say, on secondary diagonals, i.e. parallels to the main diagonal, by formulae of the form Kxx'.

Let $P_1, \ldots, P_r$ be monadic predicate symbols and a, a' elements of some domain. A sequence $P_1a, \ldots, P_ra, P_1a', \ldots, P_ra'$ of truth values can thus be encoded by K as in the example of Fig.3.22; $c \to d$ indicates that Kcd is true; elements a, a' of the given domain are represented by $(r+2)$-blocks $a = a^0, a^1, \ldots, a^{r+1}$ and $a' = a'^0, a'^1, \ldots, a'^{r+1}$ with new elements a^i, a'^i $(1 \leq i \leq r+1)$.

Let us introduce an abbreviation for the four possible truth value combinations for Kxx' and $Kx'x$:

$$\neg Kxx' \wedge \neg Kx'x, \quad Kxx' \wedge \neg Kx'x, \quad \neg Kxx' \wedge Kx'x, \quad Kxx' \wedge Kx'x.$$

We denote them by $Kxx' = 0, 1, 2, 3$ or $Fx = 0, \ldots, Fx = 3$, respectively. Formally, for $c \in \{a, a'\}$, $P_ic = 1$ is represented by $Fc^i = 1$ and $P_ic = 0$ by $Fc^i = 0$. We extend in a natural way the successor function to the new elements by $(c^i)' := c^{i+1}$, $(c^{r+1})' := c'^0$. The beginning and end of an $(r+2)$-block are represented by $Fc^0 = 0$ and $Fc^{r+1} = 2$, respectively. The 0-2-pattern for the beginning and end of an $(r+2)$-block will be ensured by

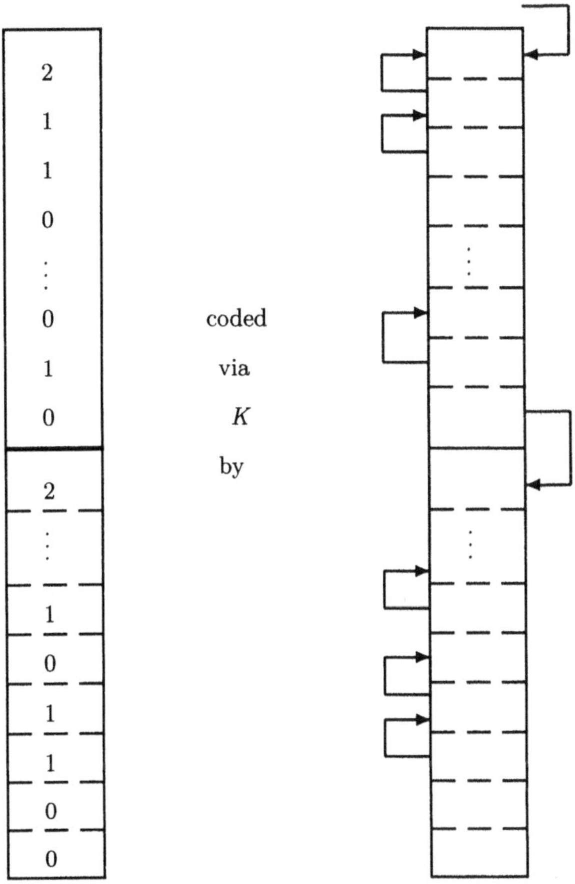

Note:
- $F_0 x = Kxx'$ and Shift-axioms $F_{i+1}x = F_i x'$
- The pattern 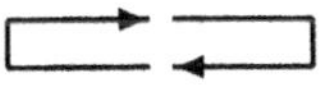 will never occur (i.e. $F_0 x \neq 3$, i.e. K is asymmetric)

$P_i a$: K-arrow from i-th box to successor box $i+1$
$\neg P_i a$: no K-arrow from i-th box to successor box $i+1$
beginning of block: no arrow between box 0 and box 1
End of block: downward arrow from successor box to box r

Figure 3.22. Encoding of $P_1 a, \ldots, P_r a, P_1 a', \ldots, P_r a'$.

corresponding modulo-$(r+2)$-axioms (see below). For later use we note the asymmetry of K in this encoding scheme; formally: $Fx \neq 3$ for each x.

To make sure that the extension of K includes $K(c^i, c^{i+1})$ and that $K(c'^0, c^{r+1})$ really encodes $P_i c$, we will guarantee that the K-relation we start with does not touch any of the two neighbour parallels of the main diagonal. That is, it satisfies for all x of the given domain the formula

$$\text{DIAG-FREE} := \neg Kxx' \wedge \neg Kx'x.$$

If we want to use this encoding for a reduction to $[\forall\exists\forall\exists^*, (0,1)]$, we have to formalize $Fc^i = j$ without applying the successor symbol more than once. The standard technique to obtain this uses auxiliary monadic predicate symbols $F_i = j$ $(0 \leq i < 2(r+2),\ 0 \leq j \leq 3)$ such that $F_i(c^0) = j$ is equivalent to $F_0(c^i) = j$; it suffices indeed to impose the following equivalence.

$$F_{i+1}\text{-SHIFT} := (F_{i+1}x = F_i x').$$

($F_{i+1}x = F_i x'$ is shorthand for $\bigvee_{j=0}^{3} F_{i+1}x = j \leftrightarrow F_i x' = j$). $F_0 = j$ now assumes the rôle of the equation $Fx = j$ above and expresses the basic coding idea, illustrated in Fig. 3.22, by

$$F_0\text{-SHIFT} := (F_0 x = Kxx').$$

This yields an encoding for the truth values $P_1 a, \ldots, P_r a, P_1 a', \ldots, P_r a'$ of r monadic predicate symbols through the sequence

$$(F_0 a^0, \ldots, F_0 a^{2r+3}) = (F_0 a^0, F_1 a^0, \ldots, F_{2r+3} a^0).$$

This is pictorially represented in Fig. 3.23. In this picture, the 0-2-pattern, signaling the beginning and end of the $(r+2)$-blocks, will be imposed by the following axioms. In presence of F_i-SHIFT axioms, they simulate the $(r+2)$-module structure $F_0 x, \ldots, F_{r+1} x$ for each x.

$$(r+2)\text{-MOD}(F_0, \ldots, F_{r+1}) := \dot{\bigvee_{0 \leq j \leq r+1}} (F_j x = 2) \wedge (F_{r+1}x = 2 \rightarrow F_0 x = 0)$$

Here $\dot{\bigvee} \psi_j$ indicates the formula expressing that ψ_j holds for precisely one j. In view of the axioms $(r+2)$-MOD$(F_0, \ldots, F_{r+1})$ and F_i-SHIFT, the formula $F_{r+2}x = 2$ means $F_0 x^{r+2} = 2$; therefore instead of $F_{r+1}x = 2$ we will also write 0-POINT(x).

Thus the problem to be solved is to express formulae $F_i t = j$ by $Ktw = j$, where w is an appropriate "witness" for F_i. Assume that a witness y for F_i has been found and consider the intended interpretation for F_i given by:

$$\begin{aligned} F_i\text{-DEF} &:= (F_{i-1}x' = Kxy) \quad \text{for } i > 0 \\ F_0\text{-DEF} &:= (Kxx' = Kxy) \end{aligned}$$

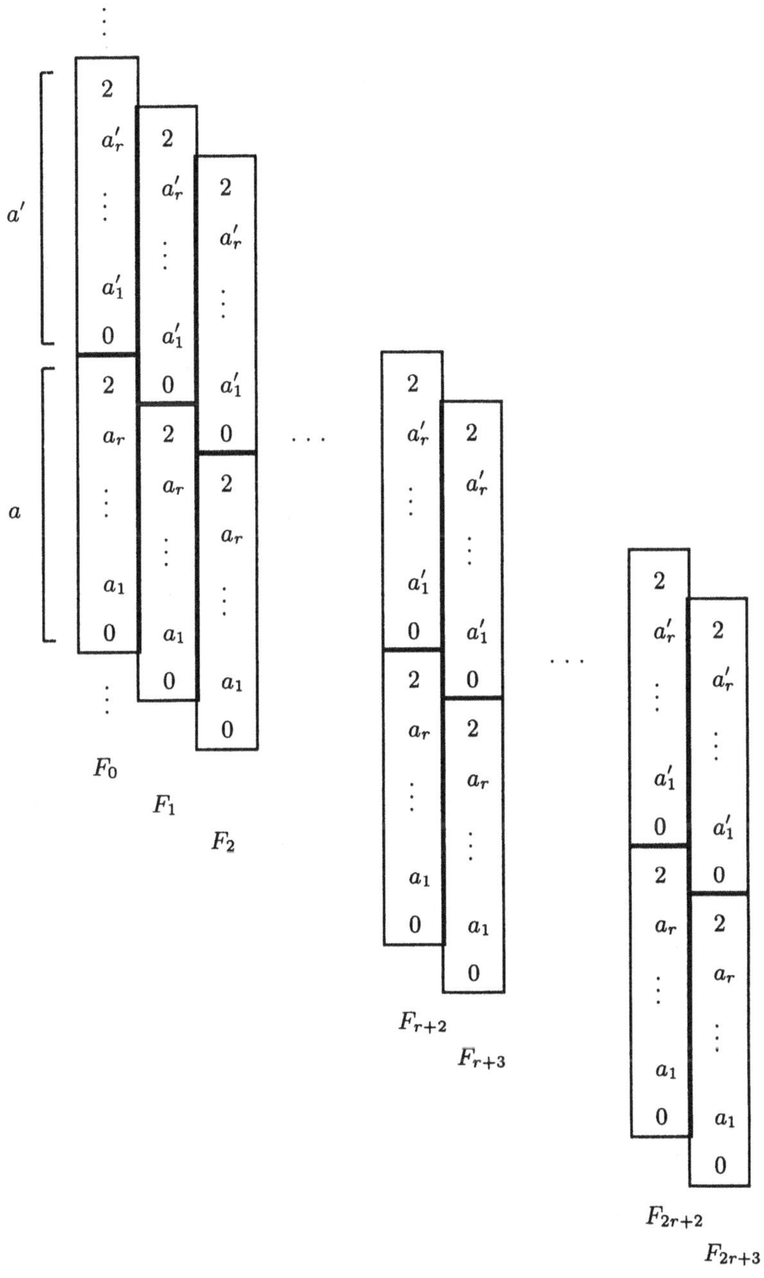

Figure 3.23. Predicate encoding (note $a_i = P_i a$, $a'_i = P_i a'$).

In view of the shift axioms, this intended interpretation can be imposed by the following F_i-axioms. In these axioms i-WITNESS is a formula which expresses the witness property needed for defining F_i through F_{i-1}:

$$\begin{aligned} F_i\text{-CODE} \ &:= \ i\text{-WITNESS}(w) \wedge \\ &\quad (\neg i\text{-WITNESS}(x) \wedge i\text{-WITNESS}(y) \rightarrow F_i\text{-DEF}) \end{aligned}$$

where w will be existentially and x, y universally quantified. F_i-SHIFT and F_i-CODE provide the desired scheme for elimination of F_i, namely by encoding F_i through F_{i-1} into K – subject to finding an F_i-witness.

If we start from an irreflexive K, we can choose

$$0\text{-WITNESS}(x) := Kxx.$$

Since, as noted above (see Fig. 3.22), the K-encoding scheme for monadic predicates does not use $F_0x = 3$, we can choose

$$1\text{-WITNESS}(x) := (F_0x = 3).$$

Since the above encoding of monadic predicates into secondary diagonals of K uses only blocks of form $0\,c_1 \ldots c_r\,2$ with $c_j \in \{0,1\}$ (see Fig. 3.22), for $i \geq 2$ we can choose sequences $2\,1 \ldots 1\,2$ of length i as witnessing property:

$$i\text{-WITNESS}(x) := (F_0x = 2) \wedge \bigwedge_{1 \leq j \leq i-2} (F_jx = 1) \wedge (F_{i-1}x = 2).$$

The preceding considerations explain the motivation for the following four conservative reduction steps to reduce an arbitrary formula $\psi_0 \in [\forall\exists\forall, (\omega, 1)]$ into a formula $\psi_4 \in [\forall\exists\forall\exists^*, (0,1)]$ which is equivalent to it and to all intermediate formulae ψ_1, ψ_2, $\psi_{3,i}$ (with respect to satisfiability and finite satisfiability).

Step 1. *(Diagonal-freeness)* An arbitrary $\psi \in [\forall\exists\forall, (\omega, 1)]$ is transformed into an equivalent formula $\psi_1 \in [\forall\exists\forall\exists^3, (\omega, 1)]$ of the form

$$\forall x \exists x' \forall y \exists u \exists v \exists w (\text{DIAG-FREE} \wedge \text{0-WITNESS-FREE} \wedge \beta_1)$$

where

$$\begin{aligned} \text{DIAG-FREE} \ &:= \ \neg Kxx' \wedge \neg Kx'x \\ \text{0-WITNESS-FREE} \ &:= \ \neg Kxx \end{aligned}$$

Thus any interpretation of ψ_1 leaves the diagonal of K and its two neighbour parallels free. Formulae containing the conjuncts DIAG-FREE and 0-WITNESS-FREE, with universally quantified variable x, are called diagonal- and 0-witness-free.

Remark. At the end of this section we indicate how this step — which presupposes the result for the Kahr class — can be replaced by a direct reduction of TM halting problems.

Step 2. *(Shift-reduced form)* Each diagonal- and 0-witness-free formula $\psi \in [\forall\exists\forall\exists^*, (\omega, 1)]$ is transformed into an equivalent formula ψ_2 with same prefix and in the same prefix-vocabulary class in *shift-reduced* form, i.e. with quantifier free conjunction

$$\beta_2 \wedge (r+2)\text{-MOD}(F_0, \ldots, F_{r+1}) \wedge \bigwedge_{i<2(r+2)} F_i\text{-SHIFT} \wedge \\ \wedge\text{-WITNESS-FREE} \wedge 1\text{-WITNESS-FREE}$$

where 1-WITNESS-FREE $:= (F_3 x \neq 0)$. Besides the unique binary predicate symbol K, the only predicate symbols occurring in ψ_2 are the monadic $F_i = j$ (for $0 \leq i < 2(r+2)$ where r is the number of monadic predicate symbols in ψ and $0 \leq j \leq 3$).

Step 3. *(F_i-elimination form)* Each shift-reduced formula

$$\psi \;:=\; \forall x \exists x' \forall y \exists u \exists v \exists w (\leq i\text{-SHIFT} \wedge \bigwedge_{j=0,1} j\text{-WITNESS-FREE} \wedge \\ \wedge (r+2)\text{-MOD}(F_0, \ldots, F_{r+1}) \wedge \beta)$$

obtained through Step 2 is transformed into an equivalent formula $\psi_{3,i} \in [\forall\exists\forall\exists^*, (4(i+1), 1)]$ in F_i-elimination form, i.e. of form

$$\forall x \exists x' \forall y \exists z_1 \cdots \exists z_n \exists z (\leq i\text{-SHIFT} \wedge \leq i\text{-WITNESS-FREE} \\ \wedge\, i\text{-PRE-WITNESS}(z)\{\text{for } i \geq 2\} \wedge \beta_3)$$

where

$$\leq i\text{-SHIFT} \;:=\; \bigwedge_{j\leq i} F_j\text{-SHIFT}$$

$$\leq i\text{-WITNESS-FREE} \;:=\; \bigwedge_{j\leq i} j\text{-WITNESS-FREE}$$

The bracketed expression $\{\text{for } i \geq 2\}$ indicates that i-PRE-WITNESS is part of this formula only if $i \geq 2$.

Formulae in F_i-elimination form guarantee the eliminability of F_i from $\psi_{3,i}$ by introducing the F_i-CODE formula and restricting the domain for interpretation of β_3 to non-i-witnesses. The existence of an i-pre-witness (see the definition below) is needed to provide an i-witness for the F_i-CODE. Note that this reduction step introduces one more existential quantifier $\exists z$.

Step 4. *(F_i-elimination)* Each formula $\psi_{3,i} \in [\forall\exists\forall\exists^*, (4(i+1), 1)]$ in F_i-elimination form is transformed into an equivalent formula $\psi_{3,i-1} \in [\forall\exists\forall\exists^*, (4i, 1)]$ in F_{i-1}-elimination form. For $i = 0$ this gives a formula $\psi_4 \in [\forall\exists\forall\exists^*, (0, 1)]$ which constitutes the final result of the whole reduction process.

Note that this reduction step introduces one more existential quantifier.

In the next section we carry out these four conservative reduction steps.

3.3.2 Reduction to Diagonal-Freeness

Here is the idea for the exclusion of diagonal pairs xx from the interpretations of K. We introduce a new element w which "witnesses" the diagonal, i.e. such that for each x, Kxx is equivalent to Kxw. We guarantee the existence of such a witness by adding a corresponding new monadic symbol W which is true for the witness and on whose successor closed complement the given formula is going to be interpreted. The same procedure applies to pairs xx' and $x'x$.

Fig. 3.24 contains the pictorial representation of this situation, in which w_1 is witness for points • (x with Kxx), w_2 for ∘ (x with Kxx') and w_3 for □ (x with $Kx'x$).

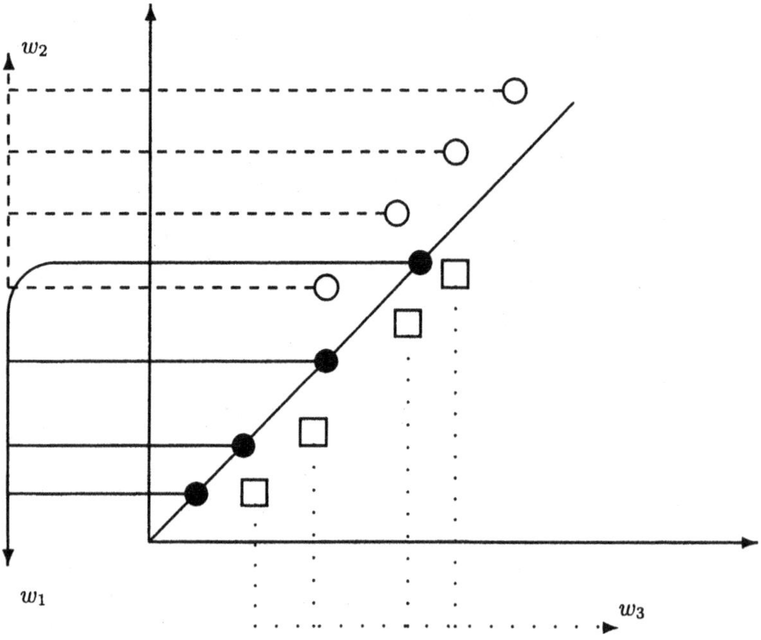

Figure 3.24. Witnessing main and secondary diagonals

Let $\psi := \forall x \exists x' \forall y \beta$ be an arbitrary formula in $[\forall\exists\forall, (\omega, 1)]$ and let W_1, W_2, W_3 be new monadic predicate symbols. Define

$$\psi_1 \quad := \quad \forall x \exists x' \forall y \exists w_1 \exists w_2 \exists w_3 \exists w (\text{DIAG-FREE} \wedge \text{0-WITNESS-FREE} \wedge \\ \wedge \bigwedge_{1 \leq i \leq 3} \text{WITNESS}(w_i) \wedge \text{CLOSURE} \wedge \mathit{Red}_1(\beta))$$

where

$$\begin{aligned}\text{WITNESS}(w_i) &:= W_iw_i \wedge \bigwedge_{i\neq j} \neg W_iw_j\\ \text{CLOSURE} &:= \bigwedge_{1\leq i\leq 3} \neg W_iw \wedge (\neg W_ix \rightarrow \neg W_ix')\end{aligned}$$

$$Red_1(\beta) := \bigwedge_{1\leq i\leq 3} \neg W_ix \wedge \neg W_iy \rightarrow \beta[Kxx/Kxw_1, Kxx'/Kxw_2, Kx'x/Kxw_3].$$

As usual $\gamma[\pi/\sigma]$ denotes the result of replacing each occurrence of π in γ by σ. Obviously $\psi_1 \in [\forall\exists\forall\exists^3, (\omega, 1)]$.

We now show that ψ is (finitely) satisfiable if and only if ψ_1 is. Indeed it is easy to check that each (finite) model of $\mathfrak{A}$ of ψ can be extended to a (finite) model $\mathfrak{B}$ of ψ_1 as follows. We introduce three new elements w_1, w_2, w_3 and put $W_i^{\mathfrak{B}} := \{w_i\}$. We set $w_i' := w_i$ and extend the given interpretation $K^{\mathfrak{A}}$ to $K^{\mathfrak{B}}$ such that:

$$\begin{aligned}\mathfrak{B} \models Kab \quad \text{iff} \quad & \mathfrak{A} \models Kab \ \&\ b \notin \{a, a'\} \ \&\ a \neq b'\\ \text{or} \quad & \mathfrak{A} \models Kaa \ \&\ b = w_1\\ \text{or} \quad & \mathfrak{A} \models Kaa' \ \&\ b = w_2\\ \text{or} \quad & \mathfrak{A} \models Ka'a \ \&\ b = w_3.\end{aligned}$$

By definition $K^{\mathfrak{B}}$ is diagonal- and 0-witness-free. Conversely it is also easy to check that each (finite) model $\mathfrak{B} \models \psi_1$ gives rise to a (finite) model $\mathfrak{A} \models \psi$. It suffices to restrict the domain to the non-witnesses (i.e. the elements outside of W_1, W_2, W_3) and to modify on this restricted (non-empty and successor closed) domain A the given interpretation of K such that

$$\begin{aligned}\mathfrak{A} \models Kab \quad \text{iff} \quad & \mathfrak{B} \models Kab\\ \text{or} \quad & a = b \ \&\ \mathfrak{B} \models Kau_1\\ \text{or} \quad & b = a' \ \&\ \mathfrak{B} \models Kau_2\\ \text{or} \quad & a = b' \ \&\ \mathfrak{B} \models Kbu_3\end{aligned}$$

Here u_i are witnesses (interpretations of w_i) for which $Red_1(\beta)$ is true. (The definition of $K^{\mathfrak{A}}$ is consistent because the u_i are pairwise distinct and different from all non-witnesses and because $K^{\mathfrak{B}}$ is diagonal- and 0-witness-free.)

3.3.3 Reduction to Shift-Reduced Form

Let $\psi := \forall x \exists x' \forall y \exists u \exists v \exists w(\text{DIAG-FREE} \wedge \text{0-WITNESS-FREE} \wedge \beta)$ be an arbitrary diagonal- and 0-witness-free formula as obtained by Step 1.

Let $P_1, \ldots, P_r$ be the monadic predicate symbols occurring in ψ. Applying the encoding idea explained above, we encode β on 0-points, replacing P_iz by $F_iz = 1$ (for $z \in \{x, y, u, v, w\}$) and P_ix' by $F_{i+r+2}x = 1$ for

$1 \leq i \leq r$. For this we need the module and shift structure – enforced by $(r+2)$-MOD$(F_0, \ldots, F_{r+1})$ and F_i-SHIFT for all $i < 2(r+2)$ – and freeness from 0-1-witnesses.

For technical reasons to be explained below we need an additional axiom which expresses the following: with respect to the K-relation to any 0-point y (i.e. truth of Kxy and/or Kyx), each 0-point is equivalent to its $r+1$ immediately preceding non-0-points:

$$\text{0-POINT}(y) \wedge \neg\text{0-POINT}(x) \rightarrow (Kxy = Kx'y).$$

This is represented pictorially in Fig. 3.25.

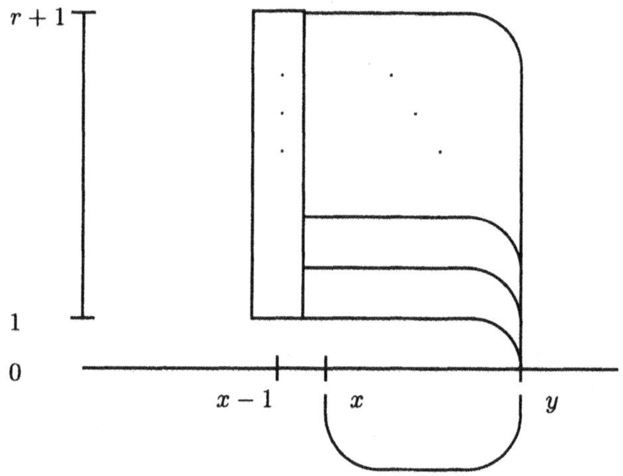

Figure 3.25. Relation to a 0-point y.

We thus define ψ_2 with the same prefix as ψ and the quantifier-free part constituted by the following conjuncts:

$$(r+2)\text{-MOD}(F_0, \ldots, F_{r+1}) \wedge \bigwedge_{i<2(r+2)} F_i\text{-SHIFT}$$

$$\text{0-WITNESS-FREE} \wedge \text{1-WITNESS-FREE}$$

$$\text{0-POINT}(u) \wedge \text{0-POINT}(v) \wedge \text{0-POINT}(w)$$

$$\text{0-POINT}(y) \wedge \neg\text{0-POINT}(x) \rightarrow (Kxy = Kx'y)$$

$$Red_2(\beta)$$

where

$$Red_2(\beta) \quad := \quad \text{0-POINT}(x) \wedge \text{0-POINT}(y) \rightarrow \beta[P_i z/(F_i z = 1), P_i x'/(F_{i+r+2} x = 1)]$$

with z standing for any variable different from x'.

We now show that ψ is (finitely) satisfiable iff ψ_2 is.

Let $\mathfrak{A} = (A, P_1, \ldots, P_r, K)$ be a model for ψ. We define the following structure $\mathfrak{B}$ which constitutes the intended interpretation of the encoding scheme explained above and which will be shown to be a model for ψ_2:

$$
\begin{aligned}
B &:= A \times \{0, 1, \ldots, r+1\} \\
(a,i)' &:= \begin{cases} (a, i+1) & \text{if } i \leq r \\ (a', 0) & \text{otherwise} \end{cases} \\
F_0^{\mathfrak{B}}(a,i) &:= \begin{cases} 0 & \text{if } i = 0 \\ 1 & \text{if } 1 \leq i \leq r \ \ \& \ \mathfrak{A} \models P_i a \\ 0 & \text{if } 1 \leq i \leq r \ \ \& \ \mathfrak{A} \models \neg P_i a \\ 2 & \text{if } i = r+1 \end{cases} \\
K^{\mathfrak{B}}(a,0)(b,0) &:= K^{\mathfrak{A}}ab \\
K^{\mathfrak{B}}(a,0)(b,i+1) &:= K^{\mathfrak{A}}ab' \qquad \text{for } i < r+1 \\
K^{\mathfrak{B}}(a,i+1)(b,0) &:= K^{\mathfrak{A}}a'b \qquad \text{for } i < r+1 \\
K^{\mathfrak{B}}(a,i)(a,i)' &:= F_0^{\mathfrak{B}}(a,i) \qquad \text{for } i \leq r+1 \\
K^{\mathfrak{B}} &\ \text{is} \ \ \text{irreflexive.}
\end{aligned}
$$

To justify the consistency of this definition we first note that the irreflexivity of $K^{\mathfrak{B}}$ is consistent with the first clause of the definition because by assumption $K^{\mathfrak{A}}$ is irreflexive, due to 0-witness-freeness in ψ. Then we have to show that the last clause of the definition is consistent with its second and third clause.

$$
\begin{aligned}
K^{\mathfrak{B}}(a,0)(a,1) &= F_0^{\mathfrak{B}}(a,0) && \text{by the last clause defining } K^{\mathfrak{B}} \\
&= 0 && \text{by definition of } F_0^{\mathfrak{B}} \\
&= K^{\mathfrak{A}}aa' && \text{by diagonal-freeness of } \psi \\
K^{\mathfrak{B}}(a,r+1)(a',0) &= F_0^{\mathfrak{B}}(a,r+1) && \text{by the last clause defining } K^{\mathfrak{B}} \\
&= 2 && \text{by definition of } F_0^{\mathfrak{B}}.
\end{aligned}
$$

Therefore $K^{\mathfrak{B}}(a,r+1)(a',0)$ is false, as is $K^{\mathfrak{A}}a'a'$ by 0-witness-freeness of ψ. $F_1^{\mathfrak{B}}, \ldots, F_{2(r+2)-1}^{\mathfrak{B}}$ are defined by, and therefore satisfy, the F_i-SHIFT axioms, as does $F_0^{\mathfrak{B}}$ by definition of $K^{\mathfrak{B}}$. By definition, the module structure required by $(r+2)$-MOD$(F_0, \ldots, F_{r+1})$ is satisfied. By definition of $F_0^{\mathfrak{B}}$, the 1-witness freeness $F_0 x \neq 3$ holds in $\mathfrak{B}$.

We now verify that the axiom

$$\text{0-POINT}(y) \wedge \neg\text{0-POINT}(x) \rightarrow (Kxy = Kx'y)$$

is satisfied in $\mathfrak{B}$. Indeed if $F_{r+1}^{\mathfrak{B}}(b,i) = 2$ and $F_{r+1}^{\mathfrak{B}}(a,j) \neq 2$, then $i = 0 \neq j \neq r+1$ (by definition of the $F_k^{\mathfrak{B}}$). Therefore from the second and the third clause of the definition of $K^{\mathfrak{B}}$ one has $\mathfrak{B} \models K(a,j)(b,0)$ iff $\mathfrak{B} \models K(a,j)'(b,0)$ and $\mathfrak{B} \models K(b,0)(a,j)$ iff $\mathfrak{B} \models K(b,0)(a,j)'$.

Since $P_i^{\mathfrak{B}}a = F_0^{\mathfrak{B}}(a,i)$ (by definition of $F_0^{\mathfrak{B}}$) $= F_i^{\mathfrak{B}}(a,0)$ (by the shift axioms) and similarly $P_i^{\mathfrak{B}}a' = F_i^{\mathfrak{B}}(a',0) = F_{i+r+2}^{\mathfrak{B}}(a,0)$, it follows that $Red_2(\beta)$ is satisfied in $\mathfrak{B}$ (where u, v, w in A are replaced by $(u,0)$, $(v,0)$, $(w,0)$ in B). Thus $\mathfrak{B} \models \psi_2$.

In the opposite direction, let $\mathfrak{B}$ be an arbitrary model of ψ_2. One obtains a model $\mathfrak{A}$ for ψ by restricting $\mathfrak{B}$ to the (by assumption non-empty) set of 0-points:

$$A := \{a \in B : \mathfrak{B} \models \text{0-POINT}[a]\}$$

on which the successor function is interpreted by defining:

$$a' := b \text{ iff } \mathfrak{B} \models (a^{r+2} = b),$$

where a^{r+2} denotes the $(r+2)$nd successor of a in $\mathfrak{B}$. The truth in $\mathfrak{B}$ of the subformula $(r+2)$-MOD$(F_0, \ldots, F_{r+1})$ guarantees that this definition is consistent; indeed a' yields the next 0-point of a in $\mathfrak{B}$.

The monadic predicate symbols P_i of ψ are interpreted by defining

$$P_i^{\mathfrak{A}} := \{a \in A : F_i^{\mathfrak{B}}a = 1\}.$$

The 0-witness-freeness is carried over from $\mathfrak{B}$ to $\mathfrak{A}$. The diagonal freeness $\mathfrak{A} \models (Kxx' = 0)$ is proved by the following equations that hold for each $a \in A$:

$$0 = F_0^{\mathfrak{B}}a = K^{\mathfrak{B}}aa^1 = K^{\mathfrak{B}}aa^2 = \cdots = K^{\mathfrak{B}}aa^{r+2}.$$

The first equality is implied by the definition of $\mathfrak{B}$ and by the fact that $\mathfrak{B} \models (r+2)$-MOD$(F_0, \ldots, F_{r+1})$. The second equality is a consequence of $\mathfrak{B} \models F_0$-SHIFT and the remaining equalities follow from the fact that

$$\mathfrak{B} \models \text{0-POINT}[a] \wedge \neg\text{0-POINT}[a^i] \rightarrow (K(a^i, a) = K(a^{i+1}, a))$$

and the symmetry of $Kxy = 0$.

Note that $\mathfrak{B} \models \neg$0-POINT$[a^i]$ by the $(r+2)$-MOD$(F_0, \ldots, F_{r+1})$ axiom of ψ_2. By symmetry of $Kxy = 0$ with $Kaa^{r+2} = 0$ we also have $\mathfrak{B} \models Ka^{r+2}a = 0$. This proves the diagonal-freeness in $\mathfrak{A}$.

By definition of $P_i^{\mathfrak{A}}$, for each 0-point a the truth of $\mathfrak{A} \models P_ia'$ is equivalent to the truth of $F_i^{\mathfrak{B}}a = 1$, and the truth of $\mathfrak{A} \models P_ia'$ equivalent to the truth of $F_i^{\mathfrak{B}}a' = 1$ and (by the shift axioms) therefore of $F_{i+r+2}^{\mathfrak{B}}a = 1$; consequently the truth of $Red_2(\beta)$ and of 0-POINT$[u]$, 0-POINT$[v]$, 0-POINT$[w]$ in $\mathfrak{B}$ implies the truth of β in $\mathfrak{A}$. This proves $\mathfrak{A} \models \psi$.

3.3.4 Reduction to F_i-Elimination Form

Let

$$\begin{aligned}\psi \;:=\; & \forall x \exists x' \forall y \exists u \exists v \exists w (\leq i\text{-SHIFT} \wedge \bigwedge_{j=0,1} j\text{-WITNESS-FREE} \wedge \\ & (r+2)\text{-MOD}(F_0, \dots, F_{r+1}) \wedge \beta)\end{aligned}$$

be an arbitrary formula in shift-reduced form with monadic predicate symbols $F_{i'} = j$ for $i' \leq i = 2(r+2)-1$ and $0 \leq j \leq 3$, as obtained by Step 2. Through this reduction step we have to find an interpretation of β on domains that contain no i'-witness (for $i' \leq i$) but admit an i-pre-witness. The latter will provide the possibility to extend such witness-free domains by an i-witness, needed for elimination of F_i by F_i-CODE in Step 4. For $i' \geq 2$ we have defined above the i'-witnessing property by intervals of F_0-values on i' successive elements of form $2\,1 \dots 1\,2$, i.e. satisfying, in the presence of $\leq i'$-SHIFT, the following equations:

$$\begin{aligned}(2, 1, \dots, 1, 2) &= (F_0 x, F_1 x, \dots, F_{i'-2} x, F_{i'-1} x) \\ &= (F_0 x, F_0 x^1, \dots, F_0 x^{i'-2}, F_0 x^{i'-1}).\end{aligned}$$

Candidates for non-i-witnesses are therefore easily obtained by restricting attention to i-intervals containing at least one 0, i.e. satisfying the formula

$$\text{ZERO}_i(x) := \bigvee_{0 \leq i' \leq i} (F_{i'} x = 0)$$

(or equivalently with $F_0 x^{i'} = 0$).

We have also to provide the means to construct from a model of $\psi_{3,i}$ an i-witness that guides the definition of F_i for a corresponding model of $\psi_{3,i-1}$; we therefore choose as pre-witnessing property for $i \geq 2$ the pattern $1 \dots 1\,2 * 1$ of F_0-values on $i+1$ successive elements, where, for reasons that will become clear below, $*$ indicates that any value is allowed. Formally:

$$i\text{-PRE-WITNESS}(z) := \bigwedge_{0 \leq j \leq i-3} (F_j z = 1) \wedge (F_{i-2} z = 2) \wedge (F_i z = 1).$$

Note that 2-PRE-WITNESS(z) := $(F_0 z = 2) \wedge (F_2 x = 1)$ and that, in presence of $\leq i$-SHIFT, i-PRE-WITNESS(z) implies i'-PRE-WITNESS(z) for $2 \leq i' < i$.

We thus define $\psi_{3,i}$ as the formula with prefix $\forall x \exists x' \forall y \exists u \exists v \exists w \exists z$ followed by the conjunction of the following formulae:

$$\leq i\text{-SHIFT} \wedge \leq i\text{-WITNESS-FREE} \wedge i\text{-PRE-WITNESS}(z)$$

$$\text{ZERO}_i(u) \wedge \text{ZERO}_i(v) \wedge \text{ZERO}_i(w)$$

$$\text{ZERO}_i(x) \wedge \text{ZERO}_i(y) \rightarrow \text{ZERO}_i(x') \wedge (r+2)\text{-MOD}(F_0, \dots, F_{r+1}) \wedge \beta)$$

We have to show that ψ is (finitely) satisfiable iff $\psi_{3,i}$ is.

Let $\mathfrak{A} = (A, F_0, \ldots, F_i, K)$ be a model of ψ. We extend $\mathfrak{A}$ by new elements b_j to make b_0 into an i-pre-witness, i.e. the b_j form a successor cycle of length $i+1$ (i.e. such that $b_j' := b_{j+1}$ for $j < i$ and $b_i' := b_0$) with associated F_0-values $1 \ldots 1211$ (i.e. $F_0(b_j) := 1$ for $j \leq i-3$ or $j > i-2$, $F_0(b_{i-2}) := 2$), This can be seen in Fig. 3.26.

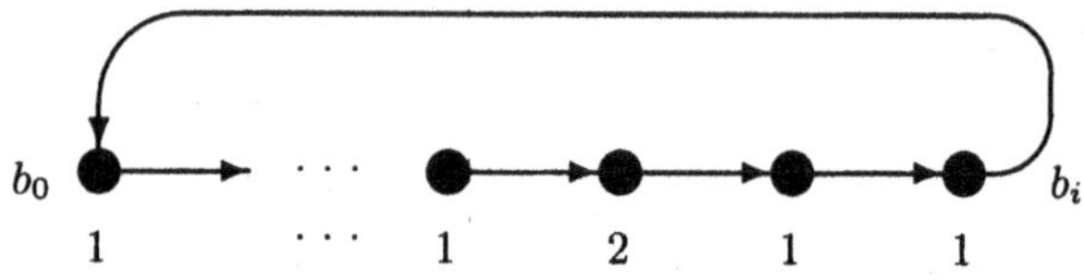

Figure 3.26. i-prewitnessing b_0.

The functions F_j with $0 < j \leq i$ and K are extended to these new elements in order to satisfy the shift axioms $\leq i$-SHIFT. To prove that this extended structure $\mathfrak{B}$ is a model of $\psi_{3,i}$, it remains to show the following:

(i) Only the new elements b_j do *not* satisfy the ZERO$_i$-predicate, and none of the b_j is a 0- or 1-witness.
(ii) $\mathfrak{B} \models \forall x \neg i'$-WITNESS$(x)$, for all i' with $2 \leq i' \leq i$.

Proof of (i). ZERO$_i$ is false for the new elements and none of them is a 1-witness because by definition the functions $F_0, \ldots, F_i$ assume only the values 1 or 2 on any b_j. No b_j is a 0-witness because the extension of K is defined to satisfy $F_0 x = Kxx'$ and therefore excludes the diagonal. For each old element $a \in A$ the formula ZERO$_i[a]$ holds in both $\mathfrak{A}$ and $\mathfrak{B}$, because due to $(r+2)$-MOD$(F_0, \ldots, F_{r+1})$, we find in the successor chain at distance $\leq r+1$ (therefore $\leq i$) a 0-POINT on which F_0 assumes the value 0.

Proof of (ii). Let $2 \leq i' \leq i$. By definition of F_0, the successor function and the shift axioms on the new elements, sequences $21 \ldots 12$ of F_0-values on successive elements have length $i+2$ and never length $i' \leq i$. Therefore no new element b_j is an i'-witness. Old elements $a \in A$ are not i'-witnesses because by $(r+2)$-MOD$(F_0, \ldots, F_{r+1})$ and the shift axioms, F_0 cannot repeat the value 2 before $r+2$ successor steps — and in between the function assumes at least once the value 0.

The opposite direction holds because by definition of $\psi_{3,i}$, the restriction of any model of $\psi_{3,i}$ to the (successor closed domain of all) elements satisfying ZERO$_i(x)$ yields a model for ψ.

3.3.5 Elimination of Monadic F_i

Let

$$\psi_{3,i} := \forall x \exists x' \forall y \exists z_1 \ldots \exists z_n \exists u (\leq i\text{-SHIFT} \wedge \leq i\text{-WITNESS-FREE} \wedge \wedge i\text{-PRE-WITNESS}(u) \wedge \beta)$$

be in F_i-elimination form. Then $\psi_{3,i-1}$ is defined to be the formula with prefix $\forall x \exists x' \forall y \exists z_1 \ldots \exists z_n \exists w \exists u$ followed by the conjunction of the formulae

$$\leq (i-1)\text{-SHIFT} \wedge \leq (i-1)\text{-WITNESS-FREE} \wedge i\text{-PRE-WITNESS}(u)$$

$$\bigwedge_{1 \leq i \leq n} \neg i\text{-WITNESS}(z_i) \wedge \neg i\text{-WITNESS}(u) \wedge F_i\text{-CODE} \wedge Red_{3,i}(\beta)$$

where

$$Red_{3,i}(\beta) := \neg i\text{-WITNESS}(x) \wedge \neg i\text{-WITNESS}(y) \rightarrow \neg i\text{-WITNESS}(x') \wedge \beta[(F_i t = j)/(Ktw = j)]$$

Here $\beta[F_i t = j/Ktw = j]$ indicates the result of replacing each subformula $F_i t = j$ in β (for any $j = 0, 1, 2, 3$ and any t) by $Ktw = j$. Note that F_i-CODE contains the conjunct i-WITNESS(w).

For $i = 0$ the formula obtained is called ψ_4 and does not contain the $(i-1)$-conjuncts; $\psi_{3,0}$ and $\psi_{3,1}$ do not contain the conjunct $0, 1$-PRE-WITNESS(u).

We have to show that $\psi_{3,i}$ is (finitely) satisfiable iff $\psi_{3,i-1}$ is.

Let $\mathfrak{A} = (A, F_0, \ldots, F_i, K)$ be a model for $\psi_{3,i}$. We distinguish three cases, following the definition of i-WITNESS for $i = 0, 1$ and $i \geq 2$.

Case $i = 0$. We extend $\mathfrak{A}$ to $\mathfrak{B}$ by adding a new element w to witness F_0 in

$$F_0\text{-CODE} = Kww \wedge (\neg Kxx \wedge Kyy \rightarrow (Kxx' = Kxy))$$

in presence of 0-SHIFT $= (F_0 x = Kxx')$.

We set $w' := w$, $\mathfrak{B} \models Kww$ and $K^{\mathfrak{B}}aw = F_0 a$ for $a \in A$. Then $\mathfrak{B} \models F_0\text{-CODE}[w]$ because $\mathfrak{A} \models F_0 x = Kxx'$ (by 0-SHIFT) and $\mathfrak{A} \models \neg Kxx$ (by 0-WITNESS-FREE). The equivalence of $F_0 a = j$ and $Kaw = j$ in this extended structure guarantees that also $Red_{3,0}(\beta)$ and therefore ψ_4 is satisfied in $\mathfrak{B}$.

Case $i = 1$. We extend $\mathfrak{A}$ by a new element w to witness F_1 in

$$F_1\text{-CODE} = (F_0 w = 3) \wedge ((F_0 x \neq 3) \wedge (F_0 y = 3) \rightarrow (F_0 x' = Kxy))$$

in presence of 1-SHIFT $= (F_1 x = F_0 x')$. We put $F_0 w = 3$ and $Kaw = F_1 a$ for all $a \in A$.

To satisfy 0-WITNESS-FREE for w we define that $Kww = 0$. Then F_1-CODE holds for w because 1-SHIFT $\wedge\ F_3 x \neq 0$ holds in $\mathfrak{A}$.

To satisfy 0-SHIFT $= (F_0x = Kxx')$ for w we introduce another new element v as successor/predecessor of w and extend K to (v, w) and (w, v) according to the definition of F_0w so that $w' := v$, $v' := w$ and $Kwv := 3$.

To satisfy 0-witness-freeness for v, we have to set $Kvv := 0$; to satisfy 0-SHIFT for v we have to set $F_0v := 3$; as a consequence of F_1-CODE and 1-SHIFT we have therefore to set also $Kav := F_1a$ for each $a \in A$.

The resulting structure $\mathfrak{B}$ satisfies 0-SHIFT, 0-WITNESS-FREE and F_1-CODE. $\mathfrak{B}$ satisfies $\mathit{Red}_{3,1}(\beta)$ because (by $\mathfrak{A} \models F_0x \neq 3$) no $a \in A$ is a 1-witness in $\mathfrak{B}$ and (by definition of K on w) $F_1a = j$ is equivalent to $Kaw = j$ for $j = 0, \ldots, 3$ and each $a \in A$. Thus $\mathfrak{B} \models \psi_{3,0}$.

Case $i \geq 2$. Truth in $\mathfrak{A}$ of the conjunct i-PRE-WITNESS in $\psi_{3,i}$ ensures that there is an i-pre-witness u in A. Truth of $\leq i$-SHIFT implies that u has an associated sequence $1 \ldots 1\, 2 * 1$ of length $i+1$ of successive F_0-values. We have to define an i-witness w – i.e. an element which has the associated F_0-value sequence $2\,1 \ldots 1\,2$ of length i; we can do this by choosing a new predecessor w of u with F_0-value 2, i.e., $w' := u$, $F_0w = 2$. To satisfy the shift axioms $\leq i-1$-SHIFT for w we extend F_j (for $1 \leq j \leq i-1$) to w and set $Kwu := 2$.

To satisfy $\neg i$-WITNESS$(x) \wedge i$-WITNESS$(y) \rightarrow (F_{i-1}x' = Kxy)$ in F_i-CODE in presence of $\leq i$-SHIFT $= (F_ix = F_{i-1}x')$ we set $Kaw := F_ia$ for each $a \in A$.

This definition is consistent with the previous definition $Kuw := 1$ (read: $Kwu := 2$) since $\mathfrak{A} \models i$-PRE-WITNESS$[u]$ ensures that $F_iu = 1$..

To satisfy $\leq i-1$-WITNESS-FREE we have to ensure $\neg j$-WITNESS$[w]$ for each $j \leq i-1$: for $j = 0$ this is obtained by defining $Kww := 0$; for $j = 1$ it follows from the definition $F_0w = 2 \neq 3$, and for $2 \leq j \leq i-1$ it follows from the i-WITNESS property of w.

Note that the model $\mathfrak{B}$ thus obtained satisfies also $i-1$-PRE-WITNESS$[u]$ for $i > 2$ (because $\mathfrak{A} \models i$-PRE-WITNESS$[u]$). (Note that in the case of $i = 2$, no 1-pre-witness is required in $\psi_{3,1}$) Since in $\mathfrak{B}$, for each $a \in A$ (i.e. non i-witness) and each $j = 0, \ldots, 3$, $Kaw = j$ is equivalent by definition to $F_ia = j$, $\mathit{Red}_{3,i}(\beta)$ will be true in $\mathfrak{B}$. Therefore $\mathfrak{B}$ is a model for $\psi_{3,i-1}$.

We now show that (finite) satisfiability of $\psi_{3,i-1}$ implies (finite) satisfiability of $\psi_{3,i}$. Let $\mathfrak{B} = (B, F_0, \ldots, F_{i-1}, K)$ be a model for $\psi_{3,i-1}$. We restrict the domain B to the (successor closed) domain $A := \{a \in B : \mathfrak{B} \models \neg i\text{-WITNESS}[a]\}$ of non i-witnesses, thus satisfying $\leq i$-WITNESS-FREE. On A we will define the function F_i satisfying $\leq i$-SHIFT to obtain a model $\mathfrak{A}$ for $\psi_{3,i}$. For this purpose let w be an i-witness satisfying F_i-CODE in $\mathfrak{B}$. Following the definition of F_i-WITNESS we again distinguish the cases $i = 0, 1$ and $i \geq 2$.

Case $i = 0, 1$. To satisfy F_0-SHIFT (for $i = 0$) or F_1-SHIFT (for $i = 1$) we have to set $F_0a := Kaa'$ and $F_1a := F_0a'$ respectively for each $a \in A$. (Note that for $i = 1$, $a \in A$ implies $a' \in A$ (since $\mathfrak{B} \models \mathit{Red}_{3,0}(\beta)$) and, by definition of A, $F_0a' \in A$.) The definition of F_0 and F_1 and the truth of F_0-CODE and F_1-CODE respectively in $\mathfrak{B}$ imply that $F_0a = j$ (for $i = 0$) or $F_1a = j$ (for $i = 1$) and $Kaw = j$ are equivalent so that $\mathfrak{A} \models \beta$ and therefore $\mathfrak{A}$ satisfies $\psi_{3,0}$ or $\psi_{3,1}$ respectively. The definition of F_1 together with truth of F_1-CODE in $\mathfrak{B}$ imply that $F_1a = j$ and $Kaw = j$.

Case $i \geq 2$. To satisfy F_i-SHIFT we have to set $F_ia = F_{i-1}a'$ for each $a \in A$. As in case $i = 1$ this definition is consistent (i.e. $a \in A$ implies $F_i(a) \in A$) and implies that $\mathfrak{A} \models \beta$. To prove that $\mathfrak{A} \models \psi_{3,i}$ it remains to exhibit an i-pre-witness u in A .

Since w is an i-witness, w' is not. We show that $\mathfrak{A} \models i$-PRE-WITNESS$[w']$. For $0 \leq j \leq i-2$ we have

$$\begin{aligned} F_jw' &= F_{j+1}w \quad (\text{by} F_{j+1}\text{-SHIFT}) \\ &= \begin{cases} 1 & \text{for } j \leq i-3 \\ 2 & \text{for } j = i-2 \end{cases} \quad (\text{since } \mathfrak{B} \models i\text{-WITNESS}[w]) \\ F_iw' &= F_{i-1}w'' \quad (\text{by definition of } F_i) \\ &= Kw'w \quad (\text{since } \mathfrak{B} \models \neg i\text{-WITNESS}[w'] \text{ and } F_i\text{-CODE}) \\ &= 1 \quad \text{because } Kww' = F_0w \quad (\text{by} \leq F_0\text{-SHIFT}) \\ &\qquad\qquad\qquad\qquad = 2 \quad (\text{by } \mathfrak{A} \models i\text{-WITNESS}[w]). \end{aligned}$$

This completes the proof of Theorem 3.3.1.

The proof we gave starts in Step 1 from the Kahr class which we already know to be a conservative reduction class. An alternative proof which is independent from the result for the Kahr class can be given by replacing Step 1 by a direct formalization of the halting problem for Turing machines through diagonal-free formulae in $[\exists^2\forall\exists\forall, (\omega, 1)]$ or $[\forall\exists^3\forall, (\omega, 1)]$ or $[\forall\exists\forall\exists^2, (\omega, 1)]$. We sketch this reduction in the following two exercises.

Exercise 3.3.1. [225] Formalize computations of deterministic Turing machine programs M with alphabet $\{0, 1\}$ by diagonal- and 0-witness-free formulae ψ_M of form

$$\forall x \forall y \beta_M \wedge \exists u \exists v \gamma.$$

The first conjunct should contain only atomic formulae Kst (for $s, t \in \{x, y, x', y'\}$) and P_is (for $s \in \{x, y, x'\}$) for one binary and finitely many monadic predicate symbols; the second conjunct should contain only formulae P_iu, P_iv. Reduce the recursively inseparable halting problems

$$H_i = \{M : (0,0)00\ldots \vdash_M (i,0)00\ldots, M \text{ a TM program}\} \quad (i = 1, 2)$$

as in the proof of Trakhtenbrot's Theorem. Show the

Reduction Property:

1. If $M \in H_1$ then ψ_M is unsatisfiable

2. If $M \in H_2$ then ψ_M is finitely satisfiable

Conclude that the class of all ψ_M is a conservative reduction class.

Sketch of a solution. One can adapt the economical description of Turing machines (given in the proof of the Church-Turing Theorem) by gluing the head position predicate H and the tape symbol predicate T_1 together into one binary predicate symbol K. (Note that $T_0 = \neg T_1$ due to the assumption that the underlying alphabet is binary.) Think about interpreting K over $(\{time, tape\} \times \mathbb{N})^2$ with $(t,n)' := (t,n')$; $K((time,t),(tape,x))$ corresponds to $H(t,x)$ and $K((tape,x),(time,t))$ corresponds to $T_1(t,x)$. This is pictorially represented in Fig. 3.27 (for the case where $H(2,1)$, $T_1(2,i)$ is true for $i = 0,3,4,5$ and false otherwise).

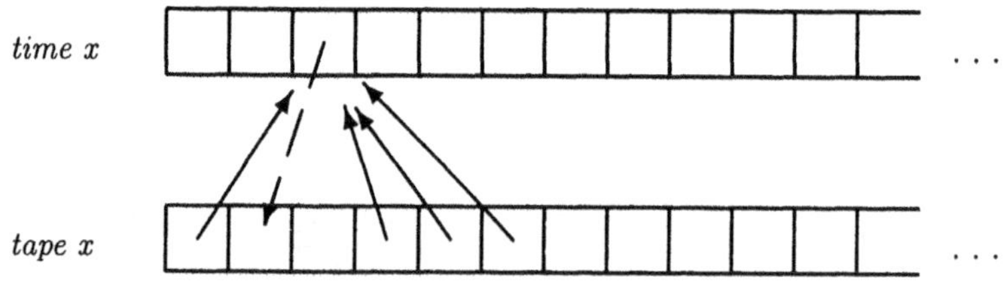

Figure 3.27. TM-description in $(\{time, tape\} \times \mathbb{N})^2$.

This interpretation satisfies $\forall x(\neg Kxx \wedge \neg Kxx' \wedge \neg Kx'x)$. To distinguish the two different uses of K, the cases $(time, x)$ and $(tape, x)$ are represented by two monadic predicates *Time* and *Tape* satisfying $\forall x((Time\, x \rightarrow Time\, x') \wedge (Tape\, x \rightarrow Tape\, x') \wedge \neg(Time\, x \wedge Tape\, x))$. To economize on applications of the successor function symbol (whose use will be reduced in the next exercise to that of a Skolem function x'), the description of printing, change of state and move of the reading head in STEP_M is split into two steps. The first describes printing and state change, the second step describes the reading head move. For this purpose new monadic predicates are used: $Move_i$ is intended to represent movement $i \in \{0,1,-1\}$ at $(time, t)$; auxiliary state predicates $I_{i'}$ are intended to represent the intermediate state reached from state i at $(time, t)$ by printing and state change. The crucial conjuncts in STEP_M are as follows, for instance for print instructions $i01pm$ of M:

– printing and state change axioms

$$Time\, x \wedge Tape\, y \wedge I_i x \wedge Kxy \wedge \neg Kyx \rightarrow I_{m'} x \wedge Move_p x \wedge Kyx'.$$

Similarly for instructions $i10pm$ replacing $\neg Kyx$ ("reading 0 in tape cell y at moment x") in the premise by Kyx ("reading 1 in tape cell y at time x") and replacing Kyx' ("reading 1 in tape cell y at moment x'") in the conclusion by $\neg Kyx'$ ("reading 0 in tape cell y at moment x'"). Analogously for instructions $ijjpm$ of M.

– move axioms

$$Time\, x \wedge Tape\, y \wedge I_{m'}\, x \wedge Move_1\, x \wedge Kxy \rightarrow Kx'y' \wedge I_m\, x'$$

There are similar axioms for left move (replacing the premise Kxy — "at moment x the reading head position is y" — by Kxy' and the conclusion $Kx'y'$ by $Kx'y$) and for idle move (replacing the conclusion $Kx'y'$ by $Kx'y$).

Thus β_M can be defined as conjunction of the adapted formula STEP$_M$ and of the corresponding formulae START, NONSTOP. For conservativity of the reduction we add $\prec_{K,M}$, see the proof of Trakhtenbrot's theorem. These formulae are written using a zero-predicate *Zero* formalized by:

$$\gamma \quad := \quad Zero\, u \wedge Time\, u \wedge Zero\, v \wedge Tape\, v$$

Exercise 3.3.2. Eliminate the use of terms y' in the formulae ψ_M obtained in the preceding exercise. This should establish the conservative reduction class property for diagonal- and 0-witness free formulae

$$\forall x \forall y \beta(x, x', y) \wedge \gamma(c, c')$$

in Skolem normal form. Therefore the classes $[\exists^2\forall\exists\forall, (\omega, 1)]$, $[\forall\exists^3\forall, (\omega, 1)]$ and $[\forall\exists\forall\exists^2, (\omega, 1)]$ — even if restricted to diagonal- and 0-witness free formulae — are conservative reduction classes.

Sketch of a solution. It suffices to replace in β_M the move axioms

$$\begin{array}{l} \text{MOVE}_1 \wedge Kxy \rightarrow Kx'y' \\ \text{MOVE}_{-1} \wedge Kxy' \rightarrow Kx'y \end{array}$$

(where MOVE$_j$:= $Time\, x \wedge Tape\, y \wedge I_{m'}\, x \wedge Move_j\, x$) by formulae which contain only terms x, y, x' and only one binary predicate symbol. Let us visualize Kab as arrow from a to b. The required shift of $a \rightarrow b$ to $a' \rightarrow b'$ (and of $a \rightarrow b'$ to $a' \rightarrow b$) has to be obtained by a series of 1-successor-step shifts—of $c \rightarrow d$ to $c' \rightarrow d$ or to $c \rightarrow d'$. To obtain such a slowing down of arrow shifts we double each element c by introducing between c and c' a new element $\bar{c}$ which is defined as new successor of c and whose successor is defined as c', pictorially represented in Fig. 3.28.

Then the shift from $a \rightarrow b$ to $a' \rightarrow b'$ can be slowed down by going through the newly created intermediate points as shown in Fig. 3.29, where the arrow index i shows the i-th step of stepwise arrow connection. Formally this can be described by introducing a modulo-2 structure through the following axiom:

$$\text{MOD-2} := \forall x((P_0 x \vee P_1 x) \wedge \neg(P_0 x \wedge P_1 x) \wedge (P_0 x \leftrightarrow P_1 x'))$$

Here, P_0 and P_1 are new monadic predicate symbols; points a satisfying $P_0 a$ are interpreted as "given" and their successors (satisfying P_1) as "new" elements. We abbreviate $Tx \wedge P_j x$ by $T_j x$ (for $T := Time, Tape$) and $I_{m'} x \wedge$

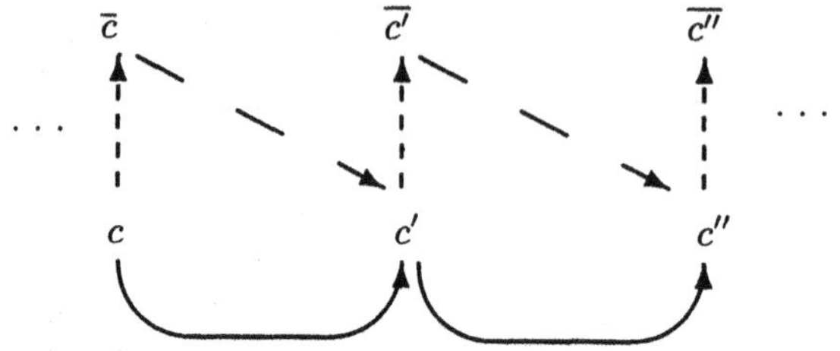

Figure 3.28. Doubling points by new successors.

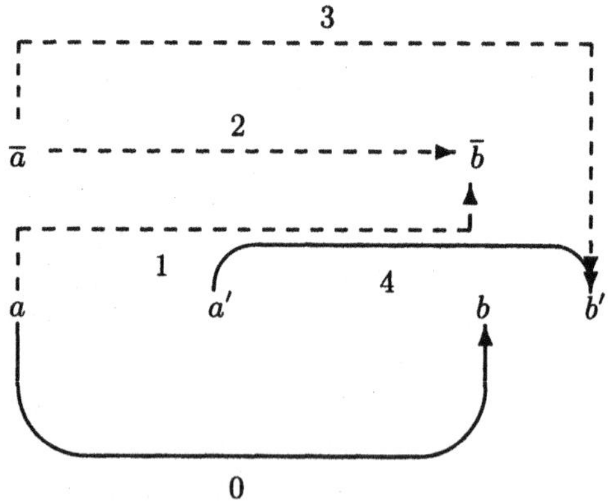

Figure 3.29. Shift of $a \to b$ to $a' \to b'$.

$Move_1 x$ by ε; then the formulae which describe the above picture in Skolem normal form are as follows:

$$\begin{array}{l} Time_0\, y \wedge Tape_0\, x \wedge \varepsilon \wedge Kyx \rightarrow Kyx' \\ Time_0\, x \wedge Tape_1\, y \wedge \varepsilon \wedge Kxy \rightarrow Kx'y \\ Time_1\, x \wedge Tape_1\, y \wedge \varepsilon \wedge Kyx \rightarrow Kyx' \\ Time_1\, x \wedge Tape_0\, y \wedge \varepsilon \wedge Kxy \rightarrow Kx'y \end{array}$$

Similar axioms formalize Fig.3.30 for the slowed down shift of $a \rightarrow b'$ to $a' \rightarrow b$.

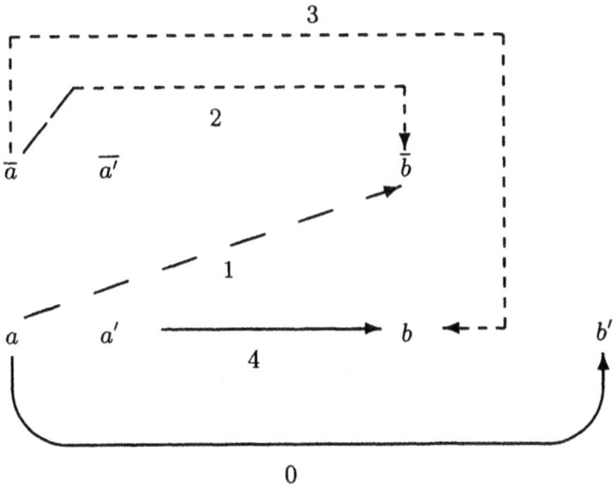

Figure 3.30. Shift of $a \rightarrow b'$ to $a' \rightarrow b$.

In presence of MOD–2, these formulae can therefore replace the above MOVE_1 and MOVE_{-1} axioms, adapting correspondingly the other conjuncts of β_M. For the axioms for printing and state change it suffices to replace $T\,t$ by $T_0\,t$ (for $T := Time, Tape$), for the idle move axioms to replace $Tape\, y$ by $Tape_0\, y$ and to add frame axioms expressing that "zero" is a 0-point ($\text{ZERO}\, x \rightarrow P_0\, x$) and that move and auxiliary state information is carried from 0-points to 1-points:

$$\begin{array}{rcll} Move_j\, x \wedge P_0\, x & \rightarrow & Move_j\, x' & \text{for } j \in \{0, 1, -1\} \\ I_{i'}\, x \wedge P_0\, x & \rightarrow & I_{i'}\, x' & \text{for auxiliary states } i' \text{ of } M. \end{array}$$

3.3.6 The Kostyrko-Genenz and Surányi Classes

Theorem 3.3.2. *The Surányi class* $[\exists^*\forall\exists\forall, (0,1)]$ *is a conservative reduction class.*

Proof. We give a conservative reduction from the class of formulae of the form

$$\psi := \exists u \exists v \forall x \exists x' \forall y(\neg Kxx \wedge \beta) \in [\exists^2\forall\exists\forall, (\omega, 1)]$$

In the two exercises of the preceding section this class has been shown to be a conservative reduction class.

The idea is to replace each monadic formula Pt by a binary formula Ktw where w is an appropriate new "witness" for P. Since the formula ψ to be reduced satisfies $\forall x \neg Kxx$, we can choose Kww as witnessing property.

Let $P_1, \ldots, P_n$ be the monadic predicate symbols occurring in ψ. Then $\varphi \in [\exists^{n+2}\forall\exists\forall, (0, 1)]$ is defined as formula with the prefix

$$\exists w_1 \ldots \exists w_n \exists u \exists v \forall x \exists x' \forall y$$

and with quantifier-free part consisting of the conjunction of the following formulae:

$\bigwedge_{1 \leq i \leq n} Kw_iw_i$	(witnessing property)
$\neg Kx'x' \wedge \neg Kuu \wedge \neg Kvv$	(closure property for non witnesses)
$\neg Kxx \wedge \neg Kyy \rightarrow \beta[P_it/Ktw_i]$	(encoding of β.)

It remains to show that ψ is (finitely) satisfiable iff φ is.

If $\mathfrak{A} = (A, P_1, \ldots, P_n, K)$ is a model for ψ, then one obtains a model $\mathfrak{B} \models \varphi$ by adding new elements $w_1, \ldots, w_n$ to A and extending K by setting

$$\mathfrak{B} \models Kaw_i \iff a = w_i \text{ or } \mathfrak{A} \models P_ia.$$

In the other direction, the restriction of each model $\mathfrak{B} \models \varphi$ to the (successor closed) set A of all a such that $\mathfrak{B} \models \neg Kaa$ yields a model for ψ by defining $P_i := \{a \in A : \mathfrak{B} \models Kaw_i\}$. □

Theorem 3.3.3. *The Kostyrko-Genenz class* $[\forall\exists^*\forall, (0, 1)]$ *is a conservative reduction class.*

Proof. We give a conservative reduction of the class of formulae of the form

$$\psi := \forall x \exists z_1 \exists z_2 \exists z_3 \forall y \ (\neg Kxx \ \wedge \ \beta) \in [\forall\exists^3\forall, (\omega, 1)].$$

The two exercises of the preceding section show that this class is a conservative reduction class.

The reduction is by induction on the number of monadic predicate symbols occurring in β. First we eliminate stepwise pairs P, Q of monadic predicate symbols by replacing Ps and Qt by Ksw and Kwt respectively. We use a new monadic predicate symbol (P, Q) that encodes P and Q via an appropriate witness w; this means that $(P, Q)w$ is true and the following "uniqueness" property (indistinguishability with respect to K) holds:

$$(P,Q)\text{-CODE} := (P,Q)y \rightarrow ((Kxw \leftrightarrow Kxy) \wedge (Kwx \leftrightarrow Kyx)).$$

These reductions preserve the irreflexivity of interpretations of K. This allows us to eliminate in Step 2 the last monadic predicate symbol R by replacing Rt by Ktw; as witnessing property we can choose Kww and the following uniqueness property (indistinguishability with respect to K):

$$R\text{-CODE} := Kyy \rightarrow (Kxw \leftrightarrow Kxy).$$

Step 1. Let $\psi := \forall x \exists z_1 \cdots \exists z_n \forall y\ (\neg Kxx \wedge \beta) \in [\forall\exists^*\forall, (r,1)]$ with $r \geq 2$ and choose two monadic predicate symbols P, Q occurring in ψ. Denote by (P,Q) a new monadic predicate symbol. Define φ as formula with prefix

$$\forall x \exists z_1 \cdots \exists z_n \exists w \forall y$$

followed by the conjunction of the formulae

$$(P,Q)w \wedge (P,Q)\text{-CODE} \qquad \text{(witnessing property)}$$

$$\bigwedge_{1 \leq i \leq n} \neg(P,Q)z_i \qquad \text{(closure property for non witnesses)}$$

$$\neg(P,Q)x \wedge \neg(P,Q)y \rightarrow \beta[Ps/Ksw, Qt/Kwt] \qquad \text{(encoding of } \beta)$$

$$\neg Kxx.$$

As usual, $\beta[Ps/Ksw, Qt/Kwt]$ indicates the result of replacing each occurrence of monadic formulae Ps, Qt in β by Ksw, Kwt respectively.

It remains to show that ψ is (finitely) satisfiable iff φ is.

A model $(A, P, Q, \ldots, K)$ for ψ is easily extended to a model $\mathfrak{B} \models \varphi$ by adding a new element w, putting $(P,Q) := \{w\}$ and extending K by

$$\mathfrak{B} \models Kaw \iff \mathfrak{A} \models Pa$$
$$\mathfrak{B} \models Kwa \iff \mathfrak{A} \models Qa$$

In the other direction, from a model $\mathfrak{B} = (B, (P,Q), \ldots, K)$ for φ we obtain a model for ψ as follows. Choose a witness w satisfying $\mathfrak{B} \models (P,Q)[w] \wedge (P,Q)$-CODE and restrict the domain to the set of non-witnesses $A = \{a \in B : \mathfrak{B} \models \neg(P,Q)a\}$. Define Pa and Qa by Kaw and Kwa respectively. The independence of this definition from the choice of w is guaranteed by the truth of (P,Q)-CODE.

Step 2. Let $\psi := \forall x \exists z_1 \cdots \exists z_n \forall y(\neg Kxx \wedge \beta) \in [\forall\exists^*\forall, (1,1)]$ with monadic predicate symbol R. Define φ as the formula with prefix

$$\forall x \exists z_1 \cdots \exists z_n \exists w \forall y$$

followed by the conjunction of:

$$Kww \wedge R\text{-CODE} \qquad \text{(witnessing property)}$$
$$\bigwedge_{1\le i\le n} \neg K z_i z_i \qquad \text{(closure property for non witnesses)}$$
$$\neg Kxx \wedge \neg Kyy \to \beta[Rt/Ktw] \qquad \text{(encoding of } \beta) .$$

Exercise 3.3.4. Show that ψ is (finitely) satisfiable iff φ is. Hint: Paraphrase Step 1, with R in place of P.

□

3.4 Historical Remarks

The study of logical decision problems dates from the earliest investigations of the first-order predicate logic. From the very beginning the vocabulary and the prefix structure of formulae in prenex normal form were used as a guide through the different ideas, methods and results which were produced in an attempt to solve Hilbert's *Entscheidungsproblem*. Already in the pioneering paper by Löwenheim [365] we find the two streams that after 50 years of intensive research led to a complete solution of the prefix-vocabulary problem for predicate logic without functions and equality [219], namely the exhibition of (a) an algorithm which solves the decision problem of a particular subclass of formulae – here monadic predicate logic – and (b) an effective reduction of the Entscheidungsproblem to the decision problem of a small class of formulae — here formulae with only binary predicates. Shortly later Skolem [477] extends Löwenheim's decidability result to the monadic predicate logic of second order [476] and uses his Normal Form Theorem to reduce the Entscheidungsproblem to the prefix class $[\forall^*\exists^*, all]$.

Löwenheim's and Skolem's reduction classes – the term itself does not appear in print before 1956 in [82], long after the unsolvability proof for the Entscheidungsproblem by Church and Turing [80, 513] – are sharpened by a chain of theorems that show how to reduce the number of predicates and quantifiers.

Herbrand and Kalmár reduce the number of predicates to just one binary predicate (see the reduction classes $[all, (0,3)]$, $[all, (0,0,1)]$ in [254] and $[all, (0,1)]$ in [295]). Gödel reduces the number of universal quantifiers in Skolem's reduction class to three (see the reduction class $[\forall^3\exists^*, all]$ in [187]). The number of existential quantifiers in Skolem's reduction class has been bound to one by Pepis, who showed the reduction class property for the class of $[\forall^* \wedge \forall\forall\exists, (1,0,1)]$-formulae and therefore for the classes $[\forall^*\exists, (1,0,1)]$ and $[\forall\forall\exists\forall^*, (1,0,1)]$ (see [420]; compare this to Ackermann's reduction class $[\forall\exists \wedge \exists\forall^*, all]$ obtained in [17] by reducing Kalmár's reduction class $[\exists^*\forall^2\exists\forall^*, (0,0,1)]$ in [292] establishing the reduction classes $[\forall\exists^2\forall^*]$,

$[\exists\forall\exists\forall^*]$ and $[\exists\forall^*\exists]$; see also the improvement to $[\forall^2\exists\forall^*, (0,1)]$ by Kalmár and Surányi in [303].) In 1943 Surányi was able to strengthen Pepis' reduction class to $[\forall^*\exists, (0,1)]$ and to bound the total number of quantifiers of either kind together with the arity of the predicates, showing the reduction class property of $[\forall^3 \wedge \forall^2\exists, (0,\omega)]$ and thereby $[\forall^3\exists, (0,\omega)]$ and $[\forall^2\exists\forall, (0,\omega)]$ [494].

Eventually, in 1959, Surányi could improve those three classes by reducing the vocabulary to $(\omega, 1)$ [498]. Our proof for the conservativity of the Suranyi reduction class $[\forall^3\exists, (\omega, 1)]$ is taken from [225].

As a consequence the only remaining prefix classes to be settled were cases in which no existential quantifier is governed by more than one universal quantifier, refining Ackermann's reduction classes $[\forall\exists \wedge \exists\forall^*, all]$, $[\exists\forall\exists\forall^*, all]$, and $[\forall\exists^2\forall^*, all]$ from [17]. In 1951 Surányi could limit the number of universal quantifiers to three, obtaining the reduction classes $[\exists\forall\exists\forall^2, (\omega, 7)]$ and $[\forall\exists^2\forall^2, (\omega, 7)]$.

The state of the art in the 50's is reported in the two books by Ackermann [18] and Surányi [498] which complement each other by covering the two sides of the medal, algorithms for solvable cases and reductions for unsolvable cases. The real breakthrough in the history of the prefix-vocabulary problem has been obtained only at he beginning of the 60's when Büchi [64] had the idea to combine (a) Turing's reduction of problems about machine computations to satisfiability/deducibility problems of logical formulae by formalizing the former through the latter, and (b) Skolem's method to reduce satisfiability of formulae to satisfiability of their normal forms over canonical domains. Büchi has applied his observation to (a dramatical simplification of) Turing's first-order formalization of Turing machines, obtained an elementary proof for the reduction class $[\exists \wedge \forall\exists\forall, (\omega, 3)]$ and therefore $[\forall\exists^2\forall, (\omega, 3)]$, $[\forall\exists\forall\exists, (\omega, 3)]$, and $[\forall\exists\forall\exists, (\omega, 3)]$; by a judicious choice of the Turing machine halting problems underlying his reduction he could even obtain, without additional difficulties, the conservativity of these classes, thus simultaneously strengthening Trakhtenbrot's Theorem of 1950 and 1953 [509, 510]. This triggered efforts to refine known reductions to conservative ones; see Gurevich's method for semi-conservative reduction [228] which is applied widely in this book.

Büchi's simple observation provided the key for a quick solution of the prefix problem and the prefix-vocabulary problem for predicate logic without functions and equality: In order to obtain reduction classes of syntactically "poor" formulae, researchers now looked for appropriate "small" computationally universal combinational problems which are reducible to the former. The first was Wang who invented the domino problem [531] a version of which immediately led to a solution of the prefix problem by establishing the reduction class $[\forall\exists\forall, (0,\omega)]$ [288]. See [258] for a simplification of this proof; our proof for the Kahr-Moore-Wang reduction class in Sect. 3.1.2 is taken from [441]. In the same year Kahr could improve the Kahr-Moore-Wang reduction class to $[\forall\exists\forall, (\omega, 1)]$, inventing more sophisticated dominoes for the asymmetric diagonal constrained domino problem. (Our proof in Sect. 3.1.3

uses a graph interpretation of $[\forall\exists\forall, (0, \omega)]$-formulae which is inspired by ideas from [287] and is taken from [441].)

In 1972 Gurevich and Koryakov [237] improved Berger's undecidability proof for the unconstrained domino problem [33] by showing that the classes of domino problems with no and with periodic solution respectively are recursively inseparable. This is what we use for the proof in Sect. 3.1.2 that $[\forall\exists\forall, (\omega, 1)]$ is a conservative reduction class. At this stage in addition to the prefix problem also the prefix-vocabulary problem for finite prefixes was solved. Using appropriate direct formalizations of machine problems, also the two simpler minimal prefix-vocabulary conservative reduction classes $[\forall\exists\forall^*, (0, 1)]$ and $[\forall\exists^*\forall, (0, 1)]$ could be established by Denton [108] and Kostyrko and Genenz [180, 317]. The previously known minimal undecidable prefix-vocabulary cases with vocabulary $(0, 1)$ and $\forall^*$ or $\exists^*$ in the prefix were the Surányi classes $[\exists^*\forall^3\exists, (0, 1)]$ and $[\exists^*\forall\exists\forall, (0, 1)]$ [498], the Kalmár-Surányi classes $[\forall^*\exists, (0, 1)]$ and $[\forall^3\exists^*, (0, 1)]$ [302] which improved the weaker results in [291, 420] where in particular one ternary predicate appeared. The difficult case $[\forall\exists\forall\exists^*, (0, 1)]$ which completed the prefix-vocabulary classification was settled by Gurevich in 1966 starting from Turing machine halting problems and using a form of existential interpretation [219]. Our proofs for all of these cases in conservative form are adapted from [219, 225, 229].

4. Undecidable Standard Classes with Functions or Equality

This chapter deals with the *Entscheidungsproblem* for formulae of full predicate logic, i.e. formulae which besides predicate symbols can contain also equality and function symbols. This problem has received little attention until the solution of the prefix-vocabulary problem for the pure predicate logic had been completed by Kahr in 1962 and Gurevich in 1966, for two reasons. On one side early algorithms for classical decidable cases turned out to work with and without equality; a notable exception is the Gödel-Kalmár-Schütte class $[\exists^*\forall^2\exists^*]$ but awareness about this grew only in the 1960's and only in 1984 it has been established that the satisfiability problem of $\exists^*\forall^2\exists^*$-formulae with equality is actually undecidable (see Sect. 4.3 below). On the other side there are simple conservative reductions from full first order logic to the pure predicate calculus (see Exercise 2.1.7 to the Church-Turing Theorem and Lemma 3.2.12 on equality as neutral congruence). These reductions became methodologically important once the proof of Church and Turing for the unsolvability of the *Entscheidungsproblem* had refined Hilbert's original problem to a classification problem. Predicate logic without equality or functions appeared as a natural choice for undecidable classes of syntactically poor formulae.

After the complete prefix-vocabulary classification of the restricted predicate logic, Gurevich observed in his Classifiability Theorem that such a solution of the classification problem is characteristic for a broad class of logics. So it is natural to look again at logic with functions and equality, this time not from the unsolvability but from the classification standpoint. Gurevich's Classifiability Theorem tells us that indeed there is a finite number of minimal undecidable prefix-vocabulary classes of the form $[\Pi, (p_1, p_2, \ldots), (f_1, f_2, \ldots)]$ or $[\Pi, (p_1, p_2, \ldots), (f_1, f_2, \ldots)]_=$ where Π is an extended prefix (word over $\forall, \exists, \forall^*, \exists^*$) and p_i, f_i are natural numbers or ω. (The presence or absence of the equality symbol as suffix indicates that the equality symbol may or may not occur.)

We prove in this chapter results which establish what these minimal undecidable classes are; we show that all of them are indeed conservative reduction classes. This is summed up by the following main theorem for the *Entscheidungsproblem* of the full predicate logic.

Theorem 4.0.1. (Reduction Classes with Functions or Equality) *The following prefix-vocabulary classes of predicate logic with functions and equality are undecidable (and indeed are conservative reduction classes):*

- *Classes with functions and equality (Gurevich 1976):*
  - $[\forall, (0), (2)]_=$
  - $[\forall, (0), (0,1)]_=$
- *Classes with functions but without equality (Gurevich 1969):*
  - $[\forall^2, (0,1), (1)]$
  - $[\forall^2, (1), (0,1)]$
- *Classes with equality but without functions (Goldfarb 1984):*
  - $[\forall^2\exists, (\omega,1), (0)]_=$
  - $[\exists^*\forall^2\exists, (0,1), (0)]_=$
  - $[\forall^2\exists^*, (0,1), (0)]_=$.

Together with the list of reduction classes in pure predicate logic as provided by Theorem 3.0.1 and the decidability results in Chap. 6 and 7 this gives a complete classification of the prefix vocabulary classes in full first-order logic with respect to decidability of the satisfiability problem.

The proof of Theorem 4.0.1 is given in Sect. 4.1–4.3 considering separately the cases of formulae with both functions and equality, with functions but without equality, and with equality but without functions. The most difficult case here is the minimal Goldfarb class $[\forall^2\exists, (\omega,1), (0)]_=$, i.e. the minimal Gödel-Kalmár-Schütte class together with equality. The following picture surveys the conservative reductions of this chapter. Note that the reductions for the case with functions and equality can be carried out for Herbrand formulae, i.e. formulae in prenex normal form whose quantifier–free part is a conjunction of atomic formulae or negations of atomic formulae.

$$\text{2-RM} \longrightarrow [\forall, (0), (\omega)]_= \cap \text{HERBRAND} \longrightarrow [\forall, (0), (2)]_= \cap \text{HERBRAND} \longrightarrow [\forall, (0), (0,1)]_= \cap \text{HERBRAND}$$

$$[\forall\exists\forall, (0,\omega)] \longrightarrow [\forall^2, (0,1), (1)] \longrightarrow [\forall^2, (1), (0,1)]$$

$$[\forall\exists\forall, (0,\omega)] \longrightarrow [\forall^2\exists^*, (\omega,\omega)]_= \longrightarrow [\forall^2\exists, (\omega,\omega)]_= \longrightarrow [\forall^2\exists, (\omega,1)]_=$$

4.1 Classes with Functions and Equality

In this section we prove the two Gurevich classes $[\forall, (0), (2)]_=$, $[\forall, (0), (0,1)]_=$ to be conservative reduction classes even with the restriction to Herbrand formulae. For expository purposes we first reduce halting problems for 2-register machines to formulae in $[\forall, (0), (\omega)]_= \cap$ KROM $\cap$ HORN and in a second step refine the reduction to Herbrand formulae. Then we show how one can encode finitely many monadic functions by two monadic functions and two monadic functions by one binary function in such a way that the propositional structure of the formulae is preserved.

Theorem 4.1.1. *The class* $[\forall, (0), (\omega)]_= \cap$KROM$\cap$HORN *is a conservative reduction class.*

Proof. In Exercise 2.1.20 to Theorem 2.1.15 it has been shown how the claim can be proved for the class $[\forall, (0), (\omega)]_= \cap$ HORN by a reduction of halting problems of 2-register machines. We refine that method here to obtain a description through formulae which use only binary disjunctions.

The idea of the proof comes from a geometrical interpretation of computations of 2-register machines in the Gaussian quadrant. Imagine the computation of a 2-register machine program M as a walk through $\mathbb{N} \times \mathbb{N}$ where M, in state i with register contents (p, q), visits point (p, q) to "colour" it with i. The colouring of (p, q) with i – i.e. the fact that (p, q) is visited by M in state i – can be expressed in terms of monadic functions $state_i$ by requiring that (p, q) is a fixed point of $state_i$, i.e.

$$state_i(p, q) = (p, q).$$

The register contents (n_1, n_2) can be encoded by terms which are built up from a single variable x with monadic functions $zero_1, zero_2, succ_1, succ_2$, by applying the zero function $zero_j$ once and the successor function $succ_j$ n_j times:

$$\underline{(n_1, n_2)} := succ_1^{n_1} zero_1 succ_2^{n_2} zero_2 x.$$

This yields the following encoding of M-configurations (i, m, n), reached by M, through equations

$$\underline{(i, m, n)}: \; state_i\underline{(m, n)} = \underline{(m, n)}.$$

To this encoding one can apply the technique of economical description of register machines explained in Sect. 2.1.1. Due to Gurevich's theorem on semi-conservative reductions (Theorem 2.1.39) it suffices to formalize two recursively inseparable halting problems H_1' and H_2'; we choose them as counterparts for register machines of the halting problems H_1 and H_2 of Turing machines used for the proof of Trakhtenbrot's Theorem:

$$H_i' = \{M : (0, 0, 0) \Rightarrow_M (i, 0, 0), M \text{ is a RM program } \}.$$

For this purpose we define now formulae GRID, STEP$_M$, START and NONSTOP such that

$$\psi_M := \forall x(\text{GRID} \land \text{STEP}_M \land \text{START} \land \text{NONSTOP})$$

is in the class $[\forall, (0), (\omega)]_= \cap$ KROM $\cap$ HORN and satisfies the following reduction properties:

(i) (*Simulation Lemma*) For all M-configurations C, D the following is true:

$$C \Rightarrow_M D \text{ implies that } \forall x(\text{GRID} \wedge \text{STEP}_M \wedge \underline{C}) \models \forall x \underline{D},$$

(ii) $M \in H_1' \implies \psi_M$ is contradictory,
(iii) $M \in H_2' \implies \psi_M$ is finitely satisfiable.

The formula GRID is the conjunction of the following equalities of terms:

$$\begin{aligned} zero_1 zero_2 x &= zero_2 zero_1 x \\ succ_1 succ_2 x &= succ_2 succ_1 x \\ zero_1 succ_2 x &= succ_2 zero_1 x \\ zero_2 succ_1 x &= succ_1 zero_2 x \end{aligned}$$

It ensures that the models of ψ_M contain a two-dimensional grid where $succ_1, succ_2$ are the successor functions along the two coordinates – i.e. $succ_1(m,n) = (m+1,n)$, $succ_2(m,n) = (m,n+1)$ – and $zero_1$ and $zero_2$ set the corresponding coordinate to 0 – i.e. $zero_1(m,n) = (0,n)$ and $zero_2(m,n) = (m,0)$.

Further, let START be the equality

$$state_0 zero_1 zero_2 x = zero_1 zero_2 x$$

and NONSTOP the inequality

$$state_1 zero_1 zero_2 x \neq zero_1 zero_2 x.$$

Finally, STEP$_M$ is defined as the conjunction of the following *instruction formulae* ε_i each of which expresses the change of $\underline{(i,m,n)}$ effected by the corresponding M-instruction $I_i = (i, a_\ell, j)$ (in state i add 1 to register l and go to state j) or $I_i = (i, s_\ell, j, k)$ (in state i go to state j if the content of the l–th register is 0, otherwise subtract 1 from it and go to state k):

For $I_i = (i, a_\ell, j) \in M$:

$$(state_i x = x \rightarrow state_j succ_\ell x = succ_\ell x).$$

For $I_i = (i, s_\ell, j, k) \in M$:

$$\begin{aligned} &(state_i zero_\ell x = zero_\ell x \rightarrow state_j zero_\ell x = zero_\ell x) \wedge \\ &\wedge(state_i succ_\ell x = succ_\ell x \rightarrow state_k x = x) \end{aligned}$$

It remains to show the reduction properties. The Simulation Lemma follows by a straightforward induction on the length of the given M-computation. Specialize the Simulation Lemma to $\underline{C}$:= START and $\underline{D}$:= ¬NONSTOP; we obtain that $M \in H_1'$ implies the truth of $\forall x \neg$NONSTOP in each model of $\forall x(\text{GRID} \wedge \text{STEP}_M \wedge \underline{(0,0,0)})$. Therefore $M \in H_1'$ implies that ψ_M is contradictory.

We now show that ψ_M has a finite model if $(0,0,0) \Rightarrow_M (1,0,0)$. Let $N-1$ be the maximal content of a register during the given M-computation from $(0,0,0)$ to $(1,0,0)$ and define $A := \{0,\ldots,N\} \times \{0,\ldots,N\}$. The following definition yields an algebra over A which clearly satisfies ψ_M:

$$\begin{aligned}
succ_1(m,n) &:= \begin{cases} (m+1,n) & \text{if } m < N \\ (m,n) & \text{if } m = N \end{cases} \\
succ_2(m,n) &:= \begin{cases} (m,n+1) & \text{if } n < N \\ (m,n) & \text{if } n = N \end{cases} \\
zero_1(m,n) &:= (0,n) \\
zero_2(m,n) &:= (m,0) \\
state_i(m,n) &= \begin{cases} (m,n) & \text{if } (0,0,0) \Rightarrow_M (i,m,n) \\ (N,N) & \text{otherwise.} \end{cases}
\end{aligned}$$

This completes the proof. □

Exercise 4.1.2. [227] Prove that $[\forall,(0),(\omega)]_=$ is a conservative reduction class by encoding the unconstrained domino problem. Hint: Associate with every domino d a function and represent the points tiled by a domino d by fixed points of that function. You will find a simpler representation of the Gaussian coordinates than the one introduced for the representation of two registers, but the disjunctions will probably violate the Horn condition and their length will increase to the number of colours of the given domino problem.

Corollary 4.1.3 (Wirsing). *The class* $[\forall,(0),(\omega)]_= \cap$ HERBRAND *is a conservative reduction class. This is even true for formulae which contain only equations and one inequality.*

Proof. For the proof of Theorem 4.1.1 we have formalized the computation $C = C_0 \xrightarrow{i_1} C_1 \xrightarrow{i_2} C_2 \ldots \xrightarrow{i_t} C_t = D$ from C to D by a sequence of implications $\underline{C} \to \ldots \to \underline{D}$ taken from STEP_M which transfer local fixed points; any fixed point of any intermediate $state_i$ at any register word $\underline{(n_1,n_2)}$ implies a fixed point of $state_j$ at the register word $\underline{(n_1',n_2')}$ obtained by executing instruction I_i on (n_1,n_2) and going to the next instruction I_j. We now express this relation by "globally" equating the encodings $\underline{C},\underline{D}$ of C and D ; in order not to loose the conservativity of the reduction we encode also the history $\underline{com}$ of the computation $C \overset{com}{\Longrightarrow}_M D$ telling us which instructions have been applied. Thus the reachability of D from C through the computation *com* will be expressed by the equality

$$\underline{C}\ start\ x = \underline{D}\ \underline{com}\ start\ x$$

where *start* is a new monadic function symbol which serves to indicate the right beginning of the coding area in a term.

With these refined definitions (including a corresponding refinement of the NONSTOP axiom, see below) we will build a formula $\forall x(\text{STEP} \wedge \text{NONSTOP})$ and prove the following reduction properties:

(i) (*Simulation Lemma*) For all M-configurations C, D the following is true: If $C \stackrel{com}{\Longrightarrow}_M D$ then

$$\forall x \text{STEP}_M \models \forall x(\underline{C}\ start\ x = \underline{D}\ \underline{com}\ start\ x).$$

(ii) $M \in H_1'$ implies that$\forall x(\text{STEP}_M \wedge \text{NONSTOP})$ is contradictory.
(iii) $M \in H_2'$ implies that$\forall x(\text{STEP}_M \wedge \text{NONSTOP})$ is finitely satisfiable.

As a consequence we keep from the previous definition of $\underline{(i, m, n)}$ only the left side of the equation, i.e. we define now $\underline{(i, m, n)}$ to be the *terms* $state_i(m, n)$.

The trace $i_1 \dots i_t$ of a computation $C_0 \stackrel{i_1 \dots i_t}{\longrightarrow} C_t$ – standing for $C_0 \stackrel{i_1}{\rightarrow} C_1 \stackrel{i_2}{\rightarrow} C_2 \dots \stackrel{i_t}{\rightarrow} C_t$ – is encoded as a sequence $\underline{i_1 \dots i_t} := (i_t) \dots (i_1)$ of the names (numbers) of the instructions which have been applied, i.e. these names $i \leq r$ are interpreted as new monadic function symbols (i). The new formula NONSTOP expressing that M does not reach the halting state 1 is

$$\underline{(0,0,0)}start\ x \neq state_1 x \wedge \text{FILL}.$$

There is no start axiom but in connection with NONSTOP the following FILL-axioms are needed which will allow us to fill in an arbitrary encoding of a computation at the right of the first occurrence of *start*. FILL is the conjunction

$$\bigwedge_{s \in S} start\ x = start\ sx$$

where $S = \{zero_1, zero_2, succ_1, succ_2\} \cup \bigcup_{I_i \in M} \{(i)\}$.

STEP_M is defined as the universal closure of the conjunction of FILL and the following instruction and migration axioms.

Addition and subtraction in the first register: These axioms are obtained from the left hand sides of the corresponding axioms in the previous proof by adding the information on the applied instruction; the instruction migration rules below will bring this information inside the encoding term.

For $I_i = (i, a_1, j) \in M$: $state_i x = state_j succ_1(i)x$.

For $(i, s_1, j, k) \in M$:

$$state_i zero_1 x = state_j zero_1(i)x \wedge state_i succ_1 x = state_k(i)x.$$

Addition in the second register. We introduce an auxiliary monadic function symbol *add* with new *migration rules*

$$add\ succ_1 x = succ_1\ add\ x \ \wedge\ add\ zero_1 x = zero_1\ succ_2 x$$

which carry the addition command from the front to the place where the second register is encoded.

For $I_i = (i, a_2, j) \in M$: $state_i x = state_j add(i) x$.

Subtraction in the second register. We introduce auxiliary monadic function symbols *succ,zero* with new migration rules which carry the information on the content of the second register to the front of the encoding term where the change of the state has to be described.

$$state_i zero\ x = state_j (i) x \wedge state_i succ\ x = state_k (i) x$$

$$zero_1 zero_2 x = zero\ zero_1 x \wedge zero_1 succ_2 x = succ\ zero_1 x$$

$$succ_1 s\ x = s\ succ_1 x \text{ for } s \in \{zero, succ\}.$$

Instruction migration rules.

$$\bigwedge_{I_i \in M} (i) s x = s(i) x \text{ for each } s \in \{zero_1, zero_2, succ_1, succ_2\}.$$

This ends the definition of STEP_M. It remains to prove the reduction properties.

Exercise 4.1.4. Prove the Simulation Lemma by induction on t.

We prove the second reduction property. Assume $M \in H_1'$ and that $\mathfrak{A} \models \forall x(\mathrm{STEP}_M \wedge \mathrm{NONSTOP})$ for some $\mathfrak{A}$. Let $C_0 = (0,0,0) \xrightarrow{i_1 \dots i_t} C_t$ be the given M-computation that terminates in state 1. From the Simulation Lemma we obtain

$$\mathfrak{A} \models \forall x(\underline{C_0}\ start\ x = \underline{C_t}\ \underline{i_1 \dots i_t}\ start\ x)$$

Note that the term on the right side is of the form $state_1 w x$ for some w. By the FILL axioms the following equality is true:

$$\mathfrak{A} \models \forall x(\underline{C_0}\ start\ x = \underline{C_0}\ start\ w x).$$

Therefore $\mathfrak{A} \models \forall x(\underline{C_0}\ start\ wx = state_1 w x)$, contradicting the axiom NONSTOP.

To prove the third reduction property we assume $M \in H_2'$ and build a finite model satisfying $\forall x \mathrm{STEP}_M \wedge \mathrm{NONSTOP}$. Let t be the successor of the length of the given computation and m be the successor of the maximal length of any register word $\overline{(n_1, n_2)}$ occurring during that computation. We interpret the formula over the domain of pairs of register and instructions words limited to length m and t resp., i.e.

$$A = \{zero_1, succ_1, zero_2, succ_2\}^{\leq m} \times \{(0), \dots, (r)\}^{\leq t}.$$

In the following definition of monadic functions over A we make use of two different elements with register words of length m, say $\mathbf{O} := (zero_1^m, \lambda)$ and $\mathbf{I} := (zero_2^m, \lambda)$ where λ denotes the empty string. For the definition we denote by $\underline{n} := succ_1^n zero_1$ the encoding of n as the contents of the first register, by reg any of the register functions $zero_1, succ_1, zero_2, succ_2$, and by $|v|$ the length of a word v.

$$\begin{aligned}
start(v,w) &:= (\lambda,\lambda)\\
reg(v,w) &:= \begin{cases}(reg\, v, w) & \text{if } |v| < m\\ \mathbf{O} & \text{otherwise}\end{cases}\\
state_i(v,w) &:= \begin{cases}\mathbf{I} & \text{if } v := \underline{(n_1,n_2)}\ ,(0,0,0) \overset{w}{\Longrightarrow} (i,n_1,n_2)\\ \mathbf{O} & \text{otherwise}\end{cases}\\
(i)(v,w) &:= \begin{cases}(v,(i)w) & \text{if } |w| < t\\ \mathbf{O} & \text{otherwise}\end{cases}\\
add(v,w) &:= \begin{cases}(\underline{n}succ_2 u, w) & \text{if } v := \underline{n}u,\ |v| < m\\ \mathbf{O} & \text{otherwise}\end{cases}\\
&= succ(v,w)\\
zero(v,w) &:= \begin{cases}(v,w) & \text{if } v := \underline{n}zero_2 u\\ \mathbf{O} & \text{otherwise}\end{cases}
\end{aligned}$$

Exercise 4.1.5. Show that this interpretation indeed satisfies the formula $\forall x(\text{STEP}_M \wedge \text{NONSTOP})$.

□

Exercise 4.1.6. Show by a counterexample that the preceding reduction does not remain conservative if the history part in the encoding is deleted.

Exercise 4.1.7. [536] Apply the proof method of the corollary to Turing instead of register machines.

Theorem 4.1.8. *The class* $[\forall,(0),(2)]_= \cap \text{HERBRAND}$ *is a conservative reduction class. Indeed there is a semi-conservative reduction of* $[\forall,(0),(\omega)]_=$ *to* $[\forall,(0),(2)]_=$ *which preserves the propositional structure.*

Proof. The idea of the proof is as follows. Let an arbitrary $\psi := \forall x\varphi \in [\forall,(0),(\omega)]_=$ be given with monadic function symbols $f_1,\dots,f_n$. Let g,h be two fresh monadic function symbols. We choose $h(g^i(a))$ as a "witness" for $f_i(a)$ where g^i denotes i iterations of g (see Fig. 4.1).

More formally we define $\psi^* := \forall x\varphi^* \in [\forall,(0),(2)]_=$ with

$$\varphi^* := \varphi[x/hx, f_1/hg, \dots, f_n/hg^n].$$

Clearly ψ^* has the same propositional structure as ψ.

Claim. *ψ^* is (finitely) satisfiable if and only if ψ is.*

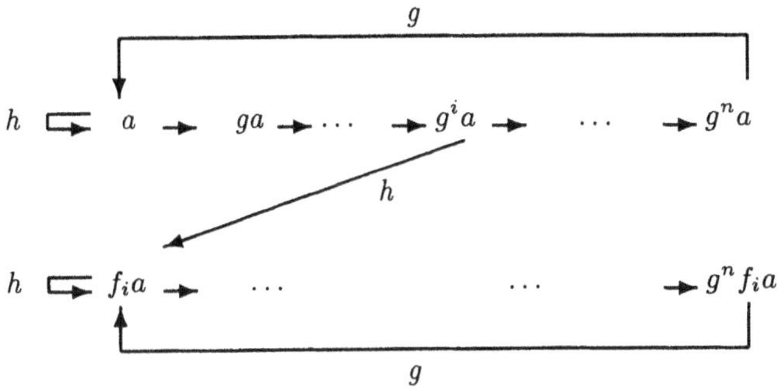

Figure 4.1. Encoding of f_i by hg^i

Clearly if $\mathfrak{B} = (B, g, h) \models \psi^*$ then $\mathfrak{A} = (A, f_1, \dots, f_n)$ with $A = h(B)$ and $f_i(a) := h(g^i(a))$ satisfies ψ.

Assume inversely that $\mathfrak{A} = (A, f_1, \dots, f_n) \models \psi$. Let $\mathfrak{B} = (B, g, h)$ be defined by $B = A \times \{0, \dots, n\}$ with functions g, h whose definition follows the illustration given in Fig. 4.1:

$$\begin{aligned} g(a,i) &:= (a, i+1) \text{ for } i < n, \quad g(a,n) := (a,0) \\ h(a,i) &:= (f_i(a), 0) \text{ for } i > 0, \quad h(a,0) := (a,0). \end{aligned}$$

This definition of hg^i in $\mathfrak{B}$ as behaving like f_i in $\mathfrak{A}$ extends to the interpretation of arbitrary terms and formulae built up from $\{f_1, \dots, f_n\}$. More formally:

Subclaim 1. Let G be the word obtained from $F \in \{f_1, \dots, f_n\}^*$ by replacing each occurrence of f_i by hg^i. For each such F and each $a \in A$ the following is true:

$$G^{\mathfrak{B}}(a, 0) = (F^{\mathfrak{A}}(a), 0).$$

Subclaim 2. For each subformula β of φ let β' be the formula obtained from β by replacing each occurrence of f_i by hg^i. For each such β and each $a \in A$ we have

$$\mathfrak{A} \models \beta[a] \iff \mathfrak{B} \models \beta'[(a, 0)].$$

Exercise 4.1.9. Prove Subclaim 1 by induction on the length of F.

Subclaim 2 follows by an induction on β. Indeed if β is of the form $(F_1 x = F_2 x)$ then β' is $(G_1 x = G_2 x)$ (where G_1, G_2 are obtained from F_1, F_2 as in Subclaim 1), and it holds that $F_1^{\mathfrak{A}}(a) = F_2^{\mathfrak{A}}(a)$ if and only if $G_1^{\mathfrak{B}}(a, 0) = G_2^{\mathfrak{B}}(a, 0)$. The inductive step is trivial.

From Subclaim 2 it follows that $\mathfrak{B} \models \psi^*$. Indeed, since $\varphi^* = \varphi'[x/hx]$ we only have to show that $\mathfrak{B} \models \varphi'[h(a, i)]$ for arbitrary $(a, i) \in B$. By Subclaim 2 we have

$$\mathfrak{B} \models \varphi'[h(a, j)] \Leftrightarrow \mathfrak{B} \models \varphi'[(f_j(a), 0)] \Leftrightarrow \mathfrak{A} \models \varphi[f_j(a)]$$

which is true since $\mathfrak{A} \models \forall x \varphi$. □

Exercise 4.1.10. Show that in Theorem 4.1.8 we can restrict attention to functions that have no fixed points, i.e., the class of all formulae

$$\forall x(gx \neq x \wedge hx \neq x \wedge \varphi) \in [\forall, (0), (2)]_=$$

is a conservative reduction class.

Theorem 4.1.11. *The class* $[\forall, (0), (0, 1)]_= \cap$HERBRAND *is a conservative reduction class. Indeed there is a semi-conservative reduction of* $[\forall, (0), (2)]_=$ *to* $[\forall, (0), (0, 1)]_=$ *which preserves the propositional structure.*

Proof. The idea of the proof is to encode the given two functions into different secondary diagonals of one binary function h, say $f(x)$ into $h(x, x')$ and $g(x)$ into $h(x', x)$. Clearly one has to ensure $x \neq x'$; this can be done for example by setting $x' = h(x, x)$.

Formally let an arbitrary $\psi := \forall x \varphi \in [\forall, (0), (2)]_=$ be given with monadic function symbols f, g. Let h be a binary function symbol and define $\psi^* := \forall x \varphi^* \in [\forall, (0), (2)]_=$ with:

$$f^* x := hxx', \quad g^* x := hx'x, \quad x' := hxx,$$

$$\varphi^* := \varphi[x/f^*x, f/f^*, g/g^*] \wedge \varphi[x/g^*x, f/f^*, g/g^*].$$

Clearly ψ^* has the same propositional structure as ψ.

Claim. ψ^* *is (finitely) satisfiable if and only if* ψ *is.*

If $\mathfrak{B} = (B, h) \models \psi^*$ we can use the above abbreviations for f^*, g^* as defining equations for functions f, g which satisfy ψ over the domain $f(A) \cup g(A)$.

Assume inversely that $\mathfrak{A} = (A, f, g) \models \psi$. Let $\mathfrak{B} = (B, h)$ be defined by $B = A \times \{0, 1, 2\}$ and by the following definition for h which realizes the above indicated encoding of f, g. Let i' be the successor of i (mod 3); we put:

$$\begin{aligned} h((a, i), (a, i)) &:= (a, i') \\ h((a, i), (a, i')) &:= (f(a), 0) \\ h((a, i'), (a, i)) &:= (g(a), 0). \end{aligned}$$

As a consequence of this definition the function $a \mapsto (a, 0)$ represents an isomorphism from $\mathfrak{A}$ to the algebra $(\{(a, 0) : a \in A\}, f^*, g^*)$ where, as indicated above, $f^*(b) = h(b, h(b, b))$ and $g^*(b) = h(h(b, b), b)$, for all $b \in B$ (Note that for each $a \in A$ and each i holds $f^*(a, i) \in \{(c, 0) : c \in A\}$.) Since by assumption $\mathfrak{A} \models \psi$ one obtains that $(\{(a, 0) : a \in A\}, f^*, g^*) \models \psi^*$. □

4.2 Classes with Functions but Without Equality

In this section we prove the two Gurevich classes $[\forall^2, (0,1), (1)]$, $[\forall^2, (1), (0,1)]$ to be conservative reduction classes. We give a conservative reduction of the Kahr-Moore-Wang class $[\forall\exists\forall, (0,\omega)]$ to $[\forall^2, (0,1), (1)]$ and of this class to $[\forall^2, (1), (0,1)]$. In the next chapter we will show that the reduction class property can be strengthened to Krom and Horn formulae for the latter class, but not for the former for which with Krom formulae a second monadic function symbol is needed.

Theorem 4.2.1 (Gurevich). *The class $[\forall^2, (0,1), (1)]$ is a conservative reduction class.*

Proof. The encoding idea for the reduction of $[\forall\exists\forall, (0,\omega)]$-formulae ψ with binary predicate symbols $R_1, \ldots, R_n$ to $[\forall^2, (0,1), (1)]$-formulae φ with only one binary predicate symbols Q is as follows. For each element a of a model $\mathfrak{A}$ which satisfies ψ, n new copies a_i for a model $\mathfrak{B}$ which satisfies φ are provided which "witness" the relation $\mathfrak{A} \models R_i ab$ through $\mathfrak{B} \models Qab_i$. The monadic function symbol $'$ allows us to define a_i as the i-th successor of a, say $a^i = a^{'\cdots'}$ with $a^0 = a$. (See Fig. 4.2):

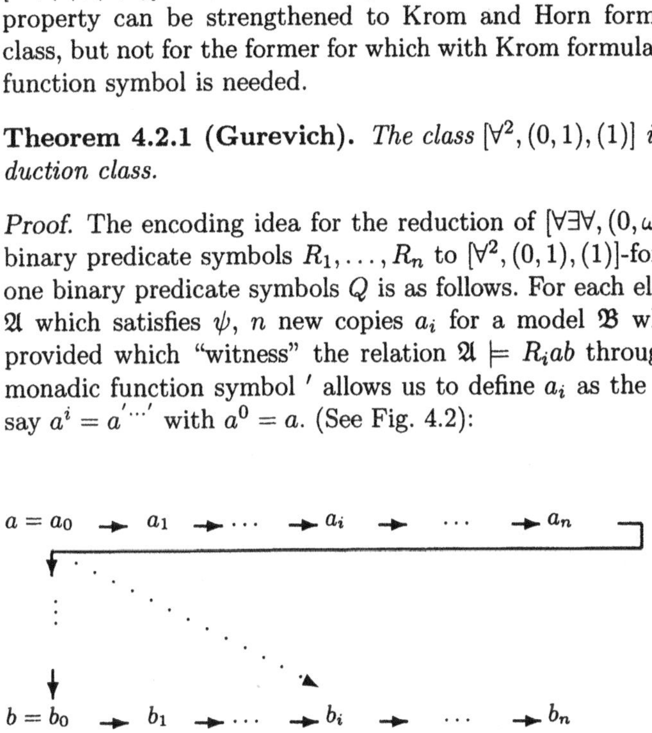

Figure 4.2. Encoding of $P_i ab$ into Qa_0b_i

Let $\forall x \forall y \beta(x, x', y)$ be the Skolem normal form of ψ and define:

$$\varphi := \forall x \forall y \text{ WITNESSES } \wedge \text{ENCODING}[\beta]$$

where

$$\begin{aligned} \text{WITNESSES} &:= \bigvee_{1 \le i \le n} Qx^i x^i \wedge \bigwedge_{i \ne j} \neg(Qx^i x^i \wedge Qx^j x^j) \\ \text{ENCODING}[\beta] &:= Qxx \wedge Qyy \rightarrow \beta^*(x, x^{n+1}, y) \\ &\qquad \text{where } \beta^* := \beta[R_i st / Qst^i]. \end{aligned}$$

Claim. *φ is (finitely) satisfiable if and only if ψ is.*

Assume that $\mathfrak{A} = (A, R_1, \ldots, R_n, ') \models \forall x \forall y \beta$. Let $\mathfrak{B}$ be defined by $B = A \times \{0, 1, \ldots, n\}$ and by the following interpretation for the function and the predicate symbol in φ:

$$(a,i)' := (a, i+1) \text{ for } i < n \qquad (a,n)' := (a', 0)$$
$$\mathfrak{B} \models Q(a,0)(a,0) \text{ for all } a \in A$$
$$\mathfrak{B} \models Q(a,0)(b,i) \text{ iff } \mathfrak{A} \models R_i ab \text{ for } 1 \leq i \leq n.$$

It is easy to check that $\mathfrak{B} \models \forall x \forall y \text{WITNESSES} \wedge \text{ENCODING}[\beta]$. Assume inversely that $\mathfrak{B} = (B, Q, ') \models \varphi$. Since $\mathfrak{B} \models$ WITNESSES the restriction of B to $\{a \in B \mid \mathfrak{B} \models Qaa\}$ is closed with respect to the new successor function defined by $a' := a^{n+1}$. Defining on this restriction $R_i ab$ to be true if and only if $\mathfrak{B} \models Qab^i$ yields a model for ψ. □

Theorem 4.2.2 (Gurevich). *The class* $[\forall^2, (1), (0,1)]$ *is a conservative reduction class. Indeed there is a semi-conservative reduction of* $[\forall^2, (0,1), (1)]$ *to* $[\forall^2, (1), (0,1)]$ *which preserves the propositional structure.*

Proof. The idea of the proof is to encode the binary relation Qxy of a given formula $\psi \in [\forall^2, (0,1), (1)]$ into a monadic predicate P by coding the two arguments x, y into one pair using the new binary function h; the given monadic function f can be represented by the diagonal of h. More formally Qxy will be replaced by $Phxy$ and fx by hxx.

Let $\psi := \forall x \forall y \beta \in [\forall^2, (0,1), (1)]$ be given with monadic function symbol f and binary predicate symbol Q. Let h be a binary function symbol and P be a monadic predicate symbol and

$$\varphi := \psi[fs/hss, Qst/Phst] \in [\forall^2, (1), (0,1)].$$

Claim. *If ψ is finitely satisfiable (unsatisfiable) then φ is finitely satisfiable (unsatisfiable).*

Assume that $\mathfrak{A} = (A, Q, f) \models \forall x \forall y \beta$ with $A = \{a_0, \ldots, a_m\}$. Since ψ is a universal formula we can assume without loss of generality that $a_{i+1} = f(a_i)$ for $i < m$ and that $f(a_m) = a_0$ (see Fig. 4.3).

It is simple to represent each element of this model as an h-pair satisfying $fx = hxx$ by interpreting h correspondingly on A:

$$h(a_i, a_i) = a_{i+1} \text{ for } i < m, \quad h(a_m, a_m) = a_0.$$

In order to satisfy $Qxy \leftrightarrow Phxy$ we need two elements to which one can map any pair (a, b) of elements in the case of truth and falsehood respectively of Qab. Choose $a = a_0$ to represent Pa and choose a new element b equivalent to a but representing $\neg Pa$. Consequently (see Fig. 4.3) we extend $\mathfrak{A}$ by setting $f(b) := f(a_0)$ and

$$Qba_i \text{ iff } Qa_0a_i, \quad Qa_ib \text{ iff } Qa_ia_0, \quad Qbb \text{ iff } Qa_0a_0.$$

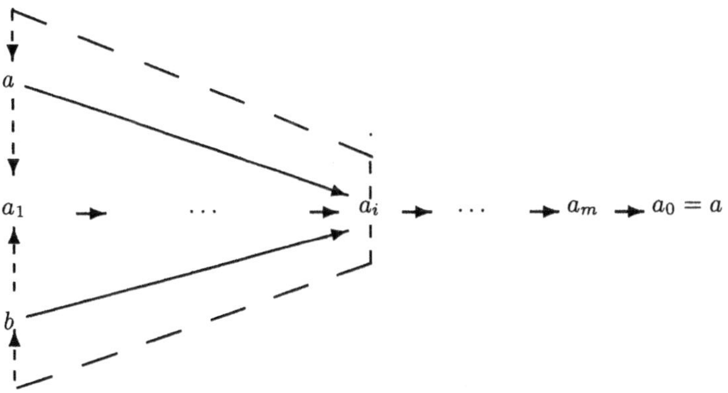

Figure 4.3. Encoding $\neg Pa$ into a new copy b of a

Clearly this extended model $\mathfrak{A}_b$ satisfies ψ and allows us to extend the above definition of h as follows to a model $\mathfrak{B}$ which satisfies the formula $\forall x \forall y(fx = hxx \land (Qxy \leftrightarrow Phxy))$:

$$h(b, b) := a_1$$
$$\mathfrak{B} \models Pa_{i+1} \text{ iff } \mathfrak{A} \models Qa_ia_i$$
$$\mathfrak{B} \models Pa_0 \text{ iff } \mathfrak{A} \models Qa_ma_m$$
$$\mathfrak{B} \models Pb \text{ iff } \mathfrak{A} \models \neg Qa_0a_0.$$

Assuming without loss of generality that $\mathfrak{A} \models Qa_ma_m$ is true we define for arbitrary $c, d \in A$:

$$h(c, d) = \begin{cases} a_0 & \text{if } \mathfrak{A} \models Qcd \\ b & \text{if } \mathfrak{A} \models \neg Qcd. \end{cases}$$

Since this model by definition satisfies $\forall x \forall y(fx = hxx \land (Qxy \leftrightarrow Phxy))$ it follows from $\mathfrak{A} \models \psi$ that $\mathfrak{B} \models \varphi$.

The other direction of the claim is easily established by taking

$$\forall x \forall y(fx = hxx \land (Qxy \leftrightarrow Phxy))$$

as defining equations for the monadic function and the binary predicate to be constructed. □

4.3 Classes with Equality but Without Functions: the Goldfarb Classes

In this section we prove that the Goldfarb class $[\forall^2\exists, (\omega, 1), (0)]_=$ – i.e. the Gödel-Kalmár-Schütte class with equality – is a conservative reduction class.

Theorem 4.3.1 (Goldfarb). *The class $[\forall^2\exists, (\omega, 1), (0)]_=$ is a conservative reduction class.*

Exercise 4.3.2. Show that for prefix classes with equality but without function symbols the only minimal unsolvable ones are the so-called minimal Goldfarb class $[\forall^2\exists]_=$ and the Kahr class $[\forall\exists\forall]$.

The following two exercises give us the two other Goldfarb classes and complete the list of minimal undecidable prefix-vocabulary classes with equality.

Exercise 4.3.3. Show that the class $[\exists^*\forall^2\exists, (0,1), (0)]_=$ is a conservative reduction class. Hint: Translate formulae $\forall x\forall y\exists z\varphi \in [\forall^2\exists, (\omega, 1), (0)]_=$ containing monadic predicate symbols $P_i (1 \leq i \leq m)$ into

$$\exists w_1 \dots \exists w_m \forall x\forall y\exists z(\neg Wz \wedge (\neg Wx \wedge \neg Wy \rightarrow \varphi[P_it/Rw_it]))$$

where $\neg Wx$ expresses that x is not a witness (i.e. different from any w_i).

Exercise 4.3.4. Show that the class $[\forall^2\exists^*, (0,1), (0)]_=$ is a conservative reduction class. Hint: First transform each formula $\forall x\forall y\exists z\varphi \in [\forall^2\exists, (\omega, 1), (0)]_=$ into an equivalent formula with irreflexive relation, e.g. into

$$\varphi' := \forall x\forall y\exists z(\neg Rxx \wedge \varphi[Rst/(Rst \vee (s = t \wedge Ds))])$$

where D is a new monadic predicate representing the diagonal. Then transform φ' containing monadic predicate symbols $P_i (1 \leq i \leq m)$ into

$$\forall x\forall y\exists z\exists w_1 \dots \exists w_m(\neg Wz \wedge \text{WITNESSES} \wedge (\neg Wx \wedge \neg Wy \rightarrow \varphi'[P_it/Rw_it])$$

where $\neg Wx$ expresses that x is not a witness (i.e. none of the w_i) and where WITNESSES stands for

$$Rw_1w_1 \wedge \bigwedge_{1\leq i\leq m} Rw_iw_{i+1} \wedge (Rw_ix \wedge Rw_iy \rightarrow x = y).$$

For expository reasons the proof of Goldfarb's Theorem is given in four steps. First we prove that the class $[\forall^2\exists^*, (\omega, \omega), (0)]_=$ contains an infinity axiom NUM whose models contain infinitely many objects which are related by a successor relation S and thus can serve as numbers for a reduction of arbitrary Kahr-Moore-Wang formulae to formulae in $[\forall^2\exists^*, (\omega, \omega), (0)]_=$. Then we apply standard techniques to improve the reduction in three steps: use of only one existential quantifier (Sect. 4.3.2), elimination of the binary predicates of the given Kahr-Moore-Wang formulae (by encoding them through pairing into monadic ones) and strengthening to a conservative reduction (Sect. 4.3.3), and finally reduction to one binary predicate by encoding the finitely many auxiliary binary predicates in NUM into just one (Sect. 4.3.4).

4.3.1 Formalization of Natural Numbers in $[\forall^2\exists^*, (\omega, \omega), (0)]_=$.

From the proof of the Church-Turing Theorem we know that in the presence of sufficiently long successor chains $\mathbf{0}, \mathbf{1}, \mathbf{2}, \ldots$ with $S\mathbf{n}+\mathbf{1}\mathbf{n}$ one can describe given Turing machine computations by formulae with two universal quantifiers ranging over the occurring time and space parameters respectively (see Sect. 2.1.1). If we can find in the Goldfarb class a description of an injective successor function S with a corresponding predicate Z representing $\mathbf{0}$, then such S-successor chains starting from $\mathbf{0}$ can serve to represent the natural numbers for describing Turing machine computations or for translating arbitrary formulae $\forall x\exists u\forall y\alpha$ of the Kahr-Moore-Wang class into equivalent formulae of $[\forall^2\exists^*, (\omega, \omega), (0)]_=$.

We will define below a satisfiable formula NUM $\in [\forall^2\exists^*, (\omega, \omega), (0)]_=$ and show that it formalizes such a unique zero element $\mathbf{0}$ and an injective successor function S generating numbers as stated in the following lemma.

Lemma 4.3.5 (Number Representation Lemma for NUM**).** *Assume that $\mathfrak{A} \models NUM$. Then Z and S are interpreted in $\mathfrak{A}$ by a unique zero-element $\mathbf{0}$ and an injective function, respectively, which together generate an infinite set $\mathbf{0}, \mathbf{1}, \mathbf{2}, \ldots$ of successive elements, i.e. satisfying $S\mathbf{i}+\mathbf{1}\mathbf{i}$, for all $i \in \mathbb{N}$.*

Corollary 4.3.6. *$[\forall^2\exists^*, (\omega, \omega), (0)]_=$ is a reduction class.*

Proof. We reduce the Kahr-Moore-Wang class $[\forall\exists\forall, (0, \omega), (0)]$ to the class $[\forall^2\exists^*, (\omega, \omega), (0)]_=$. Let $\psi := \forall x\exists u\forall y\beta(x, u, y)$ be an arbitrary formula in $[\forall\exists\forall, (0, \omega)]$. For an encoding of the successor u of x it suffices to express this relation by using the relation S formalized in NUM; therefore define φ as an appropriate prenex normal form of the conjunction of NUM with $\forall x\forall y\exists u(Sux \wedge \beta)$. Clearly $\varphi \in [\forall^2\exists^*, (\omega, \omega), (0)]_=$; it remains therefore to show that φ is satisfiable if and only if ψ is satisfiable.

If ψ is satisfiable, then by Skolem's Theorem it is satisfiable over the domain of natural numbers and therefore over the domain of natural numbers generated by a model of NUM. This implies that φ is satisfiable over such a domain.

If $\mathfrak{A} \models \varphi$, then $\mathfrak{B} \models \beta[\mathbf{m}, \mathbf{m}+\mathbf{1}, \mathbf{n}]$ for all numbers m, n where $\mathfrak{B}$ is the restriction of $\mathfrak{A}$ to the domain of numbers $\mathbf{m}$ generated by NUM in $\mathfrak{A}$. Therefore $\mathfrak{B} \models \forall x\exists u\forall y\beta$. □

Before giving the details of the definition of NUM we explain the underlying intuition. The problem to be solved is to formalize by a formula with only two universal quantifiers that one can extend each given S-successor sequence $c_0, \ldots, c_n$ (i.e. such that $(c_{i+1}, c_i) \in S$) by a new S-successor c_{n+1} of c_n. It would be easy to guarantee the uniqueness of S-successors had we three universal quantifiers allowing us to write $\forall u\forall v\forall w(Suw \wedge Svw \rightarrow u = v)$. The main idea is to use existential quantifiers to express the existence of distinguished successors as follows:

a) require each element to have a predecessor which is not a predecessor of any other element, say by a formula (see below axiom S2):

$$\forall x \forall y (x \neq y \rightarrow \exists z (Sxz \wedge \neg Syz)),$$

b) make the successor relation injective.

Condition b) means that for each S-successor $n+1$ of n and each b, $Sn{+}1b$ has to imply $b = n$. This can be guaranteed by the following construction: each S-successor $n+1$ of n which is also an S-successor of b gives rise to a chain of "Next" elements

$$(0,b) \xrightarrow{N} (1,b) \xrightarrow{N} \ldots \xrightarrow{N} (n,b)$$

such that

- the first components are in the successor relation, i.e. $Si+1i$ for each $i < n$,
- the second components are identical, namely b.

Using projection functions P_i to access the components of the elements of such chains we can guarantee $b = n$ by requiring that the S-successor of b equals the given S-successor $n+1$ of n. To achieve this goal we will write axioms which ensure that in such N-chains the S-successor of the second component is unique[1]. This will be done using auxiliary predicates S_i $(i = 0, 1, 2)$ which describe the successor of the second component and pairs containing such successors.

More formally the auxiliary predicates which we are going to use for the definition of NUM have the following intended interpretation over the natural numbers; for better readability we only indicate where the predicates are interpreted to hold, speaking about pairs of natural numbers in terms of a bijective encoding $\langle\ ,\ \rangle$:

- $P_i\langle m_1, m_2\rangle m_i$ (first and second projection)
- $N\langle m, n\rangle\langle m{+}1, n\rangle$ (Next pair, going to the successor in the first component)
- $S_0\langle m, n\rangle n+1$ (Successor of the second projection)
- $S_1\langle m, n\rangle\langle n+1, r\rangle$ (Successor of the second projection in the first component)
- $S_2\langle m, n\rangle\langle r, n+1\rangle$ (Successor of the second projection in the second component).

We define NUM as the prenex normal form with prefix $\forall x \forall y \exists z_0 \ldots \exists z_r$ of the conjunction of the following axioms **Z0** – **S2.2**.

The group of axioms **Z0** – **S2** formalizes the existence and the uniqueness of a zero element and of successors for each element; note that the uniqueness of **0** is formulated also for its rôle as a possible "component of pairs".

Z0: Zz_0 (existence of **0**).

[1] In Step 3 we will take advantage of this successor relation N between pairs for an encoding of binary predicates into monadic ones using pairing.

Z1: $Zx \wedge Zy \rightarrow x = y$ (uniqueness of **0**).
Z2: $\neg Sz_0x$ (**0** is not a successor, i.e. has no predecessor).
Z3: $\bigwedge_{i=1,2}(P_ixz_0 \wedge P_ixy \rightarrow y = z_0)$ (uniqueness of **0** as component).
S1: $\exists zSzy$ (existence of a successor z for each element y).
S2: $\neg Zx \wedge x \neq y \rightarrow \exists z(Sxz \wedge \neg Syz)$ (uniqueness of successors: each non–zero element has a unique predecessor, i.e. a predecessor which is not a predecessor of any other element).

Note that the uniqueness of successors has to be formulated here using only two universally quantified variables. This is why it is expressed as the distinctness of predecessors for distinct non-zero elements.

Exercise 4.3.7. Construct a finite model for

$$\forall x \forall y \exists z_0 \bigwedge_{0 \leq i \leq 3} \mathbf{Zi} \wedge \mathbf{S1} \wedge \mathbf{S2}.$$

The next group of axioms **N1** – **N3** ensures for each pair $\langle m, n\rangle$ the existence of a next pair (N-successor), i.e. of a pair $\langle m+1, n\rangle$ with an S-successor in the first component, with the same second component and with S-successors for the second component determined by the S_i-successors of the given pair.

N1: $\exists z(Nxz \wedge \bigwedge_{0 \leq i \leq 2}(S_ixy \rightarrow S_izy))$ (existence of an N-successor z for each x; this N-successor has the same S_0, S_1, S_2-successors (read: the same S-successor of the second component) as the given x, see Fig. 4.4)

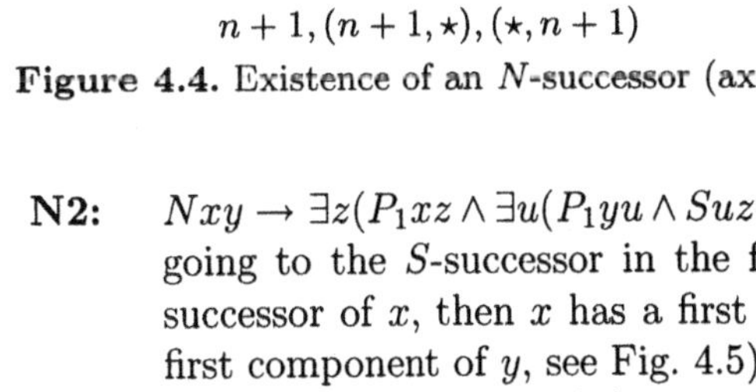

Figure 4.4. Existence of an N-successor (axiom **N1**)

N2: $Nxy \rightarrow \exists z(P_1xz \wedge \exists u(P_1yu \wedge Suz))$ (going to an N–successor means going to the S-successor in the first component, i.e. if y is an N-successor of x, then x has a first component whose successor is the first component of y, see Fig. 4.5)
N3: $Nxy \rightarrow \exists z(P_2xz \wedge P_2yz)$ (N-successors share a common second component with their predecessors, see Fig. 4.5)

Exercise 4.3.8. Construct a finite model for

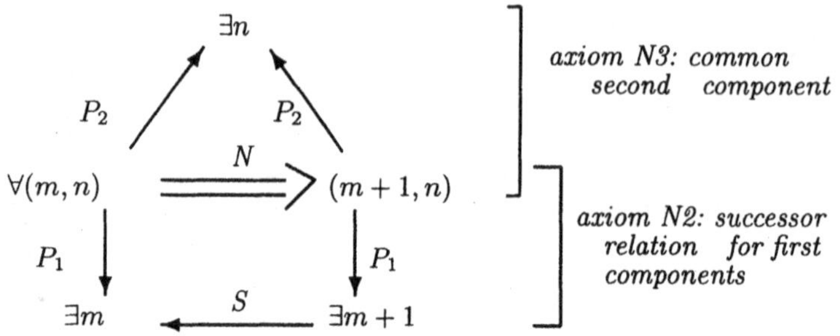

Figure 4.5. Existence of components of N-successors (axioms **N2, N3**)

$$\forall x \forall y \exists z_0 \bigwedge_{0 \le i \le 3} \mathbf{Zi} \wedge \mathbf{S1} \wedge \mathbf{S2} \wedge \bigwedge_{1 \le i \le 3} \mathbf{Ni}.$$

The S_i-axioms guarantee the uniqueness of the successor of the second component of pairs by relating the S-successor of the second component to the S-successor of the first component of the given S_i-predecessor. **Si.1** guarantees the existence of an S_i-predecessor, i.e. of a pair $(0, n)$ for each successor $n+1$ (**S0.1**) and for each positive i-th component $n+1$ (**Si.1** for $i = 1, 2$). These S_i-predecessors will be the starting points $(0, b)$ of N-chains $(0, b), \ldots, (n, b)$ in the proof that $Sn + 1b$ implies $b = n$ as explained above. The uniqueness of the successor of the second component of S_i-predecessors is guaranteed by the axioms **Si.2**. They ensure for each S_i-predecessor the existence of a first component and formalize the uniqueness of the S-successor of its second component in terms of the given S_i-successor.

S0.1: $Sxy \to \exists z(S_0zx \wedge P_1zz_0 \wedge P_2zy)$ (existence of S_0-predecessors for each S-successor: each S-successor $n+1$ has an S_0-predecessor $(0, n)$ having **0** as its first component and the predecessor n of $n+1$ as its second component, see Fig. 4.6)

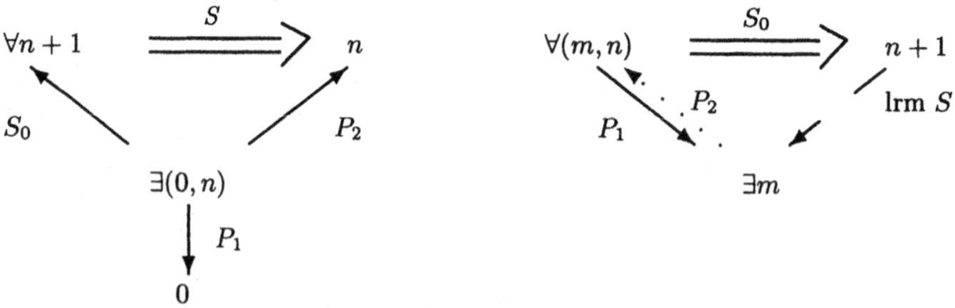

Figure 4.6. Existence of S_0-predecessors and of their first components

Si.1: $(i = 1, 2)$ $P_ixy \wedge \neg Zy \to \exists z(S_izx \wedge P_1zz_0 \wedge \exists u(P_2zu \wedge Syu))$ (existence of S_i-predecessors for positive i-th projection: each element (m_1, m_2)

with non–zero projection m_i has an S_i-predecessor $(0, m_i - 1)$ having 0 as first component and the predecessor of m_i as second component, see Fig. 4.7)

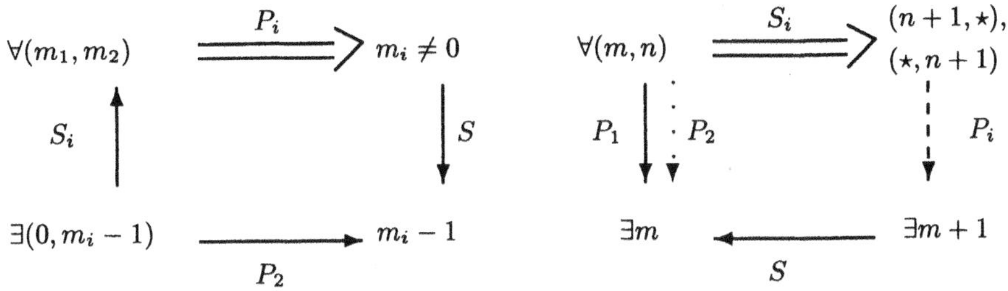

Figure 4.7. Existence of S_i-predecessors and of their first components

S0.2: $S_0xy \to \exists z(P_1xz \wedge (Syz \to P_2xz))$ (uniqueness of the S_0-successor, i.e. of the successor of the second component of an S_0-predecessor: each S_0-predecessor of y has a first component m which is also its second component if it is an S-predecessor of y, see Fig. 4.6)

Si.2: $(i = 1, 2)$ $S_ixy \to \exists z(P_1xz \wedge \exists u(Suz \wedge (P_iyu \to P_2xz)))$ (uniqueness of the successor of the second component of S_i-predecessors: each S_i-predecessor of y has a first component which comes with an S-successor and which is also its second component if it is an S-predecessor of the i-th projection of y, see Fig. 4.7)

This ends the definition of NUM.

Exercise 4.3.9. Check that the above indicated intended interpretation over the natural numbers satisfies NUM with Z interpreted as $\{0\}$ and S as the successor relation. Note that given the bijective pairing function $\langle , \rangle$, each n has a unique representation as a pair $\langle n_1, n_2 \rangle$.

We complete now and prove the Number Representation Lemma including the statement that the generated numbers are also "unique as components".

Lemma 4.3.10 (Number Representation Lemma for NUM (Cont.)). *Assume* $\mathfrak{A} \models$ NUM. *Then the domain* A *of* $\mathfrak{A}$ *contains an infinite subset* $\mathbf{0}, \mathbf{1}, \mathbf{2}, \ldots$ *such that the following properties hold:*

(Existence and Uniqueness of 0) *For all* $a \in A$, $\mathfrak{A} \models Za$ *if and only if* $a = \mathbf{0}$.

(Existence of successors) $\mathfrak{A} \models S\mathbf{n}\mathbf{n-1}$ *for all* $n > 0$.

(Injectivity of S) *For all* n *and all* $a \in A$, $\mathfrak{A} \models S\mathbf{n}a$ *implies* $n > 0$ *and* $a = \mathbf{n-1}$.

(Functionality of S) *For all* $n > 0$ *and all* $a \in A$, $\mathfrak{A} \models Sa\mathbf{n-1}$ *implies* $a = \mathbf{n}$.

(Uniqueness of numbers as component:) *For* $i = 1, 2$, *all* $a, b \in A$ *and all* n, *if* $\mathfrak{A} \models P_i a\mathbf{n}$ *and* $\mathfrak{A} \models P_i ab$, *then* $b = \mathbf{n}$.

Proof. By induction on n we construct a sequence $\mathbf{0}, \mathbf{1}, \ldots, \mathbf{n}$ satisfying the conditions of the lemma.

Induction base. Axioms **Z0**, **Z1** imply that there is a unique element $\mathbf{0} \in A$ such that $\mathfrak{A} \models Z\mathbf{0}$. Since by axiom **Z2** this element is not a successor, the injectivity claim for $n = 0$ is void. Axiom **Z3** guarantees the uniqueness of $\mathbf{0}$ as component.

Induction step. Assume the lemma for distinct elements $\mathbf{0}, \ldots, \mathbf{n}$. We will find now a new element $\mathbf{n+1} \in A$ which satisfies the statements of the lemma.

As a preparatory step we prove a sublemma which guarantees that in N-chains the first components increase – each time one successor step is taken – whereas the second components remain constant (see Fig. 4.8).

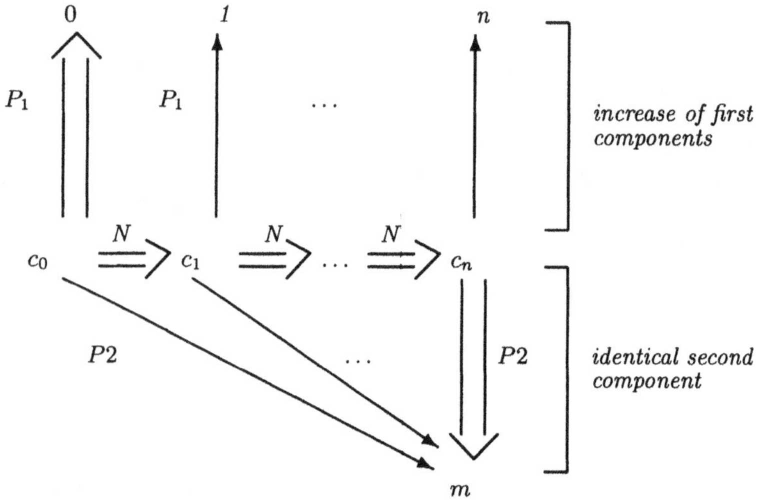

Figure 4.8. N-Chain Lemma

Sublemma 4.3.11 (N-Chain Lemma). *Let* $a, b \in A$ *and suppose* $\mathfrak{A} \models Nab$. *Then, for all* $i \leq n$, $\mathfrak{A} \models P_1 a\mathbf{i-1}$ *implies that* $\mathfrak{A} \models P_1 b\mathbf{i}$ *and* $\mathfrak{A} \models P_2 b\mathbf{i}$ *implies that* $\mathfrak{A} \models P_2 a\mathbf{i}$.

Proof. Let $i \leq n$. Since by assumption $\mathfrak{A} \models Nab$, by axiom **N2** the first component c of a changes in b to a successor d, i.e. there exist $c, d \in A$ with $\mathfrak{A} \models P_1 ac \wedge P_1 bd \wedge Sdc$. Now assume $\mathfrak{A} \models P_1 a\mathbf{i-1}$, i.e. that also $\mathbf{i-1}$ is a first component of a. Then by the uniqueness of numbers as components (induction hypothesis for $\mathbf{i}$) $c = \mathbf{i-1}$, whence $\mathfrak{A} \models Sdc$ implies by the functionality of S (induction hypothesis for $\mathbf{i-1}$) $d = \mathbf{i}$. Hence $\mathfrak{A} \models P_1 b\mathbf{i}$.

Similarly, by axiom **N3**, a and b have the same second component e, i.e. for some $e \in A$ holds $\mathfrak{A} \models P_2ae \wedge P_2be$. Now assume $\mathfrak{A} \models P_2b\mathbf{i}$, i.e. that also $\mathbf{i}$ is a second component of b. Then by the uniqueness of numbers as components (induction hypothesis for $\mathbf{i}$) $e = \mathbf{i}$, so that $\mathfrak{A} \models P_2a\mathbf{i}$. □

The next sublemma establishes the injectivity of S for $\mathbf{n}$, ensuring that no successor of $\mathbf{n}$ is a successor of anything else, as explained in the introduction to the proof.

Sublemma 4.3.12 (Injectivity Lemma for S). *Let $a, b \in A$ and suppose $\mathfrak{A} \models Sa\mathbf{n} \wedge Sab$. Then $b = \mathbf{n}$.*

Proof. Since by assumption $\mathfrak{A} \models Sab$, by axiom **S0.1** there is an S_0-predecessor $c_0 = (0, b)$ of a – i.e. for some $c_0 \in A$ holds

$$\mathfrak{A} \models S_0c_0a \wedge P_1c_0\mathbf{0} \wedge P_2c_0b.$$

Applying the N-successor axiom **N1** n times starting from c_0 yields the existence of an N-successor chain $c_0, \ldots, c_n$ of elements of A with the same S_0-successor a as c_0, i.e. satisfying $\mathfrak{A} \models Nc_ic_{i+1}$ for $0 \leq i < n$ and $\mathfrak{A} \models S_0c_ia$ for $0 \leq i \leq n$ (see Fig. 4.9).

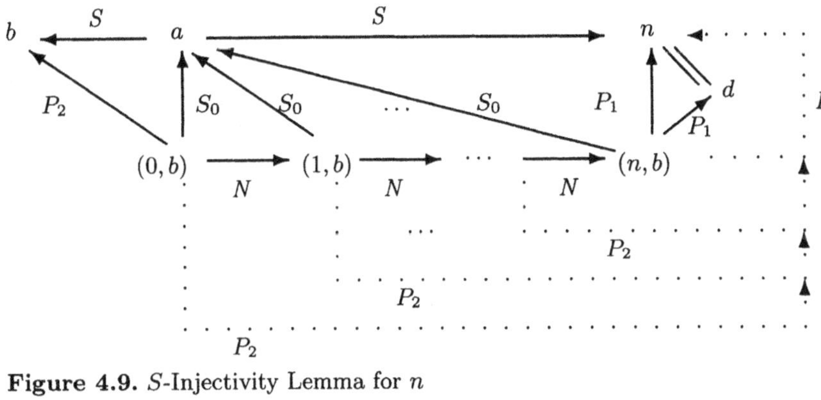

Figure 4.9. S-Injectivity Lemma for n

Since the first component of c_0 is $\mathbf{0}$ – i.e. $\mathfrak{A} \models P_1c_0\mathbf{0}$ – the first statement of the N-Chain Lemma tells us that the first component of c_n is $\mathbf{n}$, i.e. $\mathfrak{A} \models P_1c_n\mathbf{n}$. Since c_n is an S_0-predecessor of a, by the uniqueness of the successor of the second component of S_0-predecessors (axiom **S0.2**), c_n has a first projection $d \in A$ which is also its second projection if it is an S-predecessor of a, i.e. $\mathfrak{A} \models P_1c_nd \wedge (Sad \to P_2c_nd)$. But we know already that $\mathbf{n}$ is a first component of c_n, therefore by the uniqueness of numbers as components (induction hypothesis for n) we can conclude that $d = \mathbf{n}$. Since by assumption $\mathbf{n}$ is a predecessor of a one obtains that $\mathbf{n}$ is a second projection of c_n, i.e. $\mathfrak{A} \models P_2c_n\mathbf{n}$. Now we can apply the second statement of the N-Chain Lemma to the N-chain $c_0, \ldots, c_n$ and infer that c_0 has the same second component as c_n, namely $\mathbf{n}$. Since from above we know that b is a second component of c_0, by the uniqueness of numbers as components (induction hypothesis for n) we can conclude that $b = \mathbf{n}$. □

The Injectivity Lemma for S and the S-axioms imply that $\mathbf{n}$ has a unique successor in $\mathfrak{A}$, as we are going to show in the next sublemma.

Sublemma 4.3.13 (Functionality Lemma for S). *There is a unique S-successor $\mathbf{n+1} \in A$ of $\mathbf{n}$; it is distinct from each $\mathbf{i}$ with $0 \leq i \leq n$.*

Proof. By axiom **S1** there is an S-successor $a \in A$ of $\mathbf{n}$. By the uniqueness property of $\mathbf{0}$ (induction base) and axiom **Z2** it cannot be $\mathbf{0}$. By the injectivity of S (induction hypothesis for $i \leq n$) it cannot be any of $\mathbf{i}$ with $0 \leq i \leq n$.

To show the uniqueness of the $\mathbf{n}$-successor choose any $b \in A - \{a\}$. By axiom **S2** the non-zero element a has a predecessor c which is not a predecessor of the different element b, i.e. there is some $c \in A$ satisfying $\mathfrak{A} \models Sac \wedge \neg Sbc$. The injectivity lemma for S implies $c = \mathbf{n}$. Thus $\mathfrak{A} \models \neg Sb\mathbf{n}$. □

Clearly the preceding lemmas imply the statements of the Number Representation Lemma about the uniqueness of $\mathbf{0}$ and about the properties of S for $\mathbf{n+1}$. It remains to show the uniqueness of $\mathbf{n+1}$ as component.

Let $i \in \{1,2\}, a, b \in A$ and assume $\mathfrak{A} \models P_i a\mathbf{n+1} \wedge P_i ab$. We have to show that $b = \mathbf{n+1}$.

By the uniqueness of $\mathbf{0}$ and its uniqueness as component (induction hypothesis) we know that b is not the zero element, i.e. $\mathfrak{A} \models \neg Zb$. Hence a has a positive projection, namely b, and therefore by axiom **Si.1** has an S_i-predecessor $c_0 = (0, d)$ whose second component d is an S-predecessor of b, i.e. $\mathfrak{A} \models S_i c_0 a \wedge P_1 c_0 \mathbf{0} \wedge P_2 c_0, d \wedge Sbd$ for some $c_0, d \in A$. n successive applications of the N-successor axiom **N1** yield the existence of an N-successor chain $c_0, \ldots, c_n$ of elements of A with the same S_i-successor a as c_0, i.e. satisfying $\mathfrak{A} \models Nc_j c_{j+1}$ for $0 \leq j < n$ and $\mathfrak{A} \models S_i c_j a$ for $0 \leq j \leq n$ (see Fig. 4.10).

Since c_0 has first component $\mathbf{0}$ – i.e. $\mathfrak{A} \models P_1 c_0 \mathbf{0}$ – the first statement of the N-Chain Lemma tells us that c_n has first component $\mathbf{n}$. Since c_n is an S_i-predecessor of a – i.e. $\mathfrak{A} \models S_i c_n a$ – axiom **Si.2** guarantees that it has a first component $e \in A$ with a successor $e' \in A$ and that it is also its second component if the S-successor e' of e is the i-th projection of a, i.e. $\mathfrak{A} \models P_1 c_n e \wedge Se'e \wedge (P_i ae' \rightarrow P_2 c_n e)$. From the uniqueness of numbers as components (induction hypothesis for $\mathbf{n}$) and $\mathfrak{A} \models P_1 c_n \mathbf{n}$ we may infer $e = \mathbf{n}$ and thus by the Functionality Lemma for S, $e' = \mathbf{n+1}$. From the assumption $\mathfrak{A} \models P_i a\mathbf{n+1}$ it follows that $\mathfrak{A} \models P_i ae'$. Then the second clause derived above from axiom **Si.2** yields that the second component of c_n is $\mathbf{n}$, i.e. $\mathfrak{A} \models P_2 c_n \mathbf{n}$. By the N-Chain Lemma c_0 has the same second component $\mathbf{n}$ as c_n. By the uniqueness of numbers as second components (induction hypothesis for $\mathbf{n}$) we obtain from $\mathfrak{A} \models P_2 c_0 d$ (see above) that $d = \mathbf{n}$. Since b

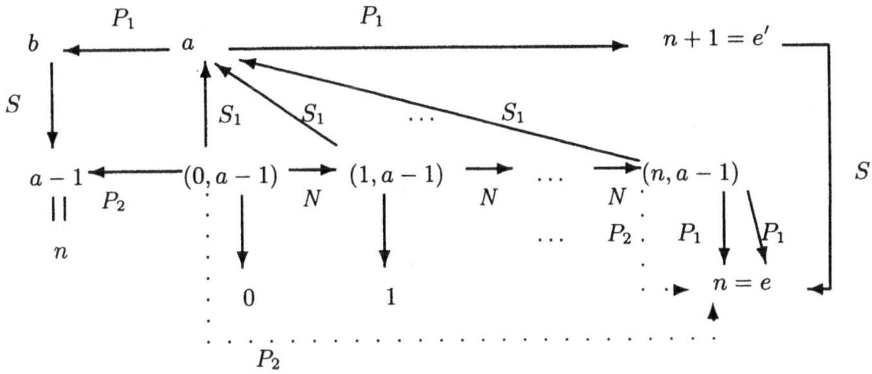

Figure 4.10. Uniqueness of $n+1$ as first component

is a successor of d (see above) the Functionality Lemma for S yields $b = \mathbf{n+1}$. □

4.3.2 Using Only One Existential Quantifier.

We refine here the formalization of natural numbers of the preceding section so as to use only one existential quantifier $\exists z$. This allows us to refine also the reduction of the Kahr-Moore-Wang class, thus obtaining the reduction class property for $[\forall^2\exists, (\omega, \omega), (0)]_=$.

An inspection of the formula NUM of the preceding section shows that nested existential quantifiers appear in two ways: a) z_0 occurs in some formulae which are in the range of a quantifier $\exists z$, b) some formulae $\exists u\beta$ occur within the range of a quantifier $\exists z$. Critical nestings of type a) are used only to express that $\mathbf{0}$ is a projection, i.e. in atomic formulae P_itz_0; critical nestings of type b) are used only to express that u is a projection and successor or predecessor of something, i.e. in formulae β of the form $P_isu \wedge Sut$ or $P_isu \wedge Stu$ or similar. Both kinds of nestings can be eliminated by the technique of formalizing auxiliary predicates $P_{i,0}$ $(i = 1, 2)$ – to encode the fixed parameter z_0 – and $P_{i,+}$ $(i = 1, 2)$ to encode one application of the successor function, i.e. such that P_itz_0 is equivalent to $P_{i,0}t$ and $P_isu \wedge Sut$ is equivalent to $P_{i,+}st$. Stated otherwise these predicates have the following intended interpretation:

- $P_{i,0}a$ if and only if $P_ia\mathbf{0}$ (0-projection),
- $P_{i,+}a\mathbf{n}$ if and only if $P_ia\mathbf{n+1}$ (successor projection).

Since the different uses of local existential quantification in NUM have now all to be handled by just one global existential quantifier which appears in the prefix $\forall x\forall y\exists z$, these uses have to be encoded into disjoint subsets of the set of pairs a, b of the given domain. For this purpose we will make the pairing function $\langle\ ,\ \rangle$ and the related component functions explicit and extend the domain of the intended interpretation by adding to the non-negative integers the set of (number) pairs. (This will also be useful in the next section where we use pairing to encode binary predicates into monadic ones.) An auxiliary (monadic) predicate I will distinguish non-negative integers from pairs of non-negative integers, another auxiliary (monadic) predicate D will represent the diagonal of the set of all pairs, i.e. these predicates have the following intended interpretation:

- Ia iff $a = n$ for some non-negative integer (natural number) n,
- Da iff $a = (n, n)$ for some natural number n.

We are now ready to define NUM' as a formula with prefix $\forall x \forall y \exists z$ whose quantifier-free part is the conjunction of the following axioms **Z0'** – **S2.2'**, **Pi.0** – **P1**. The axioms **Ax'** are obtained from the axioms **Ax** of NUM by refining the description along the lines explained above; the axioms **Pi.0** – **P1** formalize the new auxiliary predicates. For the ease of the reader we proceed again through all the conjuncts of NUM.

The axioms **Z1'** – **Z3'** for the unique zero element are obtained by reformulating their counterparts **Z1** – **Z3** in NUM as follows.

Axiom **Z1** is taken unchanged:

Z1': $Zx \wedge Zy \rightarrow x = y$ (uniqueness of the zero element).

Axiom **Z2** of NUM expresses that the zero element is not a successor. This can be reformulated – without using the existentially quantified variable for naming the zero element – by saying that no successor is a zero element. We include into **Z2'** also the condition that S ranges only over numbers.

Z2': $Sxy \rightarrow \neg Zx \wedge Ix \wedge Iy$ (no successor is zero, S holds only between numbers).

Axiom **Z3** of NUM expresses the uniqueness of the zero element as component. To avoid the use of the existentially quantified variable in $P_i x z_0$ we use the new auxiliary predicate $P_{i,0}$:

Z3': $\bigwedge_{i=1,2}(P_{i,0}x \wedge P_i xy \rightarrow Zy)$ (uniqueness of the zero element as component).

Axiom **Z0** of NUM states the existence of a zero element z_0. This existential requirement can now only be formulated using z instead of z_0 and therefore has to be kept separate from requiring – using the same z – the existence of a successor z for each x (**S1**), the existence of a unique predecessor z for each non-zero element (**S2**), etc. In other words, we have to distinguish different cases for the definition of a binary function h – the Skolem function of NUM' – which yields for each pair (a, b) some c satisfying a specific requirement. Different requirements will be encoded into different sets of pairs of the given domain.

For the encoding of the zero element we choose the non-zero part of the diagonal, i.e. the set of pairs (a, b) satisfying $\mathbf{0} \neq a = b$:

Z0': $\neg Zx \wedge x = y \rightarrow Zz$ (existence of a zero element).

To encode the **S1**-requirement of a successor $b+1$ for every given number b we choose the first column of the Gaussian quadrant, i.e. the set of pairs (a, b) satisfying $a = \mathbf{0}, b = \mathbf{n}$ for some non-negative integer n:

S1': $Zx \wedge Iy \rightarrow Szy$ (existence of a successor z for each number y).

The **S2**-requirement in NUM that each non-zero number x has a predecessor which is not a predecessor of any other element y splits now into two cases, depending on whether y is a number or a pair. In the first case we can clearly restrict attention to pairs (a, b) of numbers where $a \neq b+1$; the second case is encoded into the set of pairs (a, b) where a is a positive component of the pair b. We thus have the following two axioms refining **S2**:

S2': $\neg Zx \wedge x \neq y \wedge Ix \wedge Iy \wedge \neg Sxy \rightarrow Sxz \wedge \neg Syz$ (existence of unique predecessors: each non-zero number has a predecessor which is not the predecessor of any other number),

S3': $(i = 1, 2)$ $\neg Zx \wedge P_i yx \rightarrow Sxz \wedge P_{i,+} yz$ (existence of predecessors of positive components: each non-zero component has a predecessor).

Note that **S3'** also formalizes $P_{i,+}$ by translating the successor relation between x and z and the projection relation $P_i yx$ into the successor projection relation $P_{i,+}$ between y and z.

Axiom **N1** of NUM requires for each element x the existence of an N-successor which has the same S_i-successors y as x (for $0 \leq i \leq 2$). Clearly we need this requirement only for pairs x and can even restrict it to pairs whose components are not identical. In axiom **N2** of NUM we have only to reformulate the projection of a successor in terms of the auxiliary relation $P_{1,+}$, in **N3** we have only to replace the locally quantified z by the globally quantified z. Thus we have the following new axioms[2] **N1'** – **N3'**:

N1': $\bigwedge_{0 \leq i \leq 2}(S_i xy \wedge \neg Dx \rightarrow Nxz \wedge S_i zy)$ (for each x with an S_0, S_1, S_2-successor there exists an N-successor z with the same S_0, S_1, S_2-successor (read: the same S-successor of the second component), see Fig. 4.4).

N2': $Nyx \rightarrow P_1 yz \wedge P_{1,+} xz$ (going to an N-successor means going to the S-successor in the first component, i.e. if x is N-successor of y, then y has a first component whose successor is the first component of x, see Fig. 4.5).

N3': $Nxy \rightarrow P_2 xz \wedge P_2 yz$ (N-successors share a common second component with their predecessors, see Fig. 4.5).

[2] Comparing **N2** with **N2'** the reader will notice that we have interchanged the rôle of x and y in axiom **N2'**. The reason is that we want to handle this case as a subcase of a group of similar cases in the definition of a Skolem function for NUM' below.

The **Si.j**-axioms of NUM can easily be reformulated using $P_{i,0}, P_{i,+}$ to eliminate the occurrences of z_0 and of nested existential quantification over predecessors[3]:

S0.1': $Sxy \rightarrow S_0zx \wedge P_{1,0}z \wedge P_2zy$ (existence of S_0-predecessors for each S-successor: each S-successor $n+1$ has an S_0-predecessor $(0,n)$ having **0** as its first component and the predecessor n of $n+1$ as its second component, see Fig. 4.6).

Si.1': $(i = 1,2)$ $P_{i,+}xy \rightarrow S_izx \wedge P_{1,0}z \wedge P_2zy$ (existence of S_i-predecessors for a positive i-th projection: each element (m_1, m_2) with non-zero projection m_i has an S_i-predecessor $(0, m_i - 1)$ having 0 as its first component and the predecessor of m_i as its second component, see Fig. 4.7).

S0.2': $S_0yx \rightarrow P_1yz \wedge (Sxz \rightarrow P_2yz)$ (uniqueness of the S_0-successor, i.e. of the successor of the second projection of an S_0-predecessor: each S_0-predecessor of x has a first component m which is also its second component if it is S-predecessor of x, see Fig. 4.6).

Si.2': $(i = 1,2) S_iyx \rightarrow P_1yz \wedge (P_{i,+}xz \rightarrow P_2yz)$ (uniqueness of the successor of the second component of S_i-predecessors: each S_i-predecessor of x has a first component which comes with an S-successor and which is also its second component if it is S-predecessor of the i-th projection of x, see Fig. 4.7).

In addition to the reformulation of the axioms from NUM we have in NUM' the following new axioms which formalize the auxiliary predicates:

Pi.0: $Zx \rightarrow (P_{i,0}y \leftrightarrow P_iyx) \wedge Ix$

P1.+ $P_{1,+}xy \rightarrow \neg P_{1,0}x \wedge \neg Ix$ (if the first component of an element is positive, then it cannot be zero and this element is not a number).

Diag: $Dx \rightarrow (P_1xy \leftrightarrow P_2xy)$ (for diagonal elements the first and the second projection coincide)

P1: $Zx \wedge \neg Iy \rightarrow P_1yz$(each pair has a first component).

Note that the condition on the existence of a first component for each pair is encoded into the domain of pairs $(0, b)$ with pairs b.

This ends the definition of NUM'.

It remains to prove that the intended interpretation satisfies NUM' and that the extended version of the Number Representation Lemma holds for NUM'.

Lemma 4.3.14. *The above indicated intended interpretation satisfies NUM' where the interpretation of S_i is restricted to first arguments (m, n) which satisfy the condition $m \leq n$.*

[3] Comparing **Si.2** with **Si.2'** the reader will notice that we have interchanged the rôle of x and y in axioms **S0.2'** and **Si.2'**. The reason is that we want to handle these cases as subcase of a group of similar cases in the definition of a Skolem function for NUM' below.

Proof. We define a binary function $h : (\mathbb{N} \cup (\mathbb{N} \times \mathbb{N}))^2 \to \mathbb{N} \cup (\mathbb{N} \times \mathbb{N})$ which will serve as the interpretation of a Skolem function for NUM'. We indicate in brackets to which axioms the different cases belong. For notational convenience we denote by π_i the i-th projection function for pairs and write Rab meaning that the above defined intended interpretation of R holds between a and b.

$$h(a,b) := \begin{cases} 0 & \text{if } 0 \neq a = b \text{ (case } \mathbf{Z0'}) \\ b+1 & \text{if } a = 0 \text{ and } b \in \mathbb{N} \text{ (case } \mathbf{S1'}) \\ a-1 & \text{if } 0 \neq a \neq b, a, b \in \mathbb{N} \text{ and } a \neq b+1 \text{ (case } \mathbf{S2'}) \\ a-1 & \text{if } 0 \neq a \text{ and } \bigvee_{i=1,2} P_i ba \text{ (case } \mathbf{S3'}) \\ \pi_2 a & \text{if } Nab \text{ (case } \mathbf{N3'}) \\ \pi_1 b & \text{if } Nba \vee \bigvee_{i=0,1,2} S_i ba \vee (a = 0 \wedge b \notin \mathbb{N}) \\ & \quad \text{(cases } \mathbf{(N2'),(Si.2')} \ (i = 0,1,2), \mathbf{(P1)}) \\ (0,b) & \text{if } Sab \vee \bigvee_{i=1,2} P_{i,+} ab \text{ (cases } \mathbf{Si.1'}, i = 0,1,2) \\ (\pi_1 a + 1, \pi_2 a) & \text{if } \neg Da \wedge \bigvee_{i=0,1,2} S_i ab \text{ (case } \mathbf{N1'}) \\ \text{arbitrary} & \text{otherwise.} \end{cases}$$

Exercise 4.3.15. Check that this function h is well-defined (that is, the cases in the definition do not conflict). Check that all the axioms where z appears are covered correctly. Check that the Skolem normal form of NUM' is satisfied by the above defined intended interpretation with h as the interpretation of the Skolem function for the existential variable z of NUM'.

□

The Number Representation Lemma for NUM' asserts all the properties of the Number Representation Lemma for NUM and an additional property which verifies the intended interpretation of the predicate $P_{1,+}$.

Lemma 4.3.16 (Number Representation Lemma for NUM'). *If* $\mathfrak{A} \models$ NUM', *then the domain* A *of* $\mathfrak{A}$ *contains an infinite subset* $\mathbf{0}, \mathbf{1}, \mathbf{2}, \ldots$ *such that the following properties hold: existence and uniqueness of* $\mathbf{0}$, *existence of successors, injectivity of* S, *functionality of* S, *uniqueness of numbers as component and the following correctness property for* $P_{1,+}$. *For all* $a \in A$, $n > 0$

$$\mathfrak{A} \models P_{1,+} a \mathbf{n-1} \ \textit{implies} \ P_1 a \mathbf{n}.$$

Proof. The proof follows the lines of the proof for the Number Representation Lemma for NUM in the preceding section. We concentrate our attention on the points which do change.

Induction base. The axioms **Z0',Z1'** guarantee the existence of a unique element $\mathbf{0} \in A$ such that $\mathfrak{A} \models Z\mathbf{0}$. By **Z2'** it is not a successor (i.e. $\mathfrak{A} \models \neg S\mathbf{0}a$ for each $a \in A$), therefore the injectivity claim for $n = 0$ is void. The uniqueness of $\mathbf{0}$ as component follows from axiom **Z3'** together with the new axioms **Pi.0** which guarantee that $P_i a\mathbf{0}$ is equivalent in $\mathfrak{A}$ to $P_{i,0} a\mathbf{0}$.

Induction step (from n to $n+1$). We paraphrase the corresponding steps in the proof from NUM.

(N-Chain Lemma). Let $i \leq n$. Since by assumption $\mathfrak{A} \models Nab$, by axiom **N2'** there exists in A a first component c of a whose successor is first component of b, i.e. such that $\mathfrak{A} \models P_1ac \wedge P_{1,+}bc$. Now assume that also $\mathbf{i-1}$ is a first component of a. Then by the uniqueness of numbers as components (induction hypothesis for $\mathbf{i}$), $c = \mathbf{i-1}$, so that $\mathfrak{A} \models P_{1,+}b\mathbf{i-1}$ and therefore $\mathfrak{A} \models P_1b\mathbf{i}$ by the correctness property for $P_{1,+}$ (induction hypothesis for $i > 0$).

The proof for the second statement of the lemma is literally the same as with NUM in the preceding section, invoking **N3'** instead of **N3**.

(Injectivity Lemma for S). Since by assumption $\mathfrak{A} \models Sab$, by axiom **S0.1'** there is an S_0-predecessor $c_0 = (0, b)$ of a – i.e. for some $c_0 \in A$ holds $\mathfrak{A} \models S_0c_0a \wedge P_{1,0}c_0 \wedge P_2c_0b$. The **P1.0**-axiom implies $\mathfrak{A} \models P_1c_0\mathbf{0}$. Iterated application of the N-successor axiom **N1'** starting from c_0 yields the existence of an N-successor chain $c_1, \ldots, c_j$ of elements of A with the same S_0-successor a as c_0, i.e. satisfying $\mathfrak{A} \models Nc_ic_{i+1}$ for $0 \leq i < j$ and $\mathfrak{A} \models S_0c_ia$ for $0 \leq i \leq j$. The complication with **N1'** instead of **N1** is that we have to ensure also $\mathfrak{A} \models \neg Dc_i$ for each element in this chain and therefore have to ensure that $j = n$. We show this indirectly by deriving a contradiction from the assumption that $j < n \wedge \mathfrak{A} \models Dc_j$.

From the first statement of the N-Chain Lemma and $\mathfrak{A} \models P_1c_0\mathbf{0}$ it follows that $\mathfrak{A} \models P_1c_j\mathbf{j}$. Then $\mathfrak{A} \models Dc_j$ and the D-axiom imply $\mathfrak{A} \models P_2c_j\mathbf{j}$, so that by the second statement of the N-Chain Lemma $\mathfrak{A} \models P_2c_0\mathbf{j}$. But from above we know that P_2c_0b, therefore by the uniqueness of numbers as components (induction hypothesis for j) $b = \mathbf{j}$. The assumption $\mathfrak{A} \models Sab$ and the functionality of S (induction hypothesis for j) imply $a = \mathbf{j+1}$ in contradiction to $\mathfrak{A} \models S a\mathbf{n}$, the assumption $j < n$ and the induction hypothesis.

Therefore $j = n$ and the N-chain has length n. The rest of the proof is literally the same as with NUM in the preceding section, invoking **S0.2'** instead of **S0.2**.

(Functionality Lemma for S). By axiom **S1'** there is an S-successor $a \in A$ of $\mathbf{n}$. We prove by contradiction that such an a is unique.

Suppose $\mathfrak{A} \models Sb\mathbf{n}$ for some $b \in A - \{a\}$. In order to be able to apply axiom **S2'** we have to ensure that the premises hold in $\mathfrak{A}$. By the Injectivity Lemma for S and the induction hypothesis (injectivity of S for n) we obtain $\mathfrak{A} \models \neg Sab$. But $\mathfrak{A} \models Sa\mathbf{n}$ implies $\mathfrak{A} \models \neg Za \wedge Ia$ (axiom **Z2'**) and $\mathfrak{A} \models Sb\mathbf{n}$ implies $\mathfrak{A} \models Ib$. Therefore the premises of axiom **S2'** hold in $\mathfrak{A}$ and the proof proceeds as from NUM: **S2'** guarantees the existence of a predecessor c of a which is not a predecessor of b, i.e. of a $c \in A$ satisfying $\mathfrak{A} \models Sac \wedge \neg Sbc$. The Injectivity Lemma for S and $\mathfrak{A} \models Sa\mathbf{n}$ imply $c = \mathbf{n}$. Thus $\mathfrak{A} \models \neg Sb\mathbf{n}$, contrary to our hypothesis.

(Uniqueness of n as component in $P_{i,+}$). Let $i \in \{1,2\}$ and suppose $\mathfrak{A} \models P_{i,+}a\mathbf{n} \wedge P_{i,+}ab$ for some $a, b \in A$. We have to show that $b = \mathbf{n}$.

Axiom **Si.1'** guarantees the existence of an S_i-predecessor $c_0 \in A$ of a with first component **0** and second component b, i.e. satisfying $\mathfrak{A} \models S_i c_0 a \wedge P_{1,0} c_0 \wedge P_2 c_0 b$ and therefore (by axiom **Pi.0**) also $\mathfrak{A} \models P_1 c_0 \mathbf{0}$. As in the proof for the Injectivity Lemma for S we apply axiom **N1'** in order to get an N-successor chain $c_1, \ldots, c_j$ of elements of A with the same S_i-successor a as c_0, i.e. satisfying $\mathfrak{A} \models N c_k c_{k+1}$ for $0 \leq k < j$ and $\mathfrak{A} \models S_i c_k a$ for $0 \leq k \leq j$; as there we show $j = n$ by deriving a contradiction from the assumption $j < n \wedge \mathfrak{A} \models Dc_j$. Then the first statement of the N-Chain Lemma ensures that c_n has first projection $\mathbf{n}$, i.e. $\mathfrak{A} \models P_1 c_n \mathbf{n}$. Since c_n has the S_i-successor a, axiom **Si.2'** implies the existence of a $c \in A$ satisfying $\mathfrak{A} \models P_1 c_n c \wedge (P_{i,+} ac \rightarrow P_2 c_n c)$. By the uniqueness of numbers as first components (induction hypothesis for n) we obtain $c = \mathbf{n}$. Since $\mathfrak{A} \models P_{i,+} a\mathbf{n}$ holds by assumption we can therefore conclude that $\mathfrak{A} \models P_2 c_n \mathbf{n}$. The second statement of the N-Chain Lemma implies that $\mathfrak{A} \models P_2 c_0 \mathbf{n}$. Since $\mathfrak{A} \models P_2 c_0 b$, the uniqueness of numbers as second components (induction hypothesis for n) implies $b = \mathbf{n}$.

(Uniqueness of $\mathbf{n+1}$ as component). Let $i \in \{1,2\}, a, b \in A$ and assume $\mathfrak{A} \models P_i a \mathbf{n+1} \wedge P_i ab$. We have to show that $b = \mathbf{n+1}$.

Since $\mathfrak{A} \models \neg Z\mathbf{n+1}$ (by axiom **Z2'**), by the **Z3'**-axiom on the uniqueness of **0** as component and the assumption $\mathfrak{A} \models P_i a\mathbf{n+1}$ we obtain that $\mathfrak{A} \models \neg P_{i,0} a$. Therefore by axiom **Pi.0** we obtain from the assumption $\mathfrak{A} \models P_i ab$ that $\mathfrak{A} \models \neg Zb$. By axiom **S3'**, $\mathfrak{A} \models P_i ab \wedge \neg Zb$ implies the existence of a predecessor d of b satisfying $\mathfrak{A} \models Sbd \wedge P_{i,+} ad$; similarly there is a predecessor c of $\mathbf{n+1}$ satisfying $\mathfrak{A} \models S\mathbf{n+1}c \wedge P_{i,+} ac$. By the S-injectivity for $\mathbf{n+1}$ follows $c = \mathbf{n}$. By the uniqueness of numbers as components in $P_{i,+}$ we then obtain $d = \mathbf{n}$ from $\mathfrak{A} \models P_{i,+} ac \wedge P_{i,+} ad$. Since $\mathfrak{A} \models Sbd$, the functionality of S for $\mathbf{n+1}$ implies $b = \mathbf{n+1}$.

($P_{1,+}$-Correctness Property for $\mathbf{n+1}$). Let $\mathfrak{A} \models P_{1,+} a\mathbf{n}]$. By axiom **P1.+**, $\mathfrak{A} \models \neg P_{1,0} a \wedge \neg Ia$. By axiom **P1** there is a first component $b \in A$ of a. By axiom **P1.0** and $\mathfrak{A} \models \neg P_{1,0} a$ this first component b of a cannot be the zero element, i.e. $\mathfrak{A} \models \neg Zb$. Hence by axiom **S3'** the positive first projection b of a has a predecessor c which satisfies $\mathfrak{A} \models Sbc \wedge P_{1,+} ac$. The assumption $\mathfrak{A} \models P_{1,+} a\mathbf{n}$ implies $c = \mathbf{n}$ (by uniqueness of $\mathbf{n}$ as component in $P_{i,+}$). Hence $\mathfrak{A} \models Sbc$ implies $b = \mathbf{n+1}$ (by the functionality for $\mathbf{n+1}$) so that $\mathfrak{A} \models P_1 a\mathbf{n+1}$. □

Corollary 4.3.17. *The class* $[\forall^2\exists, (\omega, \omega), (0)]_=$ *is a reduction class.*

Proof. We refine the reduction of the Kahr-Moore-Wang class of the preceding section using NUM' instead of NUM. Let $\psi := \forall x \exists u \forall y \beta(x, u, y)$ be an arbitrary formula in $[\forall\exists\forall, (0, \omega)]$. For an encoding of the successor u of x in the reduction formula φ we have to rely upon the effect of the successor axiom **S2'** in NUM'. **S2'** ensures for each pair (m, n) with $m \neq 0$ the existence of a predecessor $m - 1$ in case $m \neq n, n+1$ (see also case **S2'** in the definition of the Skolem function h above); hence for values for x and y which are restricted by the premises of **S2'** β can be encoded by $\beta(z, x, y)$.

To this we have to add an encoding of β for the special cases $m = n$ and $m+1 = n$ by $\beta(x, y, x)$ and $\beta(x, y, y)$. This explains the definition of φ as a formula with prefix $\forall x \forall y \exists z$ whose quantifier-free part is the conjunction of the quantifier-free part of NUM' and of the following two instantiations β_1, β_2 of β:

$$\begin{aligned} \beta_1 &:= \neg Zx \wedge x \neq y \wedge Ix \wedge Iy \wedge \neg Sxy \rightarrow \beta(z, x, y) \\ \beta_2 &:= Syx \rightarrow \beta(x, y, y) \wedge \beta(x, y, x). \end{aligned}$$

Claim. *φ is satisfiable if and only if ψ is satisfiable.*

If ψ is satisfiable, then, by Skolem's Theorem, its Skolem normal form $\forall x \forall y \beta(x, x+1, y)$ is satisfiable over the domain of natural numbers and therefore over the domain of natural numbers generated by a model of NUM'. Since by the Number Representation Lemma Sab becomes true in such a model $\mathfrak{A}$ if and only if a and b are numbers satisfying $a = b+1$, $\forall x \forall y \beta_2$ is true in $\mathfrak{A}$. If the antecedent of β_1 is true for a given pair (a, b) of values for x and y, then the antecedent of axiom **S2'** is satisfied and the conclusion of **S2'** establishes that z is interpreted by the well defined predecessor $a - 1$ of a. Hence the consequent of β_1 is true in $\mathfrak{A}$. Thus φ is satisfied by $\mathfrak{A}$.

Assume $\mathfrak{A} \models \varphi$ and let $\mathfrak{A}'$ be the restriction of $\mathfrak{A}$ to the domain of numbers $\mathbf{m}$ generated by NUM' in $\mathfrak{A}$. We show that $\mathfrak{A}' \models \beta[\mathbf{m}, \mathbf{m+1}, \mathbf{n}]$ for all numbers $m, n \in \mathbb{N}$ (so that $\mathfrak{A}' \models \forall x \exists u \forall y \beta$). Note that by axiom **Z2'** all numbers generated by NUM' satisfy the predicate I. Let m, n be arbitrary natural numbers. Consider β_1 and let $n \notin \{m, m+1\}$. Then $\mathfrak{A}$ satisfies the antecedent $\neg Zx \wedge x \neq y \wedge Ix \wedge Iy \wedge \neg Sxy$ of S2' and β_1 for $x = \mathbf{m+1}, y = \mathbf{n}$ and therefore also the conclusion of both $\beta(z, x, y)$ and **S2'**. The latter means by the injectivity of S that z gets value m so that $\mathfrak{A}' \models \beta[\mathbf{m}, \mathbf{m+1}, \mathbf{n}]$.

Now consider β_2 and let $n = m + 1$. Then for $x = \mathbf{m}$ and $y = \mathbf{m+1}$ $\mathfrak{A}$ satisfies the antecedent Syx of β_2 so that its conclusion $\beta(x, y, y) \wedge \beta(x, y, x)$ is true in $\mathfrak{A}$. Therefore $\mathfrak{A}' \models \beta[\mathbf{m}, \mathbf{m+1}, \mathbf{n}]$ for $n \in \{m, m+1\}$. □

4.3.3 Encoding the Non-Auxiliary Binary Predicates.

In this section we modify the preceding reduction by encoding the binary predicates over $\mathbb{N}$ which occur in the given Kahr-Moore-Wang formula into monadic ones; we exploit the fact that the intended model for NUM' has the subdomain $\mathbb{N} \times \mathbb{N}$ and thereby allows us to use pairing of numbers. We incorporate into this refinement also what is needed to make the reduction conservative and prepare the encoding of the auxiliary binary predicates occurring in NUM' into one binary predicate which will be carried out in the next section.

More precisely we encode in this section the non-auxiliary binary predicate symbols Q occuring in the given formula ψ by monadic predicates defined on pairs of numbers, namely Qmn will be encoded into $Q(m, n)$ where for convenience we use the same letter Q to denote the encoded binary and the encoding monadic predicate. There are two problems related to the use of such an encoding by pairing. If we want to encode, for arbitrary m, n, a formula $\beta[m, n]$ in which $m+1$ also occurs, we have to find a way to represent $m + 1$ by a related pair; the N-successor relation formalized in NUM' does this job providing the N-successor $(m + 1, n)$ of (m, n). The second problem stems from the fact that we will have to encode atomic formulae of both forms Qst and Qts; this can easily be solved by introducing for each Q also an auxiliary monadic predicate symbol $\check{Q}$ to encode the converse relation of Q, i.e. with the following intended interpretation:

$$Qmn \text{ iff } Q(m,n) \quad Qnm \text{ iff } \check{Q}(m,n).$$

For the formalization of the relation between Q and $\check{Q}$ we use an auxiliary binary predicate *Link* expressing that two pairs are linked by a common first and second component respectively, i.e. with the intended interpretation

$$Link = \{((m,n),(r,m)) : m,n,r \in \mathbb{N}\}.$$

This will allow us to formalize the relation between a pair $x = (m, n)$ and its converse $y = (n, m)$ by $Link\ xy \wedge Link\ yx$ and therefore the relation between the encoding of a relation and the encoding of its converse by

$$Link\ xy \wedge Link\ yx \rightarrow (Qx \leftrightarrow \check{Q}y).$$

We prepare also the encoding of the auxiliary binary predicates $S, P_i, P_{i,+}$, S_0 and $S_1, S_2, N, Link$ occurring in NUM* which will be carried out in the next section. The predicates of the first group can be easily encoded into distinct parts of one relation R because S involves only numbers whereas the other predicates describe distinct relations between pairs and their projections or the successor or predecessor of the latter. In order to keep such an encoding free from conflicts with the encoding of the relations $S_1, S_2, N, Link$ between pairs we will use three copies for each pair (m, n) of the intended model, denoted by triples (m, n, j) with $j = 0, 1, 2$. The new component $j = 0, 1, 2$ will allow us to formulate distinct conditions serving for the encoding of $S_1, S_2, N, Link$ in an area of R which differs from the area used for the encoding of the first group of predicates. We introduce these triples already here together with two auxiliary predicates for handling them. Pairs (m, n) are encoded as triples with third component 0, i.e. they are distinguished among triples by an auxiliary monadic predicate *Basic* with the following intended interpretation:

$$Basic = \{(m,n,0) : m,n \in \mathbb{N}\}.$$

The intended interpretation of *Link* is correspondingly modified to $Link = \{((m, n, 0)(r, m, 0)) : m, n, r \in \mathbb{N}\}$. The intended interpretation of the remaining predicates is carried over in a similar way to triples (see the proof below for the satisfiability of NUM*).

In order to obtain the conservativity of the reduction one can use the idea which has been used already for the proof of Trakhtenbrot's Theorem, namely to restrict the successor axiom **S1'** of NUM' (and as a consequence also the corresponding part of the N-successor axiom **N2'**) to numbers which are not the last element of a finite domain, formalized by a new auxiliary monadic predicate *Last* whose intended interpretation in $\{0, 1, \ldots, n\}$ is $\{n\}$.

Exercise 4.3.18. Show that the reduction of the preceding section can be made conservative.

Along these lines we will refine NUM' to a finitely satisfiable formula NUM* which satisfies the following refined version of the Number Representation Lemma.

Lemma 4.3.19 (Number Representation Lemma for NUM***).** *If* $\mathfrak{A} \models$ NUM*, *then the domain* A *of* $\mathfrak{A}$ *contains a subset* $\mathbf{0}, \mathbf{1}, \mathbf{2}, \ldots$ *whose elements satisfy the properties of the Number Representation Lemma for* NUM'. *Either this set is infinite and* $\mathfrak{A} \models \neg Last\ \mathbf{n}$ *for each* n *or it is a finite set* $\{\mathbf{0}, \ldots, \mathbf{n}\}$ *such that* $\mathfrak{A} \models \neg Last\ \mathbf{m} \wedge Last\ \mathbf{n}$ *for each* $m < n$.

NUM* allows us to make the reduction of Kahr-Moore-Wang formulae $\psi \in [\forall\exists\forall, (0, \omega)]$ to formulae $\psi^* \in [\forall^2\exists, (\omega, \omega), (0)]_=$ semi-conservative (and thereby conservative by Theorem 2.1.39), i.e. such that the following two claims hold:

Claim 1. *If* ψ *is finitely satisfiable, then* ψ^* *also has a finite model.*

Claim 2. *If* ψ^* *is satisfiable, then* ψ *also has a model.*

Definition of ψ^*. Let $\psi := \forall x \exists u \forall y \beta \in [\forall\exists\forall, (0, \omega)]$ be an arbitrary formula in which binary predicates Q occur only in atomic formulae of the form Qxy, Qyx, Quy, Qyu. (From Exercise 3.1.10 we know that the class of such formulae is a conservative reduction class so that it suffices to consider such formulae.) As explained above we use the NUM*-formalization of the successor relation N to express $\beta[m, m+1, n]$ in terms of the encoding formula $\beta^*[(m, n), (m+1, n)]$ holding for the corresponding N-successors. This explains why we define ψ^* as an appropriate prenex normal form (with prefix $\forall x \forall y \exists z$) of the conjunction of NUM* and the following formula:

$$(Nxy \rightarrow \beta^*(x, y)) \wedge \bigwedge_{Q \text{ in } \beta} (Link\ xy \wedge Link\ yx \rightarrow (Qx \leftrightarrow \breve{Q}y))$$

where $\beta^*(x, y)$ is obtained from β by replacing each atomic subformula Qxy by Qx, Quy by Qy, Qyx by $\breve{Q}x$ and Qyu by $\breve{Q}y$.

Definition of NUM*. NUM* is obtained from NUM' by the following changes. **S1'** is replaced by the following formula **S1***. The restriction of the existence of successors to elements which are not the last one yields $Zx \wedge Iy \wedge \neg Last\ y \rightarrow Szy$.

Without loss of generality we assume that the finite models which we consider in this and the next section have at least two elements. We can also assume without loss of generality that the finite models which are given for a closed prenex formula whose prefix is of the form $\forall\exists\forall$ has a domain $\{0, 1, \ldots, n\}$ with successor function $i' = i + 1$ for $i < n$ and $n' = 0$. The condition $n' = 0$ is reflected in the refinement of **S1'** by formulating also the existence of a basic pair $(0, n)$, i.e. of the triple $(0, n, 0)$ if n is the last element. Thus **S1*** is defined as the conjunction of the following two formulae:

$$Zx \wedge Iy \wedge \neg Last\ y \rightarrow Szy$$

$$Zx \wedge Last\ y \rightarrow P_1zx \wedge P_2zy \wedge Basic\ z.$$

The first expresses ensures that elements which are not last have a successor, the second that last elements n generate a triple $(0, n, 0)$).

As we did for **S1'**, we now split also in the N-successor axiom**N2'** the formalization of the existence of a successor component between elements which are not last and the last element; for the latter, as explained immediately above, we will use 0 as successor. Thus **N2*** is defined as:

N2*: $Nyx \rightarrow P_1yz \wedge (\neg Last\ z \rightarrow P_{1,+}xz) \wedge (Last\ z \rightarrow P_{1,0}x)$.

Axiom **P1** expressing that each pair has a first component is changed to the condition that each pair has an N-successor (whose first component is then ensured by the new axiom **N2***), i.e. **P1** is replaced by the following axiom **P1***:

P1*: $Zx \wedge \neg Iy \rightarrow Nyz$.

We add the following formalization of the new auxiliary predicates *Last*, *Basic*, and *Link*:

Last: $Last\ x \rightarrow \neg Zx$ (the last element is different from 0).

Basic: $(Nxy \rightarrow (Basic\ x \leftrightarrow Basic\ y)) \wedge (S_0xy \rightarrow Basic\ x) \wedge (Basic\ x \rightarrow \neg Ix)$.

Link: $Basic\ x \wedge Basic\ y \wedge \neg Nxy \wedge \neg Nyx \rightarrow P_1xz \wedge (P_2yz \rightarrow Link\ xy)$ (basic elements x and y are linked if the first component of x is also the second component of y; clearly this cannot be required for triples $(m, n, 0), (m + 1, n, 0)$).

This concludes the definition of NUM*.

Lemma 4.3.20. *The above indicated intended interpretation satisfies* NUM* *over the domain* $\mathbb{N} \cup (\mathbb{N} \times \mathbb{N} \times \{0, 1, 2\})$.

Proof. The intended interpretation of the predicates $Z, S, P_i, P_{i,0}, P_{i,+}, I, D$ is independent from the third component and therefore remains as for NUM', the intended interpretation of S_0am is enriched by the condition that the third component of a is 0 (in accordance with the axiom for the predicate

Basic). The intended interpretation of N is enriched by the condition that the third component of an N-successor equals the third component of its N-predecessor (as required in the axiom for the predicate *Basic*); the intended interpretation of $S_i ab$ for $i = 1, 2$ is enriched by the condition that modulo 3 the third component of a is identical to the successor of the third component of b. (This modification of the interpretations of N, S_1, S_2 which takes into account also the third component of triples will allow us in the next section to encode these predicates into areas of a binary relation which are not in conflict with the areas where the other relations among triples are encoded.)

Therefore it suffices to refine the binary function h of the preceding section which now manipulates triples instead of pairs, i.e. $h : \mathbb{N} \cup (\mathbb{N} \times \mathbb{N} \times \{0,1,2\}))^2 \to \mathbb{N} \cup (\mathbb{N} \times \mathbb{N} \times \{0,1,2\})$. We only indicate the changes in the definition, reflecting the change of axioms in the cases **P1, Si.1', N1'** or required by the new axiom for the predicate *Link*.

$$h(a,b) := \begin{cases} (b_1+1, b_2, b_3) & \text{if } a = 0 \wedge b = (b_1,b_2,b_3) \text{ (case } \mathbf{P1^*}) \\ (0,b,0) & \text{if } Sab \text{ (case } \mathbf{S0.1'}) \\ (0,b,\pi_3 a + 1(\bmod 3)) & \text{if } \bigvee_{i=1,2} P_{i,+}ab \text{ (cases } \mathbf{Si.1'};\ i = 1,2) \\ (\pi_1 a + 1, \pi_2 a, \pi_3 a) & \text{if } \neg Da \wedge \bigvee_{i=0,1,2} S_i ab \text{ (case } \mathbf{N1'}) \\ \pi_1 b & \text{if } Basic\ a \wedge Basic\ b \wedge \neg Nab \wedge \\ & \quad \wedge \neg Nba \text{ (case } \mathbf{Link}) \\ \vdots & \end{cases}$$

Exercise 4.3.21. Check that this modified function h covers the new axioms in NUM* correctly. Check that the Skolem normal form of NUM* is satisfied by the above defined intended interpretation with h as the interpretation of the Skolem function for the existential variable z of NUM* and with *Last* interpreted as the empty set.

□

Lemma 4.3.22. *The above indicated intended interpretation can be modified so as to satisfy* NUM* *for each* $n > 0$ *over the finite domain*

$$\{0,1,\ldots,n\} \cup (\{0,1,\ldots,n\} \times \{0,1,\ldots,n\} \times \{0,1,2\}).$$

Proof. We interpret *Last* by $\{n\}$, restrict the intended interpretation described in the preceding lemma to the indicated finite domain and extend the interpretation of N by the pairs $((n,m,j),(0,m,j))$ for $0 \le j \le 3, m \le n$ in order to reflect that in this model the successor of n is 0, i.e. $n+1=0$. Correspondingly we modify the function h by setting $h(0,n) = (0,n,0)$ and $h(0,(n,m,j)) = (0,m,j)$.

Exercise 4.3.23. Check that this modified intended interpretation yields finite models for NUM*.

□

Proof. (*Claim 1.*) Assume $\mathfrak{A} \models \forall x \forall y \beta(x, x', y)$ for $\psi := \forall x \exists u \forall y \beta$ and that the domain A of $\mathfrak{A}$ is finite. Without loss of generality we can assume $A = \{0, 1, \ldots n\}$ for some $n > 0$ and that the successor function satisfies $i' = i + 1$ for $i < n$ and $n' = 0$. From the preceding lemma we know that NUM* is satisfiable over A by the intended model. We now extend this model to a model $\mathfrak{B}$ which also satisfies ψ^*. Define the interpretation of the monadic predicates $Q, \breve{Q}$ in $\mathfrak{A}$' as explained above:

$$\mathfrak{B} \models Q(m, r, j) \text{ iff } \mathfrak{A} \models Qmr,$$

$$\mathfrak{B} \models \breve{Q}(m, r, j) \text{ iff } \mathfrak{A} \models Qrm.$$

This definition clearly satisfies $\forall x \forall y (\mathit{Link}\ xy \wedge \mathit{Link}\ yx \rightarrow (Qx \leftrightarrow \breve{Q}y))$. It remains therefore to show that it satisfies $\forall x \forall y (Nxy \rightarrow \beta^*(x, y))$. Suppose $\mathfrak{B} \models Nab$ for arbitrary $a = (m, r, j), b = (m + 1, r, j)$. The definition of Q and $\breve{Q}$ implies that $\mathfrak{B} \models Q(m, r, j)$ iff $\mathfrak{A} \models Qmr$, $\mathfrak{B} \models Q(m + 1, r, j)$ iff $\mathfrak{A} \models Qm + 1r$, $\mathfrak{B} \models \breve{Q}(m, r, j)$ iff $\mathfrak{A} \models Qrm$, $\mathfrak{B} \models \breve{Q}(m + 1, r, j)$ iff $\mathfrak{A} \models Qrm + 1$. Thus from $\mathfrak{A} \models \beta[m, m + 1, r]$ and the definition of β^* we obtain that $\mathfrak{B} \models \beta^*[a, b]$. □

Proof. (*Claim 2.*) Assume $\mathfrak{B} \models \psi^*$. By the Number Representation Lemma for NUM* the domain B of $\mathfrak{B}$ contains a set of numbers $\mathit{Num} = \{0, 1, \ldots\}$ of cardinality $n_0 \leq \omega$. We define a model $\mathfrak{A}$ for ψ over Num by interpreting the predicates Q which occur in ψ as follows: $\mathfrak{A} \models Qmn$ if and only if there exists a $a \in B$ such that

$$\mathfrak{B} \models P_1am \wedge P_2an \wedge \mathit{Basic}\ a \wedge Qa.$$

We have to show that this interpretation is well-defined, namely that

$$\varepsilon(m, n) = \{a \in B : \mathfrak{B} \models \mathit{Basic}\ a \wedge P_1am \wedge P_2an\} \neq \varnothing$$

for each $m, n \in \mathit{Num}$ and that the definition of the interpretation of Q in $\mathfrak{A}$ is independent of the representatives in $\varepsilon(m, n)$.

Lemma 4.3.24 (Well-Definedness of Q). *For each $m, n \in \mathit{Num}$, each $a, b \in B$ and each predicate Q in ψ holds:*

(i) (Closure Property)

$$a \in \varepsilon(m, n) \wedge \mathfrak{B} \models Nab \implies b \in \varepsilon(m + 1, n),$$

(ii) $\varepsilon(m, n) \neq \varnothing$,
(iii) Let $a \in \varepsilon(m, n)$. Then

$$(\mathfrak{A} \models Qmn \textit{ iff } \mathfrak{B} \models Qa) \textit{ and } (\mathfrak{A} \models Qnm \textit{ iff } \mathfrak{B} \models \breve{Q}a).$$

Assuming the Well-Definedness Lemma we can show as follows that $\mathfrak{A} \models \beta[m, m+1, n]$ for all $m, n \in Num$ (and therefore $\mathfrak{A} \models \psi$). Choose any representative $a \in \varepsilon(m, n)$. As a basic element, a is not a number (by the axiom *Basic*), i.e. $\mathfrak{B} \models \neg Ia$ so that by axiom **P1*** there exists an N-successor $b \in B$ of a, i.e. satisfying $\mathfrak{B} \models Nab$. Thus $\mathfrak{B} \models \beta^*[a, b]$. This implies that $\mathfrak{A} \models \beta[m, m+1, n]$ because by the Well-Definedness Lemma the truth in $\mathfrak{A}$ of atomic formulae Qmn, Qnm in β is equivalent to the truth in $\mathfrak{B}$ of the corresponding subformulae $Qa, \breve{Q}a$ in β^* and similarly for $Qm+1n, Qnm+1$ in β and $Qb, \breve{Q}b$ in β^* (note that $b \in \varepsilon(m+1, n)$ by the closure property).

It remains to prove the Well-Definedness Lemma.

(i): Closure Property. Assume that $a \in \varepsilon(m, n)$ and $\mathfrak{B} \models Nab$. a) Since by the axiom **Basic** an element is basic if and only if its N-successor is basic, $a \in \varepsilon(m, n)$ and $\mathfrak{B} \models Nab$ imply that $\mathfrak{B} \models Basic\, b$. b) From $\mathfrak{B} \models Nab$ and axiom **N3'** we know that a and b have a common second component c, i.e. there is a $c \in B$ satisfying $\mathfrak{B} \models P_2ac \wedge P_2bc$. Since $\mathfrak{B} \models P_2an$, the uniqueness of numbers as components implies $c = n$, hence $\mathfrak{B} \models P_2bn$. c) To show that $\mathfrak{B} \models P_1bm+1$ we have to distinguish between two cases, namely whether or not $\mathfrak{B} \models \neg Last\ m$. In the first case the N-Chain Lemma can be applied to infer that $\mathfrak{B} \models P_1bm+1$ from $\mathfrak{B} \models P_1am$. In the second case axiom **P1*** provides the existence of a first component $d \in B$ of a satisfying

$$(\neg Last\ d \rightarrow P_{1,+}bd) \wedge (Last\ d \rightarrow P_{1,0}b).$$

The uniqueness of numbers as components yields $\mathfrak{B} \models d = m$; hence $\mathfrak{B} \models P_{1,0}b$. By axiom **P1.0**, $\mathfrak{B} \models P_1b0$ which was to be proved since in this case (without loss of generality) $m+1 = 0$.

(ii): We proceed by induction on m.

Base $m = 0$. There are two subcases to consider depending on whether or not $\mathfrak{B} \models \neg Last\ n$. In the first case we know that n has a successor $n+1$ which by axiom **S0.1'** has an S_0-predecessor $a \in B$ with second component n and satisfying $P_{1,0}a$, i.e. such that $\mathfrak{B} \models S_0an+1 \wedge P_{1,0}a \wedge P_2an$. By axiom **P1.0** $\mathfrak{B} \models P_1a0$ and as S_0-predecessor a is a basic element (by axiom **Basic**), i.e. $\mathfrak{B} \models Basic\ a$; hence $a \in \varepsilon(0, n)$. In the second case we have $\mathfrak{B} \models Last\ n$. As last element n generates by axiom **S1*** a basic triple $a \in B$ with first component 0 and second component n, i.e. satisfying $\mathfrak{B} \models P_1a0 \wedge P_2an \wedge Basic\ a$; hence $a \in \varepsilon(0, n)$.

For the inductive step let $a \in \varepsilon(m, n) \neq \emptyset$ and assume $\mathfrak{B} \models \neg Last\ m$. (Note that for $m = 0$ the latter is guaranteed by the axiom **Last**.) As a basic element, a is not a number (by axiom **Basic**) and therefore by axiom **P1*** has an N-successor $b \in B$. The closure property implies that $b \in \varepsilon(m+1, n)$ which is therefore not empty.

(iii): We first show the following **Auxiliary Claim:**

$$a \in \varepsilon(m, n) \wedge b \in \varepsilon(n, m) \implies \mathfrak{B} \models Link\ ab \wedge Link\ ba.$$

To prove this claim we want to use the **Link** axiom and therefore have to ensure that $\mathfrak{B} \models \neg Nab \wedge \neg Nba$. We prove this indirectly. Assume that $\mathfrak{B} \models Nab$. The closure property implies that $b \in \varepsilon(m+1, n)$. From this and $b \in \varepsilon(n, m)$ we obtain by the uniqueness of numbers as components that $m = n$ and (depending on whether $\mathfrak{B} \models \neg Last\ m$ or not) either $m + 1 = n$, or else $n = 0$. The case $m = n = m + 1$ is clearly impossible, but also the case $m = n = 0$ is impossible because (by axiom **Last**) 0 is not the last element. Similarly one refutes that $\mathfrak{B} \models Nba$. Therefore the premises of the **Link**-axiom are satisfied in $\mathfrak{B}$ and we can conclude that $\mathfrak{B} \models P_1ac \wedge (P_2bc \rightarrow Link\ ab)$ for some $c \in B$. Since $\mathfrak{B} \models P_1am$, by the uniqueness of numbers as components we conclude $c = m$; hence $\mathfrak{B} \models P_2bm$ implies that $\mathfrak{B} \models Link\ ab$. By symmetry we obtain also that $\mathfrak{B} \models Link\ ba$. This proves the claim.

Using the claim it is easy to prove the independence properties *(iii)*. Let $a \in \varepsilon(m, n)$. By definition of $\mathfrak{A}$, Qnm holds in $\mathfrak{A}$ if and only if there is some $b \in \varepsilon(n, m)$ such that $\mathfrak{B} \models Qb$. By the auxiliary claim we know that a and b are linked, i.e. $\mathfrak{B} \models Link\ ab \wedge Link\ ba$ so that the **Link**-conjunct of ψ^* implies that $\mathfrak{B} \models Qb \leftrightarrow \breve{Q}a$. Therefore $\mathfrak{A} \models Qnm$ if and only if $\mathfrak{B} \models \breve{Q}a$. This proves the second part of *(iii)*.

The first independence property of *(iii)* follows easily: If $\mathfrak{B} \models Qa$, then by definition $\mathfrak{A} \models Qmn$. For the converse suppose that $\mathfrak{A} \models Qmn$ and choose an arbitrary $b \in \varepsilon(n, m)$. The second independence property proved immediately above implies that $\mathfrak{B} \models \breve{Q}b$. Since, by the auxiliary claim, a and b are linked, the **Link**-conjunct of ψ^* implies that $\mathfrak{B} \models Qa \leftrightarrow \breve{Q}b$. Hence $\mathfrak{B} \models Qa$ as was to be shown. □

Proof. (Number Representation Lemma for NUM*.) As for NUM and NUM' the construction is by induction on n. The induction base $n = 0$ is proved in the same way as for NUM'. In addition the axiom **Last** guarantees that 0 is not the last element.

For the induction step it suffices to show that if n is not the last element (i.e. $\mathfrak{A} \models \neg Last\ n$), then it has an S-successor $n + 1 \notin \{0, \ldots, n\}$ which satisfies the conditions of the lemma.

The N-Chain Lemma and the Injectivity Lemma for S are proved as for NUM' because by assumption $\mathfrak{A} \models \neg Last\ i$ for each $i \leq n$ and therefore the invocations of **N2*** are equivalent to those of **N2'**.

The assumption that n is not the last element guarantees also that the modified axiom **S1*** can be applied and provides a successor $n+1$ the uniqueness of which can again be shown in the same way as with NUM' (Functionality Lemma for S). Also the uniqueness of numbers as components with respect to $P_i, P_{i,+}$ and the correctness property of $P_{1,+}$ can be proved in the same way as for NUM' with one modification due to the invocation of **P1*** instead of **P1**. One has to show that pairs – i.e. elements a satisfying $\mathfrak{A} \models \neg Ia$ – have a first component: axiom **P1*** guarantees the existence of an N-successor $b \in A$ of a so that axiom **N2*** provides a first component $c \in A$ for the N-predecessor a, i.e. satisfying $\mathfrak{A} \models P_1ac$. □

4.3.4 Encoding the Auxiliary Binary Predicates of NUM*.

In this section we refine the previous reduction by encoding the auxiliary binary predicates occurring in NUM* into one binary predicate R. What we have to do is to map $S, P_i, P_{i,+}, S_0$ and $S_1, S_2, N, Link$ into non-conflicting subrelations of a single relation R. On the basis of the refined interpretations which have been provided in the preceding section these predicates can easily be separated as follows: S holds only among numbers (i.e. $S \subseteq \mathbb{N} \times \mathbb{N}$), $P_i, P_{i,+}, S_0$ hold among pairs and numbers (i.e. are subsets of $(\mathbb{N} \times \mathbb{N}) \times \mathbb{N}$) and $S_1, S_2, N, Link$ hold only among pairs. The embeddings of the predicates of the groups P_1, P_2 and $P_{1,+}, P_{2,+}$ and S_0 into the single R are kept non-conflicting by their defining condition which is in terms of pairs and their projections or successors or predecessors of the latter; the embeddings of P_1 and P_2 can be kept separate from each other by encoding P_1ab into Rab and P_2ab by Rba; similarly for $P_{1,+}$ and $P_{2,+}$. Conflicts for the embeddings of S_1, S_2 and N and $Link$ into R are avoided by choosing for them different copies of the (triples which represent the) involved pairs, i.e. by using the distinguished conditions on the third component of the triples which have been introduced through the refined intended interpretation in the preceding section. The embeddings of S_1 and S_2 can be kept separate from each other by encoding S_1ab into Rab and S_2ab into Rba.

Formally we express the separating conditions in terms of monadic predicates C_i, C_i^j for $i \in \{0,1,2\}, j \in \{1,2,3\}$ with the following intended interpretation:

$$C_i m \text{ iff } m \equiv i \pmod 3, \qquad C_i^j(m_1, m_2, m_3) \text{ iff } m_j \equiv i \pmod 3.$$

These predicates allow us to express that v and w have the same remainder when divided by 3 – abbreviated $rem(v) = rem(w)$ – by the formula $\bigwedge_{i<3}(C_i v \leftrightarrow C_i w)$; similarly $rem(\pi_j(v)) = rem(w) + 1$ can be formalized by $\bigwedge_{i<3}(C_{i+1}^j v \leftrightarrow C_i w)$ where $+1$ is to be interpreted modulo 3, etc.

The encoding scheme explained above is formalized by the following definition.

Definition of ψ'. Let ψ^* be an arbitrary reduction formula as constructed in the preceding section. Define ψ' as resulting from ψ^* by replacing each atomic subformula Qvw by $Q'vw$ where the latter is defined as follows:

$$\begin{aligned}
S'vw &:= Rvw \wedge Iv \wedge Iw \\
P_1'vw &:= Rvw \wedge rem(w) = rem(\pi_1(v)) \\
P_2'vw &:= Rwv \wedge rem(w) = rem(\pi_2(v)) \\
P_{1,+}'vw &:= Rvw \wedge rem(w) + 1 = rem(\pi_1(v)) \\
P_{2,+}'vw &:= Rwv \wedge rem(w) + 1 = rem(\pi_2(v)) \\
S_0'vw &:= Rvw \wedge rem(w) = rem(\pi_2(v)) + 1 \\
S_1'vw &:= Rvw \wedge rem(\pi_3(v)) = rem(\pi_3(w)) + 1
\end{aligned}$$

$$\begin{aligned} S_2'vw &:= Rwv \wedge rem(\pi_3(v)) = rem(\pi_3(w)) + 1 \\ N'vw &:= Rvw \wedge Rwv \wedge rem(\pi_1(v)) + 1 = rem(\pi_1(w) \wedge \\ &\qquad \wedge \bigwedge\nolimits_{i=2,3} rem(\pi_i(v)) = rem(\pi_i(w)) \\ Link'vw &:= Rvw \wedge rem(\pi_3(v)) = rem(\pi_3(w)) = 0 \wedge \neg N'vw \wedge \neg N'wv. \end{aligned}$$

Clearly $\psi' \in [\forall^2\exists, (\omega, 1), (0)]_=$ so that it suffices to prove the following claims.

Claim 1. *If ψ is finitely satisfiable, then also ψ' has a finite model.*

Claim 2. *If ψ' is satisfiable, then also ψ has a model.*

Proof. Assume that $\mathfrak{A} \models \forall x \forall y \beta(x, x', y)$ for $\psi := \forall x \exists u \forall y \beta$ with finite domain A of $\mathfrak{A}$. Without loss of generality we can assume that $n+1 := 0$ (mod 3). Therefore we can assume without loss of generality that $A = \{0, 1, \dots n\}$ for some $n > 0$ and that the successor function satisfies $i' = i+1$ for $i < n$ and $n' = 0$. Let $\overline{Q}$ denote the intended interpretation of Q as defined in the previous section where it has been proved that this interpretation satisfies NUM* over $A \cup (A \times A \times \{0, 1, 2\})$. We now slightly modify and extend this interpretation to a model $\mathfrak{A}'$ (over the same domain) which satisfies ψ'.

The monadic predicates C_i, C_i^j are interpreted as indicated above. The interpretation of R in $\mathfrak{A}'$ is defined as follows:

$$\begin{aligned} R'ab &:= \overline{S}ab \vee \overline{P_1}ab \vee \overline{P_2}ba \vee \overline{P_{1,+}}ab \vee \overline{P_{2,+}}ba \vee \overline{S_0}ba \\ &\quad \vee \overline{N}ab \vee \overline{N}ba \vee \overline{S_1}ab \vee \overline{S_2}ba \vee \overline{Link}\, ab. \end{aligned}$$

Exercise 4.3.25. Show that for each binary predicate Q occuring in NUM* except *Link* and for all $a, b \in A'$ holds: $\mathfrak{A}' \models Q'ab$ if and only if $\overline{Q}ab$ holds. Hint: Distinguish the cases $a, b \in A$; $a \notin A$, $b \in A$; $a, b \notin A$.

Exercise 4.3.26. Show that for all $a, b \in A'$ holds: $\mathfrak{A}' \models Link'ab$ if and only if $\overline{Link}\, ab \wedge \neg\overline{N}ab \wedge \neg\overline{N}ba$ is true.

Exercise 4.3.27. Show that the intended interpretation still satisfies NUM* if the interpretation of *Link ab* is changed to $\overline{Link}\, ab \wedge \neg\overline{N}ab \wedge \neg\overline{N}ba$. Hint: Observe that the axiom **Link** contains $\neg Nxy \wedge \neg Nyx$ in the antecedent.

It follows from these exercises and the definition of ψ' that modifying the interpretation of *Link ab* to $\overline{Link}\, ab \wedge \neg\overline{N}ab \wedge \neg\overline{N}ba$ yields a finite model for ψ'.

Claim 2 is obvious, taking the above replacement scheme as the definition of the auxiliary relations to satisfy ψ^* and therefore ψ. □

4.4 Historical Remarks

The decision problem problem for classes $[\Pi, p, f]$ with predicates and functions but without equality has been investigated by Gurevich [223]. Our proofs in Sect 4.2 for the two minimal conservative classes $[\forall^2, (0,1), (1)]$ and $[\forall^2, (1), (0,1)]$ are taken from there. Gurevich has also studied the decision problem for classes $[\Pi, p, f]_=$ in [227] where the conservative reduction class $[\forall, (0), (\omega)]_=$ is established and then reduced to $[\forall, 0, (2)]_=$ and $[\forall, (0), (0,1)]_=$ by an appropriate encoding of finitely many unary functions by two unary or one binary function. Gurevich's proof for the class $[\forall, (0), (\omega)]$ uses an encoding of domino problems based on [237]. His reduction formulae are not Horn and contain disjunctions whose length equal the number of colours in the given domino problem. Börger [49] obtains a reduction to Horn formulae with ternary disjunctions. The improvement of this reduction to Krom and Horn formulae which appears in our proof for Theorem 4.1.1 is due to Löwen; the strengthening to Herbrand formulae is due to Wirsing [536]. His proof is adapted to register machines for our proof of Corollary 4.1.3.

The story of the $\forall^2\exists^*$ classes is somewhat unusual. In [187], Gödel proves that the class $[\exists^*\forall^2\exists^*, all]$ has the finite model property (see Sect. 6.2.3 for a proof of this result and Sect. 6.6 for the history of the class $[\exists^*\forall^2\exists^*, all]$.) In the last sentence of that paper, Gödel wrote that the same method suffices to establish the finite model property for $[\exists^*\forall^2\exists^*, all]_=$. For a long while, nobody doubted this claim. Moreover, it was even used. For example, Scott used the claim to establish the decidability of $Sat(L_2)$; see the historical remarks to Chapter 8. In the mid-1960s, Stål Aanderaa demonstrated that Gödel's criterion for the satisfiability of $\forall^2\exists^*$ formulae without equality is not sufficient for the satisfiability of $\forall^2\exists^*$ formulae with equality. There were several attempts to query Gödel himself on the subject (see [188, pp. 229–230]) but they did not clarify the matter. Eventually it became clear that the status of $[\exists^*\forall^2\exists^*, all]_=$ is open. Finally, in [192], Goldfarb has proved that $[\forall^2\exists^*, (0,1)]_=$ is a reduction class.

Two universal quantifiers correspond to quantification over ordered pairs of elements. What happens if this quantification is replaced by quantification over unordered pairs of elements? According to [194], the appropriate restriction of $[\forall^2\exists^*, (0,1)]_=$ has the finite model property.

5. Other Undecidable Cases

This chapter deals with the *Entscheidungsproblem* for classes of predicate logic formulae which are characterized not only by the vocabulary or prefix structure but also by the fine structure of the quantifier-free part of their formulae. As pointed out already occasionally in the preceding chapters, the reductions there indeed use only very special formulae to establish the conservative reduction class property for prefix-vocabulary classes; in this chapter we make some natural candidates of such refinements explicit and study their effect upon the *Entscheidungsproblem*.

Natural candidates for such further classification come from considering the propositional structure, in particular related to conjunctive and disjunctive normal form. Two outstanding examples, which originated from considerations outside reduction theory, are Horn and Krom structure. Another example comes from counting the number of conjuncts or disjuncts (for formulae in conjunctive or disjunctive normal form) or the number of atomic formulae; to find reduction classes formulated in such terms is related to finding and formalizing appropriate small universal machines. Another self-suggesting criterion for "small" reduction formulae is to look for conjunctions of "simple" subformulae which belong to decidable classes, for example to decidable prefix classes. (It is interesting to figure out in which sense every statement, of whatever complexity, can be broken into an equivalent conjunction of simple short statements.) A different syntactical classification concerns various combinations of occurrences of variables in the atomic formulae.

Gurevich's Classifiability Theorem tells us that by refining the prefix-vocabulary classification with natural additional restrictions like Krom or Horn, we can still count upon the existence of a finite number of minimal undecidable and maybe a finite number of maximal decidable classes. Nevertheless for most of the additional classification criteria which have been considered in the literature we are far from knowing what these minimal undecidable cases and their maximal decidable counterparts are.

We therefore choose for this chapter some characteristic examples which are particularly interesting from the point of view of the proof method; we abstain from an attempt to cover the subject in an exhaustive way. We will deal with additional results in numerous exercises and mention others in the references so that the reader will get a fuller picture of the state of the art.

5.1 Krom and Horn Formulae

In this section we study the *Entscheidungsproblem* for classes of Krom and Horn formulae. The notion of a *Horn* formulae originated from particular model theoretic properties of such formulae which have been studied extensively in the literature, starting with [272]; the particular computational properties of (a subclass of) Horn formulae have been recognized through the definition of Prolog (see e.g. [87]) and have led to numerous investigations of their proof theory. The notion of *Krom* formula sprang out of the Chang-Keisler normal form for predicate logic which restricts the length of disjunctions in prenex conjunctive normal forms to two or three, depending on whether the equality is present (see [75]). Herbrand [253] showed that the class HERBRAND of formulae in prenex conjunctive normal form whose conjuncts are atomic or negated atomic formulae is decidable if restricted to formulae without functions or equality; Gladstone [182] sharpened this result by showing that this class has the finite model property. (Remember that by Corollary 4.1.3 the class of Herbrand formulae with functions and equality constitutes a conservative reduction class.) Thus it was natural to start the investigation of the consequences of restricting the length of alternations, for formulae without equality, to two (see [91, 329, 331, 332, 334, 409, 437]). It turned out that both notions, of Krom and of Horn formula, have an effect upon the *Entscheidungsproblem.* We will see in this section that both the prefix and the prefix-vocabulary classification are different from the one without the restriction to Krom or Horn formulae.

We study here some representative prefix classes of Krom formulae which are known to be undecidable (and turned out to remain so even when restricted to Horn formulae). We first consider Krom formulae without equality or functions and then Krom formulae with functions or equality. On our way we deal also with the vocabulary and with the prefix-vocabulary problem for Krom classes.

5.1.1 Krom Prefix Classes Without Functions or Equality.

We concentrate our attention in this subsection on undecidable Krom prefix classes without functions or equality; along the way we collect also the information on the used vocabulary.

Remark. The vocabulary problem for Krom classes is solved by Lewis' result [348] that relational Krom sentences with a single binary predicate form a reduction class. The proof proceeds in three steps: a) the Post correspondence problem is reduced to Krom (and Horn) formulae with arbitrary nestings of function symbols (see the following exercise), b) the nestings of function symbols are eliminated, c) it is shown that the construction can be done with only one binary predicate and without function symbols. We refer the reader to the original paper for the proof of b) and c) which adapts the encoding idea from Shannon's construction of a universal Turing machine with only two internal states [465]. See also the Theorem of Lewis and Goldfarb (Theorem 5.2.2) which contains as a by-product that the class of function and quantifier free formulae in $\forall\exists\forall^* \cap$ KROM with only one predicate (whose arity is of the order of magnitude of the size of a universal 2-register machine program) is a reduction class.

Exercise 5.1.1. [348] Let $C = (v_i, w_i)_{1 \leq i \leq n}$ be a Post correspondence system. View words w over the alphabet of C as logical terms $w(c)$ where each letter of the given alphabet is represented by a monadic function symbol and c is a term, in this case an individual constant or a variable. Using this encoding formalize a binary predicate P with the following intended interpretation: $Pv(c)w(c)$ is true if and only if for some C-computation $(v_{i_1} \dots v_{i_s}, w_{i_1} \dots w_{i_s})$ (i.e. a sequence with $1 \leq i_1, \dots, i_s \leq n$) holds $v_{i_1} \dots v_{i_s} v = w_{i_1} \dots w_{i_s} w$. Define $\overline{C} := \forall x \forall y(\text{START} \wedge \text{STEP} \wedge \text{NONSTOP})$ where START $:= Pxx$, STEP $:= \bigwedge_i Pv_i(x)w_i(y) \rightarrow Pxy$, NONSTOP $:= \bigwedge_i \neg Pv_i(x)w_i(x)$. Show that C has no solution if and only if $\overline{C}$ is satisfiable.

Theorem 5.1.2. (Undecidable Krom Prefix Classes without Functions or Equality). *The following prefix classes of Krom formulae without functions or equality are undecidable and indeed are reduction classes even when restricted to Horn formulae:*

- $[\forall\exists^*\forall]$ *(Krom 1970)*
- $[\exists\forall\exists\forall], [\forall\exists^2\forall]$ *(Aanderaa and Börger 1971, Orevkov 1973)*
- $[\forall^2\exists\forall], [\forall\exists\forall^2]$ *(Lewis 1972)*

Proof. We begin with the proof for the Lewis class $[\forall^2\exists\forall, (0,0,\omega)] \cap \text{KROM} \cap \text{HORN}$ by reducing to it the halting problem of arbitrary 2-register machine programs $M = (I_i)_{0 \leq i \leq r}$. Without loss of generality we assume that M is started in the initial configuration $C_0 = (0,0,0)$ and has the only halting state 1.

We represent numbers n by special Fitch words w^n where the arbitrary Fitch word w (term built up from an individual constant using only the binary Skolem function ()) is used as relative zero. w^n is defined by $w^0 = w$ and $w^{n+1} = (w^n w)$. One can then encode each M-configuration $C = (i,p,q)$ with state i and register contents (p,q) by atomic formulae $\underline{C} = K_i w^p w^q w$ where the Fitch word w which serves as relative zero appears as third argument. The ternary relation symbols K_i thus have the following intended interpretation with respect to a given M-computation:

$K_i uvw$ is true iff there exist p, q such that $u = w^p, v = w^q, C_0 \Rightarrow_M (i,p,q)$.

Based upon this encoding we will construct a formula

$$\psi_M \in [\forall^2\exists\forall, (0,0,\omega)] \cap \text{KROM} \cap \text{HORN}$$

with Skolem normal form $\forall x \forall y \forall z \varphi_M$ so that the following holds:

Reduction Property: M, started in $C_0 = (0,0,0)$, will not halt in state 1 if and only if the formula $\forall x \forall y \forall z \varphi_M$ is satisfiable.

This will establish the claim for the first Lewis class.

As usual φ_M is defined as conjunction of formulae START $:= K_0xxx$, NONSTOP $:= \neg K_1xyz$ and STEP$_M := \bigwedge_{I_i \in M} \varepsilon_i$ with the following formulae ε_i for each instruction I_i of M:

$\varepsilon_i := K_ixzy \rightarrow K_j(xy)zy$ for addition instructions $I_i = (i, a_1, j) \in M$

$\varepsilon_i := K_izxy \rightarrow K_jz(xy)y$ for addition instructions $I_i = (i, a_2, j) \in M$

For subtraction instructions $I_i = (i, s_1, j, k)$ or $I_i = (i, s_2, j, k)$ we set respectively:

$$\varepsilon_i := (K_ixyx \rightarrow K_jxyx) \wedge (K_i(xy)zy \rightarrow K_kxzy)$$

$$\varepsilon_i := (K_iyxx \rightarrow K_jyxx) \wedge (K_iz(xy)y \rightarrow K_kzxy)$$

Exercise 5.1.3. Prove the reduction property. For the only-if direction use the above intended interpretation as model for ψ_M. For the if-direction use the following *simulation property* for each canonical model $\mathfrak{A}$ satisfying $\forall x \forall y \forall z \varphi_M$: for each t and each M-configuration C which is reached by M in t steps starting from C_0, we have that $\mathfrak{A} \models \underline{C}$ for each Fitch word w.

For the Lewis class $[\forall\exists\forall^2, (0,0,\omega)] \cap$ KROM $\cap$ HORN we give a similar reduction, changing the encoding of M-configurations $C = (i,p,q)$ to $\underline{C} = K_i(w, w + 2^p3^q, w)$. The idea is to use the first argument w as relative zero with respect to which we describe the effect of the given M-instructions I_i on the prime encoding $w + 2^p3^q$ of the register contents. The third component is used for temporary storage which allows us to give this description using only the monadic Skolem function $'$. This is nothing more than to formalize the construction given by Minsky, Shepherdson and Sturgis to simulate arbitrary register machines by 2-register machines (see [393, 469]).

An instruction $I_i = (i, o_i, j) \in M$ with an operation o_i of adding 1 in register l is simulated by a multiplication with the prime number p_l ($p_1 = 2, p_2 = 3$); i.e. one copies the double or triple of (the content of the second register encoded in) the second argument into the third one and at the end – i.e. when the second argument has become equal to the first one playing the role of the relative zero – one copies the result back into the second argument. This is formalized by the following multiplication formulae ε_i:

$$(K_iyx'z \rightarrow K_iyxz'') \wedge (K_iyyz \rightarrow K_jyzy) \text{ for } o_i = a_1,$$

$$(K_iyx'z \rightarrow K_iyxz''') \wedge (K_iyyz \rightarrow K_jyzy) \text{ for } o_i = a_2.$$

Remember that the use of the Skolem function $'$ on z can be avoided using auxiliary predicates K_i' satisfying $K_i'yzx \leftrightarrow K_iyzx'$.

For simplicity of exposition we assume without loss of generality that M executes subtraction operations only on registers not containing 0 and that the zero test is done by special operations $o_i \in \{test_1, test_2\}$. This allows us to have symmetric division formulae ε_i for instructions $I_i = (i, o_i, j) \in M$ with an operation o_i of subtracting 1 in register l where instead of multiplying we divide by two (for $l = 1$) or three (for $l = 2$):

$$(K_i yx''z \to K_i yxz') \wedge (K_i yyz \to K_j yzy) \text{ for } o_i = s_1,$$

$$(K_i yx'''z \to K_i yxz') \wedge (K_i yyz \to K_j yzy) \text{ for } o_i = s_2.$$

Instructions $I_i = (i, test_l, j, k) \in M$ to test whether the content of register l is zero are simulated by divisibility tests; keep copying the second into the third register by blocks of p_l units and at the end, before copying the result back into the second register, check whether a remainder (of 1 for $o_i = s_1$ and of 1 or 2 for $o_i = s_2$) is left (with respect to the relative zero in the first register). This is formalized by the following divisibility test formulae ε_i:

$$(K_i yx''z \to K_i yxz'') \wedge (K_i xx'z \to K_j xz'x) \wedge (K_i yyz \to K_k yzy) \text{ for } o_i = test_1.$$

For $o_i = test_2$ we have in addition to the following copying and $\neq 0$-conjuncts

$$(K_i yx'''z \to K_i yxz''') \wedge (K_i yyz \to K_k yzy)$$

also the following two conjuncts for the $= 0$-case (non divisibility):

$$(K_i xx'z \to K_j xz'x) \wedge (K_i xx''z \to K_j xz''x).$$

This completes the definition of the conjunction STEP_M of all the ε_i for $I_i \in M$. Obviously we set START $:= Kxx'x$ and NONSTOP $:= \neg K_1 xyz$.

It remains to show the reduction property. Assume that M, started in $C_0 = (0, 0, 0)$, does not halt in state 1. Then it is easy to check that the following interpretation of the ternary relation symbols K_i with respect to the given M-computation $(C_t)_{t<\infty}$ yields a canonical model for $\forall x \forall y \forall z \varphi_M$: $K_i(a, b, c)$ is true if and only if in the given computation M reaches a configuration $C_t = (i, p, q)$ such that for some number d one of the following cases holds:

	$o_i = a_1$	$b = a + 2^p 3^q - d$	$c = a + 2d$	$d \leq a + 2^p 3^q$
or	$o_i = a_2$	...	$c = a + 3d$	...
or	$o_i = s_1$	$b = a + 2^p 3^q - 2d$	$c = a + d$	$2d \leq a + 2^p 3^q$
or	$o_i = test_1$	...	$c = a + 2d$	...
or	$o_i = s_2$	$b = a + 2^p 3^q - 3d$	$c = a + d$	$3d \leq a + 2^p 3^q$
or	$o_i = test_2$	...	$c = a + 3d$	...

Exercise 5.1.4. Show the if-direction using the following *simulation property* for each canonical model $\mathfrak{A}$ satisfying $\forall x \forall y \forall z \varphi_M$: for all t, each M-configuration $C = (i, p, q)$ which is reached by M in t steps starting from C_0 and each number w, we have that $\mathfrak{A} \models K_i(w, w + 2^p 3^q, w)$.

In Chap. 2 the Aanderaa-Börger classes $[\exists\forall\exists\forall, (\omega, k)] \cap$ KROM $\cap$ HORN and $[\forall\exists\exists\forall, (\omega, k)] \cap$ KROM $\cap$ HORN have been shown to be conservative reduction classes for some natural number k of the size of a universal 2-register machine, see Corollary 2.1.16 and the Exercises 2.1.17 and 2.1.37.

For Krom's class the claim follows from the stronger result proved in the next theorem below. □

Exercise 5.1.5. [40] Refine the construction in the preceding proof by showing that $\forall^2\exists\forall \cap$ KROM $\cap$ HORN is a *conservative* reduction class. Hint (see Section 2.1.2 on Trakhtenbrot's Theorem): Use the axiomatization indicated below of a $<$-relation K, relativized to register contents. K has the following intended interpretation over $\{0, \ldots, l\}$ with respect to a given M-computation halting in state 2, where l is greater than every register content occuring in this computation: $Kuvw$ is true if and only if for some natural numbers p, q holds $u \equiv w+p \pmod{l+1}$, $v \equiv w+q \pmod{l+1}$, $p < q \leq l$ where l is greater than every register content occuring in the given halting M-computation.

$$\neg Kxxy \wedge (Kzxy \rightarrow Kz(xy)y) \wedge \bigwedge_i (K_i xzy \rightarrow Kx(xy)y) \wedge (K_i zxy \rightarrow Kx(xy)y).$$

Similarly the interpretation of K_i over $\{0, \ldots, l\}$ is refined to: $K_i uvw$ is true iff there exist p, q such that $u \equiv w + p \pmod{l+1}$, $v = w + q$ (mod $l+1$), $C_0 \Rightarrow_M (i, p, q)$.

The interpretation of the binary function ( ) is refined by setting $(vw) \equiv v+1 \pmod{l+1}$ if $w \not\equiv v+1 \pmod{l+1}$ and otherwise $(vw) \equiv v \pmod{l+1}$.

Remark. In [7] it is shown that also $\forall\exists\forall^2 \cap$KROM is a conservative reduction class.

Exercise 5.1.6. [40] Refine the above construction by showing that for some k – of the size of a universal 2-register machine – the class $[\forall^2\exists\forall, (0, \omega, k)] \cap$ KROM$\cap$HORN is a conservative reduction class. Hint: Formalize an arbitrary initial register content $(n, 0)$ by the conjunction of formulae P_0xx, $P_jxy \rightarrow P_{j+1}(xy)y$ for $0 \leq j < n$ and $P_nxy \rightarrow K_0xyy$. P_j are binary relations which have the following intended interpretation with respect to an M-computation eventually halting in state 2: P_jvw is true if and only if $v \equiv w+j \pmod{l+1}$ where l is greater than every register content occuring in the given halting M-computation.

Exercise 5.1.7. Refine the above construction by showing that for some k – of the size of a universal 2-register machine – the class $[\forall\exists\forall^2, (0, \omega, k)] \cap$ KROM $\cap$ HORN is a reduction class. For the conservativity of this class see the construction in [7].

Remark. The use of ternary relations in the reduction class $\forall\exists\forall^2 \cap$KROM is unavoidable because the class $[\forall\exists\forall^*, (\omega, \omega)] \cap$ KROM is decidable (see [41]).

It is not known whether the number of ternary relations in $[\forall\exists\forall^2, (0, \omega, k)] \cap$ KROM or $[\forall^2\exists\forall, (0, \omega, k)] \cap$ KROM can be reduced to small numbers (in particular to $k = 1$) and whether the binary relations can be replaced there by monadic predicates.

Exercise 5.1.8. [42] Prove that $[\exists^*\forall\exists\forall, (0, 0, 1)] \cap \text{KROM} \cap \text{HORN}$ is a conservative reduction class. (This improves Orevkov's reduction class

$$[\exists^*\forall\exists\forall, (0, 0, 0, 0, 0, 2, 4, 0, 1)] \cap \text{KROM}$$

in [409].) Hint: Rephrase the proof of Theorem 2.1.15 of Aanderaa and Börger.

Remark. For Krom formulae the prefix classes $\forall\exists\forall$ and $\exists^*\forall^*\exists^*$ are decidable (see Chap. 8.3). It is known that also for Horn formulae the class $\forall\exists\forall$ is decidable [190].

Exercise 5.1.9. Show that the undecidable Krom classes of Theorem 5.1.2 and the decidable Krom classes mentioned in the preceding remark constitute minimal undecidable and maximal decidable Krom prefix classes respectively and solve the prefix problem for Krom classes without equality or functions except for the classes $\forall\exists\forall\exists^k$ with $k > 0$ and $\forall\exists\forall\exists^*$ for which it is unknown whether they are decidable or not.

Theorem 5.1.10 (Rödding, Börger). $[\forall\exists^*\forall, (0, 4)] \cap \text{KROM} \cap \text{HORN}$ *is a reduction class.*

Proof. For the proof it suffices to give an effective reduction of the halting problem of arbitrary 2-register machine programs $M = (I_i)_{0 \le i \le r}$. We encode each M-configuration $C = (i, p, q)$ with state i and register contents (p, q) by the atomic formula $\underline{C} = K(state_i\underline{p}, \underline{q})$ where $\underline{m} = |^m * u$ for monadic (Skolem) functions $state_i, |, *$, a binary predicate K and arbitrary term u; think of $*u$ as denoting the empty word which stands here for the number 0. The interpretation of K is intended to satisfy the following *simulation property*, where as usual it suffices to consider M-computations which are started in the initial configuration $C_0 = (0, 0, 0)$:

$$\text{If } C_0 \Rightarrow_M (i, p, q), \text{ then } K(state_i\underline{p}, \underline{q}) \text{ is true for each } u.$$

Based upon this encoding we will construct a formula $\psi_M \in [\forall\exists^*\forall, (0, 4)] \cap$ KROM $\cap$ HORN with Skolem normal form $\forall x \forall y \varphi_M$ so that the following holds:

Reduction Property: M, started in $C_0 = (0, 0, 0)$, will not halt in state 1 if and only if the formula $\forall x \forall y \varphi_M$ is satisfiable.

This will establish the claim of the theorem.

As usual φ_M is defined as conjunction of formulae START, STEP_M, NONSTOP and an auxiliary formula AUX. For NONSTOP we assume without loss of generality that 1 is the only halting state of M; this allows us to set

$$\text{NONSTOP} := \neg K state_1 xy.$$

In START we have to formalize that the encoding $K(state_0 * u, *u)$ of $C_0 = (0,0,0)$ is true in every model of ψ_M for every possible value of u. This has to be expressed avoiding the nesting of $state_0$ and $*$. To this purpose we introduce an additional monadic function *start* – for which $K start\ x\ x$ is postulated – and an auxiliary predicate K^* which formalizes one application of the function $*$ to second arguments of K. Formally this is defined by the AUX-conjunct

$$\text{AUX.1} := K^* yx \leftrightarrow Ky * x.$$

Then START can be defined as follows:

$$\text{START} := K start\ x\ x \wedge (K^* start\ x\ y \rightarrow K^* state_0 xy).$$

In a similar way we can define the conjuncts ε_i of STEP_M which formalize instructions $I_i = (i, o_i, j, k)$ with operations o_i concerning the second register. We use another auxiliary predicate K' which formalizes one application of the function $|$ to second arguments in K, i.e. which satisfies the AUX-conjunct

$$\text{AUX.2} := K'yx \leftrightarrow Ky\,|x.$$

This allows us to set for $o_i = a_2$ (where without loss of generality $k = j$):

$$\varepsilon_i := K state_i xy \rightarrow K' state_j xy.$$

For $o_i = s_2$ we have two conjuncts, corresponding to the cases that the current value in the second register is zero or positive:

$$\varepsilon_i := (K^* state_i xy \rightarrow K^* state_j xy) \wedge (K' state_i xy \rightarrow K state_k xy).$$

It is more difficult to avoid the nesting of $state_i$ and $|$ or $*$ in the first argument position of K when formalizing the effect of an M-instruction $I_i = (i, o_i, j, k)$ on a configuration $C = (i, p, q)$ where the operation $o_i \in \{a_1, s_1\}$ concerns the first register. The idea to solve this problem can be described as follows. The transformation of $\underline{C} := K(state_i \underline{p}, \underline{q})$ into $K(state_{i'} \underline{p'}, \underline{q})$, corresponding to the successive configuration $C' = (i', p', q)$, is described by copying the state information into the second argument, then executing the operation o_i, followed by re-copying the state information back into the first argument from where the control is passed to the successive state. More precisely we have the following simulation phases.

Phase 0: We initialize the copying phase during which $state_i$ of $\underline{C}$ is stored into the second argument of K in order to bring $\underline{p}$ to the front, i.e. to make it accessible without nesting of functions. Formally we use ε_i-conjuncts $K state_i xy \rightarrow K optype_i xy$ with operation type $optype \in \{add, sub\}$ corresponding to $o_i \in \{a_1, s_1\}$ and with new functions add_i, sub_i.

Phase 1: We describe how to store the information on the state i – encoded in $K(optype_i \underline{p}, \underline{q})$ – by copying it (in the form of the sequence $*^i$) to the front of the second argument $\underline{q}$. This can be formalized by store formulae $Kadd_{l+1}xy \to K^*add_lxy$ and $Ksub_{l+1}xy \to K^*sub_lxy$ for $0 \leq l < i$. We thus can infer $K(optype_0\underline{p}, *^i|^q*u)$. This encoding works if we assume, without loss of generality, that M never comes into the situation to execute instructions concerning the first register when the second register has value $q = 0$.

Phase 2: We formalize the effect of the required operation on the first register. The case of a subtraction operation with positive register is encoded by drawing from $K(sub_0\underline{p}, *^i\underline{q})$ the conclusion $K(\underline{p}, *^i\underline{q})$ and then switching to $K(pos_0\underline{p-1}, *^i\underline{q})$ with a new function pos_0; formally this means to apply test axioms $Ksub_0xy \to Kxy$ and $K|xy \to Kpos_0xy$. For the other cases we use an auxiliary relation K^{next_0} to switch to $K(next_0\underline{p'}, |^i\underline{q})$ with a new function $next_0$; formally this means to use for subtraction instructions the axiom $Ksub_0xy \to Kxy$ as above, followed by the additional test axiom $K*xy \to K^{next_0}*xy$, and for addition instructions the addition axiom $Kadd_0xy \to K^{next_0}|xy$. K^{next_0} is formalized by the AUX-axiom

$$\text{AUX.3} := K^{next_0}xy \leftrightarrow Knext_0xy.$$

Phase 3: We describe how to load the state information $*^i$ in the result of phase 2 back from the second argument into an application of new functions $next_i$ or pos_i to the first argument. This will then allow us in the final phase to call K with the correct next state $state_j$ or $state_k$. Formally we can express this by load axioms $K^*next_lxy \to Knext_{l+1}xy$ and $K^*pos_lxy \to Kpos_{l+1}xy$ for $0 \leq l < i$ and new functions $next_l, pos_l$.

Phase 4: We transfer the control to the correct next state by the ε_i-conjuncts $K'pos_ixy \to K'state_kxy$ and $K'next_ixy \to K'state_jxy$ respectively.

This description is summarized by the following definition of ε_i for M-instructions $I_i = (i, o_i, j, k)$ and of AUX. For $o_i = a_1$ set

$$\varepsilon_i := (Kstate_ixy \to Kadd_ixy) \wedge (K'next_ixy \to K'state_jxy).$$

For $o_i = s_1$ set

$$\begin{aligned} \varepsilon_i \quad := \quad & (Kstate_ixy \to Ksub_ixy) \wedge (K'next_ixy \to K'state_jxy) \wedge \\ & (K'pos_ixy \to K'state_kxy). \end{aligned}$$

Further, let

$$\text{AUX} := \bigwedge_{1 \leq i \leq 3} \text{AUX.}i \wedge \text{STORE} \wedge \text{ADD} \wedge \text{TEST} \wedge \text{LOAD}$$

where

$$\begin{aligned}
\text{STORE} &:= \bigwedge_{0\le l<r} (Kadd_{l+1}xy \rightarrow K^*add_l xy) \wedge (Ksub_{l+1}xy \rightarrow K^*sub_l xy)\\
\text{ADD} &:= Kadd_0 xy \rightarrow K^{next_0}|xy\\
\text{TEST} &:= (Ksub_0 xy \rightarrow Kxy) \wedge (K*xy \rightarrow K^{next_0}*xy) \wedge\\
&\quad (K|xy \rightarrow Kpos_0 xy)\\
\text{LOAD} &:= \bigwedge_{0\le l<r} (K^*next_l xy \rightarrow Knext_{l+1}xy) \wedge\\
&\quad (K^*pos_l xy \rightarrow Kpos_{l+1}xy)
\end{aligned}$$

It remains to prove the reduction property. Assume that M, started in $C_0 = (0,0,0)$, will not halt in state 1. We then define the following canonical interpretation $\mathfrak{A}$ of K which satisfies $\forall x\forall y\varphi_M$. $\mathfrak{A} \models Kvw$ if and only if for some word u and some M-configuration $C_t = (i,p,q)$ and some $j \le i$ one of the following cases holds:

	$v = start\ u$	$w = u$		
or	$v = state_i\ \underline{p}$	$w = \underline{q}$		
or	$v = add_{i-j}\ \underline{p}$	$w = *^j\underline{q}$	$o_i = a_1$	
or	$v = next_{i-j}\ \underline{p+1}$	...	...	
or	$v = sub_{i-j}\ \underline{p}$	...	$o_i = s_1$	
or	$v = \underline{p}$	...	...	
or	$v = next_{i-j}\ \underline{0}$	...	...	$p = 0$
or	$v = pos_{i-j}\ \underline{p-1}$	...	...	$p > 0$

Exercise 5.1.11. Verify that $\mathfrak{A} \models \forall x\forall y\varphi_M$.

Exercise 5.1.12. Prove the if-direction of the reduction property, proving the simulation property indicated above and following the explanations given for the definition of the ε_i.

□

Exercise 5.1.13. Modify the proof to show that $[\forall\exists^*\forall, (0,4)] \cap$ KROM $\cap$ HORN is a conservative reduction class.

Exercise 5.1.14. Show that the classes $[\forall\exists^*\forall^2, (0,0,2)] \cap$ KROM $\cap$ HORN and $[\exists^2\forall\exists^*\forall^2, (0,2)] \cap$ KROM $\cap$ HORN are reduction classes.

Remark. It seems to be open whether $[\forall\exists^*\forall, (0,k)] \cap$ KROM is a reduction class for $k = 1,2,3$.

5.1.2 Krom Prefix Classes with Functions or Equality

In this section we collect what is known about Krom prefix classes with functions or equality.

Remark. In Exercise 2.1.18 (to Corollary 2.1.16 to the Aanderaa-Börger Theorem) it has been shown that $[\forall\exists\forall\exists, (\omega, k)]_= \cap$ KROM $\cap$ HORN is a reduction class for some k of the size of a universal 2-register machine. The extension of the decidable Aanderaa class to the class $[\forall\exists\forall]_= \cap$ KROM with equality is still decidable [133, page 204], whereas from the proof of the Chang-Keisler normal form in [75] it results that the extension of the Maslov class to the class $[\exists^*\forall^*\exists^*]_= \cap$ KROM is a reduction class. It seems to be unknown however whether the restriction of the Maslov class to the Gödel prefix and only one binary predicate $[\exists^*\forall^2\exists^*, (0,1)]_= \cap$ KROM is decidable. For formulae with functions but without equality we know by Theorem 4.0.1 that $[\forall^2, (0,1), (1)]$ is a reduction class, whereas its restriction to Krom formulae $[\forall^*, (\omega, \omega), (1)] \cap$ KROM is decidable (see [47]). The following exercise and theorem show that one additional monadic function suffices to get again a conservative reduction class.

Exercise 5.1.15. Infer from the construction given in Exercise 5.1.1 that $[\forall^2, (0,1), (2)] \cap$KROM$\cap$HORN is a reduction class. This improves Orevkov's reduction classes $[\forall^2, (0,1,2), (\omega)] \cap$ KROM and $[\forall^2, (k,l), (2)] \cap$ KROM for some not furthermore specified natural numbers k, l in [409]. Hint: Use that the Post correspondence problem over an alphabet with two letters is undecidable.

Theorem 5.1.16 (Börger). $[\forall^2, (0,1), (2)] \cap$ KROM $\cap$ HORN *is a conservative reduction class.*

Proof. For the proof it suffices to give an effective reduction of appropriate halting problems of arbitrary 2-register machine programs $M = (I_i)_{0 \le i \le r}$. We encode each M-configuration $C = (i, p, q)$ with state $0 \le i \le r$ and register contents (p, q) by the atomic formula $\underline{C} = K\underline{i}x\underline{(p,q)}x$ where $\underline{i} = *|^{i+1}*$ and $\underline{(p,q)} = |^p *^q *|*$ for monadic functions $|, *$ and a binary predicate K. The interpretation of K is intended to satisfy the following *Simulation Lemma*, where as usual it suffices to consider M-computations which are started in the initial configuration $C_0 = (0,0,0)$:

$$\text{If } C_0 \Rightarrow_M (i, p, q), \text{ then } K\underline{i}0\underline{(p,q)}0 \text{ is true.}$$

Based upon this encoding we will construct a formula $\psi_M \in [\forall^2, (0,1), (2)] \cap$ KROM $\cap$ HORN with Skolem normal form $\forall x \forall y \varphi_M$ so that the following *Reduction Property* holds:

1. If $C_0 = (0,0,0) \Rightarrow_M (1,0,0)$, then $\forall x \forall y \varphi_M$ is not satisfiable,
2. If $C_0 = (0,0,0) \Rightarrow_M (2,0,0)$, then $\forall x \forall y \varphi_M$ is finitely satisfiable.

This will establish the claim of the theorem.

As usual φ_M is defined as conjunction of formulae START, STEP$_M$, NONSTOP := $\neg K\underline{1}xy$ and an auxiliary formula AUX. In START we formalize that the encoding $K\underline{0}u\underline{(0,0)}u$ of $C_0 = (0,0,0)$ is true in every model of ψ_M for every possible value of u. This can be done by defining START $:= K\underline{0}x\overline{(0,0)}x$.

It is straightforward to define the conjuncts ε_i of STEP$_M$ which formalize instructions $I_i = (i, o_i, j, k)$ with operations o_i concerning the first register. For $o_i = a_1$ (where without loss of generality $k = j$) we set

$$\varepsilon_i := K\underline{i}xy \to K\underline{j}x|y.$$

For $o_i = s_1$ we have two conjuncts, corresponding to the cases that the current value in the second register is zero or positive:

$$\varepsilon_i := (K\underline{i}x * y \to K\underline{j}x * y) \wedge (K\underline{i}x|y \to K\underline{k}xy).$$

To formalize the effect of an M-instruction I_i on a configuration $C = (i, p, q)$ where the operation o_i concerns the second register we first store the content p of the first register from the second argument of $K\underline{i}x|^p *^q *| * x$ into the first one, then we execute the operation o_i on q, load the content of the first register back into the second argument and finally pass the control to the successive state. More precisely we have the following simulation phases.

Phase 0: We initialize the simulation with the copying phase during which the content p of the first register in $\underline{C}$ is stored into the first argument of K in order to bring the content q of the second register in front, i.e. to make it accessible using only a constant depth of nesting of functions. Formally we use three new combinations sim, add, sub of function nesting in ε_i-conjuncts $K\underline{i}xy \to K sim\ \underline{i}x\ optype_i y$ with operation type $optype_i \in \{add, sub\}$ corresponding to $o_i \in \{a_2, s_2\}$. We can choose for example $sim := *$, $add := *||$, $sub := **||$.

Phase 1: We describe how to store the content p of the first register in $K sim\ \underline{i}x\ optype_i\overline{(p,q)}x$ by copying it – in the form of the sequence $*^p$ – to the front of the first argument $sim\ \underline{i}x$. This can be formalized by store formulae $Kx\ optype|y \to K|x\ optype\ y$. They allow us to infer $K|^p sim\ \underline{i}x\ optype_i *^q *| * x$.

Phase 2: We formalize the effect of the required operation on the second register. The case of a subtraction operation with positive register is encoded by drawing from $K|^p sim\ \underline{i}x\ sub * *^{q-1} * | * x$ the conclusion $Kpos|^p sim\ \underline{i}x *^{q-1} *| * x$ for a new function nesting combination pos; choose for example $pos = *^7$. Formally we apply test axioms $Kx\ sub * *y \to Kpos\ x * y$. For the zero case we switch from $K|^p sim\ \underline{i}x\ sub * | * x$ to $Knext|^p sim\ \underline{i}x\ * | * x$ with a new function nesting combination $zero$; we choose $zero = *^4$. Formally we use the test axiom $Kxsub * |y \to Kzero\ x * |y$. For addition instructions we have the addition axiom $Kx\ add * y \to Knext\ * * y$ and choose $next = *^3$.

Phase 3: We describe how to load the register content p in the result of Phase 2 back from the first argument into the second argument. This will then allow us in the final phase to call K with the correct next state $\underline{j}$ or

$\underline{k}$. Formally we can express this by load axioms $Ksucc|xy \to Ksucc\ x|y$ for $succ \in \{next, zero, pos\}$.

Phase 4: We transfer the control to the correct next state by the ε_i-conjuncts $Knext\ sim\ \underline{i}xy \to K\underline{j}xy$, $Kzero\ sim\ \underline{i}xy \to K\underline{j}xy$ and $Kpos\ sim\ \underline{i}xy \to K\underline{k}xy$ respectively.

This description is summarized by the following definition of ε_i for M-instruction $I_i = (i, o_i, j, k)$ and of AUX. For $o_i = a_2$ define

$$\varepsilon_i := (K\underline{i}xy \to Ksim\ \underline{i}x\ add\ y) \wedge (Knext\ sim\ \underline{i}xy \to K\underline{j}xy).$$

For $o_i = s_2$ define

$$\begin{aligned} \varepsilon_i \quad := \quad & (K\underline{i}xy \to Ksim\ \underline{i}xsub\ y) \wedge (Kzero\ sim\ \underline{i}xy \to K\underline{j}xy) \wedge \\ & (Kpos\ sim\ \underline{i}xy \to K\underline{k}xy). \end{aligned}$$

Further, define

$$\text{AUX} := \text{STORE} \wedge \text{ADD} \wedge \text{TEST} \wedge \text{LOAD}$$

where

$$\begin{aligned} \text{STORE} &:= (Kx\ add|y \to K|x\ add\ y) \wedge (Kx\ sub|y \to K|xsub\ y) \\ \text{ADD} &:= Kx\ add * y \to Knext\ x * *y \\ \text{TEST} &:= (Kxsub * |y \to Kzero\ x * |y) \wedge (Kx\ sub * *y \to Kpos\ x * y) \\ \text{LOAD} &:= \bigwedge_{succ \in \{next, zero, pos\}} Ksucc|xy \to Ksucc\ x|y. \end{aligned}$$

Exercise 5.1.17. Prove the first claim of the reduction property. Hint: Use the following simulation property: For each canonical model $\mathfrak{A}$ satisfying $\forall x \forall y \varphi_M$ and for each t and each M-configuration C which is reached by M in t steps starting from C_0, we have that $\mathfrak{A} \models K\underline{i}u\underline{(p,q)}u$ for each value u.

Exercise 5.1.18. Prove the second claim of the reduction property. Hint: Show that the following canonical model $\mathfrak{A}$ satisfies $\forall x \forall y \varphi_M$ and can be made finite by restricting the functions to a finite domain along the lines of exercise 5.1.5. $\mathfrak{A} \models Kvw$ if and only if for some word u, some natural numbers i, p, q, k, l and some M-configuration C reached by M from C_0 one of the following cases holds:

	$v = \underline{i}u$	$w = \underline{(p,q)}u$	$C = (i,p,q)$	
or	$v = \|^k sim\ \underline{i}u$	$w = \underline{add(l,q)}u$	$C = (i, l+k, q)$	$o_i = a_2$
or	$v = next\|^k sim\ \underline{i}u$	$w = \underline{(l,q+1)}u$	...	...
or	$v = \|^k sim\ \underline{i}u$	$w = \underline{sub(l,q)}u$	...	$o_i = s_2$
or	$v = zero\|^k sim\ \underline{i}u$	$w = \underline{(l,0)}u$	$C = (i, l+k, 0)$	...
or	$v = pos\|^k sim\ \underline{i}u$	$w = \underline{(l,q)}u$	$C = (i, l+k, q+1)$	...

Note that these cases are disjoint due to the choice of the pairwise different function nesting combinations *sim,add,sub,next,zero,pos* which serve to distinguish between the different phases and cases of each simulation.

□

Exercise 5.1.19. [47] Prove the reduction class property for $[\forall^2, (1), (0,1)] \cap$ KROM∩HORN by encoding Post correspondence problems into it. Hint: For a given Post correspondence system $(v_i, w_i)_{1 \le i \le n}$ over alphabet $\{a_1, \ldots, a_m\}$ encode words over this alphabet by the Fitch word

$$(t)a_i := ({}^i tt)t) \ldots t), \qquad (t)a_0 := t$$

where t is an arbitrary Fitch word (term built up from the binary function ( ) and the variables x, y). Use parenthesis association to the left when writing $(t)a_{j_1} \ldots a_{j_r}$ for $(\ldots((t)a_{j_1})a_{j_2} \ldots)a_{j_r}$. Use as reduction formula the universal closure of the conjunction of the START formulae $P(x)v_i(x)w_i$, the STEP formulae $Pxy \to P(x)v_i(y)w_i$ and the NO-SOLUTION formula $\neg Pxx$.

Exercise 5.1.20. Show that the class $[\forall^2, (1), (0,1)] \cap$ KROM ∩ HORN is a conservative reduction class by reducing to it the class $[\forall^2, (0,1), (2)] \cap$ KROM ∩ HORN. Hint: Use the reduction method explained for the proof of Theorem 4.1.11.

Exercise 5.1.21. [275] Show that $[\forall^2, (1), (0,1)] \cap$HORN is a reduction class when restricted to formulae which are in HERBRAND except for one ternary disjunct. Hint: Use the undecidability of partial implicational propositional calculi C with only two variables. Interpret the propositional implication $\to$ as a binary function so that the finitely many axioms a_i of C become a HERBRAND formula $\mathrm{AX}_C := \bigwedge_i Pa_i$ with monadic "C-derivability" predicate P. Formalizing the *Modus Ponens* by

$$\mathrm{MP} := (Px \wedge P(x \Rightarrow y) \to Py)$$

prove that for each formula ψ of C the reduction formula $\forall x \forall y(\mathrm{AX}_C \wedge \mathrm{MP}) \to \psi$ is of the required form and is satisfiable if and only if ψ is derivable in C.

The following two exercises resume what is known about the prefix-vocabulary problem for Krom classes of formulae with functions but without equality.

Exercise 5.1.22. Show that for numbers k, l, m, n satisfying $k+l, m+n \neq 0$ and for arbitrary extended prefixes $\Pi \neq \forall^2\exists^p, \forall^2\exists^*$, for any number p, the following holds: The class $[\Pi, (k,l), (m,n)] \cap$ KROM is undecidable (and a reduction class) if and only if a) Π contains at least two universal quantifiers, b) the vocabulary contains at least one binary predicate or one binary function, i.e. $l \neq 0$ or $n \neq 0$, c) if the vocabulary does not contain any binary function, then it contains at least two monadic functions or Π contains at least one existential quantifier. Hint: Use the decidability of the purely existential case, of the Gurevich-Maslov-Orevkov class $[\exists^*\forall\exists^*, all, all]$, of the full monadic class $[all, (\omega), (\omega)]$ and of the class $[\forall^*, (\omega, \omega), (1)] \cap$ KROM (see [42]).

Exercise 5.1.23. Show that the classes $[\forall^3, (0,0,0,1), (1)] \cap$KROM$\cap$HORN and $[\exists\forall^2, (0,0,1), (1)] \cap$ KROM $\cap$ HORN are reduction classes. The maximal value of $0 \leq k < \omega$ such that $[\forall^3, (0,0,k), (1)] \cap$ KROM, $[\exists\forall^2, (\omega, k), (1)] \cap$ KROM, $[\forall\exists\forall, (\omega, k), (1)] \cap$ KROM, $[\exists^*\forall^2\exists^*, (0,k), (1)] \cap$ KROM are decidable seems to be unknown. We have the impression (but no proof) that $[\forall^2, all, (1)] \cap$ KROM and similar classes are decidable.

Also the prefix-vocabulary problem for Krom classes of formulae with functions and with equality has not yet been completely solved. Let us recall that in previous sections the following classes have been shown to be undecidable: the two classes of Herbrand formulae $[\forall, (0), (2)]_= \cap$ HERBRAND, $[\forall, (0), (0,1)]_= \cap$HERBRAND, the function free class $[\forall\exists\forall\exists, (0,k)]_= \cap$KROM$\cap$ HORN, the two equality free classes $[\forall^2, (0,1), (2)] \cap$ KROM $\cap$ HORN and $[\forall^2, (1), (0,1)] \cap$ KROM $\cap$ HORN, the equality and function free Aanderaa-Börger classes $[\exists\forall\exists\forall, (\omega, k)] \cap$ KROM $\cap$ HORN and $[\forall\exists\exists\forall, (\omega, k)] \cap$ KROM $\cap$ HORN and Lewis classes $[\forall^2\exists\forall, (0, \omega, k)]$ and $[\forall\exists\forall^2, (0, \omega, k)]$ for some k. From the decidability side we make use of the following known decidable classes of formulae with functions and with equality: the Rabin class $[all, (\omega), (1)]_=$, the existential class $[\exists^*, all, all]_=$ and Shelah's class $[\exists^*\forall\exists^*, all, (1)]_=$ (see Chap. 7).

Exercise 5.1.24. Using the preceding list of decidable and undecidable Krom classes with equality and functions, prove that the decision problem for classes $[\Pi, (p_n)_n, (f_n)_n)]_=$KROM is settled in the following cases: classes of (besides the equality) monadic formulae, of purely existential formulae, of formulae containing a binary function, of formulae with exactly one universal quantifier, of formulae with at least two functions, of formulae with at least one existential quantifier, of formulae with at least three universal quantifiers. (This holds except for determining the exact value of the number k of binary or ternary predicates for which the classes $[\exists\forall\forall, (0,k), (1)]_=$KROM, $[\forall\exists\forall, (0,k), (1)]_=$KROM, $[\forall\forall\exists, (0,k), (1)]_=$KROM are decidable.) The decision problem of the classes $[\forall\forall, (0,k), (1)]_=$KROM and $[\forall\forall, (0,0,k), (1)]_= \cap$ KROM seems to be open.

5.2 Few Atomic Subformulae

In this section we show some undecidable classes of formulae which have only a small number of atomic subformulae. Little is known about the decision problem for classes of formulae which are determined by restrictions on the number of atomic subformulae occuring in a formula, although recently some more effort has been put into this classification through investigations in the theory of logic programming where such formulae are viewed as syntactically restricted programs. Before showing some relevant proof methods we first survey what is known in the literature.

Let us start with first-order logic without functions or equality. Dreben and Goldfarb [133] proved that the class of formulae which contain at most two distinct atomic subformulae is decidable and has the finite model property. Orevkov [409] showed the undecidability of the class of formulae in prenex disjunctive normal form with two disjuncts. This result has been strengthened by Lewis and Goldfarb [354] who showed the reduction class property for the class of formulae containing five atomic subformulae where those formulae can be chosen to be in the class $\forall\exists\forall^*$ with quantifier-free part of the form $(\neg\pi_1 \wedge \pi_2 \wedge \pi_3) \vee (\neg\pi_4 \wedge \pi_5)$. They also showed that the same holds with quantifier-free part of Krom form $(\pi_1 \vee \pi_2) \wedge (\neg\pi_3 \vee \neg\pi_4) \wedge (\neg\pi_3 \vee \neg\pi_5)$ or $(\pi_1 \wedge \bigwedge_{i=3,4}(\pi_2 \rightarrow \pi_i) \wedge \neg\pi_5)$ and also for $\forall^*\exists$-formulae containing six atomic subformulae whose quantifier-free part is in conjunctive normal form with three conjuncts, namely of form $(\pi_1 \vee \pi_2 \vee \pi_3) \wedge \bigwedge_{i=5,6}(\pi_4 \rightarrow \pi_i)$. Note in this connection that any class of prenex function and equality free formulae which is determined by restricting both the number of atomic formulae and the number of universal quantifiers is reducible to a finite class of formulae, as observed by Lewis [345]. The decision problem for function and equality free formulae with only three or four atomic subformulae seems still to be open, in particular for formulae with quantifier free Krom formula of form $(\pi_1 \wedge (\pi_2 \rightarrow \pi_3) \wedge \neg\pi_4)$. We will see below that an answer to these problems can be given if functions are allowed to occur. Also open is the decision problem for function and equality free formulae with two conjuncts.

If functions are allowed to occur, then one can obtain reduction classes of formulae where both the number of atomic formulae and the number of quantifiers (and thereby also the arity of the occuring predicates and functions) can be limited by a fixed finite number. For first-order logic with functions and equality the following is known. From Gladstone's proof [182] it can be easily inferred that the class of Herbrand formulae with only one (atomic or negated atomic) subformula is decidable and has the finite model property. Wirsing [534] showed that the class of Herbrand formulae with only two subformulae, one equality and one inequality, is a conservative reduction class. In contrast Exercise 5.2.1 shows that for formulae in prenex *disjunctive* normal form the class of formulae with only two subformulae is decidable. Wirsing needed only the $[\forall^6, (0), (0,1)]_=$-fragment for his two-subformulae class and improved in [535] the prefix to $[\forall^3, (0), (0,1)]_=$ at the expense of allowing two equations besides the one inequality, and also to $[\forall, (0), (0,1)]_=$ or $[\forall, (0), (2)]_=$ allowing however three equations. In [534] Wirsing showed also that the class $[all, all, (1)]_= \cap \text{HERBRAND}$ is decidable by reducing it to S2S (see Chap. 7).

Exercise 5.2.1. [534] Show that in first-order logic with functions and equality, the class of formulae in prenex disjunctive normal form with only two subformulae is decidable. Hint: Use Gladstone's decision procedure for inequations [182] and the following equivalence, for any prefix Π: $\Pi(s \neq t \vee u \neq v)$ is satisfiable if and only if $\Pi(f(s,u) \neq f(t,v))$ is satisfiable, where f is a new binary function.

Although the original question whether in the Lewis-Goldfarbreduction class of equality and function free formulae with quantifier free part of form $(\pi_1 \wedge \bigwedge_{i=3,4}(\pi_2 \rightarrow \pi_i) \wedge \neg\pi_5)$ the number of implications can be reduced to one is still open, an answer can be given for first-order logic with functions but without equality. We show that adapting Wirsing's proof technique to Post correspondence problems yields the reduction class property for the class of prenex formulae with quantifier-free part of form $\pi \wedge (\rho \rightarrow \sigma) \wedge \neg\tau$ with atomic formulae π, ρ, σ, τ in which one predicate (of arity 4), one binary and two monadic functions occur. In the theory of logic programming this class, called the set of programs consisting of one binary Horn rule, one fact and one goal, without considering the finer analysis with respect to prefix and term structure and the arity of occurring predicates, has received some particular attention (see the discussion below).

In connection with prefix and with prefix vocabulary classes another notion of "small" formulae has been studied whether reduction formulae can be made into conjunctions of "short" formulae with "simple" prefix.

5.2.1 Few Function and Equality Free Atoms

In this section we prove Orevkov's Theorem that the class of (equality and function free) formulae in prenex disjunctive normal form with only two disjuncts is a reduction class. We prove the stronger version which has been established by Lewis and Goldfarb and where the reduction formulae have a prefix of form $\forall\exists\forall^*$ and contain only one predicate symbol (whose arity depends on the length of a universal 2-register machine program).

Theorem 5.2.2 (Lewis, Goldfarb). *The class of (function and equality free) formulae in $\forall\exists\forall^*$ with quantifier free Krom and Horn part of form*

$$\text{START} \wedge (\text{NF} \rightarrow \text{SHIFT}) \wedge (\text{NF} \rightarrow \text{PROGRAM}) \wedge \neg\text{STOP}$$

is a reduction class, where START, NF, SHIFT, PROGRAM *and* STOP *are atomic formulae.*

Proof. So far, in the reductions of halting problems of machines M to satisfiability problems of logical formulae ψ_M, we have described each possible elementary machine step (1-step transition) from a configuration C with state i *locally*, i.e. by one or more corresponding propositional formulae ε_i occurring in ψ_M. Hence the reduction formula ψ_M grows with M (or with the input n to a universal machine M). To avoid this growth we look for a *global* description of a relation $P\underline{C}\,\underline{M}$ between (the encodings of) any reachable configuration C and the whole machine program M such that every 1-step transition of M can be expressed, in terms of this relation, by comparing (the encoding of) the current configuration C with the (encoding of the) program M and by updating $\underline{C}$ accordingly to $\underline{C'}$. We have therefore to find a logical encoding $\underline{C}$ of configurations and $\underline{M}$ of programs such that this simulation can be formalized by simple propositional combinations of a few atomic formulae.

The basic idea for describing such a global relation, where $P\underline{C}\,\underline{M}$ implies $P\underline{C'}\,\underline{M}$, is a) to first rotate $\underline{M}$ until that instruction I comes in front which can be applied to C, and then b) to put C' in place of C (by applying I to the latter) and to restore $\underline{M}$.We formalize the "application of I to C" by providing in the encoding $\underline{M}$, for each configuration C, the sequence of all possible pairs (C, C'') of immediately successive M-configurations, i.e. such that $C \Rightarrow^1_M C'$.

To be more specific about the encoding details let now $M = (I_i)_{1 \le i \le r}$ be an arbitrary 2-register machine. For convenience but without loss of generality we assume that M has only instructions of form $I_i = (i, a_2, j)$ – in state i add 1 to the content of the second register and go to state j – or $I_i = (i, sc_1, j, k)$ – in state i, if the number in the first register is 0, then interchange the two registers and go to state j, otherwise subtract 1 from the content of the first register and go to state k. [1]

We want to describe M-computations by a short formulae ψ_M with prefix of form $\forall x \exists w \forall y \forall z \forall u \forall v \ldots$ which by going to the Skolem normal form provides a successor function but no explicit way to express the number 0. Therefore we adapt the idea which has been introduced already in the proof for the two Lewis classes $[\forall^2\exists\forall]$, $[\forall\exists\forall^2]$ (see Theorem 5.1.2), namely to encode register contents p as offset from a "relative" 0, say z. For the representation of M-states i we choose a form of binary code with respect to some variables, say u, v (and their instantiations by numbers a, b), namely

$$i(u,v) := \underbrace{u \ldots u}_{i-1}\; v\; \underbrace{u \ldots u}_{r-i}$$

which is intended to be used with different values a, b for u and v. This yields the following encoding of M-configurations $C = (i, p, q)$ with respect to numbers a, b, c into sequences (which will appear in the sequence of arguments of the relation P to be formalized):

$$\underline{C}(a,b,c) := i(a,b)\; p + c\; q + c\;\; c.$$

Since we will deal mainly with the case $c = 0$, we write also

[1] The proof of the Theorem of Minsky [393] and Shepherdson and Sturgis [469] implies that these restricted 2-register machines are computation universal and therefore have an unsolvable halting problem. See for example the proof in [57].

$$\underline{C}(a,b) := i(a,b)\, pq.$$

For the pairing of immediately successive configurations we have to take into account that for configurations with subtraction and register interchange state i there are two cases for immediately successive configurations to consider, depending on whether the first register in C contains 0 or not. We therefore define the following sequence of all possible pairs $step_i$ of immediately successive M-configurations, for any given pair of register contents. Let r be the only stop state of M, $1 \le i \le r-1, 1 \le j,k \le r$. Remember that x' stands for the Skolem function applied to x and that z serves as relative 0. The reader will notice that the following definition of $step_i$ rephrases the formalization of instructions of 2-register machine programs by implications ε_i which appears in the proof for the Aanderaa-Börger Theorem (Theorem 2.1.15).

For $I_i = (i, a_2, j) \in M$ we set

$$step_{2i-1} = step_{2i} = i(u,v)yx\; j(u,v)yx'.$$

For $I_i = (i, sc_1, j, k) \in M$ we set

$$step_{2i-1} = i(u,v)zy\; j(u,v)yz, \qquad step_{2i} = i(u,v)x'y\; k(u,v)xy.$$

Note that in this definition $i < r$ because no successive configuration of a configuration with stop state r needs to be considered.

Let $\mathrm{SUCC}(x,y,z,u,v) := step_1 \dots step_{2r-2}$. This sequence represents the encoding $\underline{M}$ of the given program, i.e. the sequence of the logical form of all pairs of immediately successive configurations which we may have to inspect to determine the immediate successive configuration of a given configuration C, where x, y will be instantiated according to the register contents appearing in C.

We are now almost ready for defining the reduction formula ψ_M. Let l denote the length of the encoding of $\underline{C}$ and $\underline{M}$, i.e. $l = r+3+r'm$ with the number r' of configuration pairs ($r' = 2r-2$) and their length $m = 2r+4$. It is easy to formalize the shift of $\underline{M} = step_1 \dots step_{r'}$ in $P\underline{C}(a,b,0)\underline{M}$ until that $step_k$ (for some $1 \le k \le r'$) stays in front which can be instantiated to $step_k[p,q,0,a,b] = \underline{C}(a,b)\underline{C}'(a,b)$; then the M-transition which is encoded in $step_k$ can be applied to $\underline{C}$ by copying $\underline{C}'(a,b,0)$ into the current configuration area. The shift can be obtained by shifting first $step_1$, then $step_2$, etc., i.e. by iterated shifts of m-blocks from the middle of $P\underline{C}\,\underline{M}$ to the end. This can easily be described by a formula of the following form:

$$P - - - w_1 \dots w_m \dots \to P - - - \dots w_1 \dots w_m.$$

A slightly more sophisticated trick will be used to "apply" $step_k$ to $\underline{C}$ and to restore the original program $\underline{M}$ (for use in the next simulation step). We have to formalize the condition that the state and register part $x_1 \dots x_{r+2}$ of the $\underline{C}$-section in the argument sequence $\bar{x} = x_1 \dots x_{r+2} \dots x_l$ of $P\underline{C}step_k \dots$

is identical to the first half $x_{r+4} \dots x_{2r+5}$ of $step_k = x_{r+4} \dots x_{3r+7}$; under that condition $P\underline{C}step_k \dots$ can be replaced by $Px_{2r+6} \dots x_{3r+7}z\underline{M}$, where z represents the relative zero. To do this with a small number of atoms we introduce in addition to the sequence $\bar{x}$, of length l, of universally quantified variables which appear as arguments of P, two more such sequences $\bar{y}, \bar{z}$ which are intended to be identical when they appear as second and third l-section of the argument sequence of P; this allows us to describe the required transition from $P\underline{C}step_k \dots$to $P\underline{C'}\ \underline{M}$ by an implication of the following form:

$$P\bar{x}\bar{z}\bar{z} \to Py_{2r+6} \dots y_{3r+7}z\underline{M}\ y_1 \dots y_{r+2}zy_1 \dots y_{r+2}y_{2r+6} \dots y_l\ \bar{x}.$$

This explains the definition of the reduction formula ψ_M as formula with the following Skolem normal form, where we use $\bar{x}, \bar{y}, \bar{z}$ for vectors of length l and P for a $3l$-ary relation:

$$\forall x \forall y \forall z \forall u \forall v \forall \bar{x} \forall \bar{y} \forall \bar{z} \varphi_M$$

where φ_M is the conjunction of the following formulae:

$$\begin{aligned}
\text{START} &:= P1(u,v)zzz\ \text{SUCC}(x,y,z,u,v)\ \bar{y}\bar{y} \\
\text{STOP} &:= Pr(x,x')zzz\ x_{r+4} \dots x_l\ \bar{y}\bar{y} \\
\text{SHIFT} &:= P\bar{x}\bar{z}\bar{z} \to Py_1 \dots y_{r+3}y_{r+m+4} \dots y_l y_{r+4} \dots y_{r+m+3}\bar{y}\bar{x} \\
\text{PROGRAM} &:= P\bar{x}\bar{z}\bar{z} \to \beta \ \text{ where} \\
\beta &:= Py_{2r+6} \dots y_{3r+7}z\underline{M}\ \ y_1 \dots y_{r+2}z\ y_1 \dots y_{r+2}y_{2r+6} \dots y_l\ \ \bar{x}.
\end{aligned}$$

We will show now that the following *Reduction Property* holds:

1. If ψ_M is satisfiable, then $C_0 = (1,0,0) \not\Rightarrow_M (r,0,0)$.
2. If $C_0 = (1,0,0) \not\Rightarrow_M (1,0,0)$, then ψ_M is satisfiable.

This will establish the claim of the theorem.

For the proof of the first part of the reduction property it suffices to prove the following *Simulation Lemma.*

Lemma 5.2.3 (Simulation Lemma). *Let $\mathfrak{A}$ be a model satisfying $\overline{\forall}\varphi_M$. Let t be an arbitrary natural number and C an M-configuration such that $(1,0,0) \Rightarrow_M^t C$. Then there is for all $a,b \in A$ and all numbers p,q a sequence $\bar{c}$ such that*

$$\mathfrak{A} \models P\underline{C}(a,b,0)\text{SUCC}(p,q,0,a,b)\bar{c}\,\bar{c}.$$

The lemma and the conjunct $\neg$STOP imply that if ψ_M is satisfiable, then $C_0 = (1,0,0) \not\Rightarrow_M (r,0,0)$.

The simulation can be proved by induction on t. For $t = 0$ the claim is guaranteed by the conjunct START in φ_M. For the inductive step let $(1,0,0) \Rightarrow_M^t C \to_M^1 C'$ and

$$\mathfrak{A} \models P\underline{C}(a,b,0)\text{SUCC}(p,q,0,a,b)\bar{c}\,\bar{c}$$

for some $\bar{c}$ given by the inductive hypothesis. We can apply to this formula the SHIFT-conjunct of φ_M a finite number of times, starting to instantiate $y_1 \dots y_{r+3}$ by $\underline{C}(a,b,0)$, $y_{r+4} \dots y_{r+m+3}$ by $step_1$ and $y_{r+m+4} \dots y_l$ by $step_2 \dots step_{r'}$. Iterating such SHIFT applications we obtain for each $1 \le k \le r'$:

$$\mathfrak{A} \models P\underline{C}(a,b,0) step_k \dots step_{r'} step_1 \dots step_{k-1}[p,q,0,a,b]\bar{c'}\ \bar{c'}$$

for $\bar{c'} = \underline{C}(a,b,0)\mathrm{SUCC}(p,q,0,a,b)$. We now choose k and register contents p', q' such that

$$step_k[p',q',0,a,b] = \underline{C}(a,b)\underline{C}'(a,b).$$

Then we apply the φ_M-conjunct PROGRAM by instantiating $y_{2r+6} \dots y_{3r+7}$ to $\underline{C}'(a,b)$, $y_1 \dots y_{r+2}$ to $\underline{C}(a,b)$, x, y to arbitrary $p, q$and z to 0 and $y_{3r+8} \dots y_l$ to $step_{k+1} \dots step_{r'} step_1 \dots step_{k-1}[p,q,0,a,b]$. We obtain for each p, q and some $\bar{c''}$, as was to be shown, that:

$$\mathfrak{A} \models P\underline{C}'(a,b,0)\mathrm{SUCC}(p,q,0,a,b)\bar{c''}\bar{c''}.$$

For the other direction of the reduction property assume $C_0 = (1,0,0) \not\Rightarrow_M (1,0,0)$. We define a model $\mathfrak{A}$ over the natural numbers which satisfies ψ_M.

For the definition of $\mathfrak{A}$ we will consider the natural extension $\Rightarrow_M^*$ of $\Rightarrow_M$ to integers $0, 1, -1, 2, -2, \dots$. Observe that this does not alter the halting problem $(1,0,0) \Rightarrow_M (r,0,0)$ of M because if starting in $(1,0,0)$, M at some point subtracts from 0, then the following computation will never come back to non-negative content in the first register; formally $(1,0,0) \Rightarrow_M^* (r,0,0)$ if and only if $(1,0,0) \Rightarrow_M (r,0,0)$.

Let *state* be the following function on r-tuples of natural numbers: $state(p_1,\dots,p_r) = i$ if and only if $(p_1,\dots,p_r)$ is an instance of $i(u,v)$, for $1 \le i \le r$, in which the substituents for u and v are distinct; otherwise $state(p_1,\dots,p_r) = 0$. We define now for arbitrary numbers p_i, q_i, r_i: $\mathfrak{A} \models Pp_1 \dots p_l q_1 \dots q_l r_1 \dots r_l$ if and only if

1. either $(q_1,\dots,q_l) \neq (r_1,\dots,r_l)$ or
2. the following two properties hold:
   - if $state(p_1,\dots,p_r) \neq 0$, then $(1,0,0) \Rightarrow_M^* (state(p_1,\dots,p_r), p_{r+1} - p_{r+3}, p_{r+2} - p_{r+3})$, and
   - for some $0 \le j \le 2r-3$, $p_{r+4+mj} \dots p_l p_{r+4} \dots p_{r+3+mj}$ is an instantiation of $\mathrm{SUCC}(x,y,z,u,v)$ with p_{r+3} instantiated to 0.

$\mathfrak{A} \models \overline{\forall}\neg\mathrm{STOP}$ because by definition $state(p,\dots,p,p+1) \neq 0$ and by assumption $(1,0,0) \not\Rightarrow_M^* (r,0,0)$.

$\mathfrak{A} \models \overline{\forall}\mathrm{START}$ because for every instance $Pp_1 \dots p_l q_1 \dots q_l r_1 \dots r_l$ of START either $state(p_1,\dots,p_r) = 0$ or $state(p_1,\dots,p_r) = 1$ and $p_{r+1} - p_{r+3} = 0 = p_{r+2} - p_{r+3}$. Furthermore $p_{r+4} \dots p_l$ is an instantiation of $\mathrm{SUCC}(x,y,z,u,v)$ with p_{r+3} instantiated to 0.

$\mathfrak{A} \models \overline{\forall}\mathrm{SHIFT}$ because if $\mathfrak{A} \models Pp_1 \dots p_l\bar{q}\bar{q}$, then the sequence

$$p_1 \dots p_{r+3} p_{r+m+4} \dots p_l p_{r+4} \dots p_{r+m+3}$$

is obtained from $p_1 \dots p_l$ by a shift of the arguments at the positions $r + 4, \dots, r + m + 3$ to the right end. $\mathfrak{A} \models \overline{\forall}$PROGRAM follows immediately from the definition of PROGRAM and of $\mathfrak{A}$ (remembering hat $\underline{M}$ is just another notation for SUCC(x, y, z, u, v)). □

As a corollary of (the proof of) the Theorem of Lewis and Goldfarb one obtains Orevkov's Theorem.

Theorem 5.2.4 (Orevkov). *The class of (function and equality free) formulae in prenex disjunctive normal form with only two disjuncts is a reduction class. The quantifier free part can be chosen to be in form*

$$(\text{START} \wedge \neg\text{NF}) \vee (\text{SHIFT} \wedge \text{PROGRAM} \wedge \neg\text{STOP})$$

and the prefix of form $\forall\exists\forall^*$, *where* START, NF, SHIFT, PROGRAM *and* STOP *are atomic formulae.*

Exercise 5.2.5. Infer Orevkov's Theorem from the proof of Theorem 5.2.2.

Exercise 5.2.6. [354] Show that the class of (function and equality free) formulae in $\forall\exists\forall^*$ with quantifier free Krom and Horn part of form $(\pi_1 \vee \pi_2) \wedge (\neg\pi_3 \vee \neg\pi_4) \wedge (\neg\pi_3 \vee \pi_5)$ is a reduction class. Hint: Reduce formulae

$$\forall x \exists w \forall \bar{z} (\neg P\bar{u}_1 \wedge P\bar{u}_2 \wedge P\bar{u}_3) \vee (\neg P\bar{u}_4 \wedge P\bar{u}_5)$$

obtained in Theorem 5.2.2 to formulae with prefix $\forall x \exists w \forall \bar{z} \forall \bar{z}'$, where $\bar{z}'$ is of length $5k$ for the length k of each $\bar{u}_i$, and with quantifier free part

$$(Q\bar{u}_1\bar{u}_2\bar{u}_3 \vee Q\bar{u}_4\bar{u}_5\bar{u}_5) \wedge (\neg Q\bar{v}_0\bar{v}_1\bar{v}_2 \vee \neg Q\bar{v}_1\bar{v}_3\bar{v}_4) \wedge (\neg Q\bar{v}_0\bar{v}_1\bar{v}_2 \vee \neg Q\bar{v}_2\bar{v}_3\bar{v}_4)$$

where $\bar{v}_i = z'_{ik+1} \dots z'_{(i+1)k}$. $Q\bar{p}\bar{q}\bar{r}$ interpreted by $\neg P\bar{p} \wedge P\bar{q} \wedge P\bar{r}$, vice versa $P\bar{p}$ interpreted by $\neg Q\bar{p}\bar{q}\bar{r}$ for all $\bar{q}\bar{r}$.

Exercise 5.2.7. [354] Show that the class of (function and equality free) formulae in $\forall^*\exists$ with quantifier free part of form $(\pi_1 \vee \pi_2 \vee p_3) \wedge (\neg\pi_4 \vee \neg\pi_5) \wedge (\neg\pi_4 \vee \pi_6)$ is a reduction class.

5.2.2 Few Equalities and Inequalities

In this section we prove Wirsing's Theorem which shows that the class of universal formulae with one equality and one inequality is a conservative reduction class.

Theorem 5.2.8 (Wirsing). *The class of universal formulae containing only one equality and one inequality is a conservative reduction class. Moreover this result holds for the class of formulae of form* $\overline{\forall}(s = t \wedge u \neq v) \in [\forall^6, (0), (0,1)]_=$.

Proof. We prove the result first for the class $[\forall^6, (0), (\omega, \omega)]_=$ and then apply some standard encoding of finitely many monadic and binary functions into one binary function.

For the proof of the first part it suffices to give an effective reduction of appropriate halting problems of arbitrary 2-register machine programs $M = (I_i)_{1 \le i \le r}$ to the satisfiability problem of reduction formulae ψ_M in the class considered. The general idea of the formalization of 2-register machine halting problems by (in)equalities is similar to the one used already in Chap. 4 (for the proof of Corollary 4.1.3), namely a) to equate all configurations through which M passes when started in the given initial configuration, and b) to require that the (formalization of the) start configuration is different from the (formalization of the) stop configuration. For this purpose we encode M-configurations $C = (i, p, q)$ with state $1 \le i \le r$ and register contents (p, q) by terms $\underline{C}(t) = f_i(f^p f_0(t), f^q f_0(t))$ where t is any term, f_i $(1 \le i \le r)$ are binary and f, f_0 unary functions standing for the machine states i, the successor function and the number 0 respectively. The problem is to find a way to encode all 1-step transitions of M by one single equation.

This problem can be solved by adapting the idea underlying the proof of the Theorem of Lewis and Goldfarb (Theorem 5.2.2) as follows. In Chap. 4 (Corollary 4.1.3) we have described each single M-instruction locally, by one or more equations. We now encode, for given arbitrary register contents p, q, every possible 1-step transition from $C = (i, p, q)$ to $C' = (i', p', q')$ into a pair $[C, C']$ of immediately successive configurations and collect all the pairs, for each possible M-state i of C, in one sequence determined by p, q; the one equation and the one inequality which are available for the formalization of M-computation steps are then formulated globally in terms of such sequences. $[\,]$ is a binary function.

As in the proof of the Lewis-Goldfarb Theorem, for the pairing of successive configurations we have to take into account that for some states i of M there are two possible immediately successive configurations of C to consider, namely if a subtraction and test instruction is going to be executed whose result depends on whether the given register content p, q respectively is 0 or not. We therefore define the following sequence of all possible pairs of immediately successive M-configurations, for any given pair (p, q) of register contents. Let $r - 1$ and r be the only two stop states of M, $1 \le i \le r - 2, 1 \le j, k \le r$ and let p, q be individual variables (standing for arbitrary register contents).

For $I_i = (i, a_1, j) \in M$ we set

$$step_{2i-1}(p, q) = step_{2i}(p, q) = [f_i(p, q), f_j(f(p), q)].$$

For $I_i = (i, a_2, j) \in M$ we set

$$step_{2i-1}(p, q) = step_{2i}(p, q) = [f_i(p, q), f_j(p, f(q))].$$

For $I_i = (i, s_1, j, k) \in M$ we set

$$step_{2i-1}(p,q) = [f_i(f_0(p),q), f_j(f_0(p),q)],$$
$$step_{2i}(p,q) = [f_i(f(p),q), f_k(p,q)].$$

For $I_i = (i, s_2, j, k) \in M$ we set

$$step_{2i-1}(p,q) = [f_i(p, f_0(q)), f_j(p, f_0(q))],$$
$$step_{2i}(p,q) = [f_i(p, f(q)), f_k(p,q)].$$

Let $r' = 2r-4$. The sequence $(step_i(p,q))_{1 \le i \le r'}$ represents all pairs of possibly immediately successive configurations of any given configuration with register contents p, q. For reasons which will become clear below we include also the "0-step transitions" of M which reflect the reflexivity of the reachability relation. We now use these sequences for equating any configuration $C = (i,p,q)$, if it is reachable in M from the initial configuration, with its immediately successive configuration C'. We start with the initial configuration, say x, and with $[x,x]$. Assume a sequence $x, y, [y,z], u$ has been generated where y stands for the M-configuration C_t reached in t steps and z for its immediately successive configuration C_{t+1}; we expand this sequence by the sequence $\mathrm{SUCC}(z) := z, [z,z], z, step_1(p,q), \ldots, z, step_{r'}(p,q)$ of all possible ≤ 1-step transitions of M on z, i.e. to

$$z, [z,z],\ z, step_1(p,q), \ldots, z, step_{r'}(p,q),\ x, y, [y,z], u.$$

This explains the following equation in the reduction formula $\psi_M := \forall x \forall y \forall z \forall u \forall p \forall q \varphi_M$:

$$\langle x, y, [y,z], u\rangle = \langle \mathrm{SUCC}(z), x, y, [y,z], u\rangle$$

where $\langle x_1, \ldots, x_n\rangle$ is an abbreviation for $\langle x_1, \langle x_2, \ldots \langle x_{n-1}, x_n\rangle \ldots\rangle\rangle$ with a binary function $\langle\ \rangle$. Iterated applications of this equation will allow us, at each simulation step and for each term $\langle \ldots \mathrm{SUCC}(z), rest\rangle$, to bring the encoding $\mathrm{SUCC}(z)$ for the simulation of the next M-computation step on the configuration z to the front by equating $\langle \ldots \mathrm{SUCC}(z), rest\rangle$ with a term of form $\langle \mathrm{SUCC}(z), rest'\rangle$, for some term $rest'$.

Every intermediate configuration z which is reached from the initial configuration x does appear during this process at least once at the left end of the generated sequence, so that the condition on no halt in state r with empty registers is easily expressed by the inequality of φ_M as follows. It suffices to specialize the variables x, y, z, u in the above equation to $Start(p)$ with $Start(p) := \underline{(1,0,0)}(p)$ describing the start configuration $C_0 = (1,0,0)$. Let $Begin(p) := \langle Start(p), Start(p), [Start(p), Start(p)], Start(p)\rangle$ and define the inequality of φ_M to be as follows:

$$Begin(p) \neq \langle \underline{(r,0,0)}(p), u\rangle.$$

This completes the definition of φ_M and therefore of ψ_M. It remains to prove the following *Reduction Property*:

1. If $C_0 = (1,0,0) \Rightarrow_M (r,0,0)$, then ψ_M is not satisfiable,
2. If $C_0 = (1,0,0) \Rightarrow_M (r-1,0,0)$, then ψ_M is finitely satisfiable.

By the Gurevich's Theorem on semi-conservative reductions this will establish the claim of the theorem with finitely many binary and monadic functions.

For the proof of the first part of the reduction property it suffices to show that each model of ψ_M reflects all the computations of M in the following sense.

Lemma 5.2.9 (Simulation Lemma). *Let $\mathfrak{A}$ be a model satisfying $\overline{\forall}\varphi_M$. For an arbitrary natural number t let C_t be the configuration reached by M in t steps, starting from $C_0 = (1,0,0)$. Then there is for each element $a \in A$ a sequence $rest(t)$ such that for arbitrary natural numbers p, q the following equality holds in $\mathfrak{A}$:*

$$Begin(a) = \langle \mathrm{SUCC}(\underline{C_t}(a)), rest(t)\rangle.$$

By the Simulation Lemma and the inequality $Begin(a) \neq \langle \underline{(r,0,0)}(a), u\rangle$ in ψ_M, the satisfiability of ψ_M implies $C_0 = (1,0,0) \not\Rightarrow_M (r,0,0)$.

The proof of the simulation lemma uses only the equality axiom in ψ_M and is by induction on t. For $t = 0$ the claim is established by specializing the equation as follows: instantiate x, y, z, u to $\underline{C_0}(a)$ and set $rest(0)$ to $Begin(a)$. In the inductive step we have by the inductive hypothesis that the following equality holds for some $rest(t)$ and arbitrary elements p, q in $\mathfrak{A}$.

$$Begin(a) = \langle \mathrm{SUCC}(C_t(a)), rest(t)\rangle.$$

By the assumption $C_t \Rightarrow_M C_{t+1}$ there is some $1 \leq k \leq r'$ and some p', q' such that $step_k(p', q') = [\underline{C_t}(a), \underline{C_{t+1}}(a)]$, depending on the state of M at time t and on the corresponding register contents (and on the given $a \in A$). Therefore for some u, v the right hand side of the equation can be specialized in $\mathfrak{A}$ as follows:

$$\langle \mathrm{SUCC}(\underline{C_t}(a)), rest(t)\rangle = \langle \ldots, v, \underline{C_t}(a), step_k(p', q'), u\rangle.$$

Instantiating x to v, y to $\underline{C_t}(a)$ and z to $\underline{C_{t+1}}(a)$ allows us to apply the equality in φ_M to the inner term $\langle v, \underline{C_t}(a), step_k(p', q'), u\rangle$, obtaining in $\mathfrak{A}$ for some $rest$ and arbitrary p, q the equality

$$\langle \ldots, v, \underline{C_t}(a), step_k(p', q'), u\rangle = \langle \ldots, \mathrm{SUCC}(\underline{C_{t+1}}(a)), rest\rangle.$$

Remember that $\mathrm{SUCC}(\underline{C_{t+1}}(a)$ starts with the representation $\underline{C_{t+1}}(a)$, $[\underline{C_{t+1}}(a), \underline{C_{t+1}}(a)]$ of the 0-step transition ; this allows us to iterate applications of the equality in φ_M by instantiating x successively to all the terms appearing in $\ldots$ and y, z to $\underline{C_{t+1}}(a)$ and each time u to the appropriate rest term, eliminating one after the other all the terms in $\ldots$. In this way we eventually obtain in $\mathfrak{A}$ for some $rest(t+1)$ and arbitrary p, q the desired equality

$$\langle \ldots, v, \underline{C_t}(a), step_k(p', q'), u\rangle = \langle \mathrm{SUCC}(\underline{C_{t+1}}(a)), rest(t+1)\rangle.$$

We now prove the second part of the reduction property. Assume that $C_0 = (1,0,0) \Rightarrow_M (r-1,0,0)$. We have to show that ψ_M has a finite model. The idea for the proof is to construct a model which reflects the given M-computation in the sense of the above simulation lemma; since the computation terminates, only a finite amount of information needs to be encoded. We will define an appropriate interpretation of the relativization $f_0(p)$ of the representation of 0 to make the intended model finite.

Basically what we do is to cut off the successor function, and correspondingly the intended interpretation of all the other functions, at some large enough number which majorizes all the objects needed for the encoding of the given computation. We use the following notation for this. For fixed number m, the section $\hat{f}$ of an arbitrary function f with respect to m is defined by

$$\hat{f}(x_1, \ldots, x_n) := f(min(x_1, m), \ldots, min(x_n, m)).$$

Which objects do we need for the encoding of the given M-computation so as to respect the Simulation Lemma above? First of all each number which occurs as register content during that computation and all the M-states (instruction numbers) $1 \leq i \leq r$. Let reg be 1 + the maximum of all these numbers (including r). We need the closure under each state representing function f_i to encode each reachable configuration $f_i(a, b)$ for $i, a, b < reg$. Furthermore we need the closure under the pairing function for reachable configurations and their immediate successors.

We can provide enough objects which encode such triples and pairs of triples by choosing a ternary function $H : \mathbb{N}^3 \to \mathbb{N} - \{0, 1\}$ which is strictly increasing in each argument. The configurations $f_i(a, b)$ for $i, a, b < reg$ can be encoded into H-values $\hat{H}(i, a, b)$ with respect to reg, the pairs of immediately successive configurations c, c' can be encoded into H-values $\hat{H}(r+1, c, c')$ with respect to $H(r, reg, reg)$. Therefore we need not more than $m := H(r+1, H(r, reg, reg), H(r, reg, reg))$ elements and define as domain for our model $A := \{0, \ldots, m\}$.

On this domain we define the interpretation of our functions and also the subdomains for the encoding of configurations and configuration pairs as follows, for $a, b \in A$ and $1 \leq i \leq r$.

$$\begin{aligned}
f_0(a) &:= 0, \\
f(a) &:= a \hat{+} 1, \\
f_i(a, b) &:= \hat{H}(i, a, b) \text{ with respect to } reg, \\
[a, b] &:= \hat{H}(r+1, a, b) \text{ with respect to } H(r, reg, reg),
\end{aligned}$$

$$\begin{array}{rcl} A_1 & := & \{f_i(p,q) : (1,0,0) \Rightarrow_M (i,p,q)\}, \\ A_2 & := & \{[f_i(p,q), f_j(p',q')] : (i,p,q) \Rightarrow_M^{\leq 1} (j,p',q'), p,q,p',q' \leq reg\}, \\ A_0 & := & \{1\} \cup A_1 \cup A_2, \\ \langle a,b \rangle & := & \begin{cases} 1 & \text{if } a,b \in A_0 \\ 0 & \text{otherwise.} \end{cases} \end{array}$$

From the definitions and the monotonicity of H we obtain immediately the following simple facts which hold in the model $\mathfrak{A}$.

Lemma 5.2.10. *(i) $A_0 \subset A$, i.e. encodings of register contents, instruction numbers and pairs $[f_i(p,q), f_j(p',q')]$ of immediately successive configurations which occur in the given computation are elements of A.*
(ii) A_2 and A_1 are disjoint and do not contain 1.
(iii) A_2 contains all the encodings of immediately successive configurations, i.e. for each $1 \leq i \leq r'$ and all $p,q \in A$ holds $step_i(p,q) \in A_2$.
(iv) The function $\langle\ \rangle$ maps sequences of A_0-elements to 1, i.e. if $a_1,\ldots,a_n$, $b \in A_0$, then $\langle a_1,\ldots,a_n,b\rangle = 1$.
(v) If a sequence is mapped by $\langle\ \rangle$ to 1, then its last two elements are mapped by $\langle\ \rangle$ to 1, i.e. if $\langle a_1,\ldots,a_n,b\rangle = 1$, then $\langle a_n,b\rangle = 1$.

Exercise 5.2.11. Prove this lemma.

It remains to show that the above defined model $\mathfrak{A}$ satisfies ψ_M. We first show that $\mathfrak{A}$ satisfies the equality in ψ_M. Let $a,b,c,d,p,q \in A$ be arbitrary. We show that the left side of the equation takes value 1 if and only if the right side does. We distinguish two cases.

Case 1. Assume $\langle \mathrm{SUCC}(c), a, b, [b,c], d\rangle = 1$. Since by our bracketing convention $\langle \mathrm{SUCC}(c), a, b, [b,c], d\rangle = \langle \mathrm{SUCC}(c), a, \langle b, [b,c], d\rangle\rangle$, the last statement of the lemma implies that $\langle a, b, [b,c], d\rangle = 1$, as was to be shown.

Case 2. Assume $\langle a, b, [b,c], d\rangle = 1$. We want to use the fourth statement of the lemma to conclude that also $\langle \mathrm{SUCC}(c), a, b, [b,c], d\rangle = 1$. It suffices to show that the components of SUCC(c) as well as $a, b, [b,c], d$ are elements of A_0. By our definition of (the interpretation of) $\langle\ \rangle$ the assumption implies that $a, b, [b,c], d \in A_0$. Therefore it remains to show that the components of SUCC(c) are elements of A_0. By the monotonicity of H from $[b,c] \subset A_0$ we can conclude $[b,c] \in A_2$ so that b, c have the form $b = f_{i'}(p',q'), c = f_{i''}(p'',q''$ for some numbers $i',p',q',i'',p'',q'' \leq reg$ satisfying $(i',p',q') \Rightarrow_M^{\leq 1} (i'',p'',q'')$. Since $b \in A_0$ it follows that $b \in A_1$. This implies $(1,0,0) \Rightarrow_M (i',p',q')$ and thereby $(1,0,0) \Rightarrow_M (i'',p'',q'')$ so that $c \in A_1$ and $[c,c] \in A_2$. By the third statement of the lemma for each $1 \leq i \leq r'$ and all $p,q \in A$ also the components $step_i(p,q)$ of SUCC(c) are elements of A_2.

It remains to show that $\mathfrak{A}$ satisfies the inequality in ψ_M. By the interpretation of f_0, f in $\mathfrak{A}$ and the reflexivity f $\Rightarrow_M$, we have for arbitrary $p \in A_0$ that $Start(p) \in A_1$ so that $Begin(p) = 1$ is true in $\mathfrak{A}$ by definition of $\langle\rangle$. Since by assumption M does not reach the halting configuration $(r,0,0)$, we have $f_r(0,0) \notin A_1$ and therefore by the monotonicity of H also $f_r(0,0) \notin A_0$ so that by definition of $\langle\ \rangle$ $\mathfrak{A}$ satisfies $\langle f_r(f_0(p), f_0(p)), u\rangle = 0 \neq 1$ for every $p, u \in A_0$.

We conclude the proof by reducing the class $[\forall^6, (0), (\omega, \omega)]_=$ to the class $[\forall^6, (0), (0,1)]_=$ in a way which preserves also the number of equalities and inequalities in the formulae. It suffices to reuse the encoding ideas which have been used already in Chap. 4 (see in particular the proofs of Theorem 4.1.8 and Theorem 4.1.11).

Let $\psi := \forall x_1 \cdots \forall x_n \varphi(x_1, \ldots, x_n) \in [\forall^*, (0), (\omega, \omega)]_=$. Without loss of generality we can assume that φ contains only binary functions $f_1, \ldots, f_m$. The idea is to encode the i-th binary function using the i-th secondary diagonal of one binary function. Let h be a new binary function and define the formula $\chi := \forall x_1 \cdots \forall x_n \varphi(dx_1, \ldots, dx_n) \wedge \forall x \forall y \mathrm{AUX}$ with the conjunction AUX of the following formulae:

$$\bigwedge_{1 \leq i \leq m} f_i(x,y) = dh(h(g^{i+1}x, gx), gy),$$

$$gx = h(x, x_0) \wedge dx = h(g^{m+2}x, gx).$$

Exercise 5.2.12. Show that ψ has a (finite) model if and only if χ does.

χ is equivalent to a formula in which only the binary function h appears and where the defining equations in AUX do not appear any more so that the formula is in the required class. □

Exercise 5.2.13. [533] Let HERBRAND(k) be the class of Herbrand formulae which contain at most k conjuncts. Show that $[\Pi, (0), q] \cap$HERBRAND(4) is undecidable (and a reduction class) if and only if Π contains a universal quantifier and q contains either at least two monadic functions or at least one binary function. Hint: Use Wirsing's reduction classes $[\forall, (0), (2)] \cap$ HERBRAND(4), $[\forall, (0), (0,1)] \cap$ HERBRAND(4) and the decidability of the classes $[\exists^*, all, all]_=$ ([227]) and $[all, all, (1)]_= \cap$ HERBRAND (see [533]).

The decision problems of the following classes seem still to be open: HERBRAND(3) $\cap [\forall^3, (0), (2)]$, HERBRAND(3) $\cap [\forall^k, (0), q]$ for $k = 1, 2$ and $q = (2), (0,1)$, HERBRAND(2) $\cap [\forall^6, (0), (2)]$, HERBRAND(2) $\cap [\forall^k, (0), q]$ for $1 \leq k < 6$ and $q = (2), (0,1)$.

5.2.3 Horn Clause Programs With One Krom Rule

The Krom and Horn formulae which appear in the reduction classes of previous chapters all have a Skolem normal form $\overline{\forall}(pos \wedge \bigwedge_{i \leq n}(\pi_i \rightarrow \sigma_i) \wedge neg)$ with atomic formulae π_i, σ_i and (often singleton) conjunctions pos of atomic and neg of negated (often variable-free) atomic formulae. In the logic programming interpretation, such a formula represents a "program" of binary so-called "definite clauses" $\sigma_i \leftarrow \pi_i$ together with the set pos of "facts" and the computation "goal" *neg* (a negated literal); the program computes the goal successfully if and only if the formula $\overline{\forall}(pos \wedge \bigwedge_{i\leq n}(\pi_i \rightarrow \sigma_i) \wedge neg)$ is contradictory. The decidability of the class HERBRAND implies that in order to be computation universal such programs have to contain, besides facts and goals, also some clause – representing real computation steps. By adapting the encoding idea for Wirsing's Theorem to the formalization of Post correspondence problems given in Exercises 5.1.1 and 5.1.19 we show in this section that the problem whether a program computes the goal successfully is undecidable for Krom programs with only one definite clause, one fact and a singleton goal because the class of such formulae constitutes a reduction class.

Theorem 5.2.14 (Universality of One Binary Horn Rule). *The class of prenex Krom and Horn formulae with functions containing in the quantifier free part only one implication $\rho \rightarrow \sigma$, one atomic formula and one negated atomic formula is a reduction class. The reduction class can be restricted to formulae in* $[\forall^6, (0,0,0,1), (2,1)] \cap \text{KROM} \cap \text{HORN}$.

Proof. The proof consists in a reduction of Post correspondence problems $C = (v_i, w_i)_{1\leq i\leq n}$ over a binary alphabet to formulae ψ_C in the class. We use the logical interpretation of words w as terms $w(c)$ where the letters of the alphabet are interpreted as monadic function symbols and c is an individual constant (see Exercise 5.1.1). In the Exercises 5.1.1 and 5.1.19 we formalized the application of each single correspondence pair (v_i, w_i) by a separate conjunct. The idea which leads to a description using only one implication is a simple adaptation of the idea which appeared in the proof for Wirsing's theorem, namely to encode the sequence of all possible 1-step transitions into one single term. In the context of Post correspondence problems it suffices to encode the *sequence* of simultaneous applications of all the n pairs into one implication; similarly for the start and non-stop literals.

The concatenation of words $u_1, \ldots, u_m$ to a sequence $u_1 \ldots u_m$ is expressed using a binary function symbol ( ); notationally we suppress the brackets and assume their right association, writing $s_1 \ldots s_m$ instead of $(s_1(\ldots(s_{m_1} s_m)\ldots))$ for arbitrary terms s_i.

The generation of all possible C-computations $(v_{i_1} \ldots v_{i_s}, w_{i_1} \ldots w_{i_s})$ (where $1 \leq i_1, \ldots, i_s \leq n$) can be formalized by blocks of such computations of length $1, 2, 3, \ldots$. In going from one to the next level we pass (in the same way for $u \in \{v, w\}$) from $u_{i_1} \ldots u_{i_s}$ to

$$u_1 u_{i_1} \ldots u_n u_{i_1} \quad \ldots \quad u_1 u_{i_s} \ldots u_n u_{i_s}.$$

This can be expressed with finite nesting of function symbols by using the following pattern for the stepwise generation of the elements of level $s+1$ from the sequence of all elements of level s, starting from $s = 1$; here x, x', x'', y, y', t' denote variables standing for arbitrary terms and u stands for both elements of $\{v, w\}$ representing left and right side of the computation respectively.

$$\begin{array}{llllllll} 1: & u_1 & u_2 & u_3\dots & u_n & x & & \\ 2: & & u_2 & u_3\dots & u_n & u_1u_1\dots u_nu_1 & x' & \\ 3: & & & u_3\dots & u_n & \dots & & u_1u_2\dots u_nu_2 \quad x'' \\ \vdots & & & & & & & \end{array}$$

The test whether the C-computations generated so far have led to a success reduces to the question, to be asked at each step t, whether an initial part of the first argument generated in step t coincides with an initial part of the third argument generated in that step.

The formula $\psi_C := \forall x \forall x' \forall y \forall y' \forall v \forall w \varphi_C$ which describes this process using a predicate P of arity four can be defined by the conjunction φ_C of the following formulae (compare with the formulae in Exercise 5.1.19):

$$\begin{array}{rcl} \text{START} & := & P(v_1(c)\dots v_n(c)x, x, w_1(c)\dots w_n(c)x, x) \\ \text{NO-SOLUTION} & := & \neg P(xv, w, xy, y') \\ \text{STEP} & := & \beta \rightarrow P(x, x', y, y') \qquad \text{where} \\ \beta & := & P(vx, v_1(v)\dots v_n(v)x', wy, w_1(w)\dots w_n(w)y'). \end{array}$$

Exercise 5.2.15. Prove that ψ_C is satisfiable if and only if C has no solution.

□

Remark. The quantifier-free part $\pi \wedge (\rho \rightarrow \sigma) \wedge \neg\tau$ in the reduction formulae of the preceding theorem is sharp in various respects. The so called "units" π and $\neg\tau$ cannot be both variable free because otherwise starting from π the sequence of applications of the implication $\rho \rightarrow \sigma$ either becomes periodic or produces terms of greater depth than the depth of the terms appearing in τ; therefore the class of Krom and Horn formulae with only one definite clause and with closed units (i.e. with quantifier free part $\bigwedge_i \pi_i \wedge (\rho \rightarrow \sigma) \wedge \bigwedge_j \neg\tau_j$ and closed π_i, τ_j) is decidable (see [454]). Similarly, at least one of the units has to be non-linear (i.e. to contain more than one occurrence of at least one variable) because otherwise the resulting class is decidable (see [125]). If two implications or one ternary disjunction are allowed, the resulting classes are undecidable even with the restriction to closed units (see exercise 5.2.16 and [377], sharpened in [375, 373]). Furthermore it is crucial that the premise and the conclusion of the implication are not unifiable because otherwise the class of formulae under consideration is decidable (see [539]).

Exercise 5.2.16. [454] Show that the class of Krom and Horn formulae with exactly two definite clauses and with closed units, i.e. with quantifier free part of form $\bigwedge_i \pi_i \wedge (\rho \rightarrow \sigma) \wedge (\rho' \rightarrow \sigma') \wedge \bigwedge_j \neg\tau_j$ is a reduction class. Hint: Reduce the Turing machine halting problem to this class.

Exercise 5.2.17. Show that there is a logic program $(\sigma \leftarrow \rho), \tau$ with atomic formulae ρ, σ, τ such that it is undecidable to determine for a given goal (negated atomic formula) $\neg\pi$ whether the program yields an infinite number of answer substitutions for that goal. (In [127] the corresponding question for a finite number of answer substitutions is shown to be undecidable.) Hint: Use the fact that a correspondence system is solvable if and only if it has an infinite number of solutions.

From the logic programming theory viewpoint Theorem 5.2.14 has been interpreted by its authors as a form of the Böhm-Jacopini Theorem [38], namely as expressing that a single recursion scheme

$$\text{PROC } P(s_1, \ldots, s_n) : \text{EXECUTE } P(t_1. \ldots .t_n),$$

– in logic programming notation $Ps_1 \ldots s_n \leftarrow Pt_1 \ldots t_n$ – suffices to do every computation. In the same way Wirsing's reduction class of formulae $s = t \wedge s' \neq t'$ can be interpreted as expressing that purely equational computations using one equation and one inequality can mimic every computation. Compare also Dauchet's theorem [95] that each Turing machine can be simulated by just one, a left linear, rewrite rule (whereby the termination problem of rewrite rule systems with only one, a left linear, rule is proved to be undecidable). The analysis of the proofs shows the price which is paid for this kind of computational universality, namely an infeasible complex encoding. For a more realistic analysis of the termination behaviour of logic programs with recursive predicates, such as the above P, one would expect that it is more appropriate to study the behaviour of Horn formulae without functions. But even there one has to face complexity problems. Shmueli shows in [470] a Böhm-Jacopini Theorem for pure Datalog programs (read: sets of Horn clauses in the Bernays-Schönfinkel class), namely that a single recursive predicate is sufficient. The following question, whether it is possible to eliminate recursion from a given Datalog program, is undecidable even for programs which contain only one clause [374]. This question becomes decidable (although in some cases NP-complete) only when further syntactical restrictions are considered like the one to a single linear recursive clause with a binary predicate; for more detailed references to the recent literature on the question, see the introduction to [374].

5.3 Undecidable Logics with Two Variables

Let L_k be the class of relational first-order formulae that contain only the variables $x_1, \ldots, x_k$. Logics with only a bounded number of variables are important in several branches of mathematical logic and its applications such as modal logic, finite model theory, logic of programs, model checking, database query languages and knowledge representation. Since L_3 contains the prefix class $\forall\exists\forall$, the satisfiability problem for L_k is undecidable (even for formulae without equality) for all $k \geq 3$. On the other side, we will prove in Sect. 8.1 that L_2 has the finite model property, a result that was first proved by Mortimer [396] (for L_2-sentences without equality this follows from an earlier result by Scott [459]).

Of course, interesting sentences in L_2 are not in prenex normal form; rather one uses the possibility to quantify over the same variable again and again (see Sect. 8.1 for an example). Nevertheless, the expressive power of L_2 is rather limited. Therefore one is interested in extensions of L_2 where expressiveness is enhanced by additional means like counting quantifiers, cardinality comparison or constructs like transitive closure of fixed point operators that add recursion to L_2 (see [211]).

In this section we consider two kinds of extensions of L_2, obtained a) by adding Hilbert's *choice operator* and b) by adding cardinality comparison quantifiers. We prove that the satisfiability problem for these logics is undecidable. Further undecidability results for two-variable logics were proved by Grädel, Otto and Rosen [211] We refer to Sect. 8.1 for decidability and complexity results on logics with two variables.

5.3.1 First-Order Logic with the Choice Operator

Definition 5.3.1. First-order logic with Hilbert's choice operator (the so called ε-operator) extends first-order logic with equality by an additional term building rule: If ψ is a formula and x a variable, then $\varepsilon x\psi$ is a term, read: one x such that ψ. The interpretation of ε over a given domain A is defined by a choice function $F : \mathcal{P}(A) \to A$. (Recall that a choice function satisfies the condition that $F(X) \in X$ for all non-empty X.) Thus, $\varepsilon x\psi$ is interpreted by $F(\{x : \psi(x)\})$.

This logic is also known as ε-logic [339] and is closely related to Hermes' term logic [257]. Note that the syntax of ε-logic is rather unusual since it allows to build terms from formulae (whereas in most of the usual logics, terms are built by composition of functions symbols only; they may be used to build formulae, but not the other way round).

The ε-operators plays a similar rôle as quantifiers do. In fact, existential and universal quantifiers can be easily expressed in terms of the ε-operator:

$$\begin{aligned} \exists x\psi(x) &\equiv \psi(\varepsilon x\psi) \\ \forall x\psi(x) &\equiv \psi(\varepsilon x\neg\psi) \end{aligned}$$

A sentence ψ of ε-logic is *satisfiable* if there exists a first-order structure $\mathfrak{A}$ and a choice function F on the domain of $\mathfrak{A}$ such that $(\mathfrak{A}, F) \models \psi$. Heidler [251] proved that the satisfiability problem for ε-logic is undecidable even without any function or predicate symbol besides equality. We prove here a stronger result due to Grädel, Otto and Rosen [211], imposing the additional restriction that the formulae contain only two variables.

Definition 5.3.2. Let ε-L_2 be the fragment of ε-logic consisting of the formula that contain only the two variables x, y.

It is easy to define an infinity axiom in ε-L_2. The idea is to formalize an injective but not surjective function which for every given x chooses a new element y. Let $fx := \varepsilon y(x \neq y)$ and $fy = \varepsilon x(y \neq x)$ and define

$$\text{INF} := \forall x \forall y(fx = fy \rightarrow x = y) \wedge \exists x \forall y(x \neq fy).$$

Exercise 5.3.3. Show that INF is satisfiable over A if and only if A is infinite. Hint: For given infinite domain A consider a well order $<$ without last element and interpret $f(a)$ as $\min\{b \in A : a < b\}$.

The following theorem shows that the choice operator and the equality alone, without any other function, predicate or quantifier, suffice for yielding an unsolvable decision problem for two-variable logic.

Theorem 5.3.4 (Grädel, Otto, Rosen). *The class of sentences in ε-L_2 built from the variables x, y using only equality, Boolean connectives and the ε-operator is a conservative reduction class.*

Proof. In ε-L_2 we can build the following terms:

$$\begin{aligned} c &:= \varepsilon y(y = y) \\ ft &:= \varepsilon y(y \neq t) \\ gt &:= \varepsilon y(y \neq t \wedge y \neq c) \end{aligned}$$

where t is any previously defined term without free occurrences of y.

Note that the functions f, g defined in this way can be composed arbitrarily, e.g. fgx is an abbreviation for the term

$$\varepsilon y(y \neq \varepsilon y(y \neq x \wedge y \neq \varepsilon y(y = y))).$$

Over any domain A with choice function F, these terms define an element $c = F(A) \in A$ and functions $f, g : A \rightarrow A$ such that

$$f(a) = F(A - \{a\}), \qquad g(a) = F(A - \{a, c\})$$

for all $a \in A$. If A has at least three elements, then c, f, g always satisfy the following condition, for all $a \in A$:

$$(*) \qquad f(a) \neq a,\ g(a) \neq a,\ g(a) \neq c,\ f(c) = g(c).$$

Conversely, for every domain A of cardinality greater than two, with an element c and functions $f, g : A \rightarrow A$ satisfying (*) there exists a choice function F defining c, f, g as described.

The rest is routine. One could define more functions in a similar way as f and g and then directly encode, say the domino problem along the lines of Exercise 4.1.2. A simpler way to complete the proof is by reduction from the conservative class $[\forall, (0), (2)]_=$ (see Theorem 4.1.8). As pointed out in 4.1.10 this class remains conservative when restricted to structures $\mathfrak{A} = (A, h, h')$ where the two functions h, h' have no fixed points.

Now translate any sentence $\psi = \forall x \varphi \in [\forall, (0), (2)]_=$ with unary function symbols h, h' to the sentence

$$\psi^* := \forall x (x \neq c \to (fx \neq c \wedge gx \neq c \wedge \varphi[h/f, h'/g]))$$

of ε-L_2 where $\varphi[h/f, h'/g]$ is obtained from φ by replacing the functions h and h' by f and g, respectively.

The universal quantifier can be defined in terms of the ε-operator as described above. To complete the proof it remains to show that the reduction from ψ to ψ^* preserves satisfiability and finite satisfiability.

Suppose that $\mathfrak{A} = (A, h, h') \models \psi$ where h and h' have no fixed points (and therefore $|A| \geq 2$). Let $B := A \cup \{c\}$ where c is a new element. Then there exists a choice function $F : \mathcal{P}(B) \to B$ such that the functions defined by f, g on (B, F) coincide on A with h, h'. Thus $(B, F) \models \psi^*$.

Conversely, suppose that $(B, F) \models \psi^*$. Let $A = B - \{c\}$ and let h, h' be the restrictions of the functions f, g to A (note that A is closed under f, g since $f(a) \neq c$ and $g(a) \neq c$ for all $a \in A$. Obviously, $(A, h, h') \models \psi$. □

Corollary 5.3.5 (Heidler). *The class of sentences of ε-logic without functions and predicates (besides equality) is a conservative reduction class.*

Exercise 5.3.6. Prove Heidler's Theorem by a conservative reduction from pure predicate logic with a single binary predicate – i.e. from the class $[all, (0, 1), (0)]$ – which is conservative by the results of Chap. 3. Hint: Use witnesses (new elements) w_1, w_2 to which pairs (a, b) satisfying Pab are projected. Such a projection can be formulated using the choice operator, as follows:

$$\begin{aligned} proj(x_1, \ldots, x_n) &:= \varepsilon y (\bigvee_{1 \leq i \leq n} y = x_i) \\ \pi(x, y, w_1, w_2) &:= (proj(x, y) = x \wedge x \neq y \wedge proj(x, y, w_1) = w_1) \vee \\ &\qquad (proj(x, y) = y \wedge proj(x, y, w_1) = w_2). \end{aligned}$$

Note that the formula $proj(x, y) = x$ defines a local order on pairs.

A formula ψ in the pure predicate calculus with a single, binary, predicate P is mapped to an appropriate formula φ of ε-logic by translating atoms Pxy to π and by relativizing the quantifiers to non-witnesses. Let WITNESS(x) stand for $(x = w_1 \vee x = w_2)$.

$$\varphi := \exists w_1 \exists w_2 (w_1 \neq w_2 \wedge \psi^* \wedge \exists x \neg \mathrm{WITNESS}(x))$$

where β^* is defined for subformulae β of ψ inductively as follows:

$$(Pxy)^* := \pi$$
$$(\forall x\beta)^* := \forall x(\neg\text{WITNESS}(x) \rightarrow \beta^*)$$
$$(\exists x\beta)^* := \exists x(\neg\text{WITNESS}(x) \land \beta^*).$$

Exercise 5.3.7. [251] Let $\Omega \subseteq \{=, \varepsilon, \forall, \exists\}$ and define the classes $[\Omega, p, f]$ for ε-logic in the obvious way. Show that $[\Omega, p, (0)]$ is a conservative reduction class if and only if $\{=, \varepsilon\} \subseteq \Omega$ or if $\{\varepsilon, \forall, \exists\} \cap \Omega \neq \varnothing$ and p contains at least one predicate of arity at least 2. Hint: Use Corollary 5.3.5, the conservative reduction class property for $[all, (0,1), (0)]$ and the decidability for the cases $\Omega = \{=\}$ (purely equational logic), $\Omega = \{=, \forall, \exists\}$ and $p = (\omega, 0)$ (Löwenheim's class) and for the case $\Omega = \{\varepsilon, \forall, \exists\}$ and $p = (\omega, 0)$ (see Exercise 6.2.6).

5.3.2 Two-Variable Logic with Cardinality Comparison

It is well-known that one of the major limitations of first-order logic (with respect to expressiveness) is the inability to count. For instance, there is no first-order sentence that defines parity, in the sense that its finite models are precisely the structures of even cardinality. In fact even much stronger languages like fixed point logic or the infinitary logic with bounded number of variables $L^{\omega}_{\infty\omega}$ cannot express evenness; this can be shown by a simple argument using Ehrenfeucht-Fraïssé games (see [141]). This issue is particularly relevant in finite model theory and in databases where one of the major goals is the design of logics or query languages that capture complexity classes, such as polynomial time or logarithmic space. Since counting is a computationally simple task it is natural to investigate logics with counting constructs (such as counting quantifiers, counting terms or generalized quantifiers). We refer to [209, 413] for a detailed discussion of logics with counting.

Let C_2 be the extension of L_2 by counting quantifiers of the form $\exists^{\geq n}$ and $\exists^{\leq n}$, for arbitrary $n \in \omega$. The semantics of these quantifiers is the obvious one. It is not difficult to see that C_2 contains infinity axioms. Grädel, Otto and Rosen [210] have recently proved that $Sat(C_2)$ is decidable.

But there are other forms of adding counting to a logic, for instance via *counting terms* or via the *cardinality comparison quantifier*, also called *Härtig quantifier*.

Counting Terms. In logics with counting terms we have the possibility to build for every formula φ and every variable x a term $\#_x[\varphi]$, taking cardinals as values. The free variables of φ different from x remain free in $\#_x[\varphi]$.

The semantic of $\#_x[\varphi]$ is defined as follows. As usual, the notation $\varphi(x, \bar{z})$ indicates that the free variables of φ are among $x, \bar{z}$. Given a formula $\varphi(x, \bar{z})$, a structure $\mathfrak{A}$ and a valuation $\bar{c}$ for $\bar{z}$, then the meaning of $\#_x[\varphi]$ for $\mathfrak{A}$ and $\bar{c}$ is

$$\#_x[\varphi]^{\mathfrak{A},\bar{c}} := |\{a : \mathfrak{A} \models \varphi[a, \bar{c}]\}|.$$

Fixing a logic L and any collection of basic relations Q on cardinals we can build a counting extension of L containing formulae of the form $Qt_1 \cdots t_m$ where $t_1, \ldots, t_m$ are counting terms. The simplest case is obtained by just allowing equality of counting terms.

The Härtig Quantifier. A slightly different presentation of first-order logic (or other logics) with equality of counting terms involves a particular generalized quantifier, namely the cardinality comparison quantifier I, introduced by Härtig [247]. For a survey on this quantifier, we refer to [259]; more background on generalized quantifiers can be found in [140].

The extension $L[I]$ of a logic L by the Härtig quantifier is defined by adding to L the following rule for building formulae: Given two formulae φ, ψ and two (not necessarily distinct) variables x, y, the expression

$$(Ix, y\ \varphi, \psi)$$

is a formula, saying that the set of elements x satisfying φ has the same cardinality as the set of elements y satisfying ψ.

The set of free variables of this formula is

$$\text{free}((Ix, y\ \varphi, \psi)) = (\text{free}(\varphi) - \{x\}) \cup (\text{free}(\psi) - \{y\}).$$

Formally, the semantics of such formulae can be defined by the equivalence

$$(Ix, y\ \varphi, \psi) \equiv (\#_x[\varphi] = \#_y[\psi]).$$

Thus, given a formula $(Ix, y\ \varphi(x, y, \bar{z}), \psi(x, y, \bar{z}))(x, y, \bar{z})$, a structure $\mathfrak{A}$ (of appropriate vocabulary) and valuations $a, b, \bar{c}$ for $x, y, \bar{z}$, respectively, then

$$\mathfrak{A} \models (Ix, y\ \varphi, \psi)[a, b, \bar{c}] \text{ iff } \#_x[\varphi]^{\mathfrak{A}, b, \bar{c}} = \#_y[\psi]^{\mathfrak{A}, a, \bar{c}}.$$

We prove that the satisfiability problem for $L_2[I]$ is undecidable. Note that this does not imply the undecidability of $Sat(C_2)$ since the two counting logics $L_2[I]$ and C_2 are incomparable with respect to expressive power.

The power of the Härtig quantifier in the context of L_2 results from the fact that – unlike the counting quantifiers $\exists^{\geq n}$ or $\exists^{\leq n}$ – application of H does not necessarily reduce the number of free variables. Indeed if x, y are free in $\varphi(x, y)$ and $\psi(x, y)$, then both x and y are free also in $(Ix, y\ \varphi, \psi)(x, y)$.

For instance, we can axiomatize in $L_2[I]$ the class of *regular graphs* $G = (V, E)$ by the formula

$$\psi_{\text{reg}} := \forall x \forall y (\neg Exx \wedge (Exy \rightarrow Eyx) \wedge (Ix, y\ Eyx, Exy)(x, y)).$$

Indeed $G \models \psi_{\text{reg}}$ if and only if E is irreflexive and symmetric and for all pairs of nodes u, v the number $\#_x[Evx]$ of neighbours of v is the same as the number $\#_y[Euy]$ of neighbours of u. (In fact, due to symmetry of E, we could just as well use the subformula $(Ix, y\ Exy, Exy)$ rather than $(Ix, y\ Eyx, Exy)$.)

Theorem 5.3.8 (Grädel, Otto, Rosen). *$Sat(L_2[I])$ is undecidable.*

Proof. The first step is an axiomatizing of the $\mathbb{N} \times \mathbb{N}$-grid. Let E be a binary relation symbol. We write $N_E(x)$ for $\#_y[Exy]$ and similarly $N_E(y)$ for $\#_x[Eyx]$ to specify the number of outgoing E-edges at points x, y. Clearly the statement $N_E(x) = N_E(y)$ is expressible by a formula in $L_2[I]$. We observe that the Härtig quantifier allows to express not just equality of counting terms, but also a kind of successor relation when applied to irreflexive relations.

Lemma 5.3.9. *Let C be a class of structures such that $\mathfrak{A} \models \forall x \neg Exx$ for all $\mathfrak{A} \in C$. Then there exists a formula $\psi(x, y) \in L_2[I]$ expressing on C that $N_E(y) = N_E(x) + 1$.*

Proof. Just take $\psi(x, y) := [Ix, y\ Eyx, Exy \vee (x = y)](x, y)$. □

We axiomatize the $\mathbb{N} \times \mathbb{N}$-grid by a sentence $\varphi \in L_2[I]$ containing the binary predicates H, V, E, F. The models of φ contain a copy of $\mathbb{N} \times \mathbb{N}$ with the horizontal and vertical adjacency relations H and V so that from each point $(m, n) \in \mathbb{N} \times \mathbb{N}$ we have precisely m outgoing E-edges and n outgoing F-edges.

The desired $L_2[I]$-sentence φ is the conjunction of the formulae

$$\begin{aligned}
&\forall x \exists y Hxy \wedge \forall x \exists y Vxy \\
&\forall x(\neg Exx \wedge \neg Fxx) \wedge \exists x \forall y(\neg Exy \wedge \neg Fxy) \\
&\forall x \forall y(Hxy \rightarrow (N_E(y) = N_E(x) + 1 \wedge N_F(y) = N_F(x)) \\
&\forall x \forall y(Vxy \rightarrow (N_E(y) = N_E(x) \wedge N_F(y) = N_F(x) + 1) \\
&\forall x \forall y((N_E(x) = N_E(y) \wedge N_F(x) = N_F(y)) \rightarrow x = y)
\end{aligned}$$

Obviously φ is satisfiable and all its models indeed have the properties described above. Note that φ enforces irreflexivity of E and F, so the statements $N_E(y) = N_E(x) + 1$ and $N_F(y) = N_F(x) + 1$ are expressible in $L_2[I]$.

Reducing the domino problem. Having axiomatized the grid, we can prove undecidability by the usual techniques, for instance by encoding the domino problem (see Sect. 3.1.1).

Given a domino system $\mathcal{D} = (D, H, V)$ let $\psi_{\mathcal{D}}$ be the conjunction of the grid-axiom φ with the formulae

$$\begin{aligned}
&\forall x(\bigvee_{d \in D} P_d x \wedge \bigwedge_{d \neq d'} \neg(P_d x \wedge P_{d'} y)) \\
&\forall x \forall y(Hxy \rightarrow \bigvee_{(d,d') \in H} (P_d x \wedge P_{d'} y)) \\
&\forall x \forall y(Vxy \rightarrow \bigvee_{(d,d') \in V} (P_d x \wedge P_{d'} y)).
\end{aligned}$$

Suppose that $\tau : \mathbb{N} \times \mathbb{N} \to D$ is a legal tiling. We obtain a model $\mathfrak{B} \models \psi_{\mathcal{D}}$ by taking any model $\mathfrak{A} \models \varphi$ with universe $\mathbb{N} \times \mathbb{N}$ and expanding it with the relations

$$P_d := \{(n, m) \in \mathbb{N} \times \mathbb{N} : \tau(n, m) = d\}$$

for $d \in D$. Conversely, suppose that $\mathfrak{B} \models \psi_{\mathcal{D}}$. since $\mathfrak{B} \models \varphi$ there exists for every $(m, n) \in \mathbb{N} \times \mathbb{N}$ a unique point $b_{n,m}$ in $\mathfrak{B}$ with precisely m outgoing E-edges and precisely n outgoing F-edges. We tile the point (n, m) with the unique domino $d \in D$ such that $\mathfrak{B} \models P_d[b_{n,m}]$. By the last two clauses of $\psi_{\mathcal{D}}$ this defines a legal tiling of $\mathbb{N} \times \mathbb{N}$ by $\mathcal{D}$.

This proves Theorem 5.3.8. □

Exercise 5.3.10. [211] Modify of these arguments to prove that $L_2[I]$ is in fact conservative. Hint: Instead of the infinite grid, axiomatize tori $\mathbb{Z}_s \times \mathbb{Z}_t$ (where s, t can be arbitrary natural numbers) and use the inseparability result for dominoes by Berger and Gurevich-Koryakov (Theorem 3.1.7).

Further, we can also eliminate equality. Let $L_2[I]^-$ be the set of formulae in $L_2[I]$ that do not contain the equality sign.

Theorem 5.3.11. *$L_2[I]^-$ is conservative.*

Proof. In the proof of Theorem 5.3.8, equality was used at two places. To express that $N_E(y) = N_E(x) + 1$ and to ensure that points with the same number of E- and F-neighbours coincide (see the last clause in the definition of φ)

We modify the construction as follows.

Let A, B be a unary predicate. In $L_2[I]^-$ we can express that A and B are nonempty, disjoint sets of the same cardinality such that $|A| \neq |A \cup B|$. This forces A and B to be finite.

We then add the clause $\forall x \forall y (Exy \to \neg Ay)$ and replace the condition $N_E(y) = N_E(x) + 1$ by $N_E(y) = N_E(x) + |A|$ which, given that no E-arcs has its endpoint in A, is expressible by

$$(Ix, y\ Eyx, Exy \lor Ay).$$

Now the grid can be axiomatized in a similar way as above. In any model of φ, the points $(m, n) \in \mathbb{N} \times \mathbb{N}$ are represented by elements with $m|A|$ E-neighbours and $n|A|$ F-neighbours.

The clause saying that points with the same number of E- and F-neighbours coincide is not really needed. It can be replaced by the weaker condition that such points are tiled with same domino (i.e. satisfy the same relation P_d). □

Another variant of a cardinality comparison quantifier is the *Rescher quantifier J*. With the Rescher quantifier we build formulae

$$(Jx, y\ \varphi, \psi)$$

expressing that $\#_x[\varphi] < \#_y[\psi]$. Clearly the Härtig quantifier is expressible with the Rescher quantifier since

$$(Ix, y\ \varphi, \psi) \equiv \neg(Jx, y\ \varphi, \psi) \wedge \neg(Jy, x\ \psi, \varphi).$$

Corollary 5.3.12. *$L_2[J]$ is a conservative reduction class.*

Given that $Sat(L_2[I])$ is undecidable, we may ask about $L_1[I]$, first-order logic with just one variable, extended with cardinality comparison. Of course we can restrict attention to monadic predicates, and in fact, also cardinality comparison can be reduced to statements $|F| = |G|$ for atomic predicates F, G. Indeed, every subformula $(Ix, x\ \varphi, \psi)$ has no free variable and may be replaced by

$$\forall x(F_\varphi x \leftrightarrow \varphi) \wedge \forall x(F_\psi x \leftrightarrow \psi) \wedge |F_\varphi| = |F_\psi|$$

where F_φ and F_ψ are new predicates. In this way any formula $L_1[I]$ is transformed into a formula with cardinality comparison restricted to atomic predicates which is satisfiable if and only if the original formula is.

However, note that $L_1[I]$ does not have the finite model property. Indeed the sentence

$$\forall x(Fx \rightarrow Gx) \wedge \exists x(Gx \wedge \neg Fx) \wedge |F| = |G|$$

is an infinity axiom. And since we can say that two sets F, G are infinite and of different cardinality, we neither have the Löwenheim-Skolem property. Nevertheless, it is not difficult to prove that $Sat(L_1[I])$ is decidable.

5.4 Conjunctions of Prefix-Vocabulary Classes

The Boolean operations can be extended from formulae to classes of formulae in the following pointwise way:

$$\begin{aligned}
\neg(K) &= \{\neg(\varphi) : \varphi \in K\} \\
K_1 \wedge K_2 &= \{\varphi_1 \wedge \varphi_2 : \varphi_1 \in K_1 \text{ and } \varphi_2 \in K_2\} \\
K_1 \vee K_2 &= \{\varphi_1 \vee \varphi_2 : \varphi_1 \in K_1 \text{ and } \varphi_2 \in K_2\}.
\end{aligned}$$

This section is devoted to the satisfiability and finite satisfiability problems for pointwise Boolean combinations of prefix-vocabulary classes. In Subsect. 5.4.1, we explain a quite obvious reduction of the classification problem for such Boolean combinations to the similar problem for the conjunctions of prefix vocabulary classes. In Subsect. 5.4.2, Gurevich's Classifiability Theorem (see Sect. 2.3) is generalized to cover the conjunctions of prefix-vocabulary classes. In the final subsection 5.4.3, we mention some of the known results and pose a few questions.

5.4.1 Reduction to the Case of Conjunctions

The underlying logic may be with or without equality. For any class K of formulae, let $\mathcal{P}(K)$ be one of the following three statements:

- K is decidable for satisfiability,
- K is decidable for finite satisfiability,
- K has the finite model property.

Recall that, by our conventions, every prefix-vocabulary class contains the logic constants *true* and *false*. It follows that every conjunction of prefix vocabulary classes contains a logically false sentence.

Lemma 5.4.1. *Let K_1 and K_2 be conjunctions of prefix-vocabulary classes (or any classes of formulae which contain logically false formulae). Then the following statements are equivalent:*

(i) The classes K_1, K_2 have the property $\mathcal{P}$.
(ii) The disjunction $K_1 \vee K_2$ has the property $\mathcal{P}$.

Proof. It is obvious that *(i)* implies *(ii)*. To prove that *(ii)* implies *(i)* it suffices to show that every formula in K_1 (respectively K_2) is equivalent to a formula in $K_1 \vee K_2$. But this is obvious because K_2 (respectively K_1) contains a logically false sentence. □

The $\mathcal{P}$-classification problem for pointwise Boolean combinations of prefix-vocabulary classes reduces to the $\mathcal{P}$-classification problem for conjunctions of prefix-vocabulary classes as follows: rewrite a given pointwise Boolean combination as a disjunction of conjunctions, and then use Lemma 5.4.1.

5.4.2 Another Classifiability Theorem

Again, the underlying logic may be with or without equality. For brevity, the term "class" will be restricted to mean a conjunction of prefix-vocabulary classes. We will use the terminology of Sect. 2.3. Recall the domination notion for prefix-vocabulary classes and the fact that K_1 conservatively reduces to K_2 if $K_1 \leq K_2$; see Subsect. 2.3.5 in this connection.

Definition 5.4.2. Let $X_1, \ldots, X_m, Y_1, \ldots, Y_n$ be prefix-vocabulary classes. The class $Y = Y_1 \wedge \cdots \wedge Y_n$ *dominates* the class $X = X_1 \wedge \cdots \wedge X_m$, symbolically $Y \geq X$, if there exists a one-to-one function f from $[1..m]$ into $[1..n]$ such that, for each $i = 1, \ldots, m$, Y_{fi} dominates X_i.

Exercise 5.4.3. If $X \leq Y$, then X conservatively reduces to Y.

Lemma 5.4.4. *The classes with the domination relation form a well quasi ordered set.*

Proof. Use the following facts:

1. The prefix-vocabulary classes with the domination relation form a well quasi ordered set (the Classifiability Theorem of Sect. 2.3).
2. The Finite Sequence Theorem (Subsect. 2.3.1).
3. A quasi ordered set that is the homomorphic image of a well quasi ordered set is a well quasi ordered set (Subsect. 2.3.1).

□

Definition 5.4.5. A class K is *standard* (respectively *closed*) if it is a conjunction of standard (respectively closed) prefix-vocabulary classes.

Recall that every prefix-vocabulary class is equivalent to (that is, dominates and is dominated by) a closed prefix-vocabulary class (see Subsect. 2.3.5).

Exercise 5.4.6. Every class is equivalent to a closed class.

Lemma 5.4.7. *Every closed class is a finite union of standard classes.*

Proof. We illustrate the proof on the case of the conjunction $X \wedge Y$ of two closed prefix-vocabulary classes. According to Subsect. 2.3.5, each closed prefix-vocabulary class is a finite union of standard prefix-vocabulary classes. Present X (respectively Y) as a finite union of standard prefix-vocabulary classes X_i (respectively Y_j). Then $X \wedge Y$ is the union of standard conjunctions $X_i \wedge Y_j$. □

Theorem 5.4.8 (The Classifiability Theorem for Conjunctions). *Let $\mathcal{D}$ be a collection of conjunctions of prefix-vocabulary classes and suppose that $\mathcal{D}$ is downward closed and closed under finite unions. Further, let $\mathcal{U}$ be the complement of $\mathcal{D}$. There exists a finite collection $\mathcal{M}$ of standard minimal members of $\mathcal{U}$ such that $\mathcal{U}$ is the upward closure of $\mathcal{M}$.*

Proof. Let $\mathcal{M}$ by a maximal antichain of minimal members of $\mathcal{U}$. Since the domination ordering is a well quasi ordering, $\mathcal{M}$ is finite and $\mathcal{U}$ is the upward closure of $\mathcal{M}$. According to Exercise 5.4.6, $\mathcal{M}$ can be chosen to contain only closed conjunctions. But then every member K of $\mathcal{M}$ is standard. Indeed, suppose that K is not standard. By Lemma 5.4.7, K is a finite union of standard conjunctions K_i. Since K is not standard, each $K_i < K$ and therefore belongs to $\mathcal{D}$. But then $\mathcal{D}$ contains K which is impossible. □

For example, $\mathcal{D}$ may be the collection of classes decidable for satisfiability, decidable for finite satisfiability, or having the finite model property.

5.4.3 Some Results and Open Problems

Some results on conjunctions of prefix-vocabulary classes of predicate logic have been obtained as a by-product of investigations of prefix-vocabulary classes of pure predicate formulae. We restrict attention to classes

$$[\Pi_1 \wedge \cdots \wedge \Pi_n, p] = [\Pi_1, p] \wedge \cdots \wedge [\Pi_n, p].$$

Restrictions on the Quantifier-Free Parts. One may require that the quantifier-free parts of some conjuncts have a particular form. Aanderaa proved that the class $[\exists\forall \wedge \forall\exists\forall, (\omega, \ell)] \cap \text{KROM} \cap \text{HORN}$ is a conservative reduction class for some ℓ of the size of a universal 2-register machine (see [2]). In the rest of this paragraph, we restrict attention to conjunctions of prenex formulas such that the first conjunct is of the form $\forall x_1 \cdots \forall x_j \exists y P x_1 \cdots x_j y$ where P is a $(j+1)$-ary predicate symbol. One may call conjuncts of that form *purely Skolem*.

Accordingly, restrict attention to classes K of the form $[\Pi_1 \wedge \cdots \wedge \Pi_m, p]$ or $[\Pi_1 \wedge \cdots \wedge \Pi_m, p]_=$ where Π_1 is a standard prefix set $\forall^j\exists$ and $p_j \geq 1$. The *Skolem subclass* $\mathcal{S}(K)$ of such a class K is the collection of K sentences $\varphi_1 \wedge \varphi_2 \wedge \cdots \wedge \varphi_m$ where each $\varphi_i \in [\Pi_i, p]$ and φ_1 is purely Skolem.

Pepis has proved that the Skolem subclass of $[\forall^2\exists \wedge \forall^*, (1, 0, 3)]$ is a reduction class for satisfiability [418, 419, 420]. Surányi has proved that the Skolem subclass of $[\forall^2\exists \wedge \forall^3, (0, \omega)]$ is a reduction class for satisfiability [494, 496].

On the other hand, in [318], Kostyrko gave a decision procedure for the finite satisfiability problem for the Skolem subclass of $[\forall\exists \wedge \exists^*\forall^*, (\omega, 1)]_=$. An obvious open question is whether the satisfiability problem for that subclass is decidable.

More general problems are to classify Skolem subclasses into decidable and undecidable with respect to satisfiability, finite satisfiability or the finite model property. Notice that an analogue of the Classifiability Theorem 5.4.8 holds for the Skolem subclasses. Indeed given $K_1 = [\forall^j\exists \wedge \Pi_2 \wedge \cdots \wedge \Pi_m, p]$ and $K_2 = [\forall^k\exists \wedge \Pi'_2 \wedge \cdots \wedge \Pi'_n, p']$, say that $\mathcal{S}(K_1)$ *dominates* $\mathcal{S}(K_2)$, symbolically $\mathcal{S}(K_1) \geq \mathcal{S}(K_2)$, if $i \geq j$ and $[\Pi_2 \wedge \cdots \wedge \Pi_m, p] \geq [\Pi'_2 \wedge \cdots \wedge \Pi'_n, p']$. Clearly, $\mathcal{S}(K_2)$ conservatively reduces to $\mathcal{S}(K_1)$ if $\mathcal{S}(K_2) \leq \mathcal{S}(K_1)$.

Lemma 5.4.9. *The Skolem subclasses with the domination ordering form a well quasi ordered set.*

Proof. Use the facts that the collection of wqo sets is closed under finite direct products and substructures. □

Exercise 5.4.10. Formulate an analogue of Theorem 5.4.8 for Skolem subclasses.

Pure Predicate Logic. It will be convenient to strengthen the domination ordering. Let $K = [\Pi_1 \wedge \Pi_2 \wedge \cdots \wedge \Pi_m, p]$ and $K' = [\Pi'_1 \wedge \Pi'_2 \wedge \cdots \wedge \Pi'_n, p']$. Say that K' *dominates* K_1 if p' dominates p and there exists a (not necessarily one-to-one) function f from $[1..m]$ into $[1..n]$ such that every k in the range of f satisfies a condition $C(k)$ which, for notational simplicity, we illustrate on an example where $f^{-1} = 1, 2$. In this case, the condition $C(k)$ is that the conjunction of an arbitrary Π_1 sentence and an arbitrary Π_2 sentence is equivalent to a Π'_k sentence.

Exercise 5.4.11. According to the new domination ordering, $[\exists^*\forall^* \wedge \cdots \wedge \exists^*\forall^*, all] \le [\exists^*\forall^*, all]$, $[\exists^*\forall^2\exists^* \wedge \cdots \wedge \exists^*\forall^2\exists^*, all] \le [\exists^*\forall^2\exists^*, all]$, $[\forall\exists \wedge \forall^3] \le [\forall^3\exists]$.

One can reformulate the condition $C(k)$ in syntactical terms, but we will not bother with that. Since a quasi ordered set that is the homomorphic image of a well quasi ordered set is a well quasi ordered set (Subsect. 2.3.1), classes $[\Pi_1 \wedge \Pi_2 \wedge \cdots \wedge \Pi_m, p]$ with the new domination ordering form a quasi ordered set.

Exercise 5.4.12. Formulate an analogue of Theorem 5.4.8 for classes $[\Pi_1 \wedge \Pi_2 \wedge \cdots \wedge \Pi_m, p]$ with the new domination ordering.

In the rest of this subsection, the domination ordering is new. Since every conjunction of prefix-vocabulary classes is equivalent to a closed one, and every closed conjunction is a finite union of standard ones, one may as well restrict attention to standard conjunctions $[w_1 \wedge \cdots w_m, p]$ where each w_i is a generalized prefix.

Completing the classification of prefix-vocabulary classes in [219], Gurevich also addressed the satisfiability and the finite satisfiability problems for classes $[\Pi_1 \wedge \cdots \wedge \Pi_m, p]$ of pure predicate formulae. He considered separately the case of infinite vocabulary, that is the case when $\sum_i p_i$ in infinite. In that case, the following facts are relevant.

1. The Kahr class $[\forall\exists\forall, (\omega, 1)]$ is a conservative reduction class; see Sect. 3.1. (By the way, Kahr's result was the end of chain of improvements on Büchi's theorem that the conjunction $[\exists \wedge \forall\exists\forall, (\omega, 3)]$ is a conservative reduction class [64].)
2. The conjunction $[\forall\exists \wedge \forall^3, (\omega, 1)$ is a conservative reduction class; see Corollary 3.1.19.

This gives rise to the following theorem.

Theorem 5.4.13. *Any standard conjunction $K = [w_1 \wedge \cdots w_m, p]$ with infinite $\sum_i p_i$ satisfies one of the following two conditions:*

A. K dominates at least one of the two conservative reduction classes

$$[\forall\exists\forall, (\omega, 1)], \quad [\forall\exists \wedge \forall^3, (\omega, 1)],$$

and therefore K is a conservative reduction class,

B. K is dominated by one of the classes with the finite model property

$$[all, (\omega)], \quad [\exists^*\forall^*, all], \quad [\exists^*\forall^2\exists^*, all],$$

and therefore K has the finite model property.

Proof. Assume that K does not satisfy B. Since K is not dominated by $[all, (\omega)]$, $p_2 + p_3 + \cdots \geq 1$. Since $\sum_i p_i$ is infinite, we have $p \geq (\omega, 1)$. Since K is not dominated by $[\exists^*\forall^2\exists^*, all]$, some w_i dominates $\forall\exists\forall$ or $\forall^3$. In the first case, $K \geq [\forall\exists\forall, (\omega, 1)]$ and we have finished. Assume that w_i does not dominate $\forall\exists\forall$. We have $\forall^3 \leq w_i \leq \exists^*\forall^2\exists^*$. Since K is not dominated by $[\exists^*\forall^*, all]$, some $w_j \geq \forall\exists$. But then $K \geq [\forall\exists \wedge \forall^3, (\omega, 1)]$. This is obvious if $j \neq i$. Assume that $j = i$. Then

$$K \geq [w_i, (\omega, 1)] \geq [\forall^3\exists, (\omega, 1)] \geq [\forall\exists \wedge \forall^3, (\omega, 1)].$$

□

Next Gurevich considered classes $[w_1 \wedge \cdots \wedge w_m, p]$ where $\sum_i p_i$ is finite but the prefix set $(w_1) \cup \cdots \cup (w_m)$ is infinite. (If both $\sum_i p_i$ is finite and $(w_1) \cup \cdots \cup (w_m)$ is finite, then $[w_1 \wedge \cdots \wedge w_m, p]$ has only finitely many sentences up to logical equivalence and thus is decidable for any decision problem which does not distinguish between logically equivalent sentences e.g. the satisfiability problem.) The following facts are relevant.

1. According to Denton [108], the class $[\forall\exists \wedge \forall^*, (0, 1)]$ is a conservative reduction class; a proof can be found in [320]).
2. For some integer ℓ, roughly of the size of a universal Turing machine, the classes $[\forall\exists\forall \wedge \exists^*, (\ell, 1)]$ and $[\forall\exists \wedge \forall^3 \wedge \exists^*, (\ell, 1)]$ are conservative reduction classes; see [219, 225].

Theorem 5.4.14. *Let ℓ be as above. Any standard conjunction $K = [w_1 \wedge \cdots w_m, p]$, where $\ell < \sum_i p_i < \infty$ and $(w_1) \cup \cdots \cup (w_m)$ is infinite, satisfies one of the following two conditions.*

A. K dominates at least one of the three conservative reduction classes

$$[\forall\exists \wedge \forall^*, (0, 1)], \quad [\forall\exists\forall \wedge \exists^*, (\ell, 1)], \quad [\forall\exists \wedge \forall^3 \wedge \exists^*, (k, 1)],$$

and therefore K is a conservative reduction class.

B. K is dominated by one of the three classes with the finite model property

$$[all, (\omega)], \quad [\exists^*\forall^*, all], \quad [\exists^*\forall^2\exists^*, all],$$

and therefore K has the finite model property.

Proof. The proof is similar to that of the previous theorem. Assume that K does not satisfy B. Check that $p \geq (\ell, 1)$. Again some w_i dominates $\forall\exists\forall$ or $\forall^3$, and some $w_j \geq \forall\exists$. Since $(w_1) \cup \cdots \cup (w_m)$ is infinite, some (w_k) is infinite and therefore dominates $\forall^*$ or $\exists^*$. If $w_i \geq \forall^*$, then $K \geq [\forall\exists \wedge \forall^*, (0, 1)]$. This is obvious if $k \neq j$. If $k = j$, then w_j dominates $\forall^*\exists$ or $\forall\exists\forall^*$. In either case, in the new domination order, $K \geq [w_j, (\ell, 1)] \geq [\forall\exists \wedge \forall^*, (0, 1)]$. Thus we may assume that $w_k \geq \exists^*$.

If $w_i \geq \forall\exists\forall$ then $K \geq [\forall\exists\forall \wedge \exists^*, (\ell, 1)]$. This is obvious if $k \neq i$. If $k = i$, then w_i dominates $\exists^*\forall\exists\forall$ or $\forall\exists^*\forall$ or $\forall\exists\forall\exists^*$. In all three cases, in the new domination order, $K \geq [w_i, (\ell, 1)] \geq [\forall\exists\forall \wedge \exists^*, (\ell, 1)]$.

Thus we may assume that w_i does not dominate $\forall\exists\forall$ and therefore $\forall^3 \leq w_i \leq \exists^*\forall^*\exists^*$. It is easy to see that in this case, in the new domination order, $K \geq [\forall\exists \wedge \forall^3 \wedge \exists^*, (\ell, 1)]$. □

An obvious problem is whether the conclusion of Theorem 5.4.14 remains true in the case $\ell = 0$. If not, what is the minimal appropriate ℓ?

Other Related Open Problems. Investigate the satisfiability problem, the finite satisfiability problem and the finite model property problem for conjunction of prefix-vocabulary classes in the cases when the underlying logic contains function symbols and/or equality. As far as we know, no systematic investigation of that kind has been performed (even though some conclusions can be drawn already from the formulations and proofs of the known classifications of prefix-vocabulary classes).

A related series of problems concerns zero-one laws. Recall also that every prefix-vocabulary class gives a fragment of second-order logic and gives rise to the question of the zero-one law for that fragment; see [315]. Similarly every conjunctions of prefix-vocabulary classes gives a fragment of second order logic and gives rise to the question of the zero-one law (or the limit law, the slow oscillation law, etc.) for that fragment. An appropriate version of Gurevich's Classifiability Theorem guarantees a finite solution for the resulting classification problems.

5.5 Historical Remarks

As explained in the introduction to Sect. 5.1, the notion of Krom formula sprang out from Herbrand's decidability result [253] for the class of Herbrand formulae and from the Chang-Keisler Normal Form Theorem established in [75]. Krom started the investigation of the specific logical properties of formulae with binary disjunctions [329, 330, 331, 333]. Maslov [378] showed the decidability of $\exists^*\forall^*\exists^* \cap$ KROM, Krom [331] the decidability of $\forall^* \wedge \exists^*\forall^2\exists^* \cap$ KROM. Aanderaa and Goldfarb [10] proved the finite model property for Maslov's class. Orevkov [409], Reynolds [437] and Krom [334] proved independently and with different methods that the decision problem for first-order Krom formulae is undecidable; for some years only Krom's proof was known. At the same time Cook [91] observed that the decision problem for propositional Krom formulae is not NP-complete, but decidable by a polynomial time algorithm. In [284] Cook's result is extended by showing that the unsatisfiability of Boolean Krom formulae is complete for non-deterministic logarithmic space, see [150] for a linear time algorithm. Aspvall, Plass and Tarjan [25] present a linear time algorithm for evaluating

Krom sentences of quantified propositional logic and Grädel [207] proved that this problem is also complete for nondeterministic logarithmic space.

Krom's undecidability proof proceeds by a reduction of Post's Tag systems and establishes the reduction class property for the class $[\forall\exists^*\forall, (0, \omega)] \cap$ KROM. A simpler undecidability proof has been discovered independently by Aanderaa [2] and Börger [39] (from where the register machine formalizations in Chap. 2.1 have been taken) and has been used in [2, 39, 411] to establish the conservative reduction class property for the Krom classes $[\exists\forall\exists\forall, (\omega, k)]$, $[\forall\exists^2\forall, (\omega, k)]$, $[\exists^*\forall\exists\forall, (0, k)] \cap$ KROM for some k of the size of a universal 2-register machine. Orevkov [411] contains an independent discovery of the Aanderaa-B"orger method and sharpens also Krom's undecidability result to $[\forall\exists^*\forall, (0, k)] \cap$ KROM for some k of the size of a universal 2-register machine, further improved by Rödding and Börger [442] to $[\forall\exists^*\forall, (0, 4)] \cap$ KROM $\cap$ HORN, see Theorem 5.1.10. In 1971, Aanderaa [2] shows the decidability of $\forall\exists\forall \cap$ KROM (see [12] for an elaboration of the proof) which has been extended in [6] to $[\forall\exists\forall]_= \cap$ KROM. Lewis shows in 1972 the reduction class property for $[\forall^2\exists\forall] \cap$ KROM and $[\forall\exists\forall^2] \cap$ KROM (see [12]). He introduces special counter machines which he proves to be computation universal; the proofs reported in this chapter (taken from [40]) simulate standard 2-register machines and establish for the first of the two Lewis classes also the conservativity of the reduction. For a method to prove the conservativity of the second Lewis class see [7]. Börger [41] establishes that the use of ternary predicates in the Lewis classes is necessary by proving the decidability of $[\forall^*, (\omega, \omega)] \cap$KROM, extended in [42] to the decidability of $[\forall^*, (\omega, \omega), (1)] \cap$KROM. Note that without the restriction to Krom formulae, undecidable classes can be reduced always to such classes with only binary predicates (see Chap. 3 and 4).

In [348] the class of Krom formulae with a single predicate, a binary one, is shown to be a reduction class. It is also proved that the class of prenex formulae having disjunctive normal form with only two disjuncts and having just two predicate letters, both pentadic, is a reduction class for validity, see [409, 411]. Aanderaa and Jensen [11] and Ershov [147] show that every satisfiable Krom formula without functions or equality has a recursive model. Börger [45] simplifies Aanderaa's [2] proofs about the non recursive complexity of models of simple extensions of Krom formulae and extends them to subrecursive complexity; see Sect. 2.1.3 on inseparability and model complexity.

For Krom formulae without functions but with equality, it has been shown in [6] that $[\forall\exists\forall\exists, (\omega, k)]_= \cap$KROM$\cap$HORN is a reduction class for some k of the size of a universal 2-register machine, see the Exercise 2.1.19. For Krom formulae with functions and without equality we report in this chapter the proof from [42] for the conservative reduction class $[\forall^2, (0, 1), (2)] \cap$KROM$\cap$ HORN which extends the corresponding reduction class result in [348] and for $[\exists^*\forall\exists\forall, (0, 0, 0, 0, 0, 2, 4, 0, 1)] \cap$ KROM in [409].

Theorem 5.2.2 is taken from [354], Theorem 5.2.8 from [534] (see also [535]). The idea to apply the proof technique of the Theorems of Lewis-Goldfarb and Wirsing to Post correspondence problems, for proving the theorem on the universality of binary Horn rules, appears in [244]. Another proof method, by which the same result has been established independently in [127], has found interesting applications in [374, 376].

Heidler's undecidability result (Corollary 5.3.5) appeared in [251] and is proved there also for Hermes' term logic [257]. The stronger result for two-variable logic by Grädel, Otto and Rosen and the results on logics with cardinality comparison are taken from [211]. This paper contains a number of other undecidability results for two-variable logics, including fixed-point logics and transitive closure logics.

The analysis of how much the different combinations of occurrences of variables in the reduction formulae can be restricted to only a few simple ones is discussed at length in the two books [133, 351]. A particularly interesting case where the classification along these lines has been completed satisfactorily is Kahr's reduction class of formulae of form $\forall x \exists x' \forall y \alpha \in [\forall\exists\forall, (\omega, 1)]$. The problem to classify subclasses of Kahr's class with respect to which combinations st of terms appear in atomic formulae Pst in α, has been suggested and investigated in [64, 288, 134]. Note that by Scott's Theorem [459] at least three variables are needed. Kostyrko [319] presents an interesting new proof, starting directly from Turing machines, for the result in [288] (see Exercise 3.1.10 in Chap. 3) that each such subclass is a reduction class where any three out of $xy, yx, x'y, yx'$ are allowed to occur. Aanderaa proves in his thesis [1], using the linear sampling problem, that this holds also for the combination xy, yx', xx. (For an elaboration of this proof see [13, 351].)

For interesting decidable classes obtained by restricting the form of subformulae if formulae in Skolem normal form see [478, 173, 286, 379, 380, 381, 133]. This classification line leads naturally to the study of term structure in resolution calculi based decision procedures, see [163].

Part II

Decidable Classes and Their Complexity

6. Standard Classes with the Finite Model Property

In this chapter and the next we present a complete description of the standard classes for which the satisfiability problem and the finite satisfiability problem are decidable. Together with the results in Chap. 3 and 4 this will give us a complete solution of the classification problem for standard classes. We will also investigate the complexity of decidable classes and, in most cases, present matching upper and lower complexity bounds. In addition we present a classification of the classes with the finite model property.

We first observe that there are classes whose satisfiability and finite satisfiability problems are decidable for trivial reasons and which we will exclude from further consideration. These are the (relational) classes $[\Pi, s]$ or $[\Pi, s]_=$ where Π and $s = (s_1, s_2, \ldots)$ are finite in the following sense:

- $\Pi \in \{\exists, \forall\}^*$, i.e., Π does not contain any occurrences of $\exists^*$ or $\forall^*$ and thus defines a finite set of prefixes;
- $s_i \neq \omega$ for all i and $s_i = 0$ for all but finitely many i; thus s defines (up to renaming) a finite vocabulary of relation symbols.

We call such classes *essentially finite* since their formulae are built from a (up to renaming of variables and relation symbols) finite collection K of atomic formulae. We can fix names of the variables and relation symbols and impose a linear order on K. This induces a linear order on formulae in conjunctive normal form built from the atoms of K. Every formula in an essentially finite class can be efficiently reduced to the equivalent formula whose quantifier-free part is in minimal (with respect to the chosen order) conjunctive normal form. Hence, satisfiability (and finite satisfiability) of essentially finite classes can be decided by transforming the given formulae into such a normal form, thus reducing the problem to finitely many instances, and then looking up the answer in a table. It is clear that this procedure can be implemented using only logarithmic work-space. We thus have proved:

Proposition 6.0.1. *For every essentially finite class X, the problems Sat(X) and Fin-sat(X) are decidable with logarithmic space.*

In the sequel we therefore restrict attention to standard classes $[\Pi, s, t]$ and $[\Pi, s, t]_=$ that satisfy at least one of the following conditions:

- Π contains an occurrence of $\exists^*$ or $\forall^*$;

– $t \neq 0$;
– s is not finite.

We will prove that all standard classes $[\Pi, s, t]$ or $[\Pi, s, t]_=$ that do not contain any of the sixteen conservative classes exhibited in Theorem 3.0.1 and Theorem 4.0.1 are decidable. In fact we exhibit *seven maximal standard classes for which satisfiability (and finite satisfiability) are decidable.*

Theorem 6.0.2 (Maximal Decidable Classes). *For each of the following seven classes, the satisfiability and the finite satisfiability problems are decidable:*

(1)	$[\exists^*\forall^*, all]_=$	*(Ramsey 1930)*
(2)	$[\exists^*\forall^2\exists^*, all]$	*(Gödel 1932, Kalmár 1933, Schütte 1934)*
(3)	$[all, (\omega), (\omega)]$	*(Löb 1967, Gurevich 1969)*
(4)	$[\exists^*\forall\exists^*, all, all]$	*(Gurevich 1973, Maslov-Orevkov 1972)*
(5)	$[\exists^*, all, all]_=$	*(Gurevich 1976)*
(6)	$[all, (\omega), (1)]_=$	*(Rabin 1969)*
(7)	$[\exists^*\forall\exists^*, all, (1)]_=$	*(Shelah 1977)*

The following exercise shows that this indeed gives a complete solution to the classification problem for standard classes.

Exercise 6.0.3. Prove that every standard class does either contain one of the sixteen conservative classes listed in Theorem 3.0.1 and Theorem 4.0.1, or is essentially finite, or is a subclass of one of the seven decidable classes listed in Theorem 6.0.2.

Recall that a class $X \subseteq \mathrm{FO}$ has the *finite model property* if $Sat(X) = Fin\text{-}sat(X)$, i.e. if every satisfiable formula in X has a finite model. If X is recursive and has the finite model property, then $Sat(X)$ is decidable. Indeed, by the Completeness Theorem for first-order logic and by the duality of validity and satisfiability, $Sat(X)$ is co-r.e. for every recursive set $X \subseteq \mathrm{FO}$. On the other hand, the finite model property implies that $Sat(X)$ is recursively enumerable (since $Fin\text{-}sat(X)$ is). It follows that $Sat(X)$ is recursive.

In this chapter we prove decidability for the classes (1) to (5) which in fact have the finite model property. In Chap. 7 we will then treat the remaining two classes; both of them contain infinity axioms, and we will have to use different techniques there.

While the finite model property of a recursive class X implies that $Sat(X)$ is decidable, it need not give any insight into its complexity. However, in most cases a proof of the finite model property actually determines a bound $s(n)$

such that every satisfiable formula $\psi \in X$ has a model of cardinality at most $s(|\psi|)$. In this case, we will say that X has a *small model property*. This will imply a nondeterministic upper complexity bound for $Sat(X)$ via the following proposition.

Proposition 6.0.4. *The problem whether a given prenex first-order sentence of length n with k universal quantifiers has a model of cardinality m can be decided nondeterministically in time $p(m^k n)$, for some polynomial p.*

Proof. Recall that every first-order sentence is satisfiable over the same domains as its functional form, and that transformation into functional form is easy. We can thus assume that the given sentence is of the form $\forall x_1 \cdots \forall x_k \varphi$ where φ is quantifier-free.

To check whether such a formula has a model $\mathfrak{A}$ with universe $A = \{0, \dots, m-1\}$ the decision algorithm cycles through all k-tuples $a_1, \dots, a_k \in A^k$ and, for each such tuple, guesses sufficient information about $\mathfrak{A}$ to check whether $\mathfrak{A} \models \varphi[a_1, \dots, a_k]$.

The truth value of $\varphi[a_1, \dots, a_k]$ depends on the values of at most n terms and atomic statements. Thus, for the verification that $\mathfrak{A} \models \forall x_1 \cdots \forall x_k \varphi$, at most $m^k n$ function values and truth values need to be guessed.

The verification algorithm generates and maintains a consistent list of function values and truth values of atomic statements. At every point where a function value $f(b_1, \dots, b_r)$ or a truth value $Pb_1 \cdots b_s$ is needed, it is checked whether it occurs already in the list, otherwise an appropriate value is guessed and appended to the list. Elements of the structure can be represented with $O(\log m)$ bits; therefore every entry of the list requires no more than $n \log m$ bits. Thus the list is generated with $O(n^2 m^k \log m)$ steps. The time required for the verification is a small polynomial in the length of the list, hence a polynomial in $m^k n$. □

With respect to complexity considerations it is important to keep in mind the usual convention in the theory of computation that algorithms work with strings that are composed of only finitely many distinct symbols. In particular, formulae are understood to be encoded over a finite alphabet. Hence, if X is a class of formulae that may have arbitrary many predicates, functions or variables, then these are coded (e.g. with indices in binary notation) such that a formula in X with n distinct predicates, functions or variables has length at least $cn \log n$, for some constant c that depends only on the alphabet used. To put it differently, a formula of length n has no more then $O(n/\log n)$ different predicates, functions and variables.

Here is a brief summary of this chapter. In Sect. 6.1 we discuss some techniques for proving complexity bounds for satisfiability problems. In particular we introduce a bounded variant of the domino problem that will be the main tool for most of our lower bound proofs.

In Sect. 6.2 we prove decidability and complexity results for the classical solvable cases of the decision problems. These include the class of monadic formulae, proved decidable already in 1915 by Löwenheim and the following prefix classes in pure predicate logic (without functions and equality): the Bernays-Schönfinkel class $[\exists^*\forall^*]$, the Ackermann class $[\exists^*\forall\exists^*]$ and the Gödel-Kalmár-Schütte class $[\exists^*\forall^2\exists^*]$. We will in fact prove more general decidability results, including the maximal decidable classes (1) – (3) of Theorem 6.0.2.

In Sect. 6.3 we consider sentences with only one universal quantifier. In particular we exhibit a decision procedure for the Gurevich-Maslov-Orevkov class, i.e., the $\exists^*\forall\exists^*$ prefix class in first-order logic without equality (but with arbitrary relation and function symbols), i.e. class (4) in Theorem 6.0.2.

We classify in Sect. 6.4 the standard classes of modest complexity. Here this means that the satisfiability problem is in P, NP or Co-NP. One of these classes is the existential fragment of first-order logic, i.e. class (5) of Theorem 6.0.2 (see Proposition 6.4.27 in Section 6.4.3.)

In Sect. 6.5 we present a classification of the prefix-vocabulary classes that have the finite model property and of those admitting infinity axioms.

In Sect. 6.6 we make some historical remarks concerning the results of this chapter.

6.1 Techniques for Proving Complexity Results

6.1.1 Domino Problems Revisited

We introduce a bounded domino problem that will be used in most of our lower bound proofs. It differs from the domino problems used for undecidability proofs (see Sect. 3.1.1 and Appendix A) and most of its finite variants used in the literature in two essential features:

(1) The space to be tiled is a torus $\mathbb{Z}_s \times \mathbb{Z}_t$ (where $\mathbb{Z}_s$ means the integers modulo s). This has the advantage that we don't have to verify cumbersome special conditions for borderline points.

(2) We use a more complicated initial constraint than usual. We specify the dominoes that have to be placed on the first n points in the bottom row. This makes perhaps the domino problem less elegant, but it allows a direct encoding of the input of a computation by the initial constraint for the domino problem. In fact, with any (nondeterministic) Turing machine M we can associate a fixed domino system $\mathcal{D}$ such that the language accepted by M is represented by the set of initial constraints for $\mathcal{D}$ that admit a tiling.

Definition 6.1.1. Let $\mathcal{D} = (D, H, V)$ be a domino system where D is a finite set of tiles and $H, V \subseteq D \times D$. Let $U(s,t)$ be the torus $\mathbb{Z}_s \times \mathbb{Z}_t$ and $w = w_0, \ldots, w_{n-1}$ be a n-tuple of tiles (with $n \leq s$). We say that $\mathcal{D}$ *tiles* $U(s,t)$ *with initial condition* w if there exists a mapping $\tau : U(s,t) \to D$ such that for all $(x,y) \in U(s,t)$:

(i) If $\tau(x,y) = d$ and $\tau(x+1,y) = d'$ then $(d,d') \in H$;
(ii) if $\tau(x,y) = d$ and $\tau(x,y+1) = d'$ then $(d,d') \in V$;
(iii) $\tau(i,0) = w_i$ for $0 \le i < n$.

Theorem 6.1.2 below, which may be of independent interest, establishes the correspondence between (nondeterministic) computations and domino problems on tori in rather general form, with simultaneous time and space bounds. Later we will only consider time bounds and can therefore restrict attention to tori $Z(t) := U(t,t)$.

It is convenient to consider Turing machines of some special form. We call a nondeterministic one-tape Turing machine M over alphabet Σ *simple* if it satisfies the following conditions. The alphabet of M contains Σ and at least one other symbol $\Box$ (blank). M works on a semi-infinite tape and never tries to move left from the left-most tape cell. At every stage of the computation there is some s such that the tape cells $0,\dots,s$ contain only non-blank symbols, all other tape cells contain $\Box$; in particular, to the right of a blank only other blanks may appear. Furthermore, we assume that M has a unique accepting configuration: the machine is in the unique accepting state q_a, the tape contains only blanks and the head is in position 0.

These conditions do not restrict computational power. Every language accepted in time $T(n)$ and space $S(n)$ by some one-tape nondeterministic Turing machine is accepted within the same time and space bounds by a simple Turing machine, as long as $S(n), T(n) \ge 2n$.

Theorem 6.1.2. *Let M be a simple nondeterministic one-tape Turing machine with input alphabet Σ. Then there exist a domino system $\mathcal{D} = (D,H,V)$ and a linear-time reduction which takes any input $x \in \Sigma^*$ to a word $w \in D^*$ with $|x| = |w|$ such that*

- *If M accepts x in time t_0 with space s_0 then $\mathcal{D}$ tiles $U(s,t)$ with initial condition w for all $s \ge s_0+2$, $t \ge t_0+2$;*
- *If M does not accept x, then $\mathcal{D}$ does not tile $U(s,t)$ with initial condition w for any $s,t \ge 2$.*

Proof. Let Σ' be the alphabet of M and Q its set of states; set

$$\Gamma := \Sigma' \,\dot\cup\, (Q \times \Sigma') \,\dot\cup\, \{\#, e\}.$$

The symbols $\#$ and e are used as end markers. Let $x = x_0 \cdots x_{n-1} \in \Sigma^*$ be some input, and let $s \ge s_0+2$, $t \ge t_0+2$ (if M does not accept x, let $s_0, t_0 := 0$). A configuration of M on input x can be described by a word in $C \in \Gamma^s$: $C = a_0 a_1 \cdots a_{i-1}(q a_i) a_{i+1} \cdots a_{s-2}\#$ encodes the situation that the tape stores $a_0,\dots,a_{s-2}$, the machine is in state q and scanning the ith tape cell. Thus, the accepting configuration and the initial configuration on x are encoded by

$$\begin{aligned} \mathrm{Acc} &:= (q_a\Box)\,\Box^{s-2}\,\# \\ \mathrm{Inp}(x) &:= (q_0 x_0) x_1 \cdots x_{n-1}\,\Box^{s-n-1}\,\#. \end{aligned}$$

In addition we define

$$\begin{aligned}\text{End} &:= e^{s-1}\# \\ \text{Conf} &:= \{C \in \Gamma^t : C \text{ encodes a configuration }\} \cup \{\text{End}\}\end{aligned}$$

and let Next(C) contain the encodings of those configurations that M can reach in one step from C. Next(C) is empty if $C \in \{\text{Acc, End}\}$.

The idea of the encoding is the following. Suppose that the sequence $(C^0, \dots, C^{t_0})$ represents an accepting computation of M on x. Then $C^0 = \text{Inp}(x)$, $C^{j+1} \in \text{Next}(C^j)$ and $C^{t_0} = \text{Acc}$. We extend this to a sequence $(C^0, \dots, C^{t-1})$ with $C^j = \text{Acc}$ for $t_0 \leq j < t-1$ and $C^{t-1} = \text{End}$. Let C_i^j be the i-th letter of C^j. This gives a description of the computation by a $t \times t$-table with entries from Γ. However, this table can not directly be represented by a domino tiling because the symbol C_i^{j+1} depends on the triple $C_{i-1}^j, C_i^j, C_{i+1}^j$. To overcome this problem let every domino consist of a *triple* of elements from Γ. The tiling $\tau : U(s,t) \to D$ that corresponds to the computation will then be defined by

$$\tau(i,j) = (C_{i-1}^j, C_i^j, C_{i+1}^j)$$

where addition is meant in $\mathbb{Z}_s$, i.e. we have identified the left and right borders of the configuration (or of the tape).

The definition of the domino system $\mathcal{D} = (D, H, V)$ is as follows:

$$D := \{(\alpha, \beta, \gamma) \in \Gamma^3 : \beta \in \{\Box, e\} \Longrightarrow (\gamma = \beta \vee \gamma = \#)\}$$

The horizontal adjacency relation is just a simple overlap condition:

$$H = \{(\alpha, \beta, \gamma)(\alpha', \beta', \gamma') \in D \times D : \alpha' = \beta, \beta' = \gamma\}$$

The vertical adjacency relation $V \in D \times D$ must be defined in such a way that the following condition is satisfied. Words $C \in \text{Conf}$ and $C' \in \Gamma^t$ can represent subsequent rows of the tiling, i.e.

$$[(C_{i-1}, C_i, C_{i+1})(C'_{i-1}, C'_i, C'_{i+1})] \in V$$

for all i, if and only if, one of the following holds:

(i) $C' \in \text{Next}(C)$;
(ii) $C = \text{Acc}$ and $C' \in \{\text{Acc, End}\}$;
(iii) $C = \text{End}$ and the first letter of C' is $(q_0 a)$ for some $a \in \Sigma'$.

The explicit definition of V follows well-known techniques (see e.g. [201, 203]). We only make some remarks on the nonstandard part of our construction that is necessary because we tile a torus and not a square or rectangle. The identification of the left and right borders of the tape creates no problems since we have marked the end of the configurations by #. We just have to require that above (and below) the symbol # only # may appear and that

V contains no element $(\alpha, qa, \#)(\alpha', \beta', \gamma)$ with $qa \in Q \times \Sigma'$. This precludes that the end marker # appears too far left; we can impose this condition because $s \geq s_0 + 2$. To ensure that the rows corresponding to the beginning and the end of the computation fit together, we impose special conditions on those elements of V that contain the symbol e. If $(d, d') \in V$ and at least one of the coordinates of d or d' is e then (d, d') are one of the vertically adjacent pairs in Fig. 6.1 (describing the tiling of the rows $t-2$, $t-1$ and 0) where q_a is the accepting state, q_0 the initial state and $\alpha, \beta, \gamma, \delta$ are arbitrary symbols from Σ'. Note that the domino system $\mathcal{D}$ depends only on M but not on w.

...	$(\Box, \Box, \#)$	$(\Box, \#, q_0\alpha)$	$(\#, q_0\alpha, \beta)$	$(q_0\alpha, \beta, \gamma)$	(β, γ, δ)	...
...	$(e, e, \#)$	$(e, \#, e)$	$(\#, e, e)$	(e, e, e)	(e, e, e)	...
...	$(\Box, \Box, \#)$	$(\Box, \#, q_a\Box)$	$(\#, q_a\Box, \Box)$	$(q_a\Box, \Box, \Box)$	$(\Box, \Box, \Box)$	...

Figure 6.1. Tiling of the rows $t-2$, $t-1$ and 0

Exercise 6.1.3. Define the vertical adjacency condition V explicitly.

The initial condition w is the following n-tuple of dominoes:

$$\begin{aligned} w_0 &= (\#, q_0 x_0, x_1) \\ w_1 &= (q_0 x_0, x_1, x_2) \\ &\vdots \\ w_{n-1} &= (x_{n-2}, x_{n-1}, \Box) \end{aligned}$$

If M accepts the input x then the mapping $\tau : U(s,t) \to D$ defined above is a correct tiling of $U(s,t)$ by $\mathcal{D}$ with initial condition w.

Conversely, suppose that we have a correct tiling of $U(s,t)$ with initial condition w for some $s, t \geq 2$. Then, column $s-1$ contains only dominoes of type $(\alpha, \#, \beta)$ and the dominoes at the points $(0, t-1)$ and $(0, t-2)$ are uniquely determined to be $(\#, e, e)$ and $(\#, (q_a\Box), \Box)$. Let r and q be the minimal numbers such that $\tau(0, r) = (\#, (q_a\Box), \Box)$ and $\tau(q, 0) = (\alpha, \#, \beta)$

for arbitrary α, β. For $0 \le j \le r$ and $0 \le i \le q$ and $\tau(i,j) = (\alpha,\beta,\gamma)$, set $C_i^j = \beta$ and let $C^j = C_0^j \cdots C_q^j$. Then $C^0,\ldots,C^r$ encodes an accepting computation of M on input x. □

Exercise 6.1.4. Extend this result to nondeterministic Turing machines with an arbitrary number of tapes.

Definition 6.1.5. Denote the torus $U(t,t)$ by $Z(t)$. Let $T(n)$ be a function from natural numbers to natural numbers and let $\mathcal{D}$ be a domino system. Then DOMINO$(\mathcal{D}, T(n))$ is the set of those words $w \in D^*$ for which $\mathcal{D}$ tiles $Z(T(|w|))$ with initial condition w.

We call a function $T : \mathbb{N} \to \mathbb{N}$ is *time constructible* if there exists a deterministic Turing machine making precisely $T(n)$ steps on inputs of length n (see [271, 416]).

Theorem 6.1.6. *Let $T(n)$ be a time constructible function from $\mathbb{N}$ to $\mathbb{N}$ with $T^2(dn) = o(T(n))$ for some constant $d > 0$. Then there exists a domino system $\mathcal{D}$ and a constant $c > 0$ such that*

$$\mathrm{DOMINO}(\mathcal{D}, T(n)) \notin \mathrm{NTIME}(T(cn)).$$

Proof. It is known that $\mathrm{NTIME}(T_1(n)) - \mathrm{NTIME}(T_2(n)) \neq \emptyset$ whenever T_1 is time constructible and $T_2(n+1) = o(T_1(n))$ (see [461]). Further it is known that k-tape nondeterministic Turing machines (for any k) can be simulated by one-tape nondeterministic machines with only quadratic increase of the running time [271, p. 292].

We apply these two facts as follows: Let $T_1(n) := T(dn)$. Then every problem in $\mathrm{NTIME}(T_1(n))$ is accepted by some one-tape nondeterministic Turing machine in time $T_1(n)^2$, hence also in time $T(n)$. Further, we take a constant $e > 0$, sufficiently small such that $T(e(n+1)) = o(T(dn))$ and set $T_2(n) := T(en)$. It follows that there exists a problem $B \notin \mathrm{NTIME}(T(en))$ which is accepted by a one-tape nondeterministic Turing machine in time $T(n)$.

By Theorem 6.1.2 there exists a domino system $\mathcal{D}$ and a linear-time reduction taking every input x of length n to an initial condition w also of length n such that $x \in B$ iff $w \in \mathrm{DOMINO}(\mathcal{D}, T(n))$.

Assume that $\mathrm{DOMINO}(\mathcal{D}, T(n)) \in \mathrm{NTIME}(T(cn))$ for every positive constant c. Then B could be decided in nondeterministic time $kn + T(cn)$ for some $k \in \mathbb{N}$. However, there exists a $c > 0$ such that $kn + T(cn) < T(en)$ which contradicts the assumption that $B \notin \mathrm{NTIME}(T(en))$. □

Definition 6.1.7. Let Σ and Γ be two alphabets, $A \subseteq \Sigma^*$ and $B \subseteq \Gamma^*$ two problems, and let g be a function from $\mathbb{N}$ to $\mathbb{N}$. We say that A *is polynomially reducible to B via length order $g(n)$*, in symbols $A \le_{g(n)} B$, if there exists a function $f : \Sigma^* \to \Gamma^*$ which is computable in polynomial time such that for all $x \in \Sigma^*$, $|f(x)| = O(g(|x|))$ and such that $f(A) \subseteq B$ and $f(\Sigma^* - A) \subseteq (\Gamma^* - B)$.

Theorem 6.1.8. *Let $T(n)$ be time constructible with $T^2(dn) = o(T(n))$ for some constant $d > 0$ and let A be a problem such that for all domino systems $\mathcal{D}$*

$$\mathrm{DOMINO}(\mathcal{D}, T(n)) \leq_{g(n)} A$$
respectively

Further, let $h : \mathbb{N} \to \mathbb{N}$ be such that $h(dg(n)) = O(n)$ for all d. Then there exists a constant $c > 0$ such that $A \notin \mathrm{NTIME}(T(ch(n)))$.

Exercise 6.1.9. Prove Theorem 6.1.8 from Theorem 6.1.6.

6.1.2 Succinct Descriptions of Inputs

Consider the following situation. We have a structure with successor relation S, a distinguished element 0 and a finite set of properties, e.g. unary relations P_i ($i \in \Sigma$). Suppose that, given any word $w = w_0 \cdots w_{n-1} \in \Sigma^*$, we want to express by a formula ψ_w that the condition

$$C(w) \equiv P_{w_0}(0) \wedge P_{w_1}(1) \wedge \cdots \wedge P_{w_{n-1}}(n-1)$$

is satisfied. A typical example is the encoding of the initial configuration of a Turing machine on input w or, equivalently, the initial condition of a domino problem as described in Sect. 6.1.1. The obvious way to describe it uses n variables representing the elements $0, \ldots, n-1$:

$$\exists x_0 \cdots \exists x_{n-1} \Big(x_0 = 0 \wedge \bigwedge_{i=0}^{n-2} Sx_i x_{i+1} \wedge \bigwedge_{i=0}^{n-1} P_{w_i} x_i \Big).$$

This formula has, however, length $O(n \log n)$ which is sometimes not good enough.

Therefore we introduce a technique to encode the same condition by formulae which have length $O(n)$ (but a more complicated quantifier structure):

Lemma 6.1.10. *Let Π be one of the prefix classes $[\forall\exists^*]$ or $[\forall^*]$. If the relations $y = 2x$ and $y = 2x + 1$ are available then, given $w \in \Sigma^*$, one can construct in linear time a formula $\psi_w \in \Pi$ which expresses that the condition $C(w)$ is satisfied.*

Proof. We first assume that $\Pi = [\forall\exists^*]$. The idea is to use natural numbers as labels of a binary tree such that the root is labeled by 0 and the children of a node with label x are labeled $2x$ and $2x + 1$, respectively. Thus the nodes of level i have labels $0, \ldots, 2^i - 1$.

Set $h := \lceil \log n \rceil$. We define inductively for $0 \leq i \leq h$ and $0 \leq j < 2^{h-i}$ quantifier-free formulae $\mathrm{TREE}_{i,j}(x, y_0, \ldots, y_i)$.

For all j, set

$$\mathrm{TREE}_{0,j}(x, y_0) := \begin{cases} y_0 = 0 \to P_{w_j} x & \text{if } j < n \\ x = x & \text{if } j \geq n \end{cases}$$

$$\begin{aligned}\mathrm{TREE}_{i+1,j}(x,y_0,\dots,y_{i+1}) \; := \; & ((y_{i+1}=2y_i)\wedge \mathrm{TREE}_{i,2j}(x,y_0,\dots,y_i)) \vee \\ & ((y_{i+1}=2y_i+1)\wedge \mathrm{TREE}_{i,2j+1}(x,y_0,\dots,y_i)).\end{aligned}$$

Claim. *The condition $C(w)$ is satisfied if and only if the formula*

$$\psi_w := \forall x \exists y_0 \cdots \exists y_h (y_h = x \wedge \mathrm{TREE}_{h,0}(x,y_0,\dots,y_h))$$

is true in the structure under consideration.

Indeed, for any $a \in \mathbb{N}$, the unique tuple $b_0,\dots,b_h$ that possibly satisfies the formula $\mathrm{TREE}_{h,0}(a,b_0,\dots,b_h)$ is defined by $b_h := a$, $b_{i-1} := \lfloor b_i/2 \rfloor$. If $b_0 \neq 0$ then $a \geq 2^h$ and therefore the condition $C(w)$ has nothing to do with a; clearly $\mathrm{TREE}_{h,0}[a,b_0,\dots,b_h]$ is *true* in this case. If $b_0 = 0$ then it follows by induction that $\mathrm{TREE}_{h,0}[a,b_0,\dots,b_h]$ is equivalent to $\mathrm{TREE}_{0,a}[a,b_0]$. This means that either $a \geq n$ or that $a < n$ and the condition $P_{w_a}(a)$ is satisfied.

Claim. *The length of ψ_w is $O(2^h) = O(n)$, even if the subscripts of y_i are in unary (i.e. if $|y_i| = i+1$).*

The variable y_i occurs at most $1 + 2^{h-i} + 2^{h-i+1} \leq 2^{h-i+2}$ times in ψ_w. The length of the formula is a constant multiple of the sum over the length of the variables times the number of its occurrences. This number is

$$\sum_{i=0}^{h}(i+1)2^{h-i+2} = 4\sum_{i=0}^{h}\sum_{j=0}^{h-i}2^j = 4(2^{h+2}-h-3) = O(2^h) = O(n).$$

This proves Lemma 6.1.10 for $\Pi = [\forall\exists^*]$. Only minor modifications in the definitions of $\mathrm{TREE}_{i,j}$ and ψ_w are necessary for $\Pi = [\forall^*]$:

$$\begin{aligned}\mathrm{TREE}_{i+1,j}(x,y_0,\dots,y_{i+1}) := & ((y_{i+1}=2y_i)\rightarrow \mathrm{TREE}_{i,2j}(x,y_0,\dots,y_i))\wedge \\ & ((y_{i+1}=2y_i+1)\rightarrow \mathrm{TREE}_{i,2j+1}(x,y_0,\dots,y_i))\end{aligned}$$

$$\psi_w := \forall x \forall y_0 \cdots \forall y_h (y_h = x \rightarrow \mathrm{TREE}_{h,0}(x,y_0,\dots,y_h)).$$

The proof that ψ_w has the required properties is almost verbally the same as above. □

Exercise 6.1.11. At first sight it might seem that the presence of the two binary relations $y = 2x$ and $y = 2x+1$ is a rather strong assumption. However, they are often easily axiomatizable. Prove the following: On $(\mathbb{N}, 0, S)$ where S is the successor relation, the relations $L = \{(n, 2n) : n \in \mathbb{N}\}$ and $R = \{(n, 2n+1) : n \in \mathbb{N}\}$ are axiomatizable by a formula of vocabulary $\{0, S, R, L\}$ with prefix $\forall^5$ or $\forall^2\exists^3$.

6.2 The Classical Solvable Cases

Before Church and Turing had proved the unsolvability of the *Entscheidungsproblem*, a number of mathematicians had come up with positive solutions for particular subcases. The most celebrated are the decidability results for the following classes:

$[all, (\omega)]$	(Löwenheim 1915)
$[\exists^*\forall^*, all]$	(Bernays, Schönfinkel 1928)
$[\exists^*\forall\exists^*, all]$	(Ackermann 1928)
$[\exists^*\forall^2\exists^*, all]$	(Gödel 1932, Kalmár 1933, Schütte 1934)

These are often called the *classical solvable cases* of the decision problem.

In this section, we prove that these classes are decidable and discuss their complexity.

In Sect. 6.2.1 we treat formulae of monadic vocabulary. Besides the Löwenheim class we will consider its expansions by equality – which gives the class $[all, (\omega)]_=$ – and by unary function symbols – which gives the Löb-Gurevich class $[all, (\omega), (\omega)]$, one of the maximal solvable cases.

In Sect. 6.2.2 we investigate the the class of relational $\exists^*\forall^*$-formulae.

In Sect. 6.2.3 we consider the Gödel-Kalmár-Schütte class $[\exists^*\forall^2\exists^*, all]$. We prove the finite model property of this class using a probabilistic approach due to Gurevich and Shelah [239] which dramatically simplifies the most difficult part of Gödel's original proof.

All these classes have nondeterministic exponential time complexity. More precisely, we will prove complexity upper and lower bounds of the form $\text{NTIME}(2^{cn})$ or $\text{NTIME}(2^{cn/\log n})$ and, in most cases, the bounds are sharp, i.e. upper and lower bounds differ only by the constant c.

Of course, the results on decidability and the finite model property for the Gödel-Kalmár-Schütte class are inherited by the Ackermann class. A more detailed treatment of the Ackermann class will be given in Sect. 6.3.

6.2.1 Monadic Formulae

We first fix terminology. The classes $[all, (\omega)]$ and $[all, (\omega)]_=$ are called the *Löwenheim class* and the *Löwenheim class with equality*. Formulae in these classes are called *relational monadic formulae*. By expanding the Löwenheim class with unary function symbols we obtain the *full monadic class* $[all, (\omega), (\omega)]$, which is also called the *Löb-Gurevich class*; its elements are *monadic formulae*.

We will prove that the full monadic class and the Löwenheim class with equality have the finite model property; in fact we will establish bounds on the size of minimal models of monadic formulae and thus prove a small model property that will give us good upper complexity bounds for the satisfiability problems of these classes. The Löwenheim class (with equality) has a satisfiability test of complexity $\text{NTIME}(2^{O(n/\log n)})$. For the full monadic class the complexity is slightly higher: satisfiability is decidable in $\text{NTIME}(2^{O(n)})$.

We will also prove lower complexity bounds; in fact the lower bounds apply to small subfragments of these classes and, in particular, give us a good lower bound also for the Gödel-Kalmár-Schütte class.

Decidability and Upper Complexity Bounds for the Monadic Class. We first establish a small model property for the Löwenheim class with equality.

Proposition 6.2.1. *Let ψ be a relational monadic formula, possibly with equality, of quantifier-rank q with m predicates. If ψ is satisfiable, then it has a model of cardinality at most $q2^m$.*

Proof. Let $\mathfrak{A} = (A, P_1, \ldots, P_m) \models \psi$. With every element $a \in A$ we associate a 'colour' $c(a) = c_1 \cdots c_m \in \{0,1\}^m$ where $c_i = 1$ iff $\mathfrak{A} \models P_i a$. Let $A_c \subseteq A$ be the set of elements with colour c. For every $c \in \{0,1\}^m$ we choose a set $B_c \subseteq A_c$ such that $B_c = A_c$ if $|A_c| \leq q$ and $|B_c| = q$ if $|A_c| > q$. Let $\mathfrak{B}$ be the induced substructure of $\mathfrak{A}$ with universe $B := \bigcup_{c \in \{0,1\}^m} B_c$. Obviously, $|B| \leq q2^m$.

It is easy to see that no formula with q variables distinguishes between $\mathfrak{A}$ and $\mathfrak{B}$. (For instance, it is clear that Duplicator wins the Ehrenfeucht-Fraïssé game with q moves on $\mathfrak{A}$ and $\mathfrak{B}$.) Since $\mathfrak{A} \models \psi$ and ψ has quantifier-rank q, this implies that $\mathfrak{B} \models \psi$. □

Corollary 6.2.2 (Löwenheim). *The satisfiability problem for relational monadic formulae is decidable.*

For the Löwenheim class *without equality* a somewhat stricter bound applies.

Exercise 6.2.3. Prove that every satisfiable formula in $[all, (m)]$ has a model with at most 2^m elements.

A formula of length n can have only $O(n/\log n)$ different variables and predicates. Thus we obtain the following bound on the size of a minimal model.

Corollary 6.2.4. *Every satisfiable formula of length n in $[all, (\omega)]_=$ has a model of cardinality at most $2^{O(n/\log n)}$).*

Exercise 6.2.5. [31, 477] Extend Löwenheim's decidability result to the fragment of second-order logic where all predicates, free and bound, are monadic.

Exercise 6.2.6. [251] Prove the decidability of ε-logic without equality and only monadic predicates. Here ε-logic means the extension of first-order logic with Hilbert's choice operator (see Definition 5.3.1). Hint: Extend the decidability proof for the Löwenheim class to this case.

Note that by the results in Sect. 5.3.1, ε-logic *with equality* is undecidable, even for formulae that have no relation and function symbols besides equality.

For the full monadic class we obtain a slightly higher upper bound on the size of structures that are to be checked:

Proposition 6.2.7 (Grädel). *Every satisfiable monadic formula of length n has a model of cardinality $2^{O(n)}$.*

Proof. We show that every monadic formula ψ of length n can be transformed into a formula $\varphi \in [all, (n), (0)]$ which is satisfiable over the same domains as ψ. By Exercise 6.2.3 the result follows.

Let ψ be a monadic formula containing the atom Pft (where P is a monadic predicate, f a function symbol and t a term) and let Q be a new predicate, not occuring in ψ. Then ψ is satisfiable over the same domains as

$$\psi[Pft/Qt] \wedge \forall x(Pfx \leftrightarrow Qx)$$

where $\psi[Pft/Qt]$ is obtained from ψ by replacing all atoms Pft by Qt (for arbitrary terms t). Repeated application of this transformation produces a formula

$$\psi' := \alpha \wedge \forall x\beta$$

where α does not contain any function symbols and β is a conjunction of equivalences of the form $Pfx \leftrightarrow Qx$. Let $f_1, \dots f_m$ be the function symbols in β. Observe that $\forall x\beta$ is the Skolem normal form of $\forall x \exists y_1 \cdots \exists y_m \beta[f_i x/y_i]$ which is purely relational. By the Skolem Normal Form Theorem it follows that

$$\varphi := \alpha \wedge \forall x \exists y_1 \dots \exists y_m \beta[f_i x/y_i]$$

is satisfiable over the same domains as ψ' and hence ψ. Obviously φ contains at most n predicates. □

In particular, monadic formulae have the finite model property.

Corollary 6.2.8 (Löb, Gurevich). *Sat*$[all, (\omega), (\omega)]$ *is decidable.*

A closer analysis gives an upper complexity bound.

Proposition 6.2.9 (Lewis). *The problem whether a given monadic formula of length n has a model of size s can be decided nondeterministically in $2^{O(n/\log n + \log s)}$ steps.*

Proof. We present a nondeterministic procedure which, when given a monadic formula ψ and a model size s, first guesses a structure of size s of appropriate vocabulary and then verifies by quantifier elimination that this structure is a model for ψ. Guessing a structure means writing down a word of length s for every relation and a word of length $s \log s$ for every function in ψ. This requires $O((n/\log n)s \log s)$ bits.

Then quantifiers are eliminated as follows. Begin with an innermost quantifier; assume that it is existential and that we have to eliminate it from a subformula $\exists x\varphi$ with φ quantifier-free. Transform φ into disjunctive normal form, commute the existential quantifier with the disjunctions and separate the atoms containing x from those that depend on different variables. The result is a formula

$$\bigvee_{j=1}^{r} (\tilde{\varphi}_j \wedge \exists x \varphi_j(x))$$

where each $\varphi_j(x)$ is a conjunction of literals $\pm Pfx$ (where f is a composition of at most n function symbols). Note that $r = 2^{O(n/\log n)}$ and that every subformula $\tilde{\varphi}_j \wedge \exists x\varphi_j(x)$ contains some of the at most $O(n/\log n)$ different atoms of the original formula ψ. Therefore we can evaluate every subformula $\exists x\varphi_j(x)$ and replace it by *true* or *false* in time $O(n^2 \log n \cdot s \log s)$; elimination of one quantifier requires therefore time $2^{O(n/\log n + \log s)}$. Universal quantifiers are eliminated by a dual procedure using conjunctive instead of disjunctive normal form. This procedure is repeated for every quantifier. All intermediate formulae are conjunctive or disjunctive normal forms of atoms that were originally in ψ so the same bounds as above hold for the elimination of every quantifier in ψ. Therefore the whole decision procedure takes $2^{O(n/\log n + \log s)}$ steps. □

Corollary 6.2.10 (Lewis, Grädel). *There exist constants c, d such that*

(i) $Sat[all, (\omega)]_= \in \text{NTIME}(2^{cn/\log n})$;
(ii) $Sat[all, (\omega), (\omega)] \in \text{NTIME}(2^{dn})$.

For classes of relational monadic formulae where the number of predicates is bounded by a constant, complexity is even lower. Satisfiability of a formula with m monadic predicates and q quantifiers can be decided by cycling through all structures $\mathfrak{A}$ of cardinality at most $q2^m$. The representation of such a structure takes $qm2^m$ bits, which is $O(q)$ for bounded m; on every structure the formula is evaluated by the obvious recursive procedure using additional work space $O(mq \log m \log q)$ for the representation of q elements of the model. A formula of length n has $q = O(n/\log n)$ quantifiers. Therefore deciding satisfiability takes linear space; with the Space Compression Theorem from complexity theory (see e.g. [271, pp. 288–289]) we get the following result.

Corollary 6.2.11. $Sat[all, (m)]_= \in \text{DSPACE}(n)$ *for all* $m \in \mathbb{N}$.

Exercise 6.2.12. [491] Prove that $Sat[all, (m)]_=$ is PSPACE-complete for every fixed m. Hint: Reduce QBF (the set of quantified Boolean formulae that evaluate to *true*) to $Sat[all, (0)]_=$, i.e. the first-order theory of equality. Conclude that the decision problem for every first-order theory that has a nontrivial model (in which some atomic statements are true and others false) is PSPACE-hard.

Lower Bounds for the Monadic and the Gödel-Kalmár-Schütte Class. We now prove a lower complexity bound for the Löwenheim class that matches the upper bound proved above. In fact this lower bound applies to a small subclass of monadic formulae, namely $[\forall^2\exists, (\omega)]$. In particular this class is also contained in the Gödel-Kalmár-Schütte class $[\exists^*\forall^2\exists^*, all]$, so we obtain at the same time a lower bound for the latter. We will see in Sect. 6.2.3 that this lower bound is optimal, up to the value of the constant c.

Theorem 6.2.13 (Lewis, Fürer). *There exists a constant* $c > 0$ *such that*

$$Sat[\forall^2\exists, (\omega)] \notin \text{NTIME}(2^{cn/\log n}).$$

Proof. In view of Theorem 6.1.8 it suffices to show that for every domino system $\mathcal{D} = (D, H, V)$

$$\text{DOMINO}(\mathcal{D}, 2^n) \leq_{n \log n} Sat[\forall^2\exists, (\omega)].$$

Recall that $\text{DOMINO}(\mathcal{D}, 2^n)$ is the set of all $w \in D^*$ such that $\mathcal{D}$ tiles the torus $Z(2^n) = \mathbb{Z}_{2^n} \times \mathbb{Z}_{2^n}$ with initial condition w (where $n = |w|$). We have to construct for every $w \in D^n$ a formula ψ of length $O(n \log n)$ in the prefix-vocabulary class $[\forall^2\exists^*, (\omega)]$ which is satisfiable iff $w \in \text{DOMINO}(\mathcal{D}, 2^n)$. The intended models of ψ have universe $Z(2^n)$ and encode tilings $\tau : Z(2^n) \to D$ in the following way. With any point $z = (x, y)$ of $Z(2^n)$ we associate the word $(x_0 \cdots x_{n-1} y_0 \cdots y_{n-1}) \in \{0,1\}^{2n}$ such that $x = \sum_i x_i 2^i$ and $y = \sum_i y_i 2^i$. The vocabulary of ψ consists of the monadic predicates X_i, X_i^*, Y_i, Y_i^* (for $0 \leq i < n$), N_i (for $0 \leq i \leq n$) and P_d (for $d \in D$), with the following intended interpretation:

$$\begin{aligned}
X_i z &: \; x_i = 1 \\
X_i^* z &: \; x_j = 1 \text{ for all } j < i \\
Y_i z &: \; y_i = 1 \\
Y_i^* z &: \; y_j = 1 \text{ for all } j < i \\
N_i z &: \; z = (i, 0) \\
P_d z &: \; \tau(z) = d
\end{aligned}$$

The formula ψ has the form $\forall z \exists v \alpha \wedge \forall z \forall z' \beta$ with α and β quantifier-free. $\forall z \exists v \alpha$ ensures the correct interpretations of X_i, X_i^*, Y_i and Y_i^* and N_i; it relates these predicates in the desired way among each other and states that every $z = (x, y)$ has a 'successor' v: if $x \neq 2^n - 1$ then v is its right neighbour $(x + 1, y)$, otherwise, if $z = (2^n - 1, y)$ then $v = (0, y + 1)$. To construct α

we use the following basic fact. For arbitrary elements $a = \sum_{i<n} a_i 2^i$ and $b = \sum_{i<n} b_i 2^i$ of $\mathbb{Z}_{2^n}$, we have that $b = a+1$ if and only if there exists some $j \leq n$ such that

$$a_0 = \cdots = a_{j-1} = 1,\ a_j = 0$$
$$b_0 = \cdots = b_{j-1} = 0,\ b_j = 1, \text{ and}$$
$$b_{j+1} = a_{j+1}, \ldots, b_{n-1} = a_{n-1}.$$

Thus, for every $k < n$ we can write b_k as the *exclusive or* of a_k and $\bigwedge_{i<k} a_i$.

Let α be the conjunction of the following clauses (where $\oplus$ stands for the exclusive or):

$$X_0^* z \wedge Y_0^* z$$
$$\bigwedge_{i=1}^{n-1} X_i^* z \leftrightarrow (X_{i-1}^* z \wedge X_{i-1} z)$$
$$\bigwedge_{i=1}^{n-1} Y_i^* z \leftrightarrow (Y_{i-1}^* z \wedge Y_{i-1} z)$$
$$\bigwedge_{i=0}^{n-1} X_i v \leftrightarrow (X_i z \oplus X_i^* z)$$
$$\bigwedge_{i=0}^{n-1} Y_i v \leftrightarrow \left[Y_i z \oplus (Y_i^* z \wedge X_{n-1} z \wedge X_{n-1}^* z)\right]$$
$$N_0 z \leftrightarrow \left(\bigwedge_{i=0}^{n-1} \neg X_i z \wedge \neg Y_i z \right)$$
$$\bigwedge_{i=0}^{n-1} N_i z \leftrightarrow N_{i+1} v.$$

It is easily verified that $\forall z \exists v \alpha$ axiomatizes the relations in a correct way.

The other subformula $\forall z \forall z' \beta$ asserts that the tiling is correct. The first step is the definition of two quantifier-free formulae $H(z, z')$ and $V(z, z')$ which express – given the correct axiomatization of X_i, X_i^*, Y_i, Y_i^* – that z' is the right (resp. upper) neighbour of z:

$$H(z, z') \quad := \quad \bigwedge_{i=0}^{n-1} (Y_i z' \leftrightarrow Y_i z) \wedge \bigwedge_{i=0}^{n-1} (X_i z' \leftrightarrow (X_i z \oplus X_i^* z))$$
$$V(z, z') \quad := \quad \bigwedge_{i=0}^{n-1} (X_i z' \leftrightarrow X_i z) \wedge \bigwedge_{i=0}^{n-1} (Y_i z' \leftrightarrow (Y_i z \oplus Y_i^* z))$$

With $H(z, z')$ and $V(z, z')$ at hand the tiling conditions can be expressed by the formula $\forall z \forall z' \beta$ where β is the conjunction of the following four clauses:

Every point is tiled by precisely one domino

$$\dot{\bigvee_{d \in D}} P_d z$$

(where $\dot{\bigvee}$ is an extended exclusive or)

The adjacency conditions

$$H(z,z') \rightarrow \bigvee_{(d,d') \in H} P_d z \wedge P_{d'} z'$$
$$V(z,z') \rightarrow \bigvee_{(d,d') \in V} P_d z \wedge P_{d'} z'$$

The initial condition

$$\bigwedge_{i=0}^{n-1} (N_i z \rightarrow P_{w_i} z).$$

This completes the construction of ψ. The length of ψ is $O(n \log n)$ as required. It remains to show that ψ is satisfiable if and only if $\mathcal{D}$ tiles $Z(2^n)$ with initial condition w. If a correct tiling exists, then $Z(2^n)$ with the intended interpretation of the predicates is obviously a model for ψ. Conversely, let $\mathfrak{A}$ be a model for ψ with universe A. We define the mapping $f : A \rightarrow Z(2^n)$ that associates with every $a \in A$ the point $f(a) = (x(a), y(a))$ whose binary representation coincides with the sequence of truth values $X_0^{\mathfrak{A}}a, \ldots, X_{n-1}^{\mathfrak{A}}a, Y_0^{\mathfrak{A}}a, \ldots, Y_{n-1}^{\mathfrak{A}}a$. The formula $\forall z \exists v \alpha$ ensures that this mapping is surjective. Now choose, for every point $z \in Z(2^n)$ some element $a \in f^{-1}(z)$ and define the tiling $\tau(z) = d$, where d is the unique element of D such that $\mathfrak{A} \models P_d a$. This defines a correct tiling of $Z(2^n)$ by $\mathcal{D}$ with initial condition w. □

Note that, due to the special form of ψ this proof actually gives a stronger result.

Corollary 6.2.14 (Fürer). *The satisfiability problem for sentences of the form $\forall x \forall y \alpha \wedge \forall x \exists y \beta$ where α, β are quantifier-free monadic formulae has a lower complexity bound* NTIME$(2^{cn/\log n})$*, for some constant $c > 0$.*

Remark. This lower complexity bound also applies to L_2, i.e. the class of relational first-order sentences with only two variables (see Sect. 8.1).

The first lower bound result on the monadic class was proved by Lewis [352]. It was slightly weaker than the present result since it applied to the prefix class $[\exists \wedge \forall\forall \wedge \forall\exists]$ rather than $[\forall\forall \wedge \forall\exists]$. Fürer [177] removed the leading existential quantifier using an unconstrained finite domino problem rather than a direct encoding of Turing machines. The proof presented here is due to Grädel and taken from [196].

It is not known whether the $2^{O(n)}$-upper bound for the Löb-Gurevich class is optimal in terms of nondeterministic time-complexity. But we can at least prove that the bound of Proposition 6.2.7 on the cardinality of minimal models cannot be improved in an essential way.

Proposition 6.2.15. *For every natural number n, there exists a satisfiable monadic sentence of length $O(n)$ all whose models have at least $(n+1)2^n$ elements.*

Proof. We construct sentences ψ_n whose vocabulary consists of a unary function f and monadic predicates S, C, E, Y and Y^*. The intended model consists of an f-chain $a_0, \dots, a_{n-1}$ that leads into an f-cycle of cardinality $n2^n$. This cycle is divided into 2^n segments of length n. The predicate Y defines on each segment s a natural number $m(s) < 2^n$. The formula ψ_n ensures that $m(s') = m(s) + 1 \pmod{2^n}$ for successive segments s, s'.

More precisely, the universe of the intended model $\mathfrak{A}_n \models \psi_n$ is

$$A := \{a_0, \dots, a_{n-1}\} \cup \{0, \dots, n2^n - 1\}.$$

The interpretation of f on A is given by

$$\begin{aligned} f(a_i) &:= a_{i+1} \text{ for } i < n-1 \\ f(a_{n-1}) &:= 0 \\ f(i) &:= i + 1 \pmod{n2^n}. \end{aligned}$$

The interpretations of the predicates on A are

$$\begin{aligned} S &:= \{a_0\} \\ C &:= \{0,, \dots, n2^n - 1\} \\ E &:= \{c \in C : c \equiv 0 \pmod{n}\} \\ Y &:= \{nj + k \in C : k < n, j < 2^n, \text{ the } k\text{-th bit of } j \text{ is } 1\} \\ Y^* &:= \{nj + k \in C : Y(nj) \land Y(nj+1) \land \cdots \land Y(nj+k-1)\} \end{aligned}$$

The desired sentence is $\psi_n := \exists x Sx \land \forall x \alpha$ where α is the conjunction of

$$Sx \to \neg C f^{n-1} x$$

$$C f^n x$$

$$Cx \to Cfx$$

$$Ex \to Cx$$

$$Sx \to E f^n x$$

$$E f^n x \leftrightarrow (Sx \lor Ex)$$

$$Y^* x \leftrightarrow (Ex \lor (Yx \land Y^* fx))$$

$$Y f^n x \leftrightarrow (Yx \oplus Y^* x).$$

It is easily verified that $\mathfrak{A}_n \models \psi_n$. For the converse, use the following exercise.

Exercise 6.2.16. Let $\mathfrak{B}$ be a model of ψ_n. The interpretation of E in $\mathfrak{B}$ is closed under f^n. Prove that (E, f^n) is isomorphic to $\mathbb{Z}_{2^n}, succ)$.

□

Problem. Close the gap between the lower bound $\text{NTIME}(2^{\Omega(n/\log n)})$ and the upper bound $\text{NTIME}(2^{O(n)})$ for the Löb-Gurevich class. The apparent difficulty for using the technique of Proposition 6.2.15 for pushing up the lower bound to $\text{NTIME}(2^{\Omega(n)})$ (along the lines of Theorem 6.2.13) lies in the problem of axiomatizing the grid. We can axiomatize 2^{2n} successive elements and thus have a $2^n \times 2^n$-square with horizontal adjacency condition. But how can we axiomatize vertical adjacency by a subformula of length $O(n)$? Alternatively, how can we determine in $\text{NTIME}(2^{cn/\log n})$ whether a sentence is satisfiable if its minimal models (if one exists) have cardinality 2^{cn}?

6.2.2 The Bernays-Schönfinkel-Ramsey Class

Bernays and Schönfinkel [35] proved that the satisfiability problem for relational $\exists^*\forall^*$-sentences without equality is decidable. Ramsey [435] extended this results to $\exists^*\forall^*$-sentences with equality and showed that the spectrum of every such sentence is either finite or co-finite. We prove in this section that the satisfiability problem for $\exists^*\forall^*$-sentences without functions is complete for nondeterministic exponential time, no matter whether or not the formulae contain equality.

Decidability and Upper Complexity Bounds. We first show that every relational sentence $\psi := \exists x_1 \cdots \exists x_p \forall y_1 \cdots \forall y_m \varphi$ is either unsatisfiable or has a model of cardinality at most p. There are several ways to prove this. We use a simple model-theoretic argument based on the closure of universal sentences under substructures: If $\mathfrak{A}$ is a substructure of $\mathfrak{B}$ and $\mathfrak{B}$ is a model of a prenex sentence η with only universal quantifiers, then also $\mathfrak{A} \models \eta$.

Proposition 6.2.17. *Let* $\psi := \exists x_1 \cdots \exists x_p \forall y_1 \cdots \forall y_m \varphi$ *be a satisfiable sentence in* $[\exists^*\forall^*, all]_=$. *Then* ψ *has a model with at most* $\max(1, p)$ *elements.*

Proof. Let σ be the vocabulary of ψ. Since ψ is satisfiable, there exists a σ-structure $\mathfrak{A}$ and elements $a_1, \ldots, a_p$ such that

$$\mathfrak{A} \models \forall y_1 \cdots \forall y_m \varphi[a_1, \ldots, a_p].$$

We consider $\eta := \forall y_1 \cdots \forall y_n \varphi$ as a sentence of the expanded vocabulary $\tau = \sigma \cup \{a_1, \ldots, a_p\}$ and $(\mathfrak{A}, a_1, \ldots a_p)$ as a τ-expansion of $\mathfrak{A}$. Since η is universal, it is satisfied by every substructure of $(\mathfrak{A}, a_, \ldots, a_p)$, i.e. by every τ-structure $(\mathfrak{B}, a_1, \ldots, a_p)$ where $\mathfrak{B} \subseteq \mathfrak{A}$ is a substructure of $\mathfrak{A}$ containing $a_1, \ldots, a_p$. In particular this holds for the induced substructure with universe $\{a_1, \ldots, a_p\}$ (in the case that $p = 0$, we take any substructure of cardinality one). □

Exercise 6.2.18. Give a different proof of Proposition 6.2.17 using Herbrand models. Hint: This is obvious for $\exists^*\forall^*$-sentences without equality, since the Herbrand universe H has cardinality $\max(1, p)$. To handle equality prove that a sentence in functional form with equality is satisfiable if and only if there exists an equivalence relation E on H such that ψ has a model with universe H/E.

With Proposition 6.0.4 we conclude that there exists a constant c, such that the satisfiability of a prenex sentence of length n with prefix $\exists^p\forall^m$ can be decided nondeterministically in time $t = (np^m)^c$. Note that $m = O(n/\log n)$. If we restrict attention to sentences where the number of existential quantifiers is bounded by a constant, then the time t is bounded by $2^{O(n/\log n)}$; otherwise $p = O(n/\log n)$ and $t = 2^{O(n)}$. We infer

Theorem 6.2.19 (Lewis). *There exist constants c, d such that*

(i) $Sat[\exists^*\forall^*, all]_= \in \text{NTIME}(2^{cn})$.
(ii) For all $p \in \mathbb{N}$, $Sat[\exists^p\forall^*, all]_= \in \text{NTIME}(2^{dn/\log n})$.

The same argument shows that the satisfiability problems for certain subclasses of the Bernays-Schönfinkel-Ramsey class are in P or NP.

Theorem 6.2.20. *(i) If s is finite, then* $Sat[\exists\forall^*, s]_=$ *is in* P.
(ii) $Sat[\exists\forall^*, all]_=$ *is in* NP.
(iii) For all $q \in \mathbb{N}$, $Sat[\exists^*\forall^q, all]_=$ *is in* NP.

Proof. If the given formula has only one existential quantifier, then it is either logically false or it has a model with only one element. If the vocabulary is fixed then there is a fixed list of possible structures for all formulae of the class, so the satisfiability can clearly be checked in polynomial time (in fact with logarithmic space). If the vocabulary is arbitrary the problem is equivalent to the satisfiability problem for propositional formulae (the values of the predicates on the single element of the structures are propositional variables).

If the number of universal quantifiers is bounded by a fixed $q \in \mathbb{N}$, then the claim follows immediately from Proposition 6.0.4. □

Lower Bounds.

Theorem 6.2.21 (Lewis). *There exists a constant* $c > 0$ *such that*

$$Sat[\exists^*\forall^*, all] \notin \text{NTIME}(2^{cn}).$$

Proof. We use the same basic ideas as in the proof of Theorem 6.2.13; the details of the construction are however more complicated, mainly because we want to to obtain the time bound 2^{cn} rather than $2^{cn/\log n}$. Let m be the smallest natural number with $m \log m \geq n$. We will use m-ary notation to represent a domino problem on a square of size $m^m \geq 2^n$ by a formula of length $O(m \log m) = O(n)$. More precisely we show that for any domino system $\mathcal{D}$

$$\mathrm{DOMINO}(\mathcal{D}, m^m) \leq_{m \log m} Sat[\exists^*\forall^*, all].$$

Given a word $w \in D^*$ of length n a formula ψ is constructed which is satisfiable if and only if $\mathcal{D}$ tiles $Z(m^m)$ with initial condition w. The intended model has universe $\{0, \ldots, m-1\}$; the formula ψ contains existentially quantified constants $u_0, \ldots, u_{m-1}$ which stand for the ciphers $0, \ldots, m-1$. These are used for the m-ary representation of numbers up to $m^m - 1$. Thus a point $(x, y) \in Z(m^m)$ is encoded by a $2m$-tuple $(\bar{x}, \bar{y}) = (x_{m-1}, \ldots, x_0, y_{m-1}, \ldots, y_0)$ where $x = \sum_{i=0}^{m-1} x_i m^i$ and $y = \sum_{i=0}^{m-1} y_i m^i$. The predicates in ψ together with their intended interpretations are described in the following table:

Nuv: Successor of ciphers: $u = i$ and $v = i + 1$ for some $i < m - 1$;
N^*uv: Order of ciphers: $u = i$ and $v = j$ for some $i < j$;
Euv: Equality of ciphers: $u = v = i$ for some $i \leq m - 1$;
$S\bar{x}\bar{y}$: Successor relation on $\mathbb{Z}_{m^m}$: $y = x + 1$;
$P_d\bar{x}\bar{y}$: Tiling by the domino d: $\tau(x, y) = d$.

To describe the initial condition we will need in addition:

$L\bar{x}\bar{y}$: $y = 2x$;
$R\bar{x}\bar{y}$: $y = 2x + 1$.

The formula ψ has the form $\exists u_0 \cdots \exists u_{m-1} (\alpha \land \beta \land \varphi_w)$ where α axiomatizes the predicates, β describes the tiling condition of $\mathcal{D}$ and φ_w the initial condition imposed by w.

The axiom α. The main problem in the construction of α is the axiomatization of the successor relation. This is done by defining the successor relation on numbers $< m^m$ in terms of the successor relation on numbers $< m^{m-1}$. Define α to be the formula $\forall\bar{x}\forall\bar{y}\alpha'$ where α' is the conjunction of the following formulae:

$$\bigwedge_{i=0}^{m-2} N u_i u_{i+1}$$

$$N x_0 x_1 \rightarrow N^* x_0 x_1$$

$$(N^* x_0 x_1 \land N x_1 x_2) \rightarrow (N^* x_0 x_2 \land \neg N x_0 x_2)$$

$$\neg N x_0 x_0 \land (N^* x_0 x_1 \rightarrow \neg N x_1 x_0)$$

$$\bigwedge_{i=0}^{m-1} E u_i u_i$$

$$(N^* x_0 x_1 \lor N^* x_1 x_0) \rightarrow \neg E x_0 x_1$$

$$Nx_0y_0 \rightarrow \Big(S\bar{x}\bar{y} \leftrightarrow \bigwedge_{i>0} Ex_iy_i\Big)$$

$$S\bar{x}\bar{y} \rightarrow (Nx_0y_0 \vee (Ex_0u_{m-1} \wedge Ey_0u_0))$$

$$S\bar{x}'u_{m-1}\bar{y}'u_0 \leftrightarrow \Big(Su_0\bar{x}'u_0\bar{y}' \vee \bigwedge_{i=1}^{m-1}(Ex_iu_{m-1} \wedge Ey_iu_0)\Big).$$

In the last clause, $\bar{x}'$ and $\bar{y}'$ stand for the $(m-1)$-tuple $x_{m-1}, \ldots, x_1$ and $y_{m-1}, \ldots, y_1$, respectively. These formulae determine the interpretations of the relation symbols N, N^*, E and S on the universe $\{0, \ldots, m-1\}$. Note that the combined length of all formulae is $O(m \log m)$.

The tiling formula β. With the successor relation available it is now very easy to state that there is a tiling of $Z(m^m)$ by the domino system d $=$ (D, H, V) (if the initial condition is left aside for a moment):

$$\beta \;:=\; \forall\bar{x}\forall\bar{y}\forall\bar{z}\Big(\dot{\bigvee_{d\in D}} P_d\bar{x}\bar{y} \wedge \Big(S\bar{x}\bar{z} \rightarrow$$
$$\Big[\bigvee_{(d,d')\in H}(P_d\bar{x}\bar{y} \wedge P_{d'}\bar{z}\bar{y}) \wedge \bigvee_{(d,d')\in V}(P_d\bar{y}\bar{x} \wedge P_{d'}\bar{y}\bar{z})\Big]\Big)\Big).$$

The description of the initial condition. The main difficulty concerning the initial condition is to ensure that the length of its description remains bounded by $O(n) = O(m \log m)$; the straightforward encoding of its n instances by n variables or relations would increase the length of the formula to $\Omega(n \log n)$. For this purpose we use the technique described in Sect. 6.1.2: Having the successor relation available and using the facts that

$$\begin{aligned} y = 2x \quad &\leftrightarrow \quad (x = 0 \wedge y = 0) \vee (y-2) = 2(x-1) \\ y = 2x+1 \quad &\leftrightarrow \quad (x = 0 \wedge y = 1) \vee (y-2) = 2(x-1)+1 \end{aligned}$$

it is straightforward to axiomatize the relations $L\bar{x}\bar{y}$ and $R\bar{x}\bar{y}$ by an $\forall^*$-formula of length $O(m \log m)$. Now Lemma 6.1.10 says that given an input w a universal formula φ_w of length $O(n)$ can be built in linear time which is true on the structure $\{0, \ldots, m-1\}$ if and only if the initial condition imposed by w is satisfied.

Thus we conclude that the formula

$$\psi := \exists u_0 \cdots \exists u_{m-1}(\alpha \wedge \beta \wedge \varphi_w)$$

has length $O(m \log m)$ and is satisfiable if and only if $\mathcal{D}$ tiles $Z(m^m)$ with initial condition w. This completes the proof of Theorem 6.2.21. □

Theorem 6.2.22. *There exists a constant $c > 0$ such that*

$$\mathit{Sat}[\exists^2\forall^*, \mathit{all}] \notin \mathrm{NTIME}(2^{cn/\log n}).$$

The proof is completely analogous to the previous one; there we introduced $m = O(n/\log n)$ existentially quantified constants which represented the m ciphers used for m-ary notation of numbers up to m^m. Here we have only two constants which we use to represent the numbers up to 2^m in binary notation.

6.2.3 The Gödel-Kalmár-Schütte Class: a Probabilistic Proof

We prove in this section that the satisfiability problem for the Gödel-Kalmár-Schütte class $[\exists^*\forall^*\exists^*, all]$ is decidable, and in fact is contained in the complexity class $\text{NTIME}(2^{O(n/\log n)})$. For expository reasons we first consider the case of formulae without leading existential quantifiers, i.e. the $[\forall^2\exists^*]$ prefix class.

Theorem 6.2.23 (Gödel, Schütte). *The class $[\forall^2\exists^*, all]$ has the finite model property.*

We prove Theorem 6.2.23 following Gödel's general strategy [187]: first we formulate a *necessary criterion for satisfiability* of $\forall^2\exists^*$-sentences; in a second step, we prove that this criterion is *sufficient for finite satisfiability.* In Gödel's paper the second part is a difficult and very sophisticated model construction; we will use instead a much simpler probabilistic argument due to Gurevich and Shelah [239].

In fact we will prove a more general result concerning $\forall^2\exists^*$-sentences that may contain equality, but have to satisfy a certain semantic condition: every satisfiable sentence must have a model in which no element is uniquely determined by its atomic type. As we will show below, this condition is satisfied by all sentences without equality, but there are other interesting cases as well. For instance we will conclude that the $\forall^2\exists^*$-fragment of graph theory is decidable. Also the decidability of the Ackermann class with equality is a consequence of this result (see Sect. 6.3.3).

Gödel's Criterion. Let $\psi = \forall x \forall y \exists z_1 \cdots \exists z_m \varphi(x, y, z_1, \ldots, z_m)$ be a relational first-order sentence (possibly containing equality) where φ is quantifier-free. On structures with at least two elements ψ is equivalent to

$$\forall x \forall y \exists z_1 \cdots \exists z_m \exists z'_1 \cdots \exists z'_m (x \neq y \rightarrow \varphi(x, x, z_1, \ldots, z_m) \wedge \varphi(x, y, z'_1, \ldots, z'_m)).$$

Further, on sufficiently large structures we can impose inequalities of variables by using repeatedly the equivalence

$$\exists x \exists y \alpha(x, y) \equiv \exists x \exists y ((\alpha(x, x) \vee \alpha(x, y)) \wedge x \neq y).$$

Thus we can restrict attention to sentences of the form

$$\psi := \forall x_1 \forall x_2 \exists x_3 \cdots \exists x_m (x_1 \neq x_2 \rightarrow \varphi(x_1, x_2, \ldots, x_m))$$

where $\varphi(x_1, \ldots, x_m) \models \bigwedge_{1 \leq i < j \leq m} x_i \neq x_j$, and to structures with at least m elements. We call such sentences *Gödel sentences of special form.*

Definition 6.2.24. A *k-table* of vocabulary σ is a σ-structure with universe $\{1,\dots,k\}$. Further, given a structure $\mathfrak{A}$ with a k-tuple $a_1,\dots,a_k$ of distinct elements, $T_{\mathfrak{A}}[a_1,\dots,a_k]$ is the unique k-table which is isomorphic, via the mapping $i \mapsto a_i$ (for $i = 1,\dots,k$), to the substructure of $\mathfrak{A}$ induced by $a_1,\dots,a_k$.

Definition 6.2.25. An element a of a structure $\mathfrak{A}$ is a *king* if there is no other element b of $\mathfrak{A}$ with the same 1-table, i.e. with $T_{\mathfrak{A}}[b] = T_{\mathfrak{A}}[a]$.

Lemma 6.2.26. *Let ψ be any relational first-order sentence without equality. If ψ is satisfiable then it has a model without kings.*

Proof. Suppose that $\mathfrak{A}$ is a model for ψ with universe A. Let $2\mathfrak{A}$ be the structure with universe $A \times \{0,1\}$ and relations defined in such a way that

$$2\mathfrak{A} \models R(a_1,i_1)\cdots(a_k,i_k) \iff \mathfrak{A} \models Ra_1\cdots a_k$$

for all k-ary predicates R and all $a_1,\dots,a_k \in A$ and $i_1,\dots,i_k \in \{0,1\}$. Obviously $\mathfrak{A}$ and $2\mathfrak{A}$ are indistinguishable by sentences without equality and $2\mathfrak{A}$ does not contain kings. □

Definition 6.2.27 (Gödel's Criterion). Let $\varphi(x_1,\dots,x_m)$ be quantifier-free and P,Q be non-empty sets of, respectively, 1-tables and 2-tables over σ. We say that P,Q satisfy Gödel's criterion for φ if

(1) For all $\mathfrak{B},\mathfrak{B}' \in P$, there exists a 2-table $\mathfrak{C} \in Q$ such that $T_{\mathfrak{C}}[1] = \mathfrak{B}$ and $T_{\mathfrak{C}}[2] = \mathfrak{B}'$.
(2) Every 2-table $\mathfrak{B} \in Q$ can be extended to an m-table $\mathfrak{C}$ such that
  - $T_{\mathfrak{C}}[1,2] = \mathfrak{B}$;
  - $T_{\mathfrak{C}}[i] \in P$ for all $i \in \{1,\dots,m\}$;
  - $T_{\mathfrak{C}}[i,j] \in Q$ for all distinct $i,j \in \{1,\dots,m\}$;
  - $\mathfrak{C} \models \varphi[1,\dots,m]$.

Lemma 6.2.28. *Let $\psi = \forall x_1 \forall x_2 \exists x_3 \cdots \exists x_m (x_1 \neq x_2 \rightarrow \varphi(x_1,x_2,\dots,x_m))$ be a Gödel sentence in special form. If ψ has a model without kings, then there exist non-empty sets P,Q satisfying Gödel's criterion for φ.*

Proof. Pick a model $\mathfrak{A} \models \psi$ without kings and set

$$\begin{aligned} P &:= \{T_{\mathfrak{A}}[a] : a \in A\} \\ Q &:= \{T_{\mathfrak{A}}[a,b] : a,b \in A, a \neq b\}. \end{aligned}$$

□

Exercise 6.2.29. Prove that in Lemma 6.2.28 the assumption that ψ has a model *without kings* cannot be dropped.

Sufficiency of Gödel's Criterion for Finite Satisfiability.

Theorem 6.2.30. *Let* $\psi = \forall x_1 \forall x_2 \exists x_3 \cdots \exists x_m (x_1 \neq x_2 \to \varphi(x_1, \ldots, x_m))$ *be a Gödel sentence in special form and suppose that* P, Q *satisfy the Gödel criterion for* φ. *Then* ψ *has a finite model.*

Proof. Let $P = \{\mathfrak{B}_1, \ldots, \mathfrak{B}_p\}$. For $n \geq m$ we present a probabilistic construction of an np-table $\mathfrak{A}$:

Stage 1: Every element of the universe $\{1, \ldots, np\}$ can be written in the form $ip + j$ where $0 \leq i < n, 1 \leq j \leq p$. Set $T_\mathfrak{A}[ip + j] := \mathfrak{B}_j$.

Stage 2: Let $1 \leq a < b \leq np$. By condition (1) of Gödel's criterion the set

$$\{\mathfrak{C} \in Q : T_\mathfrak{C}[1] = T_\mathfrak{A}[a], T_\mathfrak{C}[2] = T_\mathfrak{A}[b]\}$$

is non-empty. Select randomly a 2-table $\mathfrak{C}$ from this set and put $T_\mathfrak{A}[a, b] := \mathfrak{C}$.

Stage j $(3 \leq j \leq m)$*:* Define the truth value of every atomic statement $Ra_1 \cdots a_k$, for which $a_1, \ldots, a_k$ has precisely j distinct components, at random with probability 1/2.

Stages $m+1, m+2, \ldots$ *:* Set $\mathfrak{A} \models \neg Ra_1 \cdots a_k$ for every relation symbol and every tuple $a_1, \ldots, a_k$ with more than m distinct components.

Note that at stage j, the truth values of all atomic statements on j distinct elements are determined. Let S_n be the set of all np-tables that may appear with positive probability as the result of this probabilistic process. We consider S_n as a probability space with the uniform probability distribution.

Definition 6.2.31. Let $\mathfrak{A} \in S_n$, $a_1, \ldots, a_m \in \{1, \ldots, np\}$ and let $\mathfrak{C}$ be an m-table. We say that $a_3, \ldots, a_m$ *witness* $\mathfrak{C}$ *for* a_1, a_2 if

$$\mathfrak{A} \models Ra_{i_1} \cdots a_{i_k} \iff \mathfrak{C} \models Ri_1 \cdots i_k$$

for all k-ary relation symbols $R \in \sigma$ and all $i_1, \ldots, i_k \in \{1, \ldots, m\}$ such that $i_j > 2$ for at least one j. Note that for the case where σ contains only unary and binary predicates this means that

$$T_\mathfrak{A}[a_1, a_j] = T_\mathfrak{C}[i, j]$$

for all $i \neq j$ with $i > 2$ or $j > 2$.

Let s be the number of atoms $Rx_{i_1} \cdots x_{i_k}$ such that R is a k-ary predicate in σ, with $3 \leq k \leq m$, and $i_1, \ldots, i_k$ is a tuple of numbers in $\{1, \ldots, m\}$ with at least three distinct components. Further, let

$$\begin{aligned} r &:= \binom{m-2}{2} + 2(m-2) \\ q &:= |Q| \\ \varepsilon &:= \frac{1}{q^r \cdot 2^s} \end{aligned}$$

By condition (2) of Gödel's criterion there exists a function f assigning to every 2-table $\mathfrak{B} \in Q$ an appropriate m-table $f(\mathfrak{B})$. As usual, for arbitrary events E, F we write $\Pr[E \mid F]$ for the conditional probability of the event E given that the event F occurs.

Lemma 6.2.32. *Let $\mathfrak{B} \in Q, \mathfrak{C} = f(\mathfrak{B})$ and let $a_1, \ldots, a_m$ be m distinct elements of $\{1, \ldots, np\}$. Then the conditional probability*

$$\Pr[a_3, \ldots, a_m \textit{ witness } \mathfrak{C} \textit{ for } a_1, a_2 \mid T_{\mathfrak{A}}[a_i] = T_{\mathfrak{C}}[i] \textit{ for } i = 1, \ldots, m]$$

is at least ε.

Proof. There are s atoms $Rx_{i_1} \cdots x_{i_k}$ with at least three distinct variables. For each of these

$$\Pr[\mathfrak{A} \models Ra_{i_1} \cdots a_{i_k}] = 1/2.$$

The probability that $\mathfrak{A} \models Ra_{i_1} \cdots a_{i_k}$ iff $\mathfrak{C} \models Ri_1 \cdots i_k$ for all such atoms is therefore 2^{-s}.

The corresponding equivalence for all atoms with two distinct variables holds if

$$T_{\mathfrak{A}}[a_i, a_j] = T_{\mathfrak{C}}[i, j]$$

for all $i \neq j$, $\{i, j\} \neq \{1, 2\}$. There exist r such sets $\{i, j\}$. Accordingly, the probability that this is satisfied for all such i, j is $\geq q^{-r}$. For atoms with just one variable the equivalence holds by assumption. □

Let $\ell := \lfloor (n-2)/(m-2) \rfloor$; as a consequence $n \geq \ell(m-2) + 2$.

Lemma 6.2.33. *Let a_1, a_2 be distinct elements of $\{1, \ldots, np\}$. Then*

$$\Pr[\mathfrak{A} \models \neg \exists x_3 \cdots \exists x_m \varphi[a_1, a_2]] \leq (1 - \varepsilon)^{\ell}.$$

Proof. Let $\mathfrak{B}$ be any possible value for $T_{\mathfrak{A}}[a_1, a_2]$ and $\mathfrak{C} = f(\mathfrak{B})$. It suffices to show that

$$\Pr[\text{no tuple } a_3, \ldots, a_m \text{ witnesses } \mathfrak{C} \text{ for } a_1, a_2] \leq (1 - \varepsilon)^{\ell}.$$

By construction, $\mathfrak{A}$ contains at least $n-2 \geq \ell(m-2)$ pairwise distinct elements $a_{i,j} \in \{1, \ldots, np\} - \{a_1, a_2\}$ such that $T_{\mathfrak{A}}[a_{i,j}] = T_{\mathfrak{C}}[j]$ for $i = 1, \ldots, l$ and $j = 3, \ldots, m$. The ℓ events

$$\text{“}a_{i,3}, \ldots, a_{i,m} \text{ witness } \mathfrak{C} \text{ for } a_1, a_2\text{”}$$

are independent and have probability $\geq \varepsilon$. Thus the event that no tuple $a_3, \ldots, a_m$ witnesses $\mathfrak{C}$ for a_1, a_2 has probability at most $(1 - \varepsilon)^{\ell}$. □

If follows that on S_n

$$\begin{aligned}\Pr[\mathfrak{A} \models \neg\psi] &\leq \sum_{a_1 \neq a_2} \Pr[\mathfrak{A} \models \neg\exists x_3 \ldots \exists x_m \varphi[a_1, a_2]] \\ &\leq pn(pn-1)(1-\varepsilon)^\ell \leq pn(pn-1)(1-\varepsilon)^{(n-2)/(m-2)-1}\end{aligned}$$

which tends to 0 exponentially fast as n grows. Thus, for sufficiently large n, the probability that a randomly chosen $\mathfrak{A} \in S_n$ is a model for ψ is positive. Thus ψ has a finite model. □

As a consequence we obtain

Theorem 6.2.34. *Let ψ be any relational $\forall^2\exists^*$-sentence that has a model without kings. Then ψ has a finite model.*

Proof. Let $\mathfrak{A}$ be a model for ψ without kings. If $\mathfrak{A}$ is finite, nothing needs to be proved. Otherwise we transform ψ into a Gödel sentence of special form which is equivalent to ψ on all structures whose cardinality exceeds the number of variables of ψ. By Lemma 6.2.28, there exist P, Q satisfying Gödel's criterion, which by Theorem 6.2.30 implies that ψ has a finite model. □

Since every satisfiable formula without equality has a model without kings, this implies the finite model property of $[\forall^2\exists^*, all]$. Theorem 6.2.23 is proved.

The theory of directed graphs can be seen as the theory of one irreflexive binary relation. Since for an irreflexive relation, all 1-tables are identical, directed graphs are structures without kings.

Corollary 6.2.35. *The $\forall^2\exists^*$-fragment of the theory of directed graphs is decidable.*

Remark. With the method of existential interpretations (see Sect. 3.2), Gurevich [229] has shown that the $\forall^3\exists^*$-fragment of this theory is undecidable.

The Full Gödel-Kalmár-Schütte Class. We now present two methods for extending Theorem 6.2.23 to $[\exists^*\forall^2\exists^*, all]$.

The first approach generalizes Gödel's criterion to sentences with leading existential quantifiers. The same argument as above shows that we can restrict attention to the case where all variables must be interpreted by distinct elements, i.e., to sentences of the form

$$\psi := \exists x_1 \cdots \exists x_p \forall y_1 \forall y_2 \exists z_1 \cdots \exists z_t (\alpha(\bar{x}, y_1, y_2) \rightarrow \varphi(\bar{x}, y_1, y_2, \bar{z}))$$

where $\alpha(\bar{x}, y_1, y_2)$ asserts that y_1, y_2 are distinct from each other and from $x_1, \ldots, x_p$, and φ entails all remaining inequalities among distinct variables.

Definition 6.2.36 (Extended Gödel Criterion). Let $\mathfrak{A}$ be a p-table over σ and P, Q be non-empty sets of, respectively, $(p+1)$-tables and $(p+2)$-tables over σ. We say that $\mathfrak{A}, P, Q$ satisfy the extended Gödel criterion for ψ if

(0) For all $\mathfrak{B} \in P \cup Q$ we have that $T_{\mathfrak{B}}[1, \dots, p] = \mathfrak{A}$.
(1) For all $\mathfrak{B}, \mathfrak{B}' \in P$, there exists a $(p+2)$-table $\mathfrak{C} \in Q$ such that $T_{\mathfrak{C}}[1, \dots, p, p+1] = \mathfrak{B}$ and $T_{\mathfrak{C}}[1, \dots, p, p+2] = \mathfrak{B}'$.
(2) Every $(p+2)$-table $\mathfrak{B} \in Q$ can be extended to an m-table $\mathfrak{C}$ such that
  - $T_{\mathfrak{C}}[1, \dots, p+2] = \mathfrak{B}$;
  - $T_{\mathfrak{C}}[1, \dots, p, i] \in P$ for all $i \in \{p+1, \dots, m\}$;
  - $T_{\mathfrak{C}}[1, \dots, p, i, j] \in Q$ for all distinct $i, j \in \{p+1, \dots, m\}$;
  - $\mathfrak{C} \models \varphi[1, \dots, m]$.

Exercise 6.2.37. Prove that Lemma 6.2.28 and Theorem 6.2.30 generalize to the case just described, i.e., *(i)* the extended Gödel criterion holds whenever ψ has a model without kings, and *(ii)* the extended Gödel criterion implies the existence of a finite model for ψ.

Corollary 6.2.38. $[\exists^*\forall^2\exists^*, all]$ *has the finite model property.*

The same result can be established by eliminating leading existential quantifiers, and thus reducing the $[\exists^*\forall^2\exists^*]$ prefix class to the $[\forall^2\exists^*]$ class.

Exercise 6.2.39. [293, 457] Let

$$\psi := \exists x_1 \cdots \exists x_p \forall y_1 \forall y_2 \exists z_1 \cdots \exists z_m \varphi$$

be a formula of length n in the Gödel-Kalmár-Schütte class which contains r predicates of maximal arity h. Prove that ψ can be transformed in polynomial time to a formula ψ' of the form

$$\forall y_1 \forall y_2 \exists z_1 \cdots \exists z_t \varphi'$$

such that

- $t = m(p+1)^2$;
- ψ' contains at most $\min(r(p+1)^h, (p+1)^2 n / \log n)$ predicates of maximal arity h;
- If ψ has a model with universe A, then so has ψ';
- If ψ' has a model of cardinality s, then ψ has a model of cardinality at most $s + p$.

Hint: Let S be the set of substitutions π that replace both, one or none of y_1, y_2 by x-variables (i.e. any variable from $x_1, \dots, x_p$). Clearly, $|S| = (p+1)^2$. Note that for every formula α, we have the equivalence

$$\exists x_1 \cdots \exists x_p \forall y_1 \forall y_2 \alpha \equiv \exists x_1 \cdots \exists x_p \forall y_1 \forall y_2 \bigwedge_{\pi \in S} \alpha^\pi$$

where α^π is the result of applying the substitution π to α.

Let two atoms $Pu_1 \cdots u_k$ and $Pv_1 \cdots v_k$ be equivalent if they have the same predicate letter and the same x-variables at the same places. Formally $Pu_1 \cdots u_k \sim Pv_1 \cdots v_k$ if for all $i \leq k$, $j \leq p$: $u_i = x_j$ iff $v_i = x_j$. The contraction $c(\bar{u})$ of $\bar{u}$ is obtained by deleting the x-variables from u. The length of $c(\bar{u})$ is called the rank of $\bar{u}$. For every equivalence class $[P\bar{u}]$ add a predicate $Q_{[P\bar{u}]}$ whose arity is the rank of $\bar{u}$ and let

$$\psi' := \forall y_1 \forall y_2 \bigwedge_{\pi \in S} \exists z_1 \cdots \exists z_m \varphi^\pi [P\bar{u}/Q_{[P\bar{u}]}c(\bar{u})]$$

where $\varphi^\pi [P\bar{u}/Q_{[P\bar{u}]}c(\bar{u})]$ is the result of replacing every occurrence of $P\bar{u}$ in φ^π by $Q_{[P\bar{u}]}c(\bar{u})$.

Prove that (the prenex normal form of) ψ' has the required properties (see also [133, pp. 164–168].) Finally, observe that a contraction $c(\bar{u})$ may be empty, so ψ' may contain propositional variables. Show that these can be eliminated.

Exercise 6.2.40. [293, 133] Show that the method of the previous exercise can be generalized as follows: Every relational sentence ψ in the prefix class $[\exists^p \forall^m \exists^*]$ can be translated in polynomial time into a relational $[\forall^m \exists^*]$-sentence ψ' such that every model for ψ yields a model for ψ' over the same domain and, conversely, for every model for ψ' of cardinality s there exists a model for ψ of cardinality at most $s+p$. Further, ψ' has length $O((p+1)^m|\psi|)$.

Complexity Results. We now prove that the Gödel-Kalmár-Schütte class can actually be decided in $\text{NTIME}(2^{O(n/\log n)})$. This matches the lower bound established by Theorem 6.2.13.

We first restrict attention to Gödel sentences in special form. Instead of proving the upper complexity bound via a small model property, we directly investigate the complexity of verifying Gödel's criterion. It should be noted that the obvious nondeterministic procedure — to guess suitable sets P, Q of 1- and 2-tables and then to verify deterministically that P, Q satisfy conditions (1),(2) of Gödel's criterion — may require double exponential time, at least if the arities of the predicates are unbounded. Indeed, even if the vocabulary contains just one n-ary predicate, then there exist 2^{2^n} different 2-tables.

However, a refined analysis reveals that much of the data specifying a particular 2-table may be irrelevant for φ. By eliminating obsolete information we can greatly reduce the complexity and establish the desired bound.

Theorem 6.2.41. *There exists a nondeterministic algorithm which, given a quantifier-free relational formula φ of length n, decides in time $2^{O(n/\log n)}$ whether there exists a pair P, Q that satisfies Gödel's criterion for φ.*

Proof. Suppose that $\varphi(x_1, \ldots, x_m)$ has vocabulary $\sigma = \{R_1, \ldots, R_t\}$. There exist $2^t = 2^{O(n/\log n)}$ 1-tables of vocabulary σ. Thus, a set P of 1-tables can be represented by a binary word of length $2^{O(n/\log n)}$. As pointed out above, the number of 2-tables over σ may be double exponential. However, not all atomic statements are actually relevant for Gödel's criterion, and we will see that $2^{O(n/\log n)}$ bits suffice to specify enough information on Q to check that P, Q have the required properties.

Let X_φ be the set of σ-atoms $\beta(x_1, x_2)$ that contains

- all atoms $Rx_1 \cdots x_1$ and $Rx_2 \cdots x_2$ (for $R \in \sigma$), i.e., the atoms in which only one of the variables x_1, x_2 occur.
- all atoms $\beta(x_1, x_2)$ which are obtained from an atom $\beta(x_i, x_j)$ occurring in φ by substituting x_1, x_2 for x_i, x_j.

Note that $|X_\varphi| = O(n/\log n)$.

Let T_φ be the set of 2-tables over σ in which all atomic statements outside X_φ are false. Every 2-table $\mathfrak{B}$ over σ has a unique representative $\pi\mathfrak{B} \in T_\varphi$ defined by

$$\pi\mathfrak{B} \models \beta[1,2] \iff \mathfrak{B} \models \beta[1,2] \quad \text{for } \beta \in X_\varphi$$
$$\pi\mathfrak{B} \models \neg\beta[1,2] \text{ for } \beta \notin X_\varphi.$$

For every set Q of 2-tables, let $\overline{Q} := \{\pi\mathfrak{B} : \mathfrak{B} \in Q\} \subseteq T_\varphi$.

Lemma 6.2.42. *Let P be an arbitrary set of 1-tables over σ. If (P, Q) satisfies Gödel's criterion for φ, then so does $(P, \overline{Q})$.*

Proof. Condition (1) of the criterion depends only on the atomic statements on a single element. On these $\mathfrak{B}$ and $\pi\mathfrak{B}$ coincide.

Let Y_φ be the set of atoms $\beta(x_{i_1}, \dots, x_{i_k})$ that occur in φ, together with all atoms $\beta(x_i)$ or $\beta(x_i, x_j)$ obtained from some $\beta \in X_\varphi$ by substituting x_i, x_j for x_1, x_2. Note that by a rough estimate $|Y_\varphi| \le O(n/\log n) + |X_\varphi| m(m-1) = O(n^3)$.

Now suppose that for every 2-table $\mathfrak{B} \in Q$ there exists an m-table $\mathfrak{C}$ witnessing the four conditions of (2). To prove that condition (2) can be satisfied for $\pi\mathfrak{B} \in \overline{Q}$, take any m-table $\mathfrak{D}$ such that

$$\mathfrak{D} \models \beta[i_1, \dots, i_k] \iff \mathfrak{C} \models \beta[i_1, \dots, i_k]$$

for all $\beta(x_{i_1}, \dots, x_{i_k}) \in Y_\varphi$. Then, for all $i, j \in \{1, \dots, m\}$ the following hold:

- $T_{\mathfrak{D}}[i] = T_{\mathfrak{C}}[i] \in P$.
- $T_{\mathfrak{D}}[i,j] = \pi T_{\mathfrak{C}}[i,j]$. In particular $T_{\mathfrak{D}}[1,2] = \pi T_{\mathfrak{C}}[1,2] = \pi\mathfrak{B}$ and $T_{\mathfrak{D}}[i,j] \in \overline{Q}$ for all $i \neq j$.
- $\mathfrak{D} \models \varphi[1, \dots, m]$ since $\mathfrak{C}$ and $\mathfrak{D}$ coincide of all atomic statements of $\varphi[1, \dots, m]$ and $\mathfrak{C} \models \varphi[1, \dots, m]$.

This proves that $\mathfrak{D}$ witnesses condition (2) of Gödel's criterion for $\pi\mathfrak{B}$.

□

Note that we have actually proved more: it suffices to guess for each $\mathfrak{B} \in \overline{Q}$ truth values for $O(n^3)$ atomic statements of an m-table to verify condition (2).

We now describe the desired algorithm:

Guess (appropriate representations of) a set P of 1-tables and a set $Q \subseteq T_\varphi$ of 2-tables. This requires $2^{O(n/\log n)}$ bits.

To verify condition (1) of the criterion, guess for every pair $\mathfrak{B}, \mathfrak{B}' \in P$ a 2-table $\mathfrak{C} \in Q$ and check that $T_{\mathfrak{C}}[1] = \mathfrak{B}$ and $T_{\mathfrak{C}}[2] = \mathfrak{B}'$. This takes $2^{O(n/\log n)}$ steps.

To verify (2) guess, for each $\mathfrak{B} \in Q$, $O(n^3)$ truth values for all atoms in Y_φ and verify the four conditions. This takes time $2^{O(n/\log n)} n^{O(1)} = 2^{O(n/\log n)}$.

□

This proves that $Sat[\forall^2\exists^*, all] \in \mathrm{NTIME}(2^{O(n/\log n)})$. In fact, the same result holds for the full Gödel-Kalmár-Schütte class.

Theorem 6.2.43. $Sat[\exists^*\forall^2\exists^*, all] \in \mathrm{NTIME}(2^{O(n/\log n)})$.

Exercise 6.2.44. Prove Theorem 6.2.43. Hint: It is easy to see that also the extended Gödel criterion can be verified in $\mathrm{NTIME}(2^{O(n/\log n)})$. However, a little care is necessary for handling the case where constants and/or universally quantified variables assume the same values, since a straightforward translation of an arbitrary $\exists^*\forall^2\exists^*$-formula into a sentence of special form (imposing inequalities among all variables) may increase the length too much.

Remark. Theorem 6.2.43 also follows by Proposition 6.0.4, once we can establish that every satisfiable formula in the Gödel-Kalmár-Schütte class has a model of cardinality $2^{O(n/\log n)}$. Lewis [352] writes that such a bound on the model size follows from the decidability proof given in [133].

The bounds on the model size that were obtained based on Gödel's criterion are weaker. In his original paper [187], Gödel proved that every satisfiable formula $\forall x_1 \forall x_2 \exists x_1 \cdots \exists x_m \varphi$ of relational vocabulary σ has a model of cardinality N, if N satisfies the inequality

$$2(m-2)q(1+\log N) \leq 7N$$

where q is the number of different 2-tables over σ. Schütte [457, 456] proves that every formula with prefix $\exists^p\forall^2\forall^m$ containing t predicates of maximal arity h is either logically invalid or has model with

$$4^{10tm^2 2^h (p+1)^{h+4}} + p$$

elements.

Gurevich and Shelah [239] point out that their probabilistic argument yields a bound pn on the size of a minimal model for an $\forall^2\exists^m$-formula, where n satisfies $2(m-2)\log(np) \leq \varepsilon(n-m)\log e$ (where e is the basis of natural logarithms and ε is as in the proof of Theorem 6.2.30). Indeed, let $N = pn$

and suppose that this inequality holds. With $\ell > (n-2)/(m-2) - 1 = (n-m)/(m-2)$ it follows that $\log N^2 < \varepsilon\ell \log e$. Since $x \log e < -\log(1-x)$ for $0 < x < 1$ it follows that $\log N^2 + \ell \log(1-\varepsilon) < 0$ and thus $N^2(1-\varepsilon)\ell < 1$ which (at the end of the proof of Theorem 6.2.30) suffices to show that the given formula has a model of size N.

A closer analysis of these results reveals fragments of the Gödel-Kalmár-Schütte class whose satisfiability problem is in NP. For any fixed vocabulary, q is a constant; thus, there exists a constant c such that $N = cn$ satisfies Gödel's inequality. By Proposition 6.0.4 we infer

Theorem 6.2.45 (Grädel). *Let $p \in \mathbb{N}$ and s be finite. Then Sat$[\exists^p\forall^2\exists^*, s]$ is in* NP.

6.3 Formulae with One $\forall$

In this section we investigate the satisfiability problem for formula classes with one universal quantifier. We will first consider the Gurevich-Maslov-Orevkov class $[\exists^*\forall\exists^*, all, all]$ i.e. the $\exists^*\forall\exists^*$ prefix class in first-order logic without equality, but with arbitrary vocabularies of relation and function symbols. The finite model property of this class was proved by Gurevich [226]. The decidability of its satisfiability problem (but not the finite model property) has also been proved by Maslov and Orevkov [386].

The Gurevich-Maslov-Orevkov class is the unique maximal decidable prefix class without equality since the $\forall^2$-class is conservative (see Chapter 4).

We prove decidability via an alternating satisfiability test due to Grädel [204] which shows that this class is actually in DTIME$(2^{p(n)})$ for some polynomial p. We will also consider complexity results for subclasses of $[\exists^*\forall\exists^*, all, all]$ such as the Ackermann class and the monadic Ackermann class. It will turn out that most of these classes have deterministic exponential time complexity. The best lower bound known for these classes is DTIME$(2^{cn/\log n})$. It holds even for the class $[\forall\exists^2, (\omega)]$, a fragment of the monadic Ackermann class. For this class a matching upper bound will be established.

For formulae with prefix $\exists^*\forall\exists^*$ *with equality* the satisfiability problem is decidable only for relational vocabularies and for vocabularies with at most one function symbol. The relational $\exists^*\forall\exists^*$-formulae with equality form what is called the Ackermann class with equality. We prove decidability and complexity bounds for this class in Sect. 6.3.3. The latter class, $[\exists^*\forall\exists^*, all, (1)]_=$ is called the Shelah class. It has the most complicated decidability proof among all decidable standard classes and will be treated in Chap. 7.3.

To prove both upper and lower bound we use the fact – Corollary 3.5 in [73] – that for every time function $T(n) \geq \log n$

$$\text{ASPACE}(T(n)) = \bigcup_{c>0} \text{DTIME}(2^{cT(n)})$$

where ASPACE$(T(n))$ is the complexity class defined by $T(n)$ space bounded *alternating* Turing machines. For background on alternating Turing machines we refer to [30, 73].

6.3.1 A Satisfiability Test for $[\exists^*\forall\exists^*, all, all]$

Theorem 6.3.1 (Grädel). *The satisfiability problem for* $[\exists^*\forall\exists^*, all, all]$ *is in* DTIME$(2^{p(n)})$ *for some polynomial* p.

First we translate the formulae to Skolem normal form and reduce the relation symbols to a single unary predicate:

Lemma 6.3.2. *There is a polynomial time reduction taking every formula* $\psi \in [\exists^*\forall\exists^*, all, all]$ *to a formula* $\forall x\varphi$ *such that*

(i) φ *contains constants, function symbols and one monadic predicate;*
(ii) the length of φ *is linear in the length of* ψ*;*
(iii) ψ *is satisfiable if and only if* $\forall x\varphi$ *is satisfiable.*

Proof. We first bring the formula $\psi := \exists y_1 \cdots \exists y_r \forall x \exists z_1 \cdots \exists z_t \alpha$ into Skolem normal form $\forall x\beta$ where β is obtained from α by replacing every y_i by a constant c_i and every z_i by $F_i x$ where $F_1, \ldots, F_t$ are unary functions not occurring in α. We know that every formula is satisfiable if and only if its Skolem form is satisfiable.

Now choose a new monadic predicate Q and for every predicate P in β a new function F_P which has the same arity as P. Replace every atom $Pt_1 \cdots t_r$ in β by $QF_P t_1 \cdots t_r$. Let φ be the resulting formula.

If $\forall x\varphi$ has a model $\mathfrak{A}$, then $\forall x\beta$ has a model $\mathfrak{B}$ with the same universe and the same interpretation of constants and function symbols; the interpretation of a relation symbol P is defined by

$$\mathfrak{B} \models Pa_1 \cdots a_r \iff \mathfrak{A} \models QF_P(a_1, \ldots, a_r).$$

Conversely, suppose that $\mathfrak{B}$ is a model for $\forall x\beta$. It is straightforward to construct a model with two elements from a model with only one element. Therefore we may assume that $\mathfrak{B}$ has at least two distinct elements a, b. A model $\mathfrak{A}$ for $\forall x\varphi$ over the same universe (and with the same interpretation of constants) is defined in the following way:

$$\mathfrak{A} \models Qa \land \neg Qb$$

$$F_P^{\mathfrak{A}}(a_1, \ldots, a_r) := \begin{cases} a & \text{if } \mathfrak{B} \models Pa_1 \cdots a_r \\ b & \text{if } \mathfrak{B} \models \neg Pa_1 \cdots a_r. \end{cases}$$

Since the reduction consists only of some simple substitutions it is clear that it can be computed in polynomial time and that the length of $\forall x\varphi$ is linearly bounded in the length of ψ. □

For the rest of this section ψ is a sentence of the form $\forall x \varphi$ where φ is quantifier-free, contains constants $c_1, \dots, c_r$, function symbols $F_1, \dots, F_q$ and a single monadic predicate Q. The length of ψ is always denoted by n.

Algebra of Terms.

Definition 6.3.3. Let $\tau = \{F_1, \dots, F_q\}$ be the set of function symbols occurring in ψ. We define T to be the set of terms that can be formed by repeated application of these functions on the variable x and the constants (if the formula contains no constant at all we add an auxiliary constant c_1 to start the process): T is the smallest set that includes $c_1, \dots, c_r$ and x, and contains with the terms $t_1, \dots, t_r$ also the term $Ft_1 \cdots t_r$ (if F is a function symbol from τ with arity r). Let $H \subseteq T$ be the set of terms not involving x. H is the *Herbrand universe* of ψ; it is well known that every satisfiable universal formula has a model over its Herbrand universe (see Theorem 2.1.12).

Note that every term in T can be considered as a tree whose leaves are labeled by x or a constant, the other nodes are labeled by function symbols. The children of each node are ordered: if the node a is the root of a term $Ft_1 \cdots t_r$ and $a_1, \dots, a_r$ are the roots of $t_1, \dots, t_r$ (and hence the children of a), then a_1 is the first child of a and a_r the last. This can be extended to a total ordering of all nodes in a term: The root is the first element, and a node a, different from the root, precedes b if the parent of a precedes the parent of b or a is the elder child than b of the same parent. The *length* $|f|$ of f is the number of nodes in f.

Definition 6.3.4. Let f and g be two distinct terms in T. We say that (a, b) is the *minimal different pair of nodes* in f and g if, for some $i \in \mathbb{N}$,

(i) a and b are the ith nodes in f and g, respectively, with respect to the ordering defined above;
(ii) a and b have different labels;
(iii) for all $j < i$ the jth nodes in f and g have the same label.

Definition 6.3.5. For $f, g \in T$ let the product fg be the term that is obtained by replacing every occurrence of x in f by g. If $h = fg$ then f and g are, respectively, *left* and *right divisors* of h.

This product makes T a semigroup with neutral element x. $T - H$ is a sub-semigroup of T; moreover the product has the following properties:

$$(i) \quad \text{If } g \in T - H \text{ and } fg = f'g, \text{ then } f = f'.$$
$$(ii) \quad \text{If } f \in T - H \text{ and } fg = fg', \text{ then } g = g'.$$

Definition 6.3.6. A term $f \neq x$ in $T - H$ is *prime* if its only divisors are f and x.

Lemma 6.3.7. *Every $f \neq x$ in $T - H$ can be uniquely written as a product of prime terms.*

In other words, the semigroup $T - H$ is free.

Proof. It is clear that every f has at least one decomposition into prime factors. Suppose that we have two such representations and that g_1, g_2 are the rightmost factors in the two decompositions that are non-equal, i.e. $f = f_1g_1g' = f_2g_2g'$ and g_1, g_2 are non-equal primes. Then $f_1g_1 = f_2g_2$ and therefore $f_1 \neq f_2$. Let (a, b) be the minimal different pair of nodes in f_1 and f_2. One of these two nodes, say a, must be labeled by x. Let g be the subtree of f_2 whose root is b. Obviously $g \neq x$ and $g_1 = gg_2$ which contradicts the primality of g_1. □

Definition 6.3.8. We say that $f \leq g$ if f is a subterm of g (i.e. if either $f = g$ or $g = Fg_1 \cdots g_r$ and $f \leq g_i$).

Lemma 6.3.9. *If $f \neq g$ then there exists at most one term $t \in H$ such that $ft = gt$. Moreover $t \leq f$ or $t \leq g$.*

Proof. Let $f \neq g$ and $t \in H$ with $ft = gt$. Moreover, let (a, b) be the minimal pair of different nodes in f and g; one of the two nodes, say a, is labeled by x. If the subtree $g' \leq g$ whose root is b were in $T - H$ then $t \neq g't$ and therefore $ft \neq gt$. Therefore $g' = t \in H$. Thus t is uniquely determined by f and g and a subterm of g. □

Lemma 6.3.10. *Let $f, f' \in T - H$ such that neither $f = f'g$ nor $f' = fg$ for any $g \in T - H$. Then there exists at most one pair of terms $t, t' \in H$ such that $ft = f't'$. Moreover this pair has the property that at least one of t, t' is a subterm of f or f'.*

Proof. Suppose that there exists a pair t, t' such that $ft = f't'$. As in the two previous proofs we take the minimal pair (a, b) of different nodes in f and f'. Again we may suppose that a is labeled by x and that $g \leq f'$ is the subterm whose root is b. If $g \in H$ then $t = g$ is uniquely determined and a subtree of f'.

So suppose that $g \in T - H$; then $ft = f't'$ implies that $t = gt'$. The equality $ft = f't'$ defines an embedding of f into $f't'$. Since t' is a proper subterm of t, the image of f is in fact contained in f'. Consider the subtrees g' of f' whose root is the image of an x-labeled node of f; it follows that $g't' = t$. If all these g' are equal to g, then there exists an x-node c in f' not contained in any of the g' (otherwise $f' = fg$ which violates the assumptions of the Lemma). Since $t' \leq t$, the origin of c in f must be the root of a constant subterm which is equal to t', hence $t' \leq f$. But if there is a $g' \neq g$ with $g't' = t = gt'$ then, by Lemma 6.3.9, t' is uniquely determined and is a subtree of either g' or g, hence a subtree of f'. If t' is uniquely determined then t is also unique. □

Lemma 6.3.11. *If $f, f' \in T - H$ and $f' = fg$ then for all $t, t' \in H$*

$$ft = f't' \iff t = gt'.$$

Proof. Obvious. □

The Chaos and the Forest. Let $\psi = \forall x \varphi$ be an arbitrary, but fixed formula of the form specified above, and let T and H be the set of terms, respectively constant terms, defined from ψ by Definition 6.3.3.

Every term $f \in T - H$ operates on H by mapping t to ft. For $\mathcal{F} \subseteq T - H$, $M \subseteq H$ we say that M *is closed under* $\mathcal{F}$ if $ft \in M$ for all $f \in \mathcal{F}$, $t \in M$. Obviously M has to be infinite if $\mathcal{F}$ contains any element different from x. The pair $(M, \mathcal{F})$ can be considered as a directed graph with coloured edges. The nodes are the terms in M and for every node $e \in M$ and every $f \in \mathcal{F}$ there is an edge coloured f that points from e to fe.

Definition 6.3.12. Suppose that $\mathcal{F}$ includes with every element also all its prime factors and let $\mathcal{P}$ be the set of primes in $\mathcal{F}$. Then $(M, \mathcal{F})$ contains $(M, \mathcal{P})$ as a subgraph (which is obtained from $(M, \mathcal{F})$ by deleting all edges that are coloured by a non-prime). We call $(M, \mathcal{P})$ the *prime graph* of $(M, \mathcal{F})$.

In this section we will define two such sets $\mathcal{F}$ and M, such that $\mathcal{F}$ contains the terms that are relevant for ψ and M is a candidate for the universe of a model for ψ. In fact we will present a satisfiability test which defines – if it accepts ψ – a model with universe M. It will be essential that the prime graph of $(M, \mathcal{F})$ can be decomposed into two subsets C and $M - C$, where C is a finite set called the *chaos* and $M - C$ is a *forest* consisting of finitely many trees. Actually the cardinality of the chaos and the number of trees will be polynomially bounded in the length of ψ.

Definition 6.3.13. Let $\mathcal{G}$ be the set of terms $g \in T - H$ which occur in ψ (i.e. ψ contains an atom $Q(g)$). Let

$$\mathcal{F} := \{f \in T - H : (\exists g' \in T - H)(\exists g \in \mathcal{G})(fg' \leq g)\}.$$

Thus $f \in \mathcal{F}$ iff f is a left divisor of a subterm of some element of $\mathcal{G}$. By $\mathcal{P}$ we denote the set of primes in $\mathcal{F}$.

Lemma 6.3.14. *$\mathcal{F}$ has the following properties:*

(i) If $f \in \mathcal{F}$ and f' is a left divisor of f, then $f' \in \mathcal{F}$;
(ii) If $f \in \mathcal{F}$ and $f' \leq f$, then $f' \in \mathcal{F}$;
(iii) If $f \in \mathcal{F}$ then all prime divisors of f are also in $\mathcal{F}$.

Proof. Let $f \in \mathcal{F}$, $g \in \mathcal{G}$, $f' \in T - H$ and $fg' \leq g$.

(i) If f' is a left divisor of f then f' is also a left divisor of fg'.

(ii) If $f' \leq f$ then $f'g' \leq fg' \leq g$, hence $f' \in \mathcal{F}$.

(iii) If f' is a prime factor of f then there exist f_1, f_2 such that $f = f_1 f' f_2$. Now $f' f_2$ is a subterm of f, so, by *(ii)*, $f' f_2 \in \mathcal{F}$; finally since f' is a left divisor of $f' f_2$ it follows that $f' \in \mathcal{F}$. □

Definition 6.3.15. Let $E' \subseteq H$ be the smallest set of constant terms such that

(i) If ψ contains an atom $Q(e)$ then $e \in E'$.
(ii) If e is a constant subterm of some $f \in \mathcal{F}$ then $e \in E'$.
(iii) If f, f' are nonequal primes in $\mathcal{F}$ and if there exist $e, e', e'' \in H$ such that $e'' = fe = f'e'$, then e'' is in E'.
(iv) If $f \in \mathcal{F}$ and $fe \in E'$, then $e \in E'$.

Definition 6.3.16. The universe M is the closure of E' under $\mathcal{F}$. The *chaos* $C \subseteq M$ is the set of $e \in M$ for which there exist primes $f_0, f_1, \ldots, f_r \in \mathcal{P}$ such that $f_0 \neq f_1$ and $f_1 \cdots f_r e = f_0 e'$ for some $e' \in M$. More intuitively, $e \in C$ if there is a path in the prime graph $(M, \mathcal{P})$ which starts at e and eventually meets some other path. Note that the meeting point of the two paths need not be in C. Clearly $C \subseteq E'$. Finally let

$$E = E' \cup \{fe : f \in \mathcal{F}, e \in C\}.$$

E will turn out to be the part of the universe M on which the value of the relation Q will initially be guessed.

Lemma 6.3.17. *E and $\mathcal{F}$ have polynomial size. In fact there exists a $k \in \mathbb{N}$ such that*

$$\sum_{e \in E} |e| + \sum_{f \in \mathcal{F}} |f| = O(n^k)$$

where n is the length of the given formula ψ.

Proof. For $\mathcal{F}$ this follows immediately from the definition. To estimate the size of $|E|$, note that for every pair $f, f' \in \mathcal{F}$, there exists by Lemma 6.3.10 at most one triple e, e', e'' such that $e'' = fe = f'e'$. Moreover $|e''| \leq |f||f'|$. Every element of E is a product fe where f is in $\mathcal{F}$ and e is either a subterm of one of these e'', or a subterm of a $f \in \mathcal{F}$ or a subterm of a e' that occurs in ψ. The Lemma follows. □

When C is removed from M no path will ever meet any other path, so $(M - C, \mathcal{P})$ is a *forest.* The root of every tree in the forest is in E', so the number of trees is polynomially bounded; let R be the set of all roots in $(M - C, \mathcal{P})$, or to say the same thing in different words, R is the smallest set such that M is the disjoint union of C with the closure of R under $\mathcal{F}$. The tree structure of the latter component makes it comparatively easy to handle. Every element $b \in M - C$ has a unique representation $b = f_1 \cdots f_r w$ such that $w \in R$ and the f_i are prime elements of $\mathcal{F}$. We call r the *M-height* of b; if $b \in C$ we define its M-height to be 0. On the structure of C we know almost nothing, that's why we call it the chaos. Fortunately its size is polynomially bounded.

The Satisfiability Test. Using the analysis above we now present a satisfiability test for ψ. It is an alternating procedure, taking as inputs the formula $\psi := \forall x \varphi$ and the sets E, $\mathcal{F}$ together with:

(i) the decomposition of E into C, R and $E - (C \cup R)$;
(ii) the decomposition of $\mathcal{F}$ into $\mathcal{P}$ and $\mathcal{F} - \mathcal{P}$;
(iii) all true equations $fe = e'$ for $e, e' \in E, f \in \mathcal{F}$;
(iv) the prime factorization of every $f \in \mathcal{F}$. Furthermore the elements of $\mathcal{F}$ are ordered with nondecreasing height.

Note that all this data is polynomially bounded and can be computed in deterministic polynomial time from ψ because the prime factorizations of terms in $T - H$ and the solutions of equations of the form $fe = f'e'$ can be computed in polynomial time.

Informal Description: The Satisfiability Test as a Game. Informally the satisfiability test can be considered as a game for two players, the *constructor* and the *saboteur*, on the graph $(M, \mathcal{P})$. The constructor wants to build a model for ψ with universe M; he thus has to define the predicate Q on M. At every node $b \in M$ he tries to extend the definition of Q in such a way that $\varphi[b]$ is made *true*. The saboteur tries to find a point at which this is not possible. Note that to verify $\varphi[b]$ it suffices to know the values of Q on the set $E \cup \{fb : f \in \mathcal{F}\}$.

The game is played as follows:

Step 1: The constructor defines Q on E.
Step 2: $\varphi[b]$ is evaluated for every $b \in C$. If it is *false* for some b the saboteur wins. Note that $\varphi[b]$ for $b \in C$ is well-defined because $\{fb : f \in \mathcal{F}, b \in C\} \subseteq E$.
Step 3: The saboteur chooses a root $b \in R$.
Step 4, 5,...: At the currently played node b the constructor defines the values Qfb (for $f \in \mathcal{F}$) which are not yet defined. Then $\varphi[b]$ is evaluated. If it is *false* the saboteur wins. Otherwise he chooses a successor node of b in $(M, \mathcal{P})$, i.e. he chooses a prime $g \in \mathcal{P}$ and the game proceeds to gb.

In this form the game is infinite. However it turns out that if the saboteur has a winning strategy then he also has a strategy to win the game after at most $3 + 2^{|\mathcal{F}|}$ steps. Moreover the only data that must be remembered by the players are: the formula ψ; the structure of E and $\mathcal{F}$ as specified above; the values of Q on E and the current values of Qfb for $f \in \mathcal{F}$. In particular it is *not* necessary to know (the address of) the node b which is currently played. All relevant information is encoded by the pattern of the truth values for Qfb. Therefore the game needs only polynomial memory. To make the game finite we can incorporate a counter to control the number of steps: after $3+2^{|\mathcal{F}|}$ steps the game is stopped and the constructor is declared the winner.

The constructor has a winning strategy for this game if and only if ψ has a model.

Formal Description: The Satisfiability Test as an Alternating Procedure. Let $\{U_e : e \in E\}$ and $\{V_f : f \in \mathcal{F}\}$ be families of Boolean variables. During the execution of the satisfiability test truth values for Qe and Qf will be assigned to U_e and V_f; $\varphi(U, V)$ is the formula that is obtained by replacing in φ the atoms Qe by U_e and Qf by V_f. The formal description of the satisfiability test as an alternating procedure is exhibited in Fig. 6.2.

Figure 6.2. Satisfiability Test

```
Input: ψ := ∀xφ, E, F together with their structure
begin
    for all e ∈ E
        guess U_e
    for all e ∈ C
        begin
        [b := e]
        for all f ∈ F
            set V_f := U_fe
        evaluate φ(U,V); if it is false, reject.
        end
    for all e ∈ R
        begin
        [b := e]
        for all f ∈ F
            if fe = e' for some e' ∈ E, set: V_f := U_e'
            else guess V_f
        repeat 2^|F| times
            begin
            evaluate φ(U,V); if it is false, reject.
            choose universally a prime element g ∈ P and do
                begin
                [b := gb]
                for all f ∈ F
                    if fg = f' for some f' ∈ F, set V_f := V_f'
                    else guess V_f
                end
            end
        end
    accept
end
```

Remark. The assignments to b (written in square brackets) are not part of the procedure; they are comments which facilitate the following proof.

Lemma 6.3.18. *The satisfiability test can be executed by an alternating Turing machine using work space* $|E| + 2|\mathcal{F}|$.

Proof. $|E|+|\mathcal{F}|$ bits are needed to store the truth values U_e and V_f, additional $|\mathcal{F}|$ bits are used for a counter which counts the $2^{|\mathcal{F}|}$ repetitions of the loop. □

Thus, since alternating polynomial space coincides with deterministic exponential time and since the input data of the satisfiability test are computable from ψ in deterministic polynomial time, Theorem 6.3.1 is implied by

Theorem 6.3.19. *The formula $\forall x\varphi$ has a model if and only if it is accepted by the satisfiability test.*

Proof. Suppose that $\forall x\varphi$ is satisfiable. Then it has a model $\mathfrak{A}$ over the Herbrand universe H. The satisfiability test accepts $\forall x\varphi$ making the following existential guesses:

$$\begin{aligned} U_e &:= Q^{\mathfrak{A}}e \\ V_f &:= Q^{\mathfrak{A}}fb \text{ for the current value of } b. \end{aligned}$$

The other direction is more difficult. Assume that the satisfiability test accepts the formula. We choose a minimal accepting computation tree (i.e. every existential configuration has a unique successor) and use it to define a model for ψ with universe M. We have to define the interpretations of the monadic predicate Q and the functions from τ.

Let F be a function symbol in ψ with arity r. On H we have the natural interpretation

$$F^H : (t_1, \ldots, t_r) \longmapsto Ft_1 \cdots t_r.$$

On M we define:

$$F^M : (t_1, \ldots, t_r) \longmapsto \begin{cases} Ft_1 \cdots t_r & \text{if } Ft_1 \cdots t_r \in M \\ t_1 & \text{otherwise.} \end{cases}$$

This reinterpretation of the function symbols changes also the operation of terms $f \in T$ on M. For arbitrary $f \in T$ (not necessarily in $\mathcal{F}$) let f^H denote the operation of f under the old 'natural' interpretation of the function symbols, and let f^M be the new operation (if it is defined!): If $f \in M$ then f^M operates as the constant function f; if $f = x$ then f^M operates as the identity; finally if $f = F(f_1, \ldots, f_r)$ for some $F \in \tau$ and if the operations $f_1^M, \ldots, f_r^M$ are defined then for all $t \in M$:

$$f^M t = F^M(f_1^M t, \ldots, f_r^M t).$$

We want to show that the operation of $\mathcal{F}$ on M remains unchanged:

Lemma 6.3.20. *If $f \in \mathcal{F}$ then the operation f^M on M is defined and coincides with the operation f^H on M.*

Proof. If $f = x$ this is trivial. If $f = Ff_1 \cdots f_r$ then, by construction of $\mathcal{F}$ and E, each f_i is either a constant term in $E \subseteq M$ or an element of $\mathcal{F}$. By induction hypothesis we may assume that the operations f_i^M are defined and coincide with f_i^H. Thus for each $t \in M$

$$f^M t = F^M(f_1^H t, \ldots, f_r^H t).$$

Since $F(f_1^H t, \ldots, f_r^H t) = f^H t \in M$ it follows that

$$f^M t = F(f_1^H t, \ldots, f_r^H t) = f^H t.$$

□

Thus, we may give up the clumsy notation $f^M t$ and write again ft.

Recall that every $t \in M - C$ has a unique representation $t = g_1 \cdots g_r w$ where $g_1, \ldots, g_r \in \mathcal{P}$, $w \in R$ and r is the M-height of t. For every $i \geq 1$ the term $g_{i+1} \cdots g_r w$ is called a *predecessor* of t.

To define Q on M observe that every term t of M-height at most $2^{|\mathcal{F}|}$ is assigned to b exactly once in the minimal accepting computation tree. Let $V(t) = (V_f(t))_{f \in \mathcal{F}}$ be the vector of current values of V_f at that node of the computation tree at which $\psi(U, V)$ is evaluated with t being assigned to b. In particular $V_x(t)$ is the current value of V_x, i.e. of the atom Qx at this node.

Suppose that $Q^{\mathfrak{A}}$ is already defined on all elements of M with M-height smaller than r and that t has M-height r. We distinguish two cases:

(i) If $t \in C$ or if $t \in M - C$, $r \leq 2^{|\mathcal{F}|}$ and $V(t') \neq V(t'')$ for every pair of nonequal predecessors t', t'' of t, then we define

$$Q^{\mathfrak{A}} ft := V_x(t).$$

(ii) Otherwise there exists a first predecessor t' of t and a minimal term g such that gt' is also a predecessor of t and $V(t') = V(gt')$. In this case there exists a term g' such that $t = g'gt'$; we define:

$$Q^{\mathfrak{A}} t := Q^{\mathfrak{A}} g't'.$$

This implies that for any two terms $t, t' \in M - C$ such that $V(t) = V(t')$, the two trees whose roots are t and t' are isomorphic.

We have to show that this definition of $Q^{\mathfrak{A}}$ makes $\mathfrak{A}$ a model for Ψ.

Lemma 6.3.21. *For all $t, t' \in M$ with M-height at most $2^{|\mathcal{F}|}$ and for all $f, f' \in \mathcal{F}$*

$$ft = f't' \quad \Longrightarrow \quad V_f(t) = V_{f'}(t').$$

Proof. We distinguish three cases:
(i): $t, t' \in C$. Then $ft \in E$ and therefore

$$V_f(t) = V_{f'}(t') = U_{ft}.$$

(ii): $t, t' \notin C$. Then $ft = f't'$ implies that there exist primes $g_1, \ldots, g_r \in \mathcal{F}$ such that

$$f = f'g_r \cdots g_1, \qquad t' = g_r \cdots g_1 t$$

or vice versa. For every i the term $f_i := f'g_r \cdots g_{i+1}$ is in $\mathcal{F}$ because it is a left divisor of f. We claim that

$$V_{f_i}(g_i \cdots g_1 t) = V_f(t).$$

For $i = 0$ this is trivial. Assume that the equation is established for $i - 1$. If $g_i \cdots g_1 t$ is assigned to b on some path of the computation tree then the previous assignment was $g_{i-1} \cdots g_1 t$ and the universal choice was g_i. Since $f_i g_i = f_{i-1}$, it follows that

$$V_{f_i}(g_i \cdots g_1 t) = V_{f_{i-1}}(g_{i-1} \cdots g_1 t) = V_f(t).$$

Setting $i = r$ the claim follows.
(iii): $t \in C$, $t' \notin C$. Then $ft \in E$ and therefore $V_f(t) = U_{ft}$. Furthermore there exists a root $w \in R$ and $g, g' \in \mathcal{F}^*$ such that $t' = g'w$ and $w = gt$. It follows that $f = f'g'g$, hence $f'g' \in \mathcal{F}$. Since $f'g'w = ft \in E$, we conclude that $V_{f'g'}(w) = U_{ft}$. But by *(ii)* we know that $V_{f'}(t') = V_{f'g'}(w)$ and the Lemma is proved. □

For every $t \in M$ let $W(t)$ be the sequence of truth values $\{Q^{\mathfrak{A}} ft : f \in \mathcal{F}\}$.

Lemma 6.3.22. *For every $t \in M$ there exists a $t' \in M$ with M-height at most $2^{|\mathcal{F}|}$ such that $W(t) = V(t')$.*

Proof. At every t the truth value $Q^{\mathfrak{A}} t$ is defined by repeated reductions to the value of $Q^{\mathfrak{A}}$ at points with smaller M-heights until we finish at some point t' such that

$$Q^{\mathfrak{A}} t = Q^{\mathfrak{A}} t' = V_x(t').$$

We claim that $W(t) = V(t')$, i.e. that $Q^{\mathfrak{A}} ft = V_f(t')$ for every $f \in \mathcal{F}$. The proof goes by induction on the M-height of ft'.

Since t is a predecessor of ft the reduction of $Q^{\mathfrak{A}} ft$ to points with smaller M-height is analogous to the one for $Q^{\mathfrak{A}} t$ and thus leads to the point ft'. Thus, $Q^{\mathfrak{A}} ft = Q^{\mathfrak{A}} ft'$. There are two possibilities. Either the process stops there and $Q^{\mathfrak{A}} ft$ is set to $V_x(ft')$. In this case it follows by Lemma 6.3.21 that $Q^{\mathfrak{A}} ft = V_f(t')$.

Otherwise there exist a first predecessor t'' of ft' and terms g, g' such that $V(t'') = V(gt'')$ and $ft' = g'gt''$. By definition, $Q^{\mathfrak{A}} ft' = Q^{\mathfrak{A}} g't''$. From the minimality of t'' it follows that $Q^{\mathfrak{A}} t'' = V_x(t'')$. Furthermore gt'' cannot be a predecessor of t' – otherwise the reduction of $Q^{\mathfrak{A}}t$ would not finish at the point t'. Therefore gt'' lies between t' and $ft' = g'gt''$; thus g' is a left divisor of f and therefore $g' \in \mathcal{F}$. By induction hypothesis, $Q^{\mathfrak{A}}g't'' = V_{g'}(t'')$ and with help of Lemma 6.3.21 it follows that

$$Q^{\mathfrak{A}}ft = Q^{\mathfrak{A}}ft' = Q^{\mathfrak{A}}g't'' = V_{g'}(t'') = V_{g'}(gt'') = V_x(g'gt'') = \\ = V_x(ft') = V_f(t').$$

□

The satisfiability test accepts the formula $\psi \equiv \forall x\varphi$ if and only if $\varphi(U, V(b))$ is *true* for all b in the chaos and all b with M-height at most $2^{|\mathcal{F}|}$. Therefore Lemma 6.3.22 immediately implies that the relation $Q^{\mathfrak{A}}$ makes $\mathfrak{A}$ a model for ψ. □

This completes the proof of Theorem 6.3.1.

Exercise 6.3.23 (Advanced). [226] Prove that $[\exists^*\forall\exists^*, all, all]$ has the finite model property.

6.3.2 The Ackermann Class

Upper Complexity Bounds. The satisfiability test exhibited in the previous section gives also good upper bounds for the Ackermann class, i.e. the class $\exists^*\forall\exists^*$-formulae without function symbols.

Let $\exists y_1 \cdots \exists y_r \forall x \exists z_1 \cdots \exists z_t \alpha$ be a formula in $[\exists^*\forall\exists^*, all]$. When we translate it into functional form and eliminate all but one of the predicates with the procedure of Lemma 6.3.2 we obtain a formula $\forall x\varphi$ whose vocabulary consists of the monadic predicate Q, the constants $c_1, \ldots, c_r$, unary functions $g_1, \ldots, g_t$ (the Skolem functions for z_i) and the functions F_P where P is a predicate of α.

In the following, let $C_0 := \{c_1, \ldots, c_r\}$ and $\mathcal{G} = \{g_1, \ldots g_t\}$. We first consider the case where all predicates P of α (and hence all functions F_P) are monadic.

Theorem 6.3.24 (Fürer, Lewis). *There exists a constant c such that*

$$Sat[\exists^*\forall\exists^*, (\omega)] \in \text{DTIME}(2^{cn/\log n}).$$

Proof. The formula $\forall x\varphi$, obtained from a monadic Ackermann formula in the way just described, contains only unary functions and every atom in φ has one of the following three forms: $Q(F_P c_i)$, $Q(F_P x)$ or $Q(F_P g_i x)$.

In view of Lemma 6.3.18 it suffices to show that the sets E and $\mathcal{F}$ needed for the satisfiability test have cardinality $O(n/\log n)$. From the definition of $\mathcal{F}$ and E', it follows immediately that

$$\begin{aligned}\mathcal{F} &= \{gx : g \in \mathcal{G}\} \cup \{x\} \cup \{F_P x : P \text{ is a predicate of } \alpha\} \cup \\ &\quad \cup \{F_P gx : g \in \mathcal{G} \text{ and } QF_P gx \text{ occurs in } \varphi\} \\ E' &= C_0 \cup \{F_P c : c \in C_0 \text{ and } QF_P c_i \text{ occurs in } \varphi\}.\end{aligned}$$

Since all functions are unary, the chaos C is empty and therefore $E = E'$. Obviously r, t and the number of different atoms in ψ are bounded by $O(n/\log n)$. □

In the general case the functions F_P can have higher arity. Let

$$\begin{aligned}T_0 &:= C_0 \cup \{x\} \cup \{gx : g \in \mathcal{G}\} \quad \text{and} \\ T_1 &:= \{F_P t_1 \cdots t_k : P \text{ is a predicate of } \alpha \text{ and } t_1, \ldots, t_k \in T_0\}.\end{aligned}$$

The atoms in φ have the form Qf where $f \in T_1$. Let H_1 be the set of constant terms (i.e. terms without occurrences of x) in T_1. It follows that

$$\begin{aligned}\mathcal{F} &= \{gx : g \in \mathcal{G}\} \cup \{x\} \cup \\ &\quad \cup \{f \in (T_1 - H_1) : \varphi \text{ has an atom } Qfx \text{ or } Qfgx \text{ (for } g \in \mathcal{G})\}\end{aligned}$$

and thus $|\mathcal{F}| = O(n/\log n)$ also in this case. However, it is easy to give an example where the chaos C has size $\Omega((n/\log n)^2)$. Let P be a binary predicate and suppose that the atoms $Pc_i x$ and Pxz_j for $i = 1, \ldots, r$ and $j = 1, \ldots, t$ occur in α. Then $\mathcal{F}$ contains the terms $f_i = F_P c_i x$ and $f'_j = F_P x g_j x$; since $f_i g_j c_i = f'_j c_i$ the chaos C contains all constants c_i and all terms $g_j c_i$. Thus, a direct application of the satisfiability test for general Ackermann formulae would give an alternating space bound $O((n/\log n)^3)$, because $E \supseteq \mathcal{F}C$.

However, by simplifying the satisfiability test, we can improve this to a quadratic bound.

Lemma 6.3.25. *Let f and f' be distinct primes in $\mathcal{F}$ and e, e' be constant terms; if $fe = f'e'$, then either $e \in C_0$, or $e = ge'$ for some $g \in \mathcal{G}$ and $e' \in C_0$.*

Proof. If $fe = f'e'$, then, by Lemma 6.3.10 either e or e' is a subterm of f or f'. Since f and f' are in T_1, there are terms $t_1, \ldots, t_k$ and $t'_1, \ldots, t'_k \in T_0$ such that $f = F_P t_1 \cdots t_k$ and $f' = F_P t'_1 \cdots t'_k$. The only constant subterms of f and f' are those in C_0. Suppose that $e \notin C_0$; then $e' \in C_0$ and for all i, $t'_i e' \in C_0 \cup \{ge' : g \in \mathcal{G}\}$. There is at least one t_i which is not a constant term; thus $t_i e = t'_i e'$ implies that $t_i = x$ and $e = ge'$ for some $g \in \mathcal{G}$. □

Thus, whenever two edges in the graph $(M, \mathcal{P})$ finish at the same vertex, at least one of them starts at a point from C_0. Therefore truth values must initially be guessed only for the points in the set

$$E_0 = \{fc : f \in \mathcal{F}, c \in C_0\}$$

which has cardinality $O((n/\log n)^2)$.

In fact by replacing in the satisfiability test (see Fig. 6.2) the sets E, C and R by E_0, C_0 and $R_0 := E_0 - C_0$ we get a test that works for the Ackermann class and has alternating space complexity $O((n/\log n)^2)$. Thus we infer.

Theorem 6.3.26 (Fürer). *There exists a constant c such that*

$$Sat[\exists^*\forall\exists^*, all] \in \text{DTIME}(2^{c(n/\log n)^2}).$$

Lower Complexity Bounds.

Theorem 6.3.27 (Fürer, Lewis). *There is a constant $c > 0$ such that*

$$Sat[\forall\exists^2, (\omega)] \notin \text{DTIME}(2^{cn/\log n}).$$

Proof. It suffices to prove that every problem $A \in \text{ASPACE}(n)$ is reducible to $Sat[\forall\exists^2, (\omega)]$ via length order $n \log n$. Without loss of generality we make the following assumptions. A is accepted by an alternating one tape Turing machine M in space n and time $2^m - 1$ for some $m = O(n)$. All computations have exactly $2^m - 1$ steps and every configuration of M has at most two successor configurations.

If Σ is the alphabet and Q the set of states of M then every configuration of M on an input of length n is represented by a word $\#c_1 \cdots c_n\#$ where $c_i \in \Sigma \cup (Q \times \Sigma)$ and $\#$ is an end marker. Let $\Gamma := \Sigma \cup (Q \times \Sigma) \cup \{\#\}$. Thus a computation is described by a table with 2^m rows and $n+2$ columns with entries from Γ. The transition relation of M corresponds to two functions F_1, $F_2 : \Gamma^3 \to \Gamma$ such that for every entry $C_{i,j}$ of the computation table which does not belong to the first row or to the leftmost or rightmost column

$$C_{i,j+1} = F_1(C_{i-1,j}, C_{i,j}, C_{i+1,j}) \quad \text{or} \quad C_{i,j+1} = F_2(C_{i-1,j}, C_{i,j}, C_{i+1,j}).$$

The states of M are partitioned into existential, universal, accepting and rejecting states; we thus identify four subsets *Ex*, *Un*, *Acc* and *Rej* of Γ each consisting of those pairs $(q, a) \in Q \times \Sigma$ for which q is a state of the corresponding type.

For every input w a formula ψ will be constructed whose intended model has universe $\{0, \ldots, 2^m - 1\} \times \Gamma^{n+2}$. An element $x = (t, c)$ stands for the configuration $c \in \Gamma^{n+2}$ at time t; as usual we will represent t by its binary representation $t_0 \cdots t_{m-1}$ where $t = \sum_i t_i 2^i$. For every $x = (c, t)$ we will consider the computation tree $\mathcal{T}_x$ of depth $2^m - 1 - t$ whose root is the configuration c. If the configuration c occurs in a computation at time t then $\mathcal{T}_x$ is a subtree of the computation tree. The formula ψ that we are going to construct expresses that if $x = (c, 0)$ and c is the input configuration of M on w then $\mathcal{T}_x$ is an accepting computation tree.

The monadic predicates occurring in ψ together with their intended interpretations on elements $x = (t, c)$ are the following.

$P_{i,a}x:\quad c_i = a\ \ (\text{for } 0 \le i \le n+1,\ a \in \Gamma)$
$U_j x:\quad t_j = 1\ \ (\text{for } 0 \le j < m)$
$U_j^* x:\quad t_0 = \cdots = t_{j-1} = 1\ \ (\text{for } 0 \le j < m)$
$Ax:\quad$ The subtree $\mathcal{T}_x$ of the computation tree accepts.

The first step in the construction of ψ is the definition of an axiom $\forall x\alpha$ which relates the U_j^* and U_j in the desired way and asserts that the relations $P_{i,a}$ indeed encode a word $\#c\# \in \Gamma^{n+2}$; α is the formula

$$U_0^* x \wedge \bigwedge_{j=0}^{m-2} (U_{j+1}^* x \leftrightarrow (U_j^* x \wedge U_j x)) \wedge \bigwedge_i \dot{\bigvee_a} P_{i,a}x \wedge P_{0,\#}x \wedge P_{n+1,\#}x.$$

Next we specify some formulae which are needed to describe the computations of M. $\text{Exist}(x)$ expresses that c encodes an existential configuration (if it has the correct form, i.e. if it encodes indeed a configuration):

$$\text{Exist}(x) := \bigvee_i \bigvee_{a \in Ex} P_{i,a}x.$$

In an analogous way we define the formulae $\text{Univ}(x)$, $\text{Accept}(x)$ and $\text{Reject}(x)$ which describe universal, accepting and rejecting configurations. Let $x = (t, c)$ and $y = (t'c')$; we construct a formula $S(x, y)$ which says that $t' \equiv t+1 \pmod{2^m}$, and formulae $C_1(x, y)$, $C_2(x, y)$ which express that c' is a successor configuration of c via the transition function F_1 or F_2, respectively. They are defined as follows:

$$S(x,y) \ := \ \bigwedge_{j=0}^{m-1} U_j y \leftrightarrow (U_j x \oplus U_j^* x)$$

$$C_j(x,y) \ := \ \bigwedge_{i=1}^{n} \bigwedge_a \Big(P_{i,a}y \leftrightarrow \bigvee_{\substack{(b,c,d) \text{ with} \\ F_j(b,c,d)=a}} (P_{i-1,b}x \wedge P_{i,c}x \wedge P_{i+1,d}x)\Big)$$

Now we construct ψ. It has the form $\forall x(\alpha \wedge \exists y \exists z \beta)$ where α is the axiom specified above and β describes the computation tree of M; β has the following conjuncts:

Input configuration on $w = w_1 \cdots w_n$

$$\Big(\bigwedge_j \neg U_j x\Big) \rightarrow \Big(P_{1,q_0 w_1} x \wedge \bigwedge_{i=2}^{n} P_{i,w_i} x\Big)$$

Every number t has a successor (mod 2^m) *and every non-final configuration has two successor configurations*

$$S(x,y) \wedge \Big[\Big(\bigvee_j \neg U_j x\Big) \rightarrow (C_1(x,y) \wedge S(x,z) \wedge C_2(x,z))\Big]$$

Acceptance

$$\begin{aligned}
&\mathrm{Accept}(x) \to Ax;\\
&\mathrm{Reject}(x) \to \neg Ax;\\
&\mathrm{Exist}(x) \to \big(Ax \leftrightarrow (Ay \vee Az)\big);\\
&\mathrm{Univ}(x) \to \big(Ax \leftrightarrow (Ay \wedge Az)\big);\\
&\Big(\bigwedge_j \neg U_j x\Big) \to Ax.
\end{aligned}$$

It is clear that ψ has length $O(n \log n)$. The predicate A encodes the acceptance condition of M; the very last clause asserts that the computation tree is accepting. Thus ψ is satisfiable (in the structure $\{0, \dots, 2^m - 1\} \times \Gamma^{n+2}$) if and only if M accepts w. □

If we consider in the previous proof nondeterministic instead of alternating Turing machines then the accepting computation tree degenerates into a path; in this case every configuration needs to have only one successor. So we get by with only one existential quantifier and the proof of Theorem 6.3.27 gives us for free a lower bound for monadic ∀∃-formulae.

Corollary 6.3.28. *There exists a constant $c > 0$ such that*

$$Sat[\forall\exists, (\omega)] \notin \mathrm{NSPACE}(cn/\log n).$$

In particular, $Sat[\forall\exists, (\omega)]$ is PSPACE*-hard.*

6.3.3 The Ackermann Class with Equality

We first show, via the domino problem, that the Ackermann class with equality has the same lower bound as the Gödel-Kalmár-Schütte class without equality. After that we prove a corresponding result for upper complexity bounds by reducing the former class to latter.

Theorem 6.3.29 (Kolaitis, Vardi). *There exists a constant $c > 0$ such that*

$$Sat[\exists^2\forall\exists^*, all]_= \notin \mathrm{NTIME}(2^{cn/\log n}).$$

Proof. The spirit of the proof is very similar to the proof of Theorem 6.2.13. We show that for every domino system $\mathcal{D}$

$$\mathrm{DOMINO}(\mathcal{D}, 2^n) \leq_{n \log n} Sat[\exists^2\forall\exists^*, all]_=;$$

i.e. for every $w \in D^n$ we construct a formula

$$\psi := \exists u_0 \exists u_1 \forall z \exists x_0 \cdots \exists x_{n-1} \exists y_0 \cdots \exists y_{n-1} \exists z' \exists z'' \varphi$$

of length $O(n \log n)$ which is satisfiable if and only if $\mathcal{D}$ tiles $Z(2^n)$ with initial condition w. The variables u_0, u_1 represent the ciphers 0 and 1. While z ranges over all points $(x, y) \in Z(2^n)$, $(x_0, \ldots, x_{n-1}, y_0, \ldots, y_{n-1})$ will be the binary representation of (x, y) (every x_i and every y_i is equal to either u_0 or u_1), and z' and z'' represent the points $(x+1, y)$ and $(x, y+1)$ – as always modulo 2^n. The vocabulary of ψ consists of unary relations X_i, Y_i, X_i^*, Y_i^*, N_i (for $i = 0, \ldots, n-1$) and for every domino $d \in D$ two relation symbols P_d and P_d^* of arities 1 and $2n$, respectively. The intended model has universe $\{0, 1\} \dot\cup Z(2^n)$ and encodes a tiling $\tau : Z(2^n) \to D$ with the following interpretation of the relation symbols.

$$\begin{aligned} X_i z :&\quad x_i = 1 \\ X_i^* z :&\quad x_j = 1 \text{ for all } j < i \\ Y_i z :&\quad y_i = 1 \\ Y_i^*(z) :&\quad y_j = 1 \text{ for all } j < i \\ N_i z :&\quad z = (i, 0) \\ P_d z :&\quad \tau(z) = d \end{aligned}$$

The relation P_d^* 'doubles' P_d in the following sense: if $z \in Z(2^n)$ has the binary representation $(x_0, \ldots, x_{n-1}, y_0, \ldots, y_{n-1}) \in \{0, 1\}^{2n}$, then the value of $P_d^* x_0 \cdots x_{n-1} y_0 \cdots y_{n-1}$ coincides with $P_d z$.

The quantifier-free part φ of ψ is a conjunction of the following subformulae:

(1) Axioms for X_i, X_i^*, Y_i and Y_i^*:

$$X_0^* z$$

$$\bigwedge_{i=0}^{n-2} (X_{i+1}^* z \leftrightarrow (X_i^* z \land X_i z))$$

$$\bigwedge_{i=0}^{n-1} ((X_i z' \leftrightarrow (X_i z \oplus X_i^* z))$$

$$\bigwedge_{i=0}^{n-1} (X_i z'' \leftrightarrow X_i z)$$

and similarly for Y_i, Y_i^* instead of X_i, X_i^* and with z' and z'' interchanged.

(2) Axioms for $N_0, \ldots, N_{n-1}$:

$$\Big(\bigwedge_{i=0}^{n-1} \neg X_i z \land \neg Y_i z\Big) \leftrightarrow N_0 z$$

$$\bigwedge_{i=0}^{n-2} (N_{i+1} z' \leftrightarrow N_i z)$$

(3) Axioms which establish that $(x_0, \dots, x_{n-1}, y_0, \dots, y_{n-1})$ is the binary representation of z:

$$u_0 \neq u_1$$

$$\bigwedge_{i=0}^{n-1} (X_i z \to (x_i = u_1)) \wedge (\neg X_i z \to (x_i = u_0))$$

$$\bigwedge_{i=0}^{n-1} (Y_i z \to (y_i = u_1)) \wedge (\neg Y_i z \to (y_i = u_0))$$

(4) Formulae which assert that the relations P_d $(d \in D)$ encode a correct tiling:

$$P_d^* x_0 \cdots x_{n-1} y_1 \cdots y_{n-1} \leftrightarrow P_d z \qquad \text{for every } d \in D$$

$$\dot{\bigvee_{d \in D}} P_d z$$

$$\bigvee_{(d,d') \in H} (P_d z \wedge P_{d'} z')$$

$$\bigvee_{(d,d') \in V} (P_d z \wedge P_{d'} z'')$$

$$\bigwedge_{i=0}^{n-2} (N_i z \to P_{w_i} z).$$

It is clear that this formula has length $O(n \log n)$. It is also clear that the model, which is defined by a correct tiling τ in the intended way, does indeed satisfy ψ. Conversely, assume that ψ has a model $\mathfrak{A}$ with universe A and define the mapping $f : A \to Z(2^n)$ that sends every $a \in A$ to the point $(x, y) \in Z(2^n)$ whose binary representation coincides with the vector of truth values $X_0 a, \dots, X_{n-1} a, Y_0 a, \dots, Y_{n-1} a$. The axioms in part (1) assert that for every $a \in A$ with $f(a) = (x, y)$ there exist elements b, c of A such that $f(b) = (x + 1, y)$ and $f(c) = (x, y + 1)$. By repeating this argument we infer that for every $i, j < 2^n$ there is a point mapped to $(x + i, y + j)$. Hence the image of f is the whole space $Z(2^n)$. To define the tiling, choose, for every point $z \in Z(2^n)$, an arbitrary $a \in f^{-1}(z)$ and set $\tau(z) = d$ for the unique d such that $\mathfrak{A} \models P_d a$.

We have to show that this definition does not depend on the choice of a. Let $0,\ 1 \in A$ be the values that interpret the variables u_0 and u_1. Then ψ asserts that for every a there exists a $2n$-tuple $(\bar{x}(a), \bar{y}(a)) \in \{0, 1\}^{2n}$ that coincides with the vector of truth values of $X_i^{\mathfrak{A}}$ and $Y_i^{\mathfrak{A}}$ at a, and which is therefore uniquely determined by $f(a)$. Finally in (4) it is assured that $\mathfrak{A} \models P_d a$ if and only if $\mathfrak{A} \models P_d^* \bar{x}(a) \bar{y}(a)$. Thus $\tau : Z(2^n) \to D$ is a well-defined mapping. Now the formulae in (4) readily imply that τ defines a correct tiling with initial condition w; thus the Theorem follows from Theorem 6.1.8. □

We now prove that the lower bound of Theorem 6.3.29 is optimal for the Ackermann class with equality. First, we note that with the method of Exercises 6.2.39 and 6.2.40 we can reduce this class to $[\forall\exists^*, all]_=$. Since there is only one universal quantifier, the reduction increases the length only by a linear factor. But for $\forall\exists^*$-sentences we can apply the Gödel criterion since every satisfiable relational $\forall\exists^*$-sentence (even with equality!) has a model without kings. Indeed, if $\mathfrak{A}$ is a model for $\psi := \forall x \exists y_1 \cdots \exists y_m \varphi$, then so is $2\mathfrak{A}$ (defined as in the proof of Lemma 6.2.26). To see this, let $f_1, \dots, f_m$ be the Skolem functions for $y_1, \dots, y_m$ in $\mathfrak{A}$, so that $\mathfrak{A} \models \varphi[a, f_1(a), \dots, f_m(a)]$ for every a in $\mathfrak{A}$. On $2\mathfrak{A}$, put $f_i((a,j)) := (f_i(a), j)$ for all $i = 1, \dots, m$ and $j = 0, 1$. Obviously $2\mathfrak{A} \models \varphi[(a,j), f_1(a,j)), \dots, f_m(a,j))]$ for all (a,j); therefore $2\mathfrak{A} \models \psi$.

Thus the Gödel criterion is a necessary and sufficient condition for satisfiability, and by Theorem 6.2.41 we know that it can be verified in $\text{NTIME}(2^{O(n/\log n)})$.

Corollary 6.3.30. *$Sat[\exists^*\forall\exists^*, all]_= \in \text{NTIME}(2^{cn/\log n})$ for some constant c.*

Here is a different method to establish the same result. One can show that the equality sign has no influence on the cardinality of a minimal model. Since Ackermann formulae without equality are also contained in the Gödel-Kalmár-Schütte class, this implies a nondeterministic exponential upper bound.

Proposition 6.3.31. *For every $\exists^p\forall\exists^*$-formula ψ with equality which has a model of cardinality $> p$ there exists a formula φ with the same quantifier prefix and the same vocabulary as ψ, except that equality is replaced by a new binary predicate, such that φ is satisfiable and, moreover, every model of φ contains a submodel satisfying ψ.*

Proof. Let ψ be a formula with equality in prenex normal form

$$\exists x_1 \cdots \exists x_p \forall y \exists z_1 \cdots \exists z_t \alpha$$

and let $\mathfrak{A}$ be a model of ψ with universe A of cardinality at least $p+1$. This means that there exist elements $c_1, \dots, c_p$ of A and functions $f_1, \dots, f_t : A \to A$ such that for all $a \in A$:

$$\mathfrak{A} \models \alpha[c_1, \dots, c_p, a, f_1(a), \dots, f_t(a)].$$

Let V be the the set of variables $\{x_1, \dots, x_p, y, z_1, \dots, z_t\}$; it is convenient to introduce an ordering

$$x_1 \prec \cdots \prec x_p \prec y \prec z_1 \cdots \prec z_t$$

on V. Without loss of generality we may assume that every equality $v = w$ that occurs in φ is ordered, i.e. $v \preceq w$.

Every $a \in A$ defines an interpretation $I_a : V \to A$ of the variables by

$$I_a(v) = \begin{cases} c_i & \text{if } v = x_i \\ a & \text{if } v = y \\ f_i a & \text{if } v = z_i \end{cases}$$

We use this to define a *renaming* of the variables, i.e. a function $r_a : V \to V$ that maps v to the first v' (with respect to $\prec$) such that $I_a(v) = I_a(v')$. Note that $r_a(x_i)$ is independent of a. Furthermore, for every $a \in A$ and every $v \in V$:

$$r_a(v) \preceq v \quad \text{and} \quad r_a(r_a(v)) = r_a(v)$$

Independent of any model for ψ we can define the set $\mathcal{F}$ of all mappings $r : V \to V$ that satisfy this condition. Denote by α^r the formula obtained from α by replacing every variable v by $r(v)$ and equality by a new binary relation symbol E. We claim that the formula

$$\begin{aligned} \varphi &:= \exists x_1 \cdots \exists x_p \forall y \exists z_1 \cdots \exists z_t \beta \qquad \text{with} \\ \beta &:= Eyy \wedge \bigwedge_{i \neq j} \neg E x_i x_j \wedge \bigwedge_{\substack{i \leq p \\ j \leq t}} \neg E x_i z_j \wedge \\ & \bigvee_{r \in \mathcal{F}} \Big(\bigwedge_{\substack{z_i, z_j \in \mathrm{Im}(r) \\ i \neq j}} (\neg E z_i z_j \wedge \neg E y z_i) \wedge \alpha^r \Big) \end{aligned}$$

has the desired properties. ($\mathrm{Im}(r)$ denotes the image of r.)

If $\mathfrak{A}$ is a model for ψ of cardinality at least $p+1$ then it is also a model for φ with E interpreted as equality. Indeed, let $c_1, \ldots, c_p \in A$ be the constants and $f_1, \ldots, f_t$ be functions $A \to A$ such that $\mathfrak{A} \models \alpha[c_1, \ldots, c_p, a, f_1(a), \ldots, f_t(a)]$. Choose distinct elements $c'_1, \ldots, c'_p$ of A such that

$$c'_i = c_i \quad \text{if } x_i \in \mathrm{Im}(r_a) \text{ (for arbitrary } a \in A).$$

Furthermore, choose an element c distinct from $c'_1, \ldots, c'_p$ and define the Skolem functions $f'_1, \ldots, f'_t$ for $z_1, \ldots, z_t$ by

$$f'_i(a) := \begin{cases} f_i(a) & \text{if } z_i \in \mathrm{Im}(r_a) \\ c & \text{otherwise.} \end{cases}$$

Then, $\mathfrak{A} \models \beta[c'_1, \ldots, c'_p, a, f'_1(a), \ldots, f'_t(a)]$ for all $a \in A$.

Conversely, assume that $\mathfrak{B}$ is a model for φ with universe B. Thus there exist $c_1, \ldots, c_p \in B$ such that

$$\mathfrak{B} \models \forall y \exists z_1 \cdots \exists z_t \beta[c_1, \ldots, c_p].$$

Set

$$A := \{c_1, \ldots, c_p\} \cup \{b \in B : \mathfrak{B} \models \bigwedge_{i=1}^{p} \neg E c_i b\}$$

and let $\mathfrak{A}$ be the restriction of $\mathfrak{B}$ to A. Note that $\mathfrak{A}$ is also a model of φ, because, for every $a \in A$, the elements $b_1, \ldots, b_t$ such that $\mathfrak{B} \models$

$\beta[c_1, \ldots, c_p, a, b_1, \ldots, b_t]$ satisfy $\neg Ec_ib_j$ and are therefore in B. Since Eyy is a conjunct of β, it follows that $\mathfrak{A} \models \neg E[a, b]$ implies that $a \neq b$. The only atoms Evw in β that may be true in $\mathfrak{A}$ have the form Evv or Ex_iy. Therefore φ remains true in $\mathfrak{A}$ if E is replaced by equality. But the formula obtained in this way logically implies ψ. Thus, $\mathfrak{A}$ is a model for ψ. □

Corollary 6.3.32. *$Sat[\exists^p\forall\exists^*, s]_= \in$ NP for all $p \in \mathbb{N}$ and all finite s.*

Proof. We just proved that a formula in Ackermann class with equality either has a very small model, or that there exists a formula with the same quantifier prefix and the same vocabulary, except that equality is replaced by a new binary relation symbol, whose models yield models for the original formula. The new formula is in the Gödel class without equality. In the case of a fixed vocabulary and a bounded number of leading existential quantifiers there is a model of cardinality $O(n)$ (see Theorem 6.2.45). By Proposition 6.0.4 we conclude that in this case the satisfiability problem is in NP. □

6.4 Standard Classes of Modest Complexity

In this section we consider the cases with 'modest' complexity; i.e. we classify the prefix vocabulary classes whose satisfiability problem is in P, NP or Co-NP. Our description of these classes is complete, except for the case where the vocabulary consists of one unary function and equality, for which we will present an almost complete classification in Sect. 6.4.2 and identify the open problem that remains to be solved to complete the list. Most of the results of this section are due to Grädel [196].

First we dispose of the *relational* cases, i.e. the prefix vocabulary classes without function symbols.

6.4.1 The Relational Classes in P, NP and Co-NP

There are two simple cases of satisfiability problems that are solvable in polynomial time, namely any class of $\exists\forall^*$-formulae with finite vocabulary (as settled by Theorem 6.2.20) and the essentially finite classes (see Proposition 6.0.1).

Theorem 6.4.1 (The Classes in P). *Let X be a formula class which is contained in one of the classes*

(i) $[\exists\forall^*, s]_=$ for s finite;
(ii) $[\Pi, s]_=$ for Π, s finite.

Then $Sat(X)$ is decidable in polynomial time.

We will show that no other relational prefix vocabulary class is solvable in polynomial time, unless P = NP. First we determine the classes whose satisfiability problems are in NP or in Co-NP.

Recall that due to Proposition 6.0.4, the problem whether a formula of length n with p universal quantifiers has a model of size s can be decided nondeterministically in time $g(ns^p)$ for some polynomial g. Thus, if X is a formula class with a bounded number of universal quantifiers such that every satisfiable formula in X has a model of polynomial size, then $Sat(X) \in$ NP.

Theorem 6.4.2 (The Classes in NP). *Let X be contained in any of the classes*

(i) $[\exists\forall^*, all]_=$;
(ii) $[\exists^*\forall^q, all]_=$ *for* $q \in \mathbb{N}$;
(iii) $[\exists^p\forall^2\exists^*, s,]$ *for* $p \in \mathbb{N}$ *and* s *finite;*
(iv) $[\exists^p\forall\exists^*, s]_=$ *for* $p \in \mathbb{N}$ *and* s *finite;*
(v) $[\Pi_p, (q)]_=$ *for* $p, q \in \mathbb{N}$ *and* Π_p *containing at most* p *universal quantifiers.*

Then $Sat(X)$ *is in* NP.

The cases *(i)*, *(ii)*, *(iii)* and *(iv)* are settled by Theorem 6.2.20, Theorem 6.2.45 and Theorem 6.3.32; *(v)* follows immediately from Lemma 6.4.14 and Lemma 6.0.4.

The next theorem lists the minimal NP-complete classes.

Theorem 6.4.3. *Let X be one of the classes*

$$[\exists^*, (0)]_=, \qquad [\exists^*, (1)],$$
$$[\exists, (\omega)], \qquad [\forall, (\omega)].$$

Then $Sat(X)$ *is* NP*-complete.*

Proof. SAT (i.e., the satisfiability problem for propositional formulae) can be reduced to any of these classes. Let $\psi(X_1, \ldots, X_n)$ be a propositional formula. Then, ψ is satisfiable if and only if the formulae

- $\exists x \exists y_1 \cdots \exists y_n \psi[X_i/(y_i = x)]$
- $\exists x_1 \cdots \exists x_n \psi[X_i/Px_i]$
- $\exists x \psi[X_i/P_i x]$
- $\forall x \psi[X_i/P_i x]$

are satisfiable. As usual, $\psi[X_i/\alpha_i]$ denotes the formula obtained by replacing, for $i = 1, \ldots, n$, every occurrence of X_i in ψ by α_i. □

Exercise 6.4.4. Prove that every prefix vocabulary class whose satisfiability problem is in NP due to Theorem 6.4.2 is either contained in one of the classes which are in P by Theorem 6.4.1, or includes one of the NP-complete classes of Theorem 6.4.3.

Theorem 6.4.5 (The Classes in Co-NP). *Let X be a formula class contained in one of the classes*

(i) $[\exists^p\forall^*, s]_=$ *for* $p \in \mathbb{N}$ *and* s *finite;*
(ii) $[\Pi_p, (q)]_=$ *for* $p, q \in \mathbb{N}$ *and* Π_p *containing at most* p *existential quantifiers.*

Then $Sat(X)$ *is in* Co-NP.

Proof. Every satisfiable relational formula with prefix $\exists^p\forall^*$ has a model of size at most $\max(p, 1)$. Since s is finite, the list of structures that have to be checked is fixed, i.e. it does not depend on the length of the formula.

By Proposition 6.2.1 a satisfiable formula of length n whose vocabulary consists of equality and q monadic predicates has a model in which every atomic type (i.e. every sequence of truth values for the monadic predicates) is realized at most n times. Up to isomorphism, the number of structures of this form is bounded by

$$(n+1)^{2^q} = n^{O(1)} \quad \text{for fixed } q.$$

Now the theorem is implied by the following observation. Let X be a class of formulae with at most p existential quantifiers and suppose that given any $\psi \in X$, there is a list of structures $\mathfrak{A}_1, \ldots, \mathfrak{A}_k$ which can be generated in polynomial time, such that $\psi \in Sat(X)$ implies that $\mathfrak{A}_i \models \psi$ for some $i \in \{1, \ldots, k\}$. Then $Sat(X) \in$ Co-NP.

Indeed, let M be a nondeterministic algorithm which, given a formula $\psi \in X$, generates (deterministically) the structures $\mathfrak{A}_1, \ldots, \mathfrak{A}_k$ and checks (nondeterministically) for each of these whether $\mathfrak{A}_i \models \neg\psi$. By Proposition 6.0.4, each of these checks can be performed in nondeterministic polynomial time. If M finds a structure $\mathfrak{A}_i$ in which $\neg\psi$ is false, then M refutes ψ, otherwise M accepts ψ. Thus M accepts precisely the complement of $Sat(X)$. □

Theorem 6.4.6. *The satisfiability problems for each of the classes*

$$[\exists^2\forall^*, (0)]_=, \qquad [\forall^*\exists, (0)]_=, \qquad [\forall\exists\forall^*, (0)]_=$$
$$[\exists^2\forall^*, (1)], \qquad [\forall^*\exists, (1)], \qquad [\forall\exists\forall^*, (1)]$$

is Co-NP*-complete.*

Proof. By Theorem 6.4.5 the satisfiability problem for each of these classes is in Co-NP. To prove completeness, we reduce the validity problem for propositional formulae to the satisfiability problem for these classes: Let $\psi(X_1, \ldots, X_n)$ be a propositional formula. It is valid if and only if the formulae

– $\exists x \exists y \forall z_1 \cdots \forall z_n (x \neq y \wedge \psi[X_i/(z_i = x)])$

- $\exists x \exists y \forall z_1 \cdots \forall z_n((Px \oplus Py) \wedge \psi[X_i/Pz_i])$
- $\forall x \exists y(x \neq y) \wedge \forall x \forall z_1 \cdots \forall z_n \psi[X_i/(z_i = x)]$
- $\forall x \exists y(Px \oplus Py) \wedge \forall z_1 \cdots \forall z_n \psi[X_i/Pz_i]$

are satisfiable. Note that the last two formulae are in the prefix class $\forall\exists \wedge \forall^*$ which is included in both the prefix classes $\forall^*\exists$ and $\forall\exists\forall^*$. □

Thus, if X is any prefix vocabulary class, such that $Sat(X) \in$ Co-NP by Theorem 6.4.5; then either $Sat(X) \in$ P by Theorem 6.4.1, or $Sat(X)$ is Co-NP-complete. Now we show that our classification of relational prefix vocabulary classes with satisfiability problems in P, NP and Co-NP is complete.

If X is a relational prefix vocabulary class that is not contained in any of the classes in Theorems 6.4.1, 6.4.2 and 6.4.5, then X contains at least one of the following thirteen classes:

(1) $[\exists^*\forall^*, (0)]_=$

(2) $[\exists^*\forall^*, (1)]$

(3) $[\exists^2\forall^*, (\omega)]_=$

(4) $[\forall^*\exists^*, (0)]_=$

(5) $[\forall^*\exists^*, (1)]$

(6) $[\forall\exists, (\omega)]$

(7) $[\exists^*\forall\exists, (0, 1)]$

(8) $[\forall^2\exists^*, (0, 1)]_=$

(9) $[\forall^3\exists^*, (0, 1)]$

(10) $[\forall\exists^*\forall, (0, 1)]$

(11) $[\forall\exists\forall\exists^*, (0, 1)]$

(12) $[\forall^*\exists, (0, 1)]$

(13) $[\forall\exists\forall^*, (0, 1)]$

The classes (8)-(13) are conservative reduction classes. We prove that the satisfiability problem for none of the other seven classes is in NP ∪ Co-NP unless the polynomial-time hierarchy collapses to NP:

Theorem 6.4.7. *The satisfiability problems for the classes*

$$[\exists^*\forall^*, (0)]_= \qquad [\exists^*\forall^*, (1)] \qquad [\exists^2\forall^*, (\omega)]$$

are Σ_2^p-complete. In fact, $Sat(X) \in \Sigma_2^p$ for every class X of $\exists^*\forall^*$-formulae whose relations have bounded arity.

Proof. If a formula $\psi \equiv \exists x_1 \cdots \exists x_n \forall y_1 \cdots \forall y_m \varphi$ is satisfiable, then it has a model of size at most n. If ψ contains k relation symbols, each of arity at most p, then at most kn^p truth values have to guessed. Thus, the following polynomial Σ_2-algorithm decides the satisfiability of ψ:

1. **Existential step:** Guess a number $q \leq n$, and guess, for every relation P in ψ and every tuple $a \in \{0, \ldots, q-1\}^s$ (where s is the arity of P), a truth value $P[a]$. Then guess for every existentially quantified variable x_i an element $a_i \in \{0, \ldots, q-1\}$.

2. **Universal step:** Choose for every universally quantified variable y_i a value b_i.
3. If $\varphi[a_1, \ldots, a_n, b_1, \ldots, b_m]$ is true in the guessed *respectively* model, then accept, otherwise reject.

Completeness follows by reduction from $[\exists^*\forall^*] \cap$ QBF, i.e. the $\exists^*\forall^*$-subclass of quantified propositional logic. In fact, a quantified Boolean formula

$$\exists X_1 \cdots \exists X_n \forall Y_1 \cdots \forall Y_m \psi$$

is true if and only if the formulae

- $\exists z \exists z'(z \neq z' \wedge \exists x_1 \cdots \exists x_n \forall y_1 \cdots \forall y_m \psi[X_i/(x_i = z), Y_i/(y_i = z)])$
- $\exists z \exists z'(Pz \oplus Pz') \wedge \exists x_1 \cdots \exists x_n \forall y_1 \cdots \forall y_m \psi[X_i/Px_i, Y_i/Py_i]$
- $\exists z \exists z'(Qz \oplus Qz') \wedge \forall y_1 \cdots \forall y_m \psi[X_i/P_i z, Y_i/Qy_i]$

are satisfiable. □

With similar arguments we obtain

Theorem 6.4.8. *The satisfiability problems for the classes* $[\forall^*\exists^*, (0)]_=$ *and* $[\forall^*\exists^*, (1)]$ *are* Π_2^p*-complete.*

Theorem 6.4.8 is implicit in [491] and generalizes to higher levels of the polynomial time hierarchy.

Theorem 6.4.9. *Sat*$[\exists^*\forall\exists, (0,1)]$ *and Sat*$[\forall\exists, (\omega)]$ *are* PSPACE*-hard.*

Proof. For $[\forall\exists, (\omega)]$ this is settled by Theorem 6.3.28. Therefore it suffices to construct a polynomial time reduction from *Sat*$[\forall\exists, (\omega)]$ to *Sat*$[\exists^*\forall\exists, (0,1)]$:

Let α be an abbreviation for

$$Qxx \wedge \bigwedge_{i=1}^{n} \neg Qx_i x_i.$$

Then a formula $\psi := \forall y \exists z \beta$ with monadic predicates $P_1, \ldots, P_n$ is mapped to the formula

$$\varphi := \exists x_1 \cdots \exists x_n \exists x \forall y \exists z (\alpha \wedge (Qyy \rightarrow (Qzz \wedge \beta[P_i u/Qux_i])))$$

where Q is a binary relation. Obviously ψ is satisfiable, if and only if, φ is satisfiable. □

Exercise 6.4.10. Prove that these classes are in fact PSPACE-complete; furthermore, *Sat*(X) is in PSPACE for every $X \subseteq [\exists^*\forall\exists, all]$.

This completes the classification of the purely relational classes in P, NP and Co-NP.

6.4.2 Fragments of the Theory of One Unary Function

We now analyse the complexity of certain prefix classes in the theory of one unary function. In particular we will prove the following results:

- The satisfiability problem for $\exists^*\forall^*$-formulae is Σ_2^p-complete. If the number of existential quantifiers is bounded by a constant then the problem is Co-NP-complete; if we only have a constant number of universal quantifiers we get a problem that is NP-complete; if the number of both quantifiers is bounded the problem is in P.
- Satisfiability of $\exists^*\forall\exists^*$-formulae is NP-complete.
- The satisfiability problem of the $\forall^2\exists^*$ prefix class is NEXPTIME-hard.

This does not yet give a complete classification of the classes in P, NP and Co-NP. We will delineate the open problem that remains to be solved.

Configurations. To prove upper an bound on the complexity of fragments of the theory of one unary function we analyse the possible models.

Definition 6.4.11. For simplicity we refer to a structure $\mathfrak{A} = (A, f)$ with one unary function as an *algebra*. Let $B \subseteq A$ such that for all $a \in B$, there exists an $i > 0$ with $f^i(a) \in B$; we then can define the *contraction* of $\mathfrak{A}$ to B as the algebra (B, g) with $g(a) = f^i(a)$ for the minimal $i > 0$ such that $f^i(a) \in B$. We say that $\mathfrak{A}$ is an *enlargement* of $\mathfrak{B}$ if $\mathfrak{B}$ is a contraction of $\mathfrak{A}$.

Given $S \subseteq A$ we define the *closure of* S to be the set

$$\overline{S} := \{f^i(a) : i \in \mathbb{N}, a \in S\}$$

Obviously, $(\overline{S}, f)$ is also an algebra. Let $\mathfrak{A} = (A, f)$ be an algebra and $\bar{a} = (a_1, \ldots, a_k) \in A^k$; we say that $(\mathfrak{A}, \bar{a})$ is *closed* if A is the closure of $\{a_1, \ldots, a_k\}$.

Lemma 6.4.12. *Let $\psi := \forall y_1 \cdots \forall y_m \varphi(x_1, \ldots, x_k, y_1, \ldots, y_m)$ be a universal formula in the language of one unary function. If ψ is satisfiable, then there exists a closed $(\mathfrak{A}, a_1, \ldots, a_k)$ with $\mathfrak{A} \models \psi[a_1, \ldots, a_k]$.*

Proof. If $(A, f) \models \psi[a_1, \ldots, a_k]$, then also $(\overline{S}, f) \models \psi[a_1 \ldots, a_k]$ where $S = \{a_1, \ldots, a_k\}$. □

Definition 6.4.13. Given $\mathfrak{A}$ and elements $\bar{a} = (a_1, \ldots, a_k)$ we define, for every $n \in \mathbb{N}$, the *contraction operator* Z_n which maps $(\mathfrak{A}, \bar{a})$ to $(\mathfrak{B}, \bar{a})$ where $\mathfrak{B}$ is a contraction of $\mathfrak{A}$. Let

$$\begin{aligned} S &:= \{a_1, \ldots, a_k\} \\ V &:= S \cup \{a \in \overline{S} : \exists b, c \in \overline{S}(b \neq c \land fb = fc = a)\} \\ W &:= \{a \in V : f^i(a) \notin V \text{ for all } i > 0\} \\ B &:= \overline{W} \cup \bigcup_{i \leq n} f^i(V) \end{aligned}$$

Let $\mathfrak{B}$ be the contraction of $\mathfrak{A}$ to B and set

$$Z_n(\mathfrak{A}, \bar{a}) = (\mathfrak{B}, \bar{a}).$$

If $a \in V$ and $f^{r+1}(a) \in V$, but $f^i(a) \notin V$ for $i = 1, \ldots, r$, then we call the set $\{f(a), \ldots, f^r(a)\}$ a *segment.* Intuitively, $Z_n(\mathfrak{A}, \bar{a})$ is the structure obtained by restricting $\mathfrak{A}$ to the closure of $\{a_1, \ldots, a_k\}$ and by replacing every segment longer than n by a segment of length n. The following properties of Z_n follow immediately:

(i) For every n and every $(\mathfrak{A}, \bar{a})$, the contraction $Z_n(\mathfrak{A}, \bar{a})$ is closed.
(ii) The operator Z_n is idempotent, i.e. $Z_n(Z_n(\mathfrak{A}, \bar{a})) = Z_n(\mathfrak{A}, \bar{a})$.

In the sequel we write $(\mathfrak{A}, \bar{a}) \models \varphi$ to denote that $\mathfrak{A} \models \varphi[\bar{a}]$.

Proposition 6.4.14. *Let $\mathfrak{A}$ be an algebra with elements $a_1, \ldots, a_k$. Then for every quantifier-free formula $\varphi(x_1, \ldots, x_k)$ with terms of length $\leq n$*

$$(\mathfrak{A}, \bar{a}) \models \varphi \iff Z_n(\mathfrak{A}, \bar{a}) \models \varphi.$$

Proof. Every atom of φ has the form $f^r x_i = f^s x_j$ with $r, s \leq n$. Clearly the two models are indistinguishable by atoms of this form. □

In general, $Z_n(\mathfrak{A}, \bar{a})$ is infinite. However, we can define a finite *description* $Z_n^*(\mathfrak{A}, \bar{a})$ of $Z_n(\mathfrak{A}, \bar{a})$ by deleting from (the graph of) $Z_n(\mathfrak{A}, \bar{a})$ all nodes in $(\overline{W} - W)$ (i.e. the paths that go 'straight to infinity' without meeting any other path). Note, that $Z_n^*(\mathfrak{A}, \bar{a})$ uniquely determines $Z_n(\mathfrak{A}, \bar{a})$. However, $Z_n^*(\mathfrak{A}, \bar{a})$ is *not* an algebra – unless $W = \varnothing$ – because f is not defined on W.

Proposition 6.4.15. $|Z_n^*(\mathfrak{A}, \bar{a})| \leq 2kn$ *for all $\mathfrak{A}$ and all $\bar{a} = a_1, \ldots, a_k$.*

Proof. The cardinality of V is bounded by $2k$ and $|Z_n^*(\mathfrak{A}, \bar{a})| \leq n|V|$. □

Definition 6.4.16. An $[n, k]$-*configuration* is an algebra $\mathfrak{A} = (A, f)$, together with a tuple of (not necessarily distinct) elements $a_1, \ldots, a_k$, such that $Z_n(\mathfrak{A}, \bar{a}) = (\mathfrak{A}, \bar{a})$.

Proposition 6.4.17. *For every $k \geq 1$, there is a polynomial $p_k(n)$ which bounds the number of isomorphism classes of $[n, k]$-configurations .*

Proof. Up to isomorphism, an $[n, k]$-configuration $(\mathfrak{A}, \bar{a})$ is uniquely determined by the following data which we call the *map* of $(\mathfrak{A}, a)$:

- the cardinality of V, which is bounded by $2k$;
- the elements $a_1, \ldots, a_k \in V$;
- the set $W \subseteq V$;
- for every $v \in V - W$, the pair (v', i), where $v' = f^i v$ and i is the minimal positive number such that $f^i v \in V$. Clearly $i \leq n + 1$.

For fixed k the number of possible maps is polynomially bounded with respect to n. □

The $\exists^*\forall^*$ Prefix Class.

Proposition 6.4.18. *Let $\psi := \forall y_1 \cdots \forall y_m \varphi(x_1, \ldots, x_k, y_1, \ldots, y_m)$ be a satisfiable formula in the language of one unary function f with φ quantifier-free and with no term longer than n; let $q = (m+1)(n+1)$. Then, there exists a $[q,k]$-configuration which is a model for ψ.*

Proof. Let $(\mathfrak{A}, \bar{a})$ be a model of ψ with $\mathfrak{A} = (A, f)$ and $\bar{a} = (a_1, \ldots, a_k)$. We claim that $Z_q(\mathfrak{A}, \bar{a})$ is also a model for ψ. Since Z_q is idempotent, $Z_q(\mathfrak{A}, \bar{a})$ is a $[q,k]$-configuration.

Suppose that $Z_q(\mathfrak{A}, \bar{a}) = (\mathfrak{B}\, \bar{a})$ does not satisfy ψ with $\mathfrak{B} = (B, g)$. Then there exists $\bar{b} = (b_1, \ldots, b_m) \in B^m$ such that

$$(Z_q(\mathfrak{A}, \bar{a}), \bar{b}) \models \neg\varphi.$$

To prove the proposition, it suffices to construct $b'_1, \ldots, b'_m \in B$ such that also

$$(\mathfrak{A}, \bar{a}, \bar{b}') \models \neg\varphi.$$

For every b_i we distinguish two possibilities.

If b_i belongs to V, $\overline{W}$, or a segment of $Z_q(\mathfrak{A}, \bar{a})$ which was not contracted by Z_q – i.e. if b_i belongs to a part of $\mathfrak{B}$ which 'looks the same' as the corresponding part in $\mathfrak{A}$ – then set $b'_i := b_i$.

Otherwise there exist $c, d \in V$ and $r, s \in \mathbb{N}$ such that $g^r c = b_i$, $g^s b_i = d$, and $r + s = q + 1$. In $\mathfrak{A}$ we have $f^t(c) = d$ for some $t > q+1$. Let $b_{i_1}, \ldots, b_{i_\ell}$ be the elements among $b_1, \ldots, b_m$ which are in this same segment. We may assume that they are ordered in the following way. There exist $r(0), \ldots, r(\ell) \in \mathbb{N}$ such that $r(0) + \cdots + r(\ell) = q + 1$ and

$$g^{r(0)}(c) = b_{i_1}, \quad \ldots, \quad g^{r(j)}(b_{i_j}) = b_{i_{j+1}}, \quad \ldots, \quad g^{r(\ell)}(b_{i_\ell}) = d.$$

Let h be the maximal index such that $r(h) > n$; such a h exists because $\ell \le m$ and $q > (m+1)n$. Now set

$$b'_{i_j} := f^{s(j)}(c) \text{ where } s(j) = \begin{cases} r(0) + \cdots + r(j-1) & \text{for } j \le h \\ t - (r(j) + \cdots + r(\ell)) & \text{for } j > h. \end{cases}$$

From the construction of the b'_i it follows that $Z_n(\mathfrak{A}, \bar{a}, \bar{b}')$ *cong*$Z_n(\mathfrak{B}, \bar{a}, \bar{b})$ and therefore, by Proposition 6.4.14,

$$(\mathfrak{A}, \bar{a}, \bar{b}') \models \varphi \iff (Z_q(\mathfrak{A}, \bar{a}), \bar{b}) \models \varphi.$$

□

Theorem 6.4.19. *For all $k, m \in \mathbb{N}$,*

(i) $\text{Sat}[\exists^k\forall^m, (0), (1)]_= \in \text{P}$.
(ii) $\text{Sat}[\exists^*\forall^m, (0), (1)]_=$ *is* NP-complete.
(iii) $\text{Sat}[\exists^k\forall^*, (0), (1)]_=$ is Co-NP-complete.
(iv) $\text{Sat}[\exists^*\forall^*, (0), (1)]_=$ is Σ_2^p-complete.

Proof. Let $\psi := \exists x_1 \cdots \exists x_k \forall y_1 \cdots \forall y_m \varphi$ be a formula in the language of one unary function; let n be the length of the longest term in φ and let $q = (m+1)(n+1)$. Consider the following Σ_2^p-algorithm.

1. **Existential Step:** Guess (the descrip*respectively*, $k]$-configuration $(\mathfrak{A}, \bar{a})$.
2. **Universal Step:** Choose (the description of) an $[n, k+m]$-configuration $(\mathfrak{C}, \bar{a}, \bar{b})$ such that there exist $\bar{b}' = b'_1, \ldots, b'_m$ in $\mathfrak{A}$ with $Z_n(\mathfrak{A}, \bar{a}, \bar{b}') \cong (\mathfrak{C}, \bar{a}, \bar{b})$.
3. Check whether $(\mathfrak{C}, \bar{a}, \bar{b}) \models \varphi$. If yes, accept, otherwise reject.

By Proposition 6.4.15 this algorithm works in polynomial time. If the number of existential quantifiers is bounded by a fixed k then, by Proposition 6.4.17, the existential choice can be replaced by a deterministic search through all $p_k(n)$ possible configurations. If the number of universal quantifiers is bounded by a fixed m, then a similar argument applies for the universal choice of the elements $\bar{b}'$. Finally by Propositions 6.4.14 and 6.4.18 the algorithm accepts ψ if and only if ψ is satisfiable.

Note that even for the first order theory of equality, satisfiability is NP-complete for $\exists^*$-formulae and Σ_2^p-complete for $\exists^*\forall^*$-formulae (see Sect. 6.4.1. The completeness of $Sat[\forall^*, (0), (1)]_=$ in Co-NP is proved by reduction from the validity problem for Boolean formulae. A propositional formula $\psi(X_1, \ldots, X_n)$ is valid if and only if

$$\forall x \forall x_1 \cdots \forall x_n \big((fx \neq x) \wedge \psi[X_i/(x_i \neq x)]\big)$$

is satisfiable. □

The $\exists^*\forall\exists^*$ Prefix Class. The Ackermann prefix class in the theory of one unary function is NP-hard because it contains the pure existential theory of equality. (The NP-completeness of this theory is proved in Sect. 6.4.1). We establish the NP-completeness by presenting a nondeterministic satisfiability test which accepts precisely the satisfiable $\exists^*\forall\exists^*$-formulae in the theory of one unary function.

Satisfiability Test

Input: $\psi := \exists x_1 \cdots \exists x_k \forall y \exists z_1 \cdots \exists z_r \varphi$
set $q = (r+1)(n+1)$ where $n = |\varphi|$
guess (the description of) a $[q, k]$-configuration $(\mathfrak{B}, \bar{a})$
initialize $X := \{(\mathfrak{B}, \bar{a}, b) : b \in \mathfrak{B}, Z_q(\mathfrak{B}, \bar{a}, b) = (\mathfrak{B}, \bar{a}, b)\}$
initialize $Y := \varnothing$
while $X \neq Y$ do
    begin
    choose $(\mathfrak{B}, \bar{a}, b) \in X$
    guess (the description of) a $[q, k+1+r]$-configuration $(\mathfrak{C}, \bar{a}, b, \bar{e})$ which
        is an enlargement of $(\mathfrak{B}, \bar{a}, b)$
    check whether $Z_n(\mathfrak{C}, \bar{a}, b, \bar{e}) \models \varphi$. If not, reject.

set $Y = Y - \{(\mathfrak{B}, \bar{a}, b)\}$
set $X = X \cup \{Z_q(\mathfrak{C}, \bar{a}, b') : b' \in \mathfrak{C}\}$
end
accept

By Proposition 6.4.17, the test runs in polynomial time.

Theorem 6.4.20. *A formula $\psi := \exists x_1 \cdots \exists x_k \forall y \exists z_1 \cdots \exists z_r \varphi$ in the language of one unary function is satisfiable if and only if it is accepted by the satisfiability test.*

Proof. Assume that ψ is satisfiable. This means that there exist an algebra $\mathfrak{A} = (A, f)$, elements $a_1, \ldots, a_k$ of A and functions $g_1, \ldots, g_r : A \to A$ such that for all $b \in A$

$$(\mathfrak{A}, \bar{a}, b, \bar{g}(b)) \models \varphi.$$

Then the satisfiability test accepts ψ making the following guesses:

- $(\mathfrak{B}, \bar{a}) = Z_q(\mathfrak{A}, \bar{a})$
- $(\mathfrak{C}, \bar{a}, b, \bar{e}) = Z_q(\mathfrak{A}, \bar{a}, b, \bar{g}(b))$.

Indeed, by Proposition 6.4.14 all instances of φ that are checked by the test are satisfied.

The converse is more complicated. Assuming that the satisfiability test accepts ψ we will inductively define a model $\mathfrak{A}$. The construction of $\mathfrak{A}$ proceeds in stages. At every stage $i > 0$ an enlargement $\mathfrak{A}_i$ of $\mathfrak{A}_{i-1}$ is defined. Finally we set $\mathfrak{A} = \bigcup_{i \in \mathbb{N}} \mathfrak{A}_i$. After stage i is completed we enumerate the (countably many) elements of $\mathfrak{A}_i - \mathfrak{A}_{i-1}$ in some arbitrary way as $c_{i,0}, c_{i,1}, \ldots$. We turn this into an enumeration of all elements of $\mathfrak{A}$ by choosing a bijection $h : \mathbb{N} \times \mathbb{N} \to \mathbb{N}$ with $h(j, k) \geq j$ and by defining: $b_{h(j,k)} = c_{j,k}$. Let X^* be the set X at the end of an accepting computation of the satisfiability test on input ψ.

Stage 0: Initially the satisfiability test guesses a $[q, k]$-configuration $(\mathfrak{B}, \bar{a})$. Set $\mathfrak{A}_0 := \mathfrak{B}$. Clearly, $Z_q(\mathfrak{A}_0, \bar{a}, b_0) \in X^*$.

Stage $i+1$: Let $b := b_i$. There exists a unique pair (j, k) such that $h(j, k) = i$. Thus, $b = c_{j,k} \in \mathfrak{A}_i$ because $i \geq j$. Let $(\mathfrak{B}, \bar{a}, b) = Z_q(\mathfrak{A}_i, \bar{a}, b)$. By induction we may assume that $(\mathfrak{B}, \bar{a}, b) \in X^*$. This means that during the execution of the satisfiability test an enlargement $(\mathfrak{C}, \bar{a}, b, \bar{e})$ of $(\mathfrak{B}, \bar{a}, b)$ was guessed which satisfies φ.

Lemma 6.4.21. *There exists an enlargement $\mathfrak{A}_{i+1}$ of $\mathfrak{A}_i$, containing $e_1, \ldots, e_r$, such that $Z_n(\mathfrak{A}_{i+1}, \bar{a}, b, \bar{e}) \cong Z_n(\mathfrak{C}, \bar{a}, b, \bar{e})$. Moreover, for every element b' of $\mathfrak{A}_{i+1}$, $Z_q(\mathfrak{A}_{i+1}, \bar{a}, b')$ is in X^*.*

Proof. Let $\mathfrak{A}_i = (A, f)$, $\mathfrak{B} = (B, g)$ and $\mathfrak{C} = (C, g')$. Clearly, $B \subseteq A$ and $B \subseteq C$. Without loss of generality we may assume that $A \cap C = B$. Moreover

$$Z_q(\mathfrak{A}_i, \bar{a}, b) = (\mathfrak{B}, \bar{a}, b) \text{ and } Z_q(\mathfrak{C}, \bar{a}, b, \bar{e}) = (\mathfrak{C}, \bar{a}, b, \bar{e}).$$

Recall from Definition 6.4.13 that the operation of Z_q gives us the sets $V \subseteq B$ and $V' \subseteq C$. Let $s \leq r$ be the number of those elements among $e_1, \dots, e_r$ which are not in B. Then (the graph of) $\mathfrak{C} - \mathfrak{B}$ has at most s connected components. Furthermore every connected component D of $\mathfrak{C} - \mathfrak{B}$ is either also a connected component of $\mathfrak{C}$, or there exists a unique $v \in V' \cap B$ such that $g'(d) = v$ for some $d \in D$. We call v the *target* of the component D. This implies that that every element of $(V' - V) \cap B$ belongs either to $\{e_1, \dots, e_r\}$ or is a target of a connected component of $\mathfrak{C} - \mathfrak{B}$. Thus

$$|(V' - V) \cap B| \leq r.$$

Next we define an embedding $\sigma : B \to A$. If $u \in B$ belongs to V or $\overline{W}$ or to a segment of $\mathfrak{B}$ which was left invariant by the contraction Z_q then set $\sigma(u) = u$. On the other hand, let $\{g(v), g^2(v), \dots, g^q(v)\}$ be a contracted segment; its corresponding segment in $\mathfrak{A}_i$ has the form $\{f(v), f^2(v), \dots, f^{q+t}(v)\}$ for some $t > 0$. Since $q = (r+1)(n+1)$ there exists a minimal $j \leq q - (n+1)$ such that $g^{j+1}(v), \dots, g^{j+n}(v) \notin V'$. Now set, for $i = 1, \dots, q$

$$\sigma(g^i(v)) = \begin{cases} f^i(v) & \text{if } i \leq j \\ f^{i+t}(v) & \text{if } i > j \end{cases}$$

Now we can define $\mathfrak{A}_{i+1} = (A', f')$:

$$\begin{aligned} A' &:= A \cup C \\ f'(u) &= \begin{cases} f(u) & \text{if } u \in A \\ g'(u) & \text{if } u,\ g'(u) \in (C - B) \\ \sigma(g'(u)) & \text{if } u \in (C - B) \text{ and } g'(u) \in B \end{cases} \end{aligned}$$

We have to show that $Z_n(\mathfrak{A}_{i+1}, \bar{a}, b, \bar{e}) \simeq Z_n(\mathfrak{C}, \bar{a}, b, \bar{e})$. By construction V' is the same set in $\mathfrak{A}_{i+1}$ and in $\mathfrak{C}$ and every segment (with respect to V') in $\mathfrak{A}_{i+1}$ corresponds to a unique segment in $\mathfrak{C}$. Let $S \subseteq A'$ be a segment in $\mathfrak{A}_{i+1}$; then either $S \subseteq A$ or $S \subseteq (C - B)$. If $S \subseteq (C - B)$ then the corresponding segment in $\mathfrak{C}$ is also S. If $S \subseteq A$ then the corresponding segment S' in $\mathfrak{C}$ lies entirely in B; it is part of a possibly larger segment $S'' = \{f(v), \dots, f^s(v)\}$ of $\mathfrak{B}$ (with respect to V). The embedding σ of B into A has the property that $\sigma(S') = S$, unless $s > q$ and S' begins with the 'breaking point' $g^{j+1}(v)$ of S''. But this means that $\{g^{j+1}, \dots, g^{j+n}\} \subseteq S'$ and, since $\sigma(S') \subseteq S$, $|S'|, |S| \geq n$.

Thus, in every case, corresponding segments S and S' of $\mathfrak{A}_{i+1}$ and $\mathfrak{C}$ have the property that $\max(|S|, n) = \max(|S'|, n)$. But this implies $Z_n(\mathfrak{A}_{i+1}, \bar{a}, b, \bar{e}) \simeq Z_n(\mathfrak{C}, \bar{a}, b, \bar{e})$.

This proves the first part of the Lemma. From the construction of $\mathfrak{A}_{i+1}$ it immediately follows for all $b' \in B$, that $Z_q(\mathfrak{A}_{i+1}, \bar{a}, b') = Z_q(\mathfrak{A}_i, \bar{a}, b')$ which, by induction, is in X^*. If $b' \in (C - B)$, then $Z_q(\mathfrak{A}_{i+1}, \bar{a}, b') = Z_q(\mathfrak{C}, \bar{a}, b')$ which is put into X after the satisfiability test has disposed of $(\mathfrak{C}, \bar{a}, b)$. Thus $Z_q(\mathfrak{A}_{i+1}, \bar{a}, b') \in X^*$ for all $b \in A'$. □

It remains to prove that $\mathfrak{A} \models \psi$. Take an arbitrary element b of $\mathfrak{A}$; then $b = b_i$ for some i. We show that there exist $e_1, \dots, e_r$ such that $(\mathfrak{A}, \bar{a}, b, \bar{e}) \models \varphi$. By Lemma 6.4.21 the substructure $\mathfrak{A}_{i+1}$ of $\mathfrak{A}$ contains elements $e_1, \dots, e_r$ such that $Z_n(\mathfrak{A}_{i+1}, \bar{a}, b, \bar{e}) \models \varphi$. Since $\mathfrak{A}$ is an extension of $\mathfrak{A}_{i+1}$, $Z_n(\mathfrak{A}, \bar{a}, b, \bar{e}) \simeq Z_n(\mathfrak{A}_{i+1}, \bar{a}, b, \bar{e})$. By Proposition 6.4.14 this implies that $(\mathfrak{A}, \bar{a}, b, \bar{e}) \models \varphi$. □

We now can infer

Corollary 6.4.22. *$Sat[\exists^*\forall\exists^*, (0), (1)]_=$ is* NP*-complete.*

Corollary 6.4.23. *For all $k, m \in \mathbb{N}$, $Sat[\exists^k\forall\exists^m, (0), (1)]_=$ is in P.*

Proof. Replace the existential guesses in the satisfiability test by a deterministic search through all possible configurations. From Proposition 6.4.17, it follows that, for fixed k and m, the resulting procedure works in polynomial time. □

The $\forall^2\exists^*$ Prefix Class. We now show that the class of $\forall^2\exists^*$-formulae in the theory of one unary function is hard for nondeterministic exponential time. This is proved by a polynomial reduction from $Sat[\forall^2\exists^*, (\omega)]$ to $Sat[\forall^2\exists^*, (0), (1)]_=$. The reduction is given using the method of existential interpretation.

First we construct for every $n \in \mathbb{N}$ a formula ψ_n in the language of one unary function. Let

$$D_i(x) := (f^{i+1}x = f^i x) \land (f^i x \neq f^{i-1}x).$$

Then $\psi_n := \forall x \forall y \exists z \exists z_1 \cdots \exists z_n \varphi_n$ where φ_n is the conjunction of

$$f^{2n+2}x = f^{2n+1}x$$

$$D_{2n+1}(z) \land f^{2n+1}z = f^{2n+1}x$$

$$(D_{2n+1}(x) \land D_{2n+1}(y) \land f^{2n+1}x = f^{2n+1}y) \rightarrow (x = y)$$

$$\bigwedge_{i=1}^{n}\Big((fz_i = f^{2i}z \land z_i \neq f^{2i-1}z) \lor (fz_i = f^{2i+1}z \land z_i \neq f^{2i}z)\Big)$$

$$(D_{2i}(x) \land D_{2i-1}(y) \land f^{2i}x = f^{2i-1}y) \rightarrow (f^{2(n-i)+1}z = x \lor f^{2(n-i)+2}z = y).$$

(The last formula has to be included for $i = 1, \dots, n$.)

An algebra $\mathfrak{A} = (A, f)$ is a model of ψ_n if and only if every connected component C of $\mathfrak{A}$ has the following form. There exist unique elements a, $b \in C$ such that $f(b) = b$, $f^{2n}(a) \neq b$ and $f^{2n+1}(c) = b$ for all $c \in C$, i.e. there is a unique chain of $2n + 2$ nodes in C with first element a and last element b; there is no element mapped to a and no element besides a is mapped to $f(a)$. Moreover for $i = 1, \dots, n$, precisely one of $f^{2i}(a)$ and $f^{2i+1}(a)$ has an element outside this chain that is mapped to it. We call a the *leader* of C, and let $L \subseteq A$ be the set of leaders of $\mathfrak{A}$. Clearly L is definable by the formula

$\delta(x) := (f^{2n+1}x \neq f^{2n}x)$. On L we introduce n monadic predicates by the formulae $\pi_1(x), \ldots, \pi_n(x)$ where

$$\pi_i(x) := \exists u(fu = f^{2i}x \wedge u \neq f^{2i-1}x).$$

Note that also the negation of π_i is definable by an existential formula, namely

$$\tilde{\pi}_i(x) := \exists u(fu = f^{2i+1}x \wedge u \neq f^{2i}x).$$

Thus, a model $\mathfrak{A} \models \psi_n$ uniquely determines a structure $\mathfrak{A}^* = (L, Q_1, \ldots, Q_n)$ for the language of n monadic predicates. Conversely, for every structure $\mathfrak{B}$ of this vocabulary, there exists an algebra $\mathfrak{A}$ such that $\mathfrak{A}^* \cong \mathfrak{B}$.

Now, let χ be a formula in the pure monadic predicate calculus, with predicates $P_1, \ldots, P_n$. We map it to a formula η in the language of one unary function by relativizing every quantifier to δ and by replacing the predicates P_i by the formulae π_i and $\tilde{\pi}_i$; positive occurrences of P_i are substituted by π_i, negative occurrences by $\tilde{\pi}_i$. If χ is an $\forall^2\exists^*$-formula, then so is ψ'; therefore, also the formula $\psi_n \wedge \eta$ is equivalent to an $\forall^2\exists^*$-formula. Moreover, this formula is computable in polynomial time from χ. Finally the analysis of the models of ψ_n implies that χ is satisfiable if and only if $\psi_n \wedge \eta$ is satisfiable.

Since $Sat[\forall^2\exists^*, (\omega), 0]$ – in fact already $Sat[\forall^2\exists, (\omega), 0]$ – is hard for nondeterministic exponential time, by Theorem 6.2.13 we infer

Theorem 6.4.24. *$Sat[\forall^2\exists^*, (0), (1)]_=$ is hard for* NEXPTIME *via polynomial-time reductions.*

Finite Prefix Classes. We show that there are prefix classes in the theory of one unary function which are defined by simple finite prefixes and whose satisfiability problems are NP-hard:

Theorem 6.4.25. *Let X contain all formulae of the theory of one unary function with prefix $\forall\exists \wedge \forall^3$. Then $Sat(X)$ is* NP*-hard.*

Proof. We will reduce 3-SAT to $Sat(X)$. First we define, for every $n \in \mathbb{N}$, an axiom α_n in the language of one unary function, which satisfies the following properties:

(i) If $\mathfrak{A}$ is a model of α_n, then every connected component of $\mathfrak{A}$ is a k-cycle for some $k \leq 2n$;
(ii) For every $k \leq n$ there exists a $2k$-cycle in $\mathfrak{A}$ if and only if there exists no $(2k-1)$-cycle in $\mathfrak{A}$;
(iii) α_n has the form $\forall x \forall y \beta_n(x, y) \wedge \forall x \exists u \delta_n(x, u)$ where β_n and δ_n are quantifier-free;
(iv) There is a Turing machine which, given n, constructs α_n in time $O(n^2)$.

For every $i \in \mathbb{N}$, let

$$C_i(x) := (f^i x = x) \wedge \bigwedge_{1 \le j < i} (f^j x \neq x).$$

$C_i[a]$ is true if and only if a is a member of an i-cycle. Then let $\beta_n(x, y)$ be the conjunction of the following clauses.

$$\bigvee_{i=1}^{2n} f^i x = x$$

$$\neg C_{2i}(x) \vee \neg C_{2i-1}(y) \qquad \text{for } i = 1, \ldots, n.$$

Let $\delta_n(x, u)$ be a tautology for $n = 1$ and, for $n > 1$, be the conjunction of

$$(C_{2i-1}(x) \vee C_{2i}(x)) \to (C_{2i+1}(u) \vee C_{2i+2}(u)) \qquad \text{for } i = 1, \ldots, n-1$$

$$(C_{2n-1}(x) \vee C_{2n}(x)) \to (C_1(u) \vee C_2(u)).$$

Obviously $\alpha_n := \forall x \forall y \beta_n \wedge \forall x \exists u \delta_n$ has the required properties. Now, let $\psi(X_1, \ldots, X_n)$ be a propositional formula in 3-CNF, i.e.

$$\psi := \bigwedge_{i=1}^{k} Y_{i_1} \vee Y_{i_2} \vee Y_{i_3} \qquad \text{where } Y_{i_j} \in \{X_1, \ldots, X_n, \neg X_1, \ldots, \neg X_n\}.$$

The reduction maps ψ to the formula

$$\varphi \quad := \quad \alpha_n \wedge \forall x \forall y \forall z \Big(\bigwedge_{i=1}^{k} F_{i_1}(x) \vee F_{i_2}(y) \vee F_{i_3}(z) \Big) \qquad \text{where}$$

$$F_{i_j}(v) \quad := \quad \begin{cases} \neg C_{2k}(v) & \text{if } Y_{i_j} \text{ is } X_k \\ \neg C_{2k-1}(v) & \text{if } Y_{i_j} \text{ is } \neg X_k \end{cases}$$

Obviously the transformation from ψ to φ is computable in polynomial time. We claim that $\psi \in$ 3-SAT if and only if φ is satisfiable.

Suppose that ψ is satisfied by the assignment $(\varepsilon_1, \ldots, \varepsilon_n) \in \{0, 1\}^n$. Then let $\mathfrak{A}$ be the f- structure that consists of n connected components $Z_1, \ldots, Z_n$, where Z_k is a $(2k-1)$-cycle if $\varepsilon_k = 1$ and otherwise, Z_k is a $2k$-cycle. Clearly, $\mathfrak{A} \models \alpha_n$. Moreover, for every i, there is at least one literals Y_{i_j} that is made true by the assignment. If Y_{i_j} is X_k then $\varepsilon_k = 1$ and $\mathfrak{A}$ contains no $2k$-cycle; therefore, for all $a \in A$, $\mathfrak{A} \models \neg C_{2k}[a]$. Otherwise, Y_{i_j} is $\neg X_k$, i.e. $\mathfrak{A}$ contains no $(2k-1)$-cycle and thus, for arbitrary a, $\mathfrak{A} \models \neg C_{2k-1}[a]$. This implies that $\mathfrak{A} \models \varphi$.

Conversely, suppose that $\mathfrak{A} \models \varphi$. For $k = 1, \ldots, n$, let $\varepsilon_k = 1$ if $\mathfrak{A}$ contains a $2k - 1$-cycle, $\varepsilon_k = 0$, otherwise. Since α_n holds in $\mathfrak{A}$, $\varepsilon_k = 1$ is equivalent with the non-existence of a $2k$-cycle in $\mathfrak{A}$. By analogous reasoning as above it follows that $\psi(\varepsilon_1, \ldots, \varepsilon_n)$ is true as a propositional formula. □

The exact classification of the complexities of the finite prefix classes in the theory of one unary function remains open. The same holds for the $\exists^*\Pi$-prefix classes where Π is finite.

Problem: Determine the finite prefixes $\Pi \in \{\exists, \forall\}^*$ for which

(i) $Sat[\Pi, (0), (1)]_=$ is in P, NP or Co-NP;
(ii) $Sat[\exists^*\Pi, (0), (1)]_=$ is in P, NP or Co-NP.

Up to this problem the classification of all prefix classes in the first order theory of one unary function whose satisfiability problems are in P, NP or Co-NP is complete.

6.4.3 Other Functional Classes

We now describe the standard classes of 'modest' complexity whose vocabulary contains at least two functions, or a function and a relation, or a function of arity greater than one.

We prove the following result:

Theorem 6.4.26. *Let X be a prefix vocabulary class whose vocabulary contains at least two functions, or a function and a relation, or a function of arity greater than one. Then Sat(X) is* NP*-hard. Further, Sat(X) is* NP*-complete if and only if X includes only existential formulae. Otherwise, Sat(X) is at least* PSPACE*-hard.*

Note that the classes $[\forall, (0), (2)]_=$ and $[\forall, (0), (0, 1)]_=$ are unsolvable. Thus Theorem 6.4.26 is immediately implied by the following three propositions:

Proposition 6.4.27. *The satisfiability problem for existential first order formulae, i.e. $Sat[\exists^*, all, all]_=$, is* NP*-complete.*

Note that $[\exists^*, all, all]_=$ is one of the maximal solvable classes.

Proof. Given an existential formula $\exists x_1 \cdots \exists x_k \psi$, let T be the set of terms that occur in ψ, possibly as a subterm of another term. This means that for every term $s = fs_1 \cdots s_r$ appearing in ψ, T contains not only s, but also $s_1, \ldots, s_r$ and all their subterms. In particular, T contains all variables of ψ. Let $T = \{t_1, \ldots, t_m\}$. We transform ψ to the formula

$$\exists t_1 \cdots \exists t_m (\psi' \wedge \bigwedge_{\substack{s = fs_1 \cdots s_r \\ s \in T}} s = fs_1 \cdots s_r)$$

where ψ' is the same formula as ψ except that the terms in ψ are considered as variables in ψ'. In the new formula all atoms are of the form $Pz_1 \cdots z_r$, $z_1 = z_2$ and $fz_1 \cdots z_r = z_s$ where all the z_i are variables. A satisfiable formula of this form has a model of size at most $k + m$. Thus there is a straightforward procedure for deciding satisfiability of existential formulae in nondeterministic polynomial time.

Moreover this problem is certainly at least as hard as satisfiability of propositional formulae and therefore NP-complete. □

Proposition 6.4.28. *If X contains any of the three classes*

$$[\exists, (1), (1)] \qquad [\exists, (0), (2)]_= \qquad [\exists, (0), (0, 1)]_=$$

then $Sat(X)$ is NP*-hard.*

Proof. A propositional formula $\psi(X_1, \dots, X_n)$ is in SAT if and only if the following formulae are satisfiable:

- $\exists x \psi[X_i/Pf^{i-1}x]\psi$
- $(\exists x)\psi[X_i/(gf^{i-1}x = x)]$
- $(\exists x)\psi[X_i/(ft_i x = x)]$ where the t_i are terms defined as: $t_0 = x$ and $t_{i+1} = fxt_i$.

□

Proposition 6.4.29. *The class $Sat[\forall, (1), (1)]$ is* PSPACE*-hard.*

Proof. Let A be a problem in PSPACE. Without loss of generality we make the following assumptions. A is decided by a Turing machine N in space n^k and time 2^{n^k}. After acceptance N remains in the accepting configuration. Let Σ be the alphabet and Q the set of states of the Turing machine. A configuration of N is encoded by a word of length n^k over the alphabet $\Gamma = \Sigma \cup (Q \times \Sigma)$ and the transition function of N is described by function $\delta : \Gamma^3 \to \Gamma$ (the jth symbol of the successor configuration of c is determined by applying δ to the $(j-1)^{\text{th}}$, j^{th} and $(j+1)^{\text{th}}$ symbols of c). If we use instead of Γ the alphabet $\{0, 1\}$ then, for some constant d, the encoding of a configuration has length dn^k and the computational behaviour of N is described by a function δ from $\{0, 1\}^{3d}$ to $\{0, 1\}^d$.

Set $m := dn^k$. We want to describe computations by N on inputs of length n by structures $\mathfrak{A} = (A, f, P)$ (with unary function f and monadic predicate P on the universe A) in the following way. Given any element a of A, the set $\{f^i(a) : i \in \mathbb{N}\}$ is a homomorphic image of the natural numbers with successor, and P defines an infinite binary word on this set. We want this word to be divided into segments of length $4m + 3$ which have the form

$$110b_0 0b_1 0 \cdots b_{m-1} 0c_0 0c_1 0 \cdots c_m 0.$$

This is asserted by the formula $\forall x \alpha$ where

$$\alpha := \bigvee_{i=0}^{4m+2} (Pf^i x \wedge Pf^{i+1} x) \wedge ((Px \wedge Pfx) \to \bigwedge_{i=1}^{2m+1} \neg P(f^{2i} x)).$$

Given a structure $\mathfrak{A}$ which satisfies $\forall x\alpha$, every $a \in A$ with $\mathfrak{A} \models (Pa \wedge Pf(a))$ is the leading element of such a segment. It defines a natural number $b(a) < 2^m$ with binary representation $b_0 \cdots b_{m-1}$ and a word $c(a) = c_0, \ldots, c_{m-1} \in \{0,1\}^*$ via

$$\begin{aligned} b_i = 1 \quad &\text{iff} \quad \mathfrak{A} \models Pf^{2i+3}(a) \\ c_i = 1 \quad &\text{iff} \quad \mathfrak{A} \models Pf^{2i+2m+3}(a). \end{aligned}$$

A computation is encoded by a sequence of 2^m such segments; if a_i is the leading element of the ith segment then $b(a_i) = i$ and $c(a_i)$ encodes the ith configuration of the computation.

It is not difficult to construct from the transition function δ quantifier free formulae of length polynomial in m which express the following properties:

$\text{Next}(x,y)$:	$c(y)$ encodes the successor configuration of $c(x)$.
$\text{Acc}(x)$:	$c(x)$ encodes an accepting configuration.
$\text{Inp}_w(x)$:	$c(x)$ is the initial configuration on input w.
$\text{Succ}(x,y)$:	$b(y) = b(x) + 1 \pmod{2^m}$.
$\text{First}(x)$:	$b(x) = 0$.
$\text{Last}(x)$:	$b(x) = 2^m - 1$.

Given an input w of length n construct the formula

$$\begin{aligned} \psi \;:=\; & (Px \wedge Pfx) \rightarrow \big(\text{Succ}(x, f^{4m+3}x) \wedge (\text{First}(x) \rightarrow \text{Inp}_w(x)) \wedge \\ & (\neg\text{Last}(x) \rightarrow \text{Next}(x, f^{4m+3}x)) \wedge (\text{Last}(x) \rightarrow \text{Acc}(x))\big). \end{aligned}$$

N accepts w if and only if the formula $\forall x(\alpha \wedge \psi)$ is satisfiable. This proves the Proposition. □

So we now have, except for the open problem in Sect. 6.4.2, a complete classification of the prefix vocabulary classes whose satisfiability problems are in P, NP or Co-NP.

6.5 Finite Model Property vs. Infinity Axioms

The question whether a formula class has the finite model property or whether it contains infinity axioms is interesting by itself. Thus it is desirable to have a classification of the prefix vocabulary classes with respect to this question. Gurevich's Classifiability Theorem implies that also in this case, a finite classification exists. We present an almost complete such classification here.

We have seen that in many cases, the satisfiability problem (and the finite satisfiability problem) for a prefix-vocabulary class is decidable if and only if the class has the finite model property. However, if both function symbols and equality are present, then we may have decidable classes with infinity axioms. The maximal ones are the Rabin class $[all, (\omega), (1)]_=$ and the Shelah class $[\exists^*\forall\exists^*, all, (1)]_=$ (see Chap. 7). It should also be noted that the essentially finite classes – i.e. classes with finite prefix and finite relational vocabulary – cannot be excluded from consideration. These classes are trivial for algorithmic questions like decidability of the satisfiability problem, but they may of course contain infinity axioms. In fact, any particular infinity axiom has finite vocabulary and finite prefix; therefore every relational reduction class has essentially finite subclasses with infinity axioms.

Example. The sentences

$$\begin{aligned}\varphi &:= \forall x\exists y\forall z(\neg Rxx \wedge Rxy \wedge (Ryz \rightarrow Rxz))\\ \psi &:= \forall x\forall y\forall z\exists u(\neg Rxx \wedge Rxu \wedge (Rxy \wedge Ryz \rightarrow Rxz))\end{aligned}$$

are infinity axioms in the essentially finite classes $[\forall\exists\forall, (0,1)]$ and $[\forall^3\exists, (0,1)]$.

We will present the classification by two theorems. The first gives a list of maximal classes with the finite model property, the second gives a list of minimal classes with infinity axioms. The two lists almost exhaust all prefix vocabulary classes. We will at the end of this section discuss the few cases that remain open.

Theorem 6.5.1 (Maximal Classes with Finite Model Property). *The following nine classes have the finite model property.*

(1)	$[\exists^*\forall^*, all]_=$	*(Ramsey 1930)*
(2)	$[\exists^*\forall^2\exists^*, all]$	*(Gödel 1932, Schütte 1934)*
(3)	$[\exists^*\forall\exists^*, all]_=$	*(Ackermann 1928)*
(4)	$[all, (\omega)]_=$	*(Löwenheim 1915)*
(5)	$[all, (\omega), (\omega)]$	*(Löb 1967, Gurevich 1969)*
(6)	$[\exists^*\forall\exists^*, all, all]$	*(Gurevich 1973)*
(7)	$[\exists^*, all, all]_=$	*(Gurevich 1976)*
(8)	$[\forall^*, (\omega), (1)]_=$	*(Ash 1975)*
(9)	$[\exists^*\forall, all, (1)]_=$	*(Grädel 1996)*

The finite model property of the classes (1) to (7) has already been established in the previous sections of this chapter. The finite model property of class (8) was proved by Ash using rather sophisticated machinery (Ramsey's Theorem, Vaught's test and decomposition theorems for the monadic theory of linear orderings). We present here an elementary proof.

Theorem 6.5.2 (Ash). *The class $[\forall^*, (\omega), (1)]_=$ has the finite model property.*

Proof. Let $\psi = \forall x_1 \cdots \forall x_k \varphi$ be a formula in the given class, where φ is quantifier-free with monadic predicates $P_1, \ldots, P_t$, and a unary function f. Further let m be such that φ contains only terms $f^i x_j$ with $i < m$.

Given structures $\mathfrak{A}, \mathfrak{B}$ of appropriate vocabulary and elements a, b of $\mathfrak{A}$ and $\mathfrak{B}$, respectively, we write $(\mathfrak{A}, a) \sim_r (\mathfrak{B}, b)$ to denote that for all monadic predicates P_i and all $j \leq r$

$$\mathfrak{A} \models P_i f^j(a) \iff \mathfrak{B} \models P_i f^j(b).$$

Further, given k-tuples $\bar{a} = a_1, \ldots, a_k$ and $\bar{b} = b_1, \ldots, b_k$ of elements of $\mathfrak{A}$ and $\mathfrak{B}$, respectively, we write $(\mathfrak{A}, \bar{a}) \cong_r (\mathfrak{B}, \bar{b})$ if for all $i, j \leq k$ and $h, \ell \leq r$

(i) $(\mathfrak{A}, a_i) \sim_r (\mathfrak{B}, b_i)$,
(ii) $f^h(a_i) = f^\ell(a_j)$ iff $f^h(b_i) = f^\ell(b_j)$.

Since all atoms of φ have the form $P_i f^h x_j$ or $f^h x_i = f^\ell x_j$ for $h, \ell < m$ it follows that $\mathfrak{A} \models \varphi[\bar{a}]$ iff $\mathfrak{B} \models \varphi[\bar{b}]$ whenever $(\mathfrak{A}, \bar{a}) \cong_m (\mathfrak{B}, \bar{b})$.

Suppose now that $\mathfrak{A} \models \psi$ and consider any substructure $\mathfrak{B} \subseteq \mathfrak{A}$ that is generated by a single element b, i.e. a substructure with universe $B = \{f^i(b) : i < \omega\}$. Since $\mathfrak{A} \models \psi$ and ψ is universal it follows that $\mathfrak{B} \models \psi$. If B is finite, we are done. Otherwise $(B, f) \cong (\omega, succ)$.

Clearly, for every $r < \omega$, there are only finitely many $\sim_r$-equivalence classes. Thus, there exist $c \in B$ and a natural number $n > km$ such that $(\mathfrak{B}, c) \sim_{km} (\mathfrak{B}, f^n(c))$. Let $\mathfrak{C}$ be the substructure of $\mathfrak{B}$ generated by c and let $\mathfrak{D}$ be the structure with universe $D = \{f^i(c) : 0 \leq i < n\}$, with $f(f^{n-1}(c)) := c$ and $P_i^{\mathfrak{D}} := P_i^{\mathfrak{C}} \cap D$. Obviously $\mathfrak{C} \models \psi$. We claim that also $\mathfrak{D} \models \psi$. To see this we take any $\bar{d} = d_1, \ldots, d_k \in D$ and show that we can find $\bar{c} = c_1, \ldots, c_k \in C$ such that $(\mathfrak{C}, \bar{c}) \cong_m (\mathfrak{D}, \bar{d})$.

Without loss of generality, we can assume that $d_1 = f^{h_1}(c), \ldots, d_k = f^{h_k}(c)$ with $0 \leq h_1 \leq h_2 \leq \cdots \leq h_k < n$. We distinguish two cases.

Case 1: If d_1 is not reachable from d_k, in the sense that $\mathfrak{D} \models (f^j(d_k) \neq d_1)$ for all $j < m$ we can take $\bar{c} = \bar{d}$.

Case 2: Otherwise $\mathfrak{D} \models (f^j(d_k) = d_1)$ for some $j < m$. Take the maximal $s \leq k$ such that $h_i - h_{i-1} < m$ for $i = 2, \ldots s$. Note that $h_s < sm \leq km$. Now set

$$c_i := \begin{cases} f^n(d_i) & \text{for } i = 1, \ldots, s \\ d_i & \text{for } i = s+1, \ldots, k. \end{cases}$$

It is easily verified that $(\mathfrak{C}, \bar{c}) \cong_m (\mathfrak{D}, \bar{d})$ in both cases. Since $\mathfrak{C} \models \varphi[\bar{c}]$ for all $\bar{c}$ it follows that $\mathfrak{D} \models \varphi[\bar{d}]$. But $\bar{d}$ was chosen arbitrarily, so we have proved that $\mathfrak{D}$ is a finite model of ψ. □

It remains to prove the finite model property of class (9) which is a fragment of the Shelah class. In fact it turns out to be a maximal fragment with the finite model property.

Theorem 6.5.3 (Grädel). *The class* $[\exists^*\forall, all, (1)]_=$ *has the finite model property.*

Proof. We can replace leading existential quantifiers by constants and thus assume that we have a sentence $\psi = \forall x\varphi$ where φ is a quantifier-free formula whose vocabulary consists of constants $c_1, \dots, c_k$, one unary function symbol f and a finite number of relation symbols of arbitrary arity. Suppose that $\mathfrak{A} \models \psi$ and let $\mathfrak{B} \subseteq \mathfrak{A}$ be the substructure generated by the constants, i.e. the substructure with universe $B = \{f^i(c_j) : i < \omega, j = 1, \dots, k\}$. (If ψ has no constants, take the substructure generated by an arbitrary element c_1 of $\mathfrak{A}$.) Since ψ is universal, $\mathfrak{B} \models \psi$.

For elements a, b of $\mathfrak{B}$, let $a \sim_m b$ denote that $\mathfrak{B} \models \eta[a] \leftrightarrow \eta[b]$ for all atomic formula $\eta(x)$ with one variable and terms $f^i x$ and $f^i c_j$ with $i < m$.

The $\{f\}$-reduct (B, f) of $\mathfrak{B}$ can be seen as an infinite directed graph, with an arc from a to b iff $f(a) = b$. If B is finite we are done. Otherwise, let K be an infinite (weakly) connected component of this digraph. Obviously the number of $\sim_m$-equivalence classes is finite, so there exist $b \in K$ and a natural number $n > m$ such that $b \sim_m f^n(b)$ and $\{f^{n+i}(b) : i < \omega\} \cap \{c_1, \dots, c_k\} = \varnothing$. By identifying $f^n(b)$ with b, the component K is changed to a finite one. By repeating this operation for all infinite components of $\mathfrak{B}$, we obtain a finite structure $\mathfrak{C}$. For every element c of $\mathfrak{C}$ there is an element b of $\mathfrak{B}$ such that $b \sim_m c$. Since $\mathfrak{B} \models \varphi[b]$ for all b, this implies that $\mathfrak{C} \models \varphi[c]$. Thus $\mathfrak{C}$ is a finite model of $\forall x\varphi$. □

This completes the proof of Theorem 6.5.1. We now consider minimal classes that admit infinity axioms.

Theorem 6.5.4 (Minimal Classes with Infinity Axioms). *Each of the following classes contains infinity axioms.*

(1)	$[\forall^3\exists, (0,1)]$	(6)	$[\forall^2, (1), (0,1)]$
(2)	$[\forall\exists\forall, (0,1)]$	(7)	$[\forall, (0), (2)]_=$
(3)	$[\forall^2\exists, (\omega,1)]_=$	(8)	$[\forall, (0), (0,1)]_=$
(4)	$[\forall^2\exists^*, (0,1)]_=$	(9)	$[\exists\forall^2, (0), (1)]_=$
(5)	$[\exists^*\forall^2\exists, (0,1)]_=$	(10)	$[\forall\exists, (0), (1)]_=$

We have exhibited infinity axioms for the classes (1) and (2) in the example above. The classes (3) – (8) are conservative (see Theorem 4.0.1). It remains to show that the classes (9) and (10), i.e. the $\exists\forall^2$ and $\forall\exists$ prefix classes in the logic of one unary function admit infinity axioms.

Proposition 6.5.5. *There exist infinity axioms in the first-order logic of one unary function with prefix* $\exists\forall^2$ *and* $\forall\exists$*, respectively.*

Proof. The desired infinity axioms are

$$\begin{aligned}\varphi &:= \exists x\forall y\forall z(fy \neq x \land (fy = fz \rightarrow y = z))\\ \psi &:= \forall x\exists y(f^2y = fx \land fy \neq x)\end{aligned}$$

Indeed, φ states that f is injective but not surjective. The argument for ψ is only slightly more complicated. A model for ψ is given by the infinite binary tree $T = (\{0,1\}^*, f)$ with $f(\lambda) = \lambda$ and $f(w0) = f(w1) = w$ for all $w \in \{0,1\}^*$. Now suppose that $\mathfrak{A} = (A, f)$ is a finite model of ψ. Call $a \in A$ *cyclic* if $f^m(a) = a$ for some $m < \omega$, and *acyclic* otherwise. Since $\mathfrak{A}$ is finite it contains a cyclic element a; but then there exists $b \in A$ such that $f^2(b) = f(a)$ and $f(b) \neq a$. In particular, b is acyclic. By the finiteness of $\mathfrak{A}$ there exists a maximal number $n < \omega$ such that $f^n(c) = f(a)$ for some acyclic element c. However, ψ implies that there exists an element d with $f^2(d) = f(c)$ and therefore $f^{n+1}(d) = f(a)$ which contradicts the maximality of n. Thus ψ admits no finite models. □

Theorem 6.5.1 and Theorem 6.5.4 give an almost complete classification of the prefix vocabulary with the finite model property The only cases that are open are some essentially finite subclasses of the Goldfarb class.

Indeed, for every prefix-vocabulary class X one of the following holds:

1. X is included in one of the classes (1) – (9) of Theorem 6.5.1 (and thus has the finite model property), or
2. X contains one of the classes (1) – (10) of Theorem 6.5.4 (and thus has infinity axioms), or
3. X is an essentially finite subclass of the $\exists^*\forall^2\exists^*$-class with equality and without functions (the Goldfarb class). More precisely, X has the form $[\exists^k\forall^2\exists^m, (p_1, p_2, \ldots)]_=$ such that $k \geq 0, m \geq 1$, all p_i are finite and at least one of $p_2, p_3, \ldots$ is positive (but only finitely many of them are). Goldfarb [192] exhibits infinity axioms for some of these classes, such as $[\forall^2\exists, (3,9)]_=$ and $[\forall^2\exists^{17}, (1,7)]_=$, but a complete classification is not known. In particular, it seems to be unknown whether $[\forall^2\exists, (0,1)]_=$ has infinity axioms.

6.6 Historical Remarks

The study of decidable cases of the *Entscheidungsproblem* started with Löwenheim's ground-breaking paper [365] where he established the decidability and finite model property of the monadic predicate calculus (even with equality). In the same paper he showed that first-order logic contains infinity axioms and proved that the validity problem for first-order logic can be reduced to the validity problem of the pure predicate calculus with only binary predicates. Skolem [477] and Behmann [31] extended Löwenheim's decidability result to the fragment of second-order logic where all predicates, free and bound, are monadic. Bernays and Schönfinkel proved that a monadic formula with n predicates (without equality) is valid if and only if it is valid over a domain of cardinality 2^n and thus established a small model property.

The first decidability results for formula classes with non-monadic predicates were due to Bernays and Schönfinkel [35] in 1928 and concerned the $\exists^*\forall^*$ and the $\forall\exists$ prefix classes in the pure predicate calculus with relation symbols of arbitrary arity. Ackermann [16] extended the latter result to the class $\exists^*\forall\exists^*$ of relational sentences with at most one universal quantifier. Other proofs of the same result were given by Skolem [478] and Herbrand [254]. Ramsey [435] proved that the satisfiability problem for relational $\exists^*\forall^*$-sentences is decidable also in the predicate calculus with equality. In fact he proved that the spectrum of every $\exists^*\forall^*$-sentence without function symbols is either finite or co-finite. To prove this result he developed his famous combinatorial theorem and initiated was is now called Ramsey theory, a still very active subfield of combinatorics.

In 1932 – 1934, Gödel [186], Kalmár [293] and Schütte [457, 456] independently discovered decision procedures for the class of $\exists^*\forall^2\exists^*$-sentences in pure predicate logic. Gödel [187] and Schütte also established the finite model property of this class. The original proof relies on an ingenious and very complicated model construction. In 1984, Gurevich and Shelah found a simpler proof that replaces Gödel's explicit combinatorial construction by a probabilistic argument (see Sect. 6.2.3).

The results obtained later by Surányi [494] and Kahr, Moore and Wang [288] that the prefix classes $\forall^3\exists$ and $\forall\exists\forall$ in pure predicate logic are reduction classes showed that the Bernays-Schönfinkel class $[\exists^*\forall^*]$ and the Gödel-Kalmár-Schütte class $[\exists^*\forall^2\exists^*]$ are the two maximal decidable prefix classes in pure predicate logic which are decidable for satisfiability (and finite satisfiability).

At the end of his paper [187] Gödel claims, without substantiation, that his method to show the finite model property for the $\forall^2\exists^*$ class suffices to show the same result also in the presence of equality. Only in the 1960's examples were discovered showing that Gödel's criterion is not sufficient for satisfiability of $\forall^2\exists^*$-sentences with equality. In 1984 Goldfarb proved the undecidability of the $\forall^2\exists$-class and thus completed the classification of the decidable and undecidable prefix classes with equality.

Thus, in predicate logic with equality (but without function symbols), the $\exists^*\forall\exists^*$-sentences and the $\exists^*\forall^*$-sentences form the two maximal decidable decidable prefix classes.

Formula classes with function symbols had been excluded from consideration for a long time. It was only in 1954, at the very end of his book [18], in fact on the last four lines, that Ackermann suggested to investigate the decision problem for formulae with both predicates and functions. However, such a study was initiated only in the late 1960s when Löb [363] and Gurevich [223] proved that first-order logic without equality and with only monadic predicate and function symbols (the class $[all, (\omega), (\omega)]_=$) has the finite model property. In a series of papers Gurevich then classified decidable and undecidable standard classes with function symbols [223, 226, 227]. The most difficult case in first-order logic without equality is the class of $\exists^*\forall\exists^*$-sentences with arbitrary vocabulary of relation and function symbols. Gurevich proved in [226] that this class has the finite model property and therefore is decidable for satisfiability. The decidability result (but not the finite model property) also follows from the fact that the derivability problem for the dual class $[\forall^*\exists\forall^*, all, all]$ is decidable; this had been announced by Orevkov in [407] and proved by Maslov and Orevkov in [386] who cite Gurevich's proof (accepted for publication in 1968, but published only in 1973). Their method is proof theoretical and quite different from that of Gurevich. See also [395] for a proof of this result.

Thus, the $\exists^*\forall\exists^*$-sentences form the unique maximal decidable prefix class in first-order logic without equality, whereas the $\forall^2$-sentences form a reduction class (even with very restricted vocabulary, such as one unary relation and one binary function, or one binary relation and one unary function) [223].

For full first-order logic (with equality, arbitrary functions and relation symbols) only the existential prefix class (which readily reduces to the propositional case) is decidable. In fact Gurevich proved that the classes $[\forall, (0), (2)]_=$ and $[\forall, (0), (0,1)]_=$ are reduction classes. This leaves two maximal decidable standard classes, both of which contain infinity axioms, namely the Rabin class $[all, (\omega), (1)]_=$ and the Shelah class $[\exists^*\forall\exists^*, all, (1)]_=$ (see the historical remarks in Sect. 7.4)

The study of complexity results for decidable cases of the decision problem originates in the work of Lewis [352] and Fürer [175, 177] who determined upper and lower complexity bounds for the classical solvable cases. Specifically they proved that the Löwenheim class, the Bernays-Schönfinkel class and the Gödel-Kalmár-Schütte class have nondeterministic exponential time complexity, and that the Ackermann class has deterministic exponential time complexity (for details see Sect. 6.2).

The complexity of the Ackermann class with equality $[\exists^*\forall\exists^*, all]_=$ was first determined by Kolaitis and Vardi [314]. The proof presented here was found independently, but later, by Grädel [196].

Grädel [200, 204] investigated the complexity of classes with function symbols; in particular he strengthened Gurevich's decision procedures for the classes $[\exists^*\forall\exists^*, all, all]$ and $[all, (\omega), (\omega)]$ and proved essentially optimal complexity bounds. In his Habilitationsschrift [196] Grädel wrote a unified presentation of complexity results for decidable cases of the decision problem which has been the basis of a large part of this chapter. In particular, he developed there the particular bounded domino problem (see Sect. 6.1.1) that we used for most of the lower complexity bounds. Other finite variants of domino problems had been devised earlier by Lewis and Papadimitriou [355], Harel [246], Savelsbergh and van Emde Boas [448], Chlebus [78] and Grädel

[201, 203, 206]; they were used to prove lower complexity bounds for various systems of propositional logic and decision problems in mathematical theories. In [196], Grädel also classified the standard classes whose satisfiability problems are in P, NP or Co-NP (see Sect. 6.4).

A useful survey on solvable and unsolvable decision problems in mathematical logic is given by Grigorieff [216]. The reader can find there a large amount of related material that we don't cover in this book (like decidability and complexity results for mathematical theories).

7. Monadic Theories and Decidable Standard Classes with Infinity Axioms

Not all decidable prefix-vocabulary classes in first-order logic have the finite model property. In fact, among the seven maximal decidable standard classes as given by Theorem 6.0.2 the following two contain infinity axioms:

- $[all, (\omega), (1)]_=$, i.e. first-order logic with equality, one unary function and monadic predicates.
- The Shelah class, i.e. the class $[\exists^*\forall\exists^*, all, (1)]_=$ of prenex first-order sentences with at most one universal quantifier, at most one unary function symbol and arbitrary relation symbols, with equality, but without function symbols of arity > 1.

Indeed, we can formulate infinity axioms with very modest quantifier prefixes even in the first-order theory of one unary function (see Proposition 6.5.5). Since the finite model property does not hold, we need different methods than in the previous chapter to establish decidability. It should also be noted that for classes with infinity axioms, satisfiability and finite satisfiability are two different problems. It is conceivable that one is decidable but not the other. However, it turns out that for the classes studied here, both satisfiability and finite satisfiability can be proved decidable by essentially the same arguments.

These results, in our exposition, rely on Rabin's famous result that S2S, the monadic theory of the infinite binary tree, is decidable. This is one of the most important decidability theorems for mathematical theories and has numerous applications in several areas of mathematics and computer science. We prove this result in Sect. 7.1. The proof, due essentially to Gurevich and Harrington [236], replaces the most complicated parts of Rabin's paper – notably the Complementation Theorem for tree automata based on an induction on countable ordinals – by simpler arguments based on the determinacy of certain games. The crucial ingredient is the Forgetful Determinacy Theorem; it is presented and proved in Sect. 7.1.4.

In Sect. 7.2 we show that the monadic theory of one unary function, like many other monadic theories, can be interpreted in S2S and thus proved to be decidable. It is a simple consequence of this result that the satisfiability problem and the finite satisfiability problem for the standard class $[all, (\omega), (1)]_=$ in first-order logic are decidable. However, the complexity is enormous. We show that even the first-order theory of one unary function is not elementary recursive, i.e. its time complexity exceeds any constant number of iterations of the exponential function.

In Sect. 7.3 we prove the decidability of the Shelah class $[\exists^*\forall\exists^*, all, (1)]_=$ by reducing it to the $\exists^*\forall\exists^*$-fragment of the monadic theory of one unary function. This may be the hardest decidability proof in the book.

In Sect. 7.4, we give relevant historical remarks.

7.1 Automata, Games and Decidability of Monadic Theories

7.1.1 Monadic Theories

Monadic (second-order) logic [1] is the extension of first-order logic that allows quantification over monadic predicates. Predicates and functions of any arity may appear in monadic formulae, but they may not be quantified over. Monadic theories are useful in various branches of mathematical logic and its applications. Often they have a reasonable level of expressiveness sufficient to formalize interesting features but modest enough to be manageable. For more information on monadic theories, we refer to [231].

Let σ be an arbitrary vocabulary of relation and function symbols. *Monadic formulae* of vocabulary σ are built from first-order σ-formulae and atomic formulae $Z_i(t)$ involving monadic predicate variables Z_i by means of negation, conjunction, disjunction, existential and universal quantification over individual variables, and existential and universal quantification over the set variables Z_i. There is an infinite supply of monadic predicate variables of course. A *monadic σ-sentence* is a formula of the monadic logic over σ (that is a monadic σ-formula) without free occurrences of any individual or set variables.

Definition 7.1.1 (Definable Relations). Let $\mathfrak{A}$ be a σ-structure with universe A and $\psi(x_1, \ldots, x_k, Z_1, \ldots, Z_m)$ be a formula with free individual variables $x_1, \ldots, x_k$ and free predicate variables $Z_1, \ldots, Z_m$. The *relation defined by ψ on $\mathfrak{A}$* is

$$\psi^{\mathfrak{A}} := \{(\bar{a}, \bar{S}) \in A^k \times \mathcal{P}(A)^m : \mathfrak{A} \models \psi[\bar{a}, \bar{S}]\}$$

where $\mathcal{P}(A)$ stands for the power set of A. Let L be a logic, say first-order or monadic second-order logic. We say that a given relation $R \subseteq A^k \times \mathcal{P}(A)^m$ is *definable* by L on $\mathfrak{A}$ if there exists a formula $\psi(x_1, \ldots, x_k, Z_1, \ldots, Z_m)$ of L such that $R = \psi^{\mathfrak{A}}$. A function is said to be definable if its graph is.

[1] We will often the adjective 'monadic' as an abbreviation for 'monadic second-order'. Thus monadic formulae are monadic second-order formulae, monadic theory is monadic second-order theory, etc.

Definition 7.1.2 (Monadic Theories). The *monadic theory* $\mathrm{Th}_{\mathrm{mon}}(\mathfrak{A})$ of a σ-structure $\mathfrak{A}$ is the set of monadic σ-sentences ψ such that $\mathfrak{A} \models \psi$. The monadic theory of a class $\mathcal{C}$ of σ-structures is $\mathrm{Th}_{\mathrm{mon}}(\mathcal{C}) := \bigcap_{\mathfrak{A}\in\mathcal{C}} \mathrm{Th}_{\mathrm{mon}}(\mathfrak{A})$. The *weak monadic theory* of a structure or a class of structures is defined in a similar way, but predicate variables Z_i range only over the finite subsets of the universe.

The *monadic theory of the binary tree* is the monadic theory of the structure $T^2 := (\{0,1\}^*, succ_0, succ_1)$ over the set of binary words, with the successor functions $succ_0(w) = w0$ and $succ_1(w) = w1$. The monadic theory of T^2 is also called S2S (an acronym for *(monadic) second-order theory of two successors*); the weak monadic theory of T^2 is denoted WS2S.

Sometimes a richer vocabulary for T^2 is used, including besides the two successor functions also the constant λ (for the empty word), and the prefix relation $<$. However, it is easy to see that λ and $<$ are definable on T^2; hence these modifications do not change the expressive power of S2S.

Exercise 7.1.3. Express λ and $<$ in T^2.

Example 7.1.4. Here are some further examples of definable relations on T^2. To enhance readability, we freely use abbreviations like $x \leq y$, $Y \subseteq Z$, $Y = Z$ and write $x0, x1$ rather that $succ_0(x), succ_1(x)$.

1. The lexicographical ordering $\prec$ on $\{0,1\}^*$ is definable by
$$x < y \lor \exists z(z0 \leq x \land z1 \leq y).$$
2. The formula
$$\mathrm{CHAIN}(Z) := \forall x \forall y (Zx \land Zy \rightarrow x < y \lor x = y \lor y < x)$$
defines the class of chains, i.e. the class of subsets of $\{0,1\}^*$ on which $<$ is a total order.
3. The class of infinite chains is defined by
$$\mathrm{INFCHAIN}(Z) := \mathrm{CHAIN}(Z) \land \forall x (Zx \rightarrow \exists y (x < y \land Zy)).$$
4. The formula
$$\begin{aligned}\mathrm{FIN}(Z) \quad := \quad & \neg\exists X \exists Y (\forall y (Xy \leftrightarrow \exists x (y \leq x \land Zx)) \land \\ & Y \subseteq X \land \mathrm{INFCHAIN}(Y))\end{aligned}$$
expresses that the prefix-closure X of Z does not contain an infinite chain. By König's Lemma this is true if and only if Z is finite.

The last example shows that WS2S can be interpreted in S2S.

The *sequential calculus* is the monadic logic of the structure $(\omega, succ)$. The monadic theory of $(\omega, succ)$ is also called S1S, for monadic (second-order) theory of one successor.

7.1.2 Automata on Infinite Words and the Monadic Theory of One Successor

We assume that the reader is familiar with the basic notions and results on finite automata and regular sets of finite words, as provided by most introductory textbooks on the theory of computation.

We consider here automata on infinite objects, such as infinite words and infinite trees. We only present the concepts and results that we need for the decidability theorems we are interested in. For more information on this exciting subject, we refer the reader to the excellent survey article by W. Thomas [507] and the references given there.

Let Σ be a finite non-empty alphabet. We denote by Σ^ω the set of *ω-sequences* (or *ω-words*) $\alpha = \alpha_0\alpha_1\cdots$ over Σ, or equivalently, the set of functions $\alpha : \omega \to \Sigma$. As usual, for $U, V \subseteq \Sigma^*$ we write UV for the set of words $uv \in \Sigma^*$ with $u \in U, v \in V$ and U^ω for the set of ω-words $\alpha = u_0u_1u_2\cdots$ obtained by infinite concatenation of words $u_i \in U$.

Definition 7.1.5. A *Büchi automaton* over the alphabet Σ is of the form $A = (S, S_0, T, E)$ where S is a finite set of *states*, $S_0 \subseteq S$ is the set of *initial states*, $T : S \times \Sigma \to \mathcal{P}(S)$ is the *transition function* and $E \subseteq S$ is the set of *final states*. A triple (s, a, s') such that $s' \in T(a, s)$ is called a *transition* of A. A *run* of A on an ω-word α is an ω-sequence $s_0s_1\cdots$ of states such that $s_0 \in S_0$ and $s_{n+1} \in T(s_n, \alpha_n)$ for all $n \in \omega$. The automaton A *accepts* α if there exists a run of A on α that contains some state $s \in E$ infinitely often. Let

$$L(A) = \{\alpha \in \Sigma^\omega : A \text{ accepts } \alpha\}$$

be the ω-language accepted by A.

A set $L \subseteq \Sigma^\omega$ is called *Büchi recognizable* or *ω-regular* if there exists a Büchi automaton A such that $L = L(A)$.

Let $A = (S, S_0, T, E)$ be a Büchi automaton with states $s, s' \in S$ and let $w = w_0\cdots w_{n-1} \in \Sigma^*$. We write $s \xrightarrow{w} s'$ if there exists a sequence $s_0\cdots s_n$ of states with $s_0 = s$, $s_{i+1} \in T(s_i, w_i)$ (for $i = 0, \ldots, n-1$) and $s_n = s'$. Clearly, the sets

$$W_{ss'} := \{w \in \Sigma^* : s \xrightarrow{w} s'\}$$

are regular sets of finite words.

Lemma 7.1.6. $L(A) = \bigcup_{s_0 \in S_0, s \in E} W_{s_0 s}(W_{ss})^\omega.$

The ω-language recognized by A is non-empty if and only if there exists an initial state s_0 and a final state s of A such that W_{s_0s} and $W_{ss} - \{\lambda\}$ are non-empty. Clearly, the existence of a reachable final state which is located in a loop of A is effectively decidable.

Corollary 7.1.7 (Emptiness Problem for Büchi Automata). *There is an algorithm, which for every given Büchi automaton A decides whether $L(A) = \varnothing$.*

Remark. It is known that the emptiness problem for Büchi automata is complete for NLOGSPACE (see [518]), and the universality problem for Büchi automata (given A, decide whether $L(A) = \Sigma^\omega$) is PSPACE-complete (see [475]).

Next, we prove some simple closure properties for ω-regular languages.

Lemma 7.1.8. *(i) If $U \subseteq \Sigma^*$ is regular, then U^ω is ω-regular.*
(ii) If $U \subseteq \Sigma^*$ is regular and $L \subseteq \Sigma^\omega$ is ω-regular, then UL is ω-regular.
(iii) If L, L' are ω-regular, then so are $L \cup L'$ and $L \cap L'$.

Proof. (i): Since $U - \{\lambda\}$ is regular if U is, and $(U - \{\lambda\})^\omega = U^\omega$ we can assume that U does not contain the empty word. Let A be a finite automaton recognizing U such that no transition of A leads into its initial state s_0. A Büchi automaton B recognizing U^ω is constructed from A by adding transitions (s, a, s_0) for any transition (s, a, f) of A into a final state f and by declaring s_0 as the single final state of B.

(ii): Let A be a finite automaton recognizing U and B be a Büchi automaton recognizing L. Without loss of generality we assume that A and B have no state in common. We obtain a Büchi automaton C recognizing UL by taking the union of the two automata and adding transitions (s, a, s_0) for every transition (s, a, f) into a final state f of A and every initial state s_0 of B.

(iii): Let $A = (S, S_0, T, E)$ and $A' = (S', S_0', T', E')$ be Büchi automata recognizing L and L', respectively, such that $S \cap S' = \varnothing$. Then $B = (S \cup S', S_0 \cup S_0', T \cup T', E \cup E')$ is a Büchi automaton recognizing $L \cup L'$. A Büchi automaton for $L \cap L'$ has the form $C = (S \times S' \times \{0, 1, 2\}, S_0 \times S_0' \times \{0\}, T'', E'')$ where

$$T''(ss'i, a) := T(s, a) \times T'(s', a) \times \{j\}$$

$$\text{where} \begin{cases} j = 1 & \text{if } i = 0 \text{ and } s \in E \\ j = 2 & \text{if } i = 1 \text{ and } s' \in E' \\ j = 0 & \text{if } i = 2 \\ j = i & \text{otherwise.} \end{cases}$$

Thus, a run of C simulates in parallel a run of A and a run of A'. Initially the third component of the state is 0. When some $s \in E$ is reached in the first component, the third component is set to 1 until a final state $s' \in E'$ is reached in the second component. Then the third component is set to 2 and in the next step back to 0. Thus C reaches infinitely often a state with third component 2 if and only if both A and A' reach final states infinitely often. Hence by setting $E'' := S \times S' \times \{2\}$ we obtain the desired result. □

Note that these closure properties are effective, i.e. the desired Büchi automaton recognizing, say, the union or intersection of two ω-regular languages L, L' can be effectively constructed from the Büchi automata for L and L'.

The following theorem is an immediate consequence of Lemma 7.1.6 and Lemma 7.1.8.

Theorem 7.1.9 (Büchi). *An ω-language $L \subseteq \Sigma^\omega$ is Büchi recognizable if and only if it can be represented as a finite union of sets UV^ω where $U, V \subseteq \Sigma^*$ are regular sets. Further, one can even assume that $VV \subseteq V$.*

A more difficult problem is the closure under complementation. On finite words, the closure under complementation of the regular languages follows immediately from the result of Rabin and Scott that every language recognizable by a finite automaton can also be recognized by a deterministic one. However, the corresponding result for Büchi automata fails and the closure of the ω-regular languages under complementation has to be established by more sophisticated arguments.

Exercise 7.1.10 (Deterministic Büchi Automata). A Büchi automaton is deterministic if it admits for every state s and every symbol a precisely one transition (s, a, s'). In contrast to the situation for automata over finite words, it is not the case that every Büchi automaton is equivalent to a deterministic one. To see this show that for $\Sigma = \{0, 1\}$, the ω-language $\Sigma^* 1^\omega$ of ω-words with only finitely many occurrences of 0 (which is obviously ω-regular) cannot be recognized by any deterministic Büchi automaton. However, there is an obvious deterministic Büchi automaton recognizing the complement of $\Sigma^* 1^\omega$, i.e., the set of all binary ω-sequences with infinitely many occurrences of 0. It thus follows that the class of ω-languages recognizable by *deterministic* Büchi automata is *not* closed under complementation.

To prove that the ω-regular languages are closed under complementation, we will (as in Büchi's original proof) use Ramsey's Theorem, in the version for countable sets.

Given a set X, we denote by $[X]^k$ the set of k-element subsets of X.

Theorem 7.1.11 (Ramsey). *For every finite set M, every $k \in \omega$ and every function $f : [\omega]^k \to M$ there exists an infinite set $X \subseteq \omega$ such that f maps all sets in $[X]^k$ to the same element of M.*

For $k = 1$ this is the pigeonhole principle. We will need the case where $k = 2$. Proofs of Ramsey's Theorem can be found in [270, pp. 538–539] and [76, pp. 168–169].

Theorem 7.1.12 (Büchi). *The class of ω-regular languages is closed under complementation.*

Proof. Given a Büchi automaton $A = (S, S_0, T, E)$ over Σ, we introduce a congruence relation $\sim_A$ over Σ^*. (Here, a congruence relation is an equivalence relation that is compatible with concatenation.) The equivalence classes of $\sim_A$ are regular sets. We will then show that both $L(A)$ and its complement can be represented as finite unions of sets UV^ω where U and V are equivalence classes with respect to $\sim_A$. By Theorem 7.1.9 this implies that $\Sigma^\omega - L(A)$ is ω-regular.

Let $s \xrightarrow{w,E} s'$ denote that there exists a run of A on w from state s to state s' such that at least one of the states in the run (including s and s') belongs to E. For words $u, v \in \Sigma^*$, let $u \sim_A v$ if for all states s, s' of A

$$s \xrightarrow{u} s' \Leftrightarrow s \xrightarrow{v} s' \quad \text{and} \quad s \xrightarrow{u,E} s' \Leftrightarrow s \xrightarrow{v,E} s'.$$

Lemma 7.1.13. *(i) The relation $\sim_A$ is a congruence relation of finite index over Σ^*. (The index is the number of equivalence classes.)*
(ii) Each $\sim_A$-class is regular.

Proof. The proof of *(i)* is straightforward. To see *(ii)*, let

$$W_{ss'}^E := \{w \in \Sigma^* : s \xrightarrow{w,E} s'\}.$$

Obviously, the sets $W_{ss'}^E$ are regular. Further the $\sim_A$-class $[w]$ of any $w \in \Sigma^*$ is the intersection of the sets $W_{ss'}$, $W_{ss'}^E$, $\Sigma^* - W_{ss'}$ and $\Sigma^* - W_{ss'}^E$ that contain w. □

We say that a congruence relation $\sim$ over Σ^* *saturates* an ω-language $L \subseteq \Sigma^\omega$ if for any pair U, V of $\sim$-equivalence classes

$$UV^\omega \cap L \neq \varnothing \quad \Longrightarrow \quad UV^\omega \subseteq L.$$

Note that if $\sim$ saturates L then it also saturates its complement.

Proposition 7.1.14. *(i) Let A be a Büchi automaton. Then $\sim_A$ saturates $L(A)$.*
(ii) Let $\sim$ be any congruence relation over Σ^* of finite index. Then, for every ω-word α there exist $\sim$-classes U, V (even with $VV \subseteq V$) such that $\alpha \in UV^\omega$.

Theorem 7.1.12 is an immediate consequence of this proposition. Indeed, since $\sim_A$ saturates $L(A)$ (and hence saturates $\Sigma^\omega - L(A)$) and since $\sim_A$ is of finite index it follows that

$$\Sigma^\omega - L(A) = \bigcup\{UV^\omega : U, V \text{ are } \sim_A\text{-classes}, UV^\omega \cap L(A) = \varnothing\}.$$

Hence, $\Sigma^\omega - L(A)$ is a finite union of ω-regular sets and thus ω-regular.

It remains to prove Proposition 7.1.14. Let $U, V \subseteq \Sigma^*$ be $\sim_A$-equivalence classes and $\alpha \in UV^\omega \cap L(A)$. Thus $\alpha = uv_1v_2\cdots$ with $u \in U$ and $v_i \in V - \{\lambda\}$.

Further, since A accepts α, there exists an ω-sequence $s_0 s_1 s_2 \cdots$ of states such that $s_0 \in S_0$,

$$s_0 \xrightarrow{u} s_1 \xrightarrow{v_1} s_2 \xrightarrow{v_2} s_3 \xrightarrow{v_3} \cdots$$

and, since the run leads to acceptance, $s_i \xrightarrow{v_i, E} s_{i+1}$ is true for infinitely many i. We have to show that every other ω-word $\beta = u'v_1'v_2' \cdots \in UV^\omega$ is also accepted by A. Since U, V are $\sim_A$-classes, we have $u \sim_A u'$ and $v_i \sim_A v_i'$ for all i. Therefore

$$s_0 \xrightarrow{u'} s_1 \xrightarrow{v_1'} s_2 \xrightarrow{v_2'} s_3 \xrightarrow{v_3'} \cdots$$

and $s_i \xrightarrow{v_i', E} s_{i+1}$ for infinitely many i. Hence $\beta \in L(A)$.

To prove *(ii)* we use Ramsey's Theorem. A congruence $\sim$ and an ω-word α induce a function $f : [\omega]^2 \to \Sigma^*/\sim$ mapping $\{i, j\}$ (with $i < j$) to the $\sim$-class of the word $\alpha(i, j) = \alpha_i \cdots \alpha_{j-1}$. Since $\sim$ is of finite index we can apply Ramsey's Theorem and infer that there exists an infinite set $X \subseteq \omega$ such that all words $\alpha(k, \ell) := \alpha_k \cdots \alpha_{\ell-1}$ for $k, \ell \in X, k < \ell$ are $\sim$-equivalent. In particular we have an infinite sequence $i_0 < i_1 < i_2 < \cdots$ of indices such that all segments $\alpha(i_j, i_{j+1})$ belong to the same $\sim$-class. Let V be this class and U be the $\sim$-class of the segment $\alpha(0, i_0)$. Then $\alpha \in UV^\omega$. This completes the proof of Theorem 7.1.12. □

Corollary 7.1.15. *Given a Büchi automaton A one can effectively construct a Büchi automaton $\overline{A}$ recognizing the complement of $L(A)$.*

Proof. By Lemma 7.1.13, the $\sim_A$-classes are regular sets and thus can be defined e.g. by regular expressions. Compute these regular expressions using the languages $W_{ss'}$ and $W_{ss'}^E$. For each pair U, V of $\sim_A$ classes, check whether $UV^\omega \cap L(A) = \emptyset$. Finally, construct a Büchi automaton recognizing the appropriate finite union of such sets UV^ω. □

We now are ready to prove the decidability of the sequential calculus. We associate with every tuple $V_1, \ldots, V_n$ of subsets of ω an ω-sequence $S(V_1, \ldots, V_n)$ over the alphabet $\{0,1\}^n$. Let $c_V : \omega \to \{0,1\}$ be the characteristic function of a subset V of ω. Then $S(V_1, \ldots, V_n) = s_0 s_1 s_2 \cdots$ where $s_m \in \{0,1\}^n$ indicates which of the sets $V_1, \ldots, V_m$ contain the number m. (It is convenient to write each letter of $\{0,1\}^n$ as a column of n bits, so that an ω-sequence $\alpha \in (\{0,1\}^n)^\omega$ can be seen as a binary matrix with n rows and ω columns.)

Theorem 7.1.16. *For every formula $\psi(X_1, \ldots, X_n)$ in the monadic logic of one successor, one can effectively construct a Büchi automaton A_ψ over the alphabet $\{0,1\}^n$ such that for all $V_1, \ldots, V_n \subseteq \omega$*

$$(\omega, succ) \models \psi[V_1, \ldots, V_n] \Leftrightarrow A_\psi \text{ accepts } S(V_1, \ldots, V_n).$$

Proof. We reformulate the sequential calculus in a formally first-order language with the binary predicates $\subseteq$ and *Succ*. The variables range over subsets of ω; the interpretation of $\subseteq$ is the usual one and $Succ(U, V)$ holds if and only $U = \{m\}$ and $V = \{m+1\}$ for some $m \in \omega$. Obviously, every formula of the sequential calculus can effectively be rewritten in this form.

The proof now proceeds by induction over ψ. For atomic formulae the construction is easy. For $X \subseteq Y$ we need an automaton that accepts precisely the ω-words over $\{0,1\}^2$ that do not contain the letter 10. This is achieved by an automaton with a single state s (which is initial and final) and transitions (s, a, s) for all $a \neq 10$. For the atom $Succ(X, Y)$ the desired automaton has three states s_0, s_1, s_2, where s_0 is the unique initial state and 2 the unique final state; the transitions are

$$(s_0, 00, s_0),\ (s_0, 10, s_1),\ (s_1, 01, s_2),\ (s_2, 00, s_2).$$

For formulae formed by disjunction, conjunction or negation we use that the ω-regular languages are effectively closed under Boolean operations. Finally, let $\psi(\bar{Y}) = \exists X \varphi(X, \bar{Y})$. By induction hypothesis one can construct a Büchi automaton $A_\varphi = (S, S_0, T, E)$ recognizing the ω-sequences $S(U, \bar{V})$ that satisfy $\varphi[U, \bar{V}]$. The desired automaton for $\psi(\bar{Y})$ – which, intuitively, on $S(\bar{V})$ "guesses" an additional component U and simulates A_φ on $S(U, \bar{V})$ – is $A_\psi = (S, S_0, T', E)$ where $T'(s, \bar{a}) = T(s, \bar{a}0) \cup T(s, \bar{a}1)$. □

We immediately get Büchi's Theorem.

Corollary 7.1.17 (Decidability of S1S). *The monadic theory of* $(\omega, succ)$ *is decidable.*

7.1.3 Tree Automata, Rabin's Theorem and Forgetful Determinacy

Let Σ be a non-empty finite alphabet. A Σ-*tree* is given by a labeling function $F : \{0,1\}^* \to \Sigma$ assigning to every node of the infinite binary tree a letter from Σ.

Definition 7.1.18. A Σ-*tree automaton* is a quadruple $A = (S, T, T_0, \mathcal{E})$ where S is a finite set of *states*, $T : S \times \{0,1\} \times \Sigma \to \mathcal{P}(S)$ is the *transition table*, $T_0 : \Sigma \to \mathcal{P}(S)$ the *initial table* and $\mathcal{E} \subseteq \mathcal{P}(S)$ is the *set of final collections of states.* We assume that $T_0(a)$ and $T(s, d, a)$ are non-empty for all $a \in \Sigma$, $d \in \{0,1\}$ and $s \in S$.

It is convenient to view automata from a game-theoretic point of view [236]. In order to describe when the automaton A accepts a Σ-tree F, we introduce a game $\Gamma(A, F)$ between the automaton A and another player called Pathfinder. The game starts at the root λ of the tree where the automaton chooses a state $s_0 \subseteq T_0(F(\lambda))$. The players alternate; at odd positions

Pathfinder chooses a direction $d \in \{0,1\}$, at even positions the automaton selects a state $s \in S$. The players thus define an infinite sequence $s_0 d_0 s_1 d_1 s_2 \cdots$, called a *play*. The choices of the automaton are restricted by the condition that

$$s_{n+1} \in T(s_n, d_n, F(d_1 \cdots d_n)).$$

The automaton wins the play $s_0 d_0 s_1 d_1 s_2 \cdots$ if the the collection of states that appear infinitely often in the play is a final collection, i.e. is contained in $\mathcal{E}$. Otherwise, Pathfinder wins. We say that the automaton accepts F if it has a winning strategy for $\Gamma(A, F)$. We write $L(A)$ for the set of Σ-trees accepted by the Σ-tree automaton A.

Exercise 7.1.19. Let $\Sigma = \{0,1\} \times \{0,1\}$. Construct a tree automaton A that accepts a Σ-tree F if and only if F avoids the label 01, i.e. $F(w) \neq 01$ for all nodes w. Further construct Σ-tree automata A_0, A_1 such that A_i accepts F if and only there exists a node $w \in \{0,1\}^*$ such that $F(w) = 10$, $F(wi) = 01$ and $F(v) = 00$ for any other node v.

Exercise 7.1.20. Given a Σ_1-tree automaton A and a Σ_2-tree automaton B construct a $(\Sigma_1 \cup \Sigma_2)$-tree automaton accepting $L(A) \cup L(B)$. Further, given a $(\Sigma_1 \times \Sigma_2)$-tree automaton A construct a Σ_1-tree automaton A that accepts a Σ_1-tree F if and only if there exists a Σ_2-tree G such that A accepts (F, G).

The *node* of a game position p is the string Node(p) of even letters in p, i.e. the node of the binary tree that is currently played. Note that if p is a position where the automaton makes the next move then Node(p) = Node(ps) for every successive position ps (where $s \in S$). For every node $v \in \{0,1\}^*$ and every Σ-tree F, the v-residue of F is the Σ-tree F_v with $F_v(w) = F(vw)$.

The *latest appearance record* LAR(p) at position p is intuitively the list of states (without repetitions) in the order of their latest appearance. We give an inductive definition of LAR. It does not really matter what is the LAR of the empty string; it is convenient to define it as a list of all states (without repetition) in some order. Moves by Pathfinder do not change the LAR, i.e. LAR(pd) =LAR(p) for any odd position p and $d \in \{0,1\}$. If $p = qs$ for some even position q and $s \in S$, then LAR(p) $= rs$ where r is the result of removing s from LAR(q).

A *strategy* of either player in $\Gamma(A, F)$ is a function that assigns to each position of that player a set of legal moves from that position. The Forgetful Determinacy Theorem states that one of the players has a winning strategy for $\Gamma(A, F)$ that is 'forgetful' in the sense that it only depends on the residual game from the given position and on its latest appearance record (and not on the entire history).

Theorem 7.1.21 (Forgetful Determinacy for Tree Automata). *One of the players has a strategy f for winning $\Gamma(A, F)$ that satisfies the following condition: If p and q are positions from which the winner makes moves, such that LAR(p) = LAR(q) and the Node(p)-residue and the Node(q)-residue of F coincide, then $f(p) = f(q)$.*

We will prove (a slightly more general version of) the Forgetful Determinacy Theorem in the next section. But first we derive some interesting consequences and show in particular, that the Forgetful Determinacy Theorem gives a simple proof for the decidability of S2S.

Theorem 7.1.22 (Emptiness Problem for Tree Automata). *There exists an algorithm that, given a Σ-tree automaton A, decides whether there exists a Σ-tree accepted by A.*

Proof. We first reduce the claim to the case where the alphabet contains only one letter. Let $A = (S, T, T_0, \mathcal{E})$ be a Σ-tree automaton for an arbitrary alphabet Σ. Then let B be the $\{0\}$-tree automaton $(S, T', T_0', \mathcal{E})$ with $T_0'(0) := \bigcup_{a \in \Sigma} T_0(a)$ and $T'(s, i, 0) = \bigcup_{a \in \Sigma} T(s, i, a)$. Obviously B accepts the unique $\{0\}$-tree F if and only if A accepts some Σ-tree.

By the Forgetful Determinacy Theorem, one of the players has a forgetful winning strategy for $\Gamma(B, F)$. List all forgetful strategies $f_1, \ldots, f_m$ for the automaton and $g_1, \ldots, g_n$ for Pathfinder. Since the plays eventually become periodic, one can effectively check each f_i against each g_j and thus determine whether B accepts F. □

Theorem 7.1.23 (Complementation Theorem for Tree Automata). *One can effectively construct from each Σ-tree automaton A a Σ-tree automaton $\overline{A}$ accepting exactly the Σ-trees rejected by A.*

Proof. Let F be a Σ-tree and g be any (not necessarily winning) forgetful strategy for Pathfinder. Without loss of generality, we may assume that g is deterministic. Indeed, if g allows Pathfinder both moves from p, just refine g to $g(p) = 0$. Let R be the set of *a priori* possible latest appearance records of A, i.e. the set of all lists of states containing each state at most once. The strategy g for Pathfinder can be considered as a function $g : \{0,1\}^* \times R \to \{0,1\}$.

Let Δ be the set of all functions $h : R \to \{0,1\}$. We encode the strategy g by the Δ-tree G with

$$G(w)(r) = g(w, r)$$

i.e. every node w is labeled by the function assigning to each LAR r the move that Pathfinder makes from the node w with LAR r according to strategy g. Combining the labels of a Σ-tree F and a Δ-tree G gives a $(\Sigma \times \Delta)$-tree which may be denoted (F, G).

Lemma 7.1.24. *Given A, one can effectively construct a $(\Sigma \times \Delta)$-tree automaton B such that the following holds: Pathfinder wins $\Gamma(A, F)$ via the forgetful strategy represented by G if and only if B wins all plays of the game $\Gamma(B, (F, G))$ (i.e., B wins by every conceivable strategy).*

Proof. We construct a tree automaton B whose states are the latest appearance records of A together with a new state *win*. The rôle of this additional state is to force Pathfinder in the new game $\Gamma(B,(F,G))$ to play along the strategy G. If she ever deviates from this strategy the state *win* will be entered and not be left anymore. Of course the singleton set $\{win\}$ will be a winning collection for B.

Let $A = (S,T,T_0,E)$. For every nonempty LAR $r \in R$, let $\ell(r)$ be the rightmost (that is the last) member of the list r. For every LAR $r \in R$, let $u(r,s)$ be the updated LAR obtained from r by removing s and appending it at the end (so that $\ell(u(r,s)) = s$). Every node in the tree (F,G) has a label ah where $a \in \Sigma$ and $h : R \to \{0,1\}$ assigns a direction to each conceivable LAR of A.

The desired automaton is of the form $B = (R \cup \{win\}, T', T_0', \mathcal{E}')$; the transition table T', the initial table T_0' and the set of final collections $\mathcal{E}'$ are defined as follows.

To ensure that the win-state is never left we put $T'(win,d,ah) := win$ for all $a \in \Sigma$, $h \in \Delta$ and $d \in \{0,1\}$. For $r \in R$, let

$$T'(r,d,ah) := \begin{cases} \{win\} & \text{if } h(r) \neq d \\ \{u(r,s) : s \in T(\ell(r),d,a)\} & \text{if } h(r) = d. \end{cases}$$

Intuitively, in terms of the game $\Gamma(B,(F,G))$, this means that the automaton assumes the win-state whenever Pathfinder at a node labeled ah does not choose the direction $d = h(r)$ prescribed by G. If Pathfinder chooses $h(r)$ (i.e. simulates the strategy G) then the automaton B simulates a move of the old automaton, namely he picks one of the states $s \in T(\ell(r),d,a)$, and the new state of B is the one obtained by updating r accordingly.

To define the initial table of B we identify a state with the list consisting of just that state and put $T_0'(ah) := T_0(a)$.

A set $R_0 \subseteq R \cup \{win\}$ is in $\mathcal{E}'$ if either $win \in R_0$ or the set $\{\ell(r) : r \in R_0\}$ is *not* a final collection of A. Thus, intuitively the plays of $\Gamma(B,(F,G))$ won by the automaton correspond to plays of $\Gamma(A,F)$ won by Pathfinder along strategy G.

We claim that B has the desired properties. Indeed, suppose Pathfinder wins $\Gamma(A,F)$ with strategy G. Then B wins $\Gamma(B,(F,G))$ by any strategy: If Pathfinder sticks to strategy G (i.e. at any position with state r and label ah she chooses the direction $h(r)$) then B wins because the final collection of states thus produced corresponds to the sequence of LAR's of a play of $\Gamma(A,F)$ won by Pathfinder. On the other side if Pathfinder ever deviates from G, then B wins because the win-state is assumed.

Conversely, suppose that the automaton A wins $\Gamma(A,F)$ with a strategy f against Pathfinder's strategy G. It suffices to show that B has a strategy f' that loses when the pathfinder plays along the strategy G in $\Gamma(B,(F,G))$. The desired losing strategy f' of B is obtained by utilizing the winning strategy f of A. It is losing because f wins against G and because the winning conditions for B essentially complement the winning conditions for A. □

Lemma 7.1.25. *For every tree automaton B one can effectively construct a tree automaton C which accepts a tree H if only if B wins all plays of $\Gamma(B, H)$.*

Proof. Let $B = (S, T, T_0, \mathcal{E})$ be a Σ-tree automaton. The condition that B wins all plays of $\Gamma(B, H)$ means that each path $d_0 d_1 d_2 \cdots \in \{0,1\}^\omega$ through the infinite binary tree satisfies the following condition where the ω-sequence $H(\lambda)H(d_0)H(d_0 d_1) \cdots \in \Sigma^\omega$ is the sequence of labels along this path:

(*) For all sequences $s_0 s_1 \cdots \in S^\omega$ such that $s_0 \in T(H(\lambda))$ and $s_{n+1} \in T(s_n, d_n, H(d_0 \cdots d_n))$ the collection of states that occur infinitely often in $s_0 s_1 \cdots$ belongs to $\mathcal{E}$.

Condition (*) is expressible by an S1S-formula $\varphi(X, \bar{Y})$ with a free set variable X to encode the given sequence $d_0 d_1 d_2 \cdots$ and a tuple $\bar{Y}$ of set variables to encode the label sequence $H(\lambda)H(d_0)H(d_0 d_1) \cdots$ (see the exercise below).

By Theorem 7.1.16, there exists a Büchi automaton $B' = (S', S'_0, T', E')$ over the alphabet $\{0,1\} \times \Sigma$ that accepts a pair of sequences $d_0 d_1 d_2 \cdots$ and $H(\lambda)H(d_0)H(d_0 d_1) \cdots$ if and only if they satisfies condition (*).

We use B' to construct a Σ-tree automaton C which wins $\Gamma(C, H)$ if and only if for every path $d_0 d_1 \cdots$ chosen by Pathfinder, the induced sequence $H(\lambda), d_0 H(d_0), d_1 H(d_0 d_1), \ldots$ is accepted by the Büchi automaton B'. In other words, C accepts H if and only if B wins all conceivable plays of $\Gamma(B, H)$.

The desired $C = (S'', T'', T''_0, \mathcal{E}'')$ with

$$
\begin{aligned}
S'' &:= S' \\
T''(s, d, a) &:= T'(s, da) \\
T''_0(a) &:= \bigcup_{s \in S'_0} \bigcup_{i \in \{0,1\}} T'(s, ia) \\
\mathcal{E}'' &:= \{X \subseteq S' : X \cap E \neq \varnothing\}.
\end{aligned}
$$

Let us justify the definition of $T''(a)$. One can view a sequential (Büchi) automaton as a tree automaton playing against a sequential "pathfinder" that does not make any choices. There is, however, a little discrepancy in the way sequential and tree automata start. A sequential automaton starts without looking at the label of zero, then the "pathfinder" makes a move (from say -1 to 0), and only then the automaton reacts to the label of zero. On the other hand, the very first move of a tree automaton depends on the label of the root. Thus the first move of a tree automaton corresponds to the first two moves of sequential automaton (with an additional move of the sequential "pathfinder" from -1 to 0 that does not correspond to any move of the tree pathfinder). □

Exercise 7.1.26. Construct explicitly an S1S-formula expressing statement $(*)$ above. Hint: A sequence $H(\lambda), d_0H(d_0), d_1H(d_0d_1), \ldots$ is encoded by subsets $D \subseteq \omega$ and $H_a \subseteq \omega$ (for $a \in \Sigma$) namely $D = \{n < \omega : d_n = 1\}$ and $H_a = \{n < \omega : H(d_0 \cdots d_{n-1}\}) = a\}$. Let $\bar{H} = (H_a)_{a \in \Sigma}$. The desired formula has the form $\psi(X, \bar{Y})$ such that $\psi[D, \bar{H}]$ is true if and only if D and $\bar{H}$ encode a sequence satisfying $(*)$. To construct ψ, proceed as follows: Let $\psi(X, \bar{Y}) := \forall \bar{Z} \varphi(X, \bar{Y}, \bar{Z})$ where φ states that if $\bar{Z}$ encodes a state sequence $s_0s_1 \ldots$ such that X and $\bar{Z}$ constitute a legal play of $\Gamma(B, H)$ then the collection of states s such that for all $i \in \omega$ there exists a $j > i$ with $s_j = s$ belongs to $\mathcal{E}$.

We now finish the proof of the Complementation Theorem. Given a Σ-tree automaton A, use Lemma 7.1.24 and Lemma 7.1.25 to construct a $\Sigma \times \Delta$ automaton C that accepts a tree (F, G) if and only if Pathfinder wins $\Gamma(A, F)$ with the strategy encoded by G.

Let D be the Σ-tree automaton which, on every Σ-tree F guesses a Δ-tree G and simulates C on (F, G). More formally if $C = (S, T, T_0, E)$ is any $\Sigma \times \Delta$ automaton then let $D = (S, T', T_0', E)$ be the Σ-tree automaton with $T'(s, d, a) := \bigcup_{d \in \Delta} T(s, d, ab)$ and $T_0'(a) = \bigcup_{b \in \Delta} T_0(ab)$.

A Σ-tree F is accepted by D iff there exists a Δ-tree G such that C accepts (F, G). But his means that D accepts precisely the trees rejected by A. □

We now are ready to prove that S2S is decidable.

Theorem 7.1.27 (Rabin's Tree Theorem). *The monadic theory of the infinite binary tree is decidable.*

Proof. As in the case of S1S, it is convenient to reformulate S2S in a (formally first-order) language with binary predicates $\subseteq$, $Succ_0$ and $Succ_1$, with variables ranging over subsets of $\{0, 1\}^*$, with the obvious interpretation of $\subseteq$ and with $Succ_i(U, V)$ if and only $U = \{w\}$ and $V = \{wi\}$ for some $w \in \{0, 1\}^*$.

Obviously, every monadic second-order formula on the language of two successor function can be translated into an equivalent formula of this form.

Let $\Sigma = \{0, 1\}$ and Σ_n be the n-fold Cartesian product of Σ. Every tuple $V_1, \ldots, V_n$ of subsets of the infinite binary tree yields a Σ_n-tree $T(V_1, \ldots, V_n)$ that labels each $w \in \{0, 1\}^*$ with $c_{V_1}(w), \ldots, c_{V_n}(w)$ where c_V is the characteristic function of V.

Theorem 7.1.28. *With every* S2S-*formula* $\psi(X_1, \ldots, X_n)$ *one can effectively associate a* Σ_n*-tree automaton* A_ψ *such that for all* $V_1, \ldots, V_n \subseteq \{0, 1\}^*$

$$T^2 \models \psi[V_1, \ldots, V_n] \iff A_\psi \textit{ accepts } T(V_1, \ldots, V_n).$$

Proof. By induction over φ. The construction is easy for atomic ψ (see Exercise 7.1.19). For $\psi = \varphi \vee \eta$ and $\psi = \exists X \varphi$, use Exercise 7.1.20. For $\psi = \neg\varphi$ use the Complementation Theorem. □

Thus a formula ψ is a theorem of S2S if and only if the set of tree accepted by the automaton $A_{\neg\psi}$ is empty. Since $A_{\neg}\psi$ can be effectively constructed from ψ and since the emptiness problem for tree automata is decidable, this proves Rabin's Tree Theorem. □

7.1.4 The Forgetful Determinacy Theorem for Graph Games

In this section we formulate and prove a more general version of the Forgetful Determinacy Theorem. Our presentation follows Zeitman's proof [546] which in turn simplifies (and generalizes in some respect) the proof given by A. Yakhnis and V. Yakhnis [540]. More information on the background and history of forgetful determinacy is given in Sect. 7.4.

Graph Games. Let *MOVE* be a finite alphabet. An *arena* A is a coloured bipartite multi-digraph (a directed graph in which parallel edges are allowed) that satisfies the following conditions:

- The vertices are divided into *east* vertices and *west* vertices. Edges go only from east to west vertices or from west to east vertices.
- There is a distinguished vertex called the *start vertex* of A. Every vertex is reachable from the start vertex, and there is at least one outgoing edge from each vertex.
- The edges of A are labeled by elements of *MOVE* in such a way that no two outgoing edges from the same vertex have the same label.
- There is a finite set S of colours that partition the set of vertices. (Zeitman [546] allows multiply coloured vertices but this is unnecessary for our purposes here.) We write C^s for the set of vertices with colour s.

A game on A is played by two players, Mr. 0 and Mr. 1, who alternately choose an outgoing edge from the current vertex, thus defining an infinite path (possibly revisiting some vertices) through A. A *position* p is a finite directed path through A from the start vertex; it is uniquely described by a word in $MOVE^*$. We identify p with the appropriate word. The labels on the edges leading out of the last vertex of a position p are the possible moves at position p. A *play* in A is an ω-sequence $P \in MOVE^\omega$ such that all of its finite prefixes are positions. We denote the set of plays over A by $PLAY(A)$. This set is divided into two complementary parts: the *winning sets* for Mr. 0 and for Mr. 1. Mr. δ wins a play if the play belongs to his winning set. We will only consider winning sets that are Boolean combinations of sets $[C^s]$, where $[C^s]$ is the set of plays that pass infinitely often through a vertex of colour s.

Definition 7.1.29. A *graph game* is a triple $\Gamma = (A, \varepsilon, W_\varepsilon)$, where A is an arena, $\varepsilon \in \{0,1\}$ (denoting the player who goes first), and $W_\varepsilon \subseteq PLAY(A)$ is a Boolean combination of sets $[C^s]$, the winning set for player ε.

It is Mr. ε's turn to move at the vertices whose orientation (east or west) coincides with that of the start vertex; call the set of these vertices V_ε. The set of remaining vertices will be denoted $V_{1-\varepsilon}$. Thus, for each δ, it is Mr. δ's turn to move at vertices $v \in V_\delta$.

Definition 7.1.30. A *forgetful strategy* f for Mr. δ in Γ is a function $f : V_\delta \to \mathcal{P}(MOVE)$ assigning to every vertex $v \in V_\delta$ a non-empty set of possible moves (i.e. labels of outgoing edges) from v. The strategy is "forgetful" in the sense that it depends only on v and not on how v was reached.

The *latest appearance record* LAR of a position is an ordering of the colours defined inductively. The LAR of the start vertex is an ordering whose last colour is that of the start vertex. If a position q is obtained from a position p by adjoining an edge to a vertex of colour s, then LAR of q is obtained from LAR of p by moving s to the last place. The colouring of an arena A is *forgetful* if any two positions ending at the same vertex have the same LAR. In this case we can speak of the LAR at vertex v.

Theorem 7.1.31 (Forgetful Determinacy). *Let $\Gamma = (A, \varepsilon, W_\varepsilon)$ be any graph game with a forgetful colouring of the arena A. Then one of the players has a forgetful strategy winning Γ.*

Before we prove this more general version of the Forgetful Determinacy Theorem, we prove that it implies the version for tree automata (Theorem 7.1.21). A game $\Gamma(A, F)$ between a tree automaton A and Pathfinder on the Σ-tree F gives a graph game whose alphabet comprises the states of A and (some names for) the two directions left and right. A starts the game and chooses a state according to its initial table, then Pathfinder chooses (the name for) a direction, then A chooses a state and so on. All positions p where A makes a move have the same default colour. If A chooses a state s at p then the colour of position ps is s. This colouring is forgetful in a trivial way.

The Forgetful Determinacy Theorem for tree automata states that one of the players has a winning strategy that is the same for any two positions with the same LAR and the same residual game.

To prove this, we define an equivalence relation on the vertices of an arbitrary graph game $\Gamma = (A, \varepsilon, W_\varepsilon)$, that relates vertices having the same LAR and the same future. This will allow us to define a factor game and thus to show that the Forgetful Determinacy Theorem gives in fact a forgetful strategy that assigns the same value to any two equivalent vertices.

Consider an arena A with a forgetful colouring. We say that two vertices v, w of A are equivalent, $v \sim w$, if the following conditions hold where C^s is the collection of vertices of colour s:

- v and w both are east vertices or both are west vertices;
- $\mathrm{LAR}(v) = \mathrm{LAR}(w)$;
- the same sequences of moves are possible from v and w;

– furthermore, for any colour s, the same sequences of moves lead from v or w to C^s.

Exercise 7.1.32. Prove that the equivalence relation $\sim$ has the following properties:

– if $v \sim w$ then, for all $s \in S$, $v \in C^s$ iff $w \in C^s$;
– if $v \sim w$ and if there is an edge labeled by μ from v to a vertex v', then there is an edge labeled by μ from w to a vertex w' such that $w \sim w'$.

Corollary 7.1.33 (Strong Forgetful Determinacy). *Let $\Gamma = (A, \varepsilon, W_\varepsilon)$ be a graph game in which the colouring of A is forgetful. Then one of the players has a forgetful strategy f winning Γ such that $f(v) = f(w)$ for all vertices v, w with $v \sim w$.*

Proof. We define a factor game $\Gamma/\!\sim\, = (A/\!\sim, \varepsilon, W_\varepsilon)$. The arena $A/\!\sim$ has as its vertex set the set of $V/\!\sim$ of equivalence classes $[v]$ of vertices $v \in V$. The new start vertex is the equivalence class of the original start vertex. There is an edge labeled by $\mu \in MOVE$ from $[v]$ to $[w]$ in $A/\!\sim$ if there is an outgoing edge labeled μ from a vertex in $[v]$ to a vertex in $[w]$. A vertex $[v]$ of $A/\!\sim$ is coloured by $s \in S$ if one (and hence all) of the vertices in $[v]$ has colour s; we write $C^s/\!\sim$ for the set of equivalence classes $[v]$ with colour s. Also, $[v]$ is a east vertex if the vertices in $[v]$ are east; otherwise $[v]$ is a west vertex.

Obviously, if p is a position that ends at v in A, then p is also a position in $A/\!\sim$ that ends at $[v]$. Furthermore $PLAY(A) = PLAY(A/\!\sim)$. The set $[C^s]$ of plays that pass infinitely often through a vertex with colour s is the same for A and $A/\!\sim$. Thus also W_ε coincides on both arenas.

Note that the colouring of $A/\!\sim$ is forgetful. By the Forgetful Determinacy Theorem, there exists a forgetful winning strategy g for one of the players in $\Gamma/\!\sim$. The forgetful strategy f for Γ defined so that $f(v) := g([v])$ is a winning strategy that has the same value at any two vertices of A related by $\sim$. □

The Forgetful Determinacy Theorem for tree automata is a special case of this corollary.

Proof of the Forgetful Determinacy Theorem. Let δ be 0 or 1. From now on, f denotes a forgetful strategy for Mr. δ, and g denotes a forgetful strategy for Mr. $(1-\delta)$. We say that a play P is *consistent* with with f and g after position $p = \mu_0 \cdots \mu_n$ if every position $p_m = \mu_0 \cdots, \mu_m$, $m \geq n$, in P satisfies the following condition where v is the last vertex of p_m: $\mu_{m+1} \in f(v)$ if $v \in V_\delta$, and $\mu_{m+1} \in g(v)$ if $v \in V_{1-\delta}$. We say that f *wins Γ from vertex v against g* if for all positions p that end at v, all plays consistent with f and g after p are in W_δ. We say that f *wins Γ* for Mr. δ if f wins Γ from the start vertex against the strategy for Mr. $(1-\delta)$ that allows any possible move at each vertex in $V_{1-\delta}$. A strategy f' for Mr. δ is a *refinement* of f if $f'(v) \subseteq f(v)$ for all $v \in V_\delta$.

The set of *winning vertices* for Mr. δ with restrictions f and g, denoted $\mathrm{Win}(\Gamma, \delta, f/g)$, is the set of vertices v such that Mr. δ has a refinement of f that wins Γ from v against g. Note that playing a winning refinement against g from a vertex in $\mathrm{Win}(\Gamma, \delta, f/g)$ keeps the play within the set of winning vertices. Indeed, suppose that f' is a refinement of f that wins against g at v, and that w is a vertex reached by a play that is consistent with f' and g after position p that ends at v. Then w is in $\mathrm{Win}(\Gamma, \delta, f/g)$ because any play consistent with f' and g after a position ending at w differs from a play consistent with f' and g after p only with respect to a finite prefix, and thus is also in W_δ since membership in this set depends only on the set of colours that are reached infinitely often.

The Forgetful Determinacy Theorem is a consequence of the following theorem.

Theorem 7.1.34. *Let $\Gamma = (A, \varepsilon, W_\varepsilon)$ be a graph game with a forgetful colouring, f a forgetful strategy for Mr. δ , and g a forgetful strategy for Mr. $(1-\delta)$ in Γ. Then Mr. δ has a refinement of f that wins against g at every vertex in $\mathrm{Win}(\Gamma, \delta, f/g)$, and Mr. $(1-\delta)$ has a refinement of g that wins against f at every vertex in $\mathrm{Win}(\Gamma, 1-\delta, g/f)$. Furthermore, every vertex of A is either in $\mathrm{Win}(\Gamma, \delta, f/g)$ or in $\mathrm{Win}(\Gamma, 1-\delta, g/f)$.*

The Forgetful Determinacy Theorem follows by taking for f and g the trivial strategies that allow at each vertex the set of all possible moves from that vertex. Since the start vertex of A is either in $\mathrm{Win}(\Gamma, \delta, f/g)$ or in $\mathrm{Win}(\Gamma, 1-\delta, g/f)$, there either is a forgetful strategy that wins Γ for Mr. δ, or there is one that wins Γ for Mr. $(1-\delta)$.

To prove Theorem 7.1.34 we first consider the presentation of the winning sets W_δ and $W_{(1-\delta)}$. Any intersection of sets $[C^s]$ and complements $[C^s]^c$ of such sets will be called a *term*. Clearly, the sets W_δ and $W_{(1-\delta)}$ can be expressed as unions of terms. Let $\beta \in \{0,1\}$ be such that W_β has the fewest of such terms.

Lemma 7.1.35. *Either W_β or its complement can be expressed in the form*

$$W = (U_1 \cup [B_1]) \cap \cdots \cap (U_m \cup [B_m]) = ([B_1] \cap \cdots \cap [B_m]) \cup \left(\bigcup_{i=1}^{m} U_i \right)$$

where each B_i is a union of sets C^s and the sets U_i (the derived winning sets) are unions of terms. Furthermore, each U_i is empty, or has fewer terms than W_β, or the complement of U_i has the same number of terms as W_β and contains a term in which no set of the form $[C^s]^c$ occurs (that is, every member of that term has the uncomplemented form $[C^s]$).

Proof. Note that $\bigcup_{j \in I} [C^j] = [\bigcup_{j \in I} C^j]$ for any $I \subseteq S$. If W_β contains a term in which no $[C^s]$ set is complemented, say

$$W_\beta = ([C^{s_1}] \cap [C^{s_2}] \cdots \cap [C^{s_k}]) \cup U',$$

then W_β can easily be expressed in the desired manner with each $U_i = U'$. Otherwise W_β contains some number of terms, say m, each of which contains some sets of the form $[C^s]^c$. Let B_i be the union of the sets C^s that appear in complemented form in the ith term. If there are uncomplemented $[C^s]$ sets in the same term, let V_i be their intersection; otherwise let V_i be all of $PLAY(A)$. Then

$$W_\beta = \bigcup_{i=1}^{m}(V_i \cap [B_i]^c) = \bigcup_{i=1}^{m}((V_i \cup (\bigcup_{j\neq i}(V_j \cap [B_j]^c))) \cap [B_i]^c).$$

Define U_i as the complement of the set $V_i \cup \left(\bigcup_{j\neq i}(V_j \cap [B_j]^c)\right)$. Each U_i set is empty, or its complement has m terms and contains one, namely V_i, with no complemented sets. Also $W_{(1-\beta)} = (U_1 \cup [B_1]) \cap \cdots \cap (U_m \cup [B_m]) = ([B_1] \cap \cdots \cap [B_m]) \cup \left(\bigcup_{i=1}^{m} U_i\right)$. □

Clearly, Theorem 7.1.34 is true if one of the winning sets is empty, that is, has no terms. Otherwise the winning set with the fewest terms or its complement can be expressed in the form described by Lemma 7.1.35. By induction on the minimum number of terms in either of the winning sets one can show that Theorem 7.1.34, is implied by the following lemma.

Lemma 7.1.36. *Let A be an arena with a forgetful colouring and suppose that*

$$W_\delta = (U_1 \cup [B_1]) \cap \cdots \cap (U_m \cup [B_m]) = ([B_1] \cap \cdots \cap [B_m]) \cup (\bigcup_{i=1}^{m} U_i)$$

where each B_i is a union of sets C^s. If Theorem 7.1.34 holds for each game Γ^i obtained from the given game Γ by replacing the winning set for Mr. δ with U_i, then it also holds for given game Γ (in which W_δ is the winning set for Mr. δ).

Exercise 7.1.37. Prove Theorem 7.1.34 from Lemma 7.1.36. Hint: Use induction on the minimum number of terms in either of the winning sets.

To prove Lemma 7.1.36, we introduce the notion of the *rank* of a vertex (with respect to a given player δ, a set of vertices X, and forgetful strategies f and g). A vertex v has rank $k > 0$ if the following two conditions hold:

- Player δ has a refinement of f that allows him to reach a vertex in X within k moves from v as long as his opponent plays according to the strategy g.
- Mr. $(1-\delta)$ has a strategy refining g that keeps Mr. δ from reaching X from v in fewer than k moves as long as Mr. δ plays f.

More formally, the rank is defined inductively as follows. Given any forgetful strategy f, a vertex w is called an *f-successor* of a vertex $v \in V_\delta$ if there exists a move $\mu \in f(v)$ that labels an edge from v to w. Recall that V_δ is the set of vertices from which player δ makes moves.

Then, $\text{rank}(X, \delta, f/g)(v) = 1$ if either $v \in V_\delta$ and there exists a move $\mu \in f(v)$ from v into X, or $v \in V_{1-\delta}$ and all moves $\mu \in g(v)$ lead from v into X. If v is a vertex that is not of rank $1, \ldots, k$, let $\text{rank}(X, \delta, f/g)(v) = k+1$ if either $v \in V_\delta$ and there exists an f-successor w of v with $\text{rank}(X, \delta, f/g)(w) = k$; or if $v \in V_{1-\delta}$ and all g-successors of v are ranked and the maximal value of $\text{rank}(X, \delta, f/g)(w)$ for g-successors w of v is k. Further, we say that $\text{rank}(X, \delta, f/g)(v) = 0$ if $v \in X$ and $\text{rank}(X, \delta, f/g)(v) \neq k$ for all $k > 0$.

Also let $\text{Dom}(X, \delta, f/g)$ be the set of vertices v where $\text{rank}(X, \delta, f/g)(v)$ is defined, and $\text{Dom}^+(X, \delta, f/g)$ be the set of vertices where this rank is positive.

We next define two basic strategies, one for each player. Here and below, when defining a strategy for a given player at vertex v, we always assume it is that player's turn to move at v. The basic strategy for Mr. δ is to decrease the rank. Let $\mathit{decrease}(X, \delta, f/g)(v)$ be $f(v)$ if $v \notin \text{Dom}^+(X, \delta, f/g)$; otherwise $\mathit{decrease}(X, \delta, f/g)(v)$ is the set of $\mu \in f(v)$ that label an edge from v to a vertex w with $\text{rank}(X, \delta, f/g)(w) < \text{rank}(X, \delta, f/g)(v)$ or with $w \in X$. The basic strategy for Mr. $(1-\delta)$, is to avoid the set $\text{Dom}(X, \delta, f/g)$; thus let $\mathit{avoid}(X, 1-\delta, g/f)$ be the strategy that assigns to a vertex v the set of $\mu \in g(v)$ that label an outgoing edge from v to a vertex outside of $\text{Dom}(X, \delta, f/g)$ if this set is not empty, and is the same as $g(v)$ otherwise. Note that any play from a vertex $v \notin \text{Dom}^+(X, \delta, f/g)$, that is consistent with f and $\mathit{avoid}(X, 1-\delta, g/f)$ remains out of $\text{Dom}^+(X, \delta, f/g)$ and in fact remains out of $\text{Dom}(X, \delta, f/g)$ after the first move (v itself may belong to $\text{Dom}(X, \delta, f/g)$).

We now define a monotone operator on sets of vertices of A such that the set of winning vertices for Mr. δ is the greatest fixed point of that operator. For $1 \leq i \leq m$ and X a set of the vertices of A, define

$$F_i(X) := \text{Dom}^+(X \cap B_i, \delta, f/g) \cup \text{Win}(\Gamma^i, \delta, f/g^{iX})$$

where $g^{iX} = \mathit{avoid}(X \cap B_i, 1-\delta, g/f)$. Further, let $H(X) := \bigcap_{1 \leq i \leq m} F_i(X)$.

We have to show that each F_i, and therefore H, is indeed monotone.

Lemma 7.1.38. $X \subseteq Y \implies F_i(X) \subseteq F_i(Y)$.

Proof. Let $v \in F_i(X)$ and $v \notin \text{Dom}^+(Y \cap B_i, \delta, f/g)$. Since

$$\text{Dom}^+(X \cap B_i, \delta, f/g) \subseteq \text{Dom}^+(Y \cap B_i, \delta, f/g),$$

it follows that $v \notin \text{Dom}^+(X \cap B_i, \delta, f/g)$. Therefore, $v \in \text{Win}(\Gamma^i, \delta, f/g^{iX})$. Let f' be the refinement of f that wins against g^{iX} at v. Since $v \notin \text{Dom}^+(Y \cap B_i, \delta, f/g)$, any play from vertex v that is consistent with f' and g^{iY} is also consistent with f' and g^{iX}, so f' wins against g^{iY} at v. Thus, $v \in \text{Win}(\Gamma^i, \delta, f/g^{iY})$, that is, $v \in F_i(Y)$. □

Define the sequence X_α (where α ranges over ordinals) as follows. Let X_0 be the entire vertex set of A. For $\alpha > 0$, let

$$X_\alpha := H(X_{<\alpha}) \text{ where } X_{<\alpha} := \bigcap_{\beta<\alpha} X_\beta.$$

Since H is monotone, this sequence is non-increasing, and therefore there is a first ordinal η such that $X_\eta = X_{\eta+1}$, which is the greatest fixed point of H. Let $P = X_\eta$. Then

$$P = \bigcap_{1\le i\le m} (\mathrm{Dom}^+(P \cap B_i, \delta, f/g) \cup \mathrm{Win}(\Gamma^i, \delta, f/g^{iP})).$$

Lemma 7.1.39 (Strategy for Mr. $(1-\delta)$). *There is a forgetful strategy G for Mr. $(1-\delta)$ that refines g and wins Γ against f from all vertices not in P.*

Proof. Fix α and i and consider the set $\mathrm{Win}(\Gamma^i, 1-\delta, g^{iX_{<\alpha}}/f)$. By the assumption that Theorem 7.1.34 holds for each game Γ^i, there is a forgetful strategy $g^{\alpha i}$ refining $g^{iX_{<\alpha}}$ and winning Γ^i against f from all vertices in $\mathrm{Win}(\Gamma^i, 1-\delta, g^{iX_{<\alpha}}/f)$. The strategy G coincides with g on vertices in P. For every vertex $v \notin P$ take the smallest ordinal α such that $v \notin X_\alpha$. Then $v \in X_{<\alpha} - H(X_{<\alpha})$. Thus, there exists a smallest i such that $v \notin F_i(X_{<\alpha})$. We associate the pair $\langle\alpha, i\rangle$ with v. Then $v \notin \mathrm{Dom}^+(X_{<\alpha} \cap B_i, \delta, f/g)$ and $v \notin \mathrm{Win}(\Gamma^i, \delta, f/g^{iX_{<\alpha}})$. By the assumption for Γ^i, $v \in \mathrm{Win}(\Gamma^i, 1-\delta, g^{iX_{<\alpha}}/f)$. Let $G(v) = g^{\alpha i}(v)$.

Consider a play from $v \notin P$ that is consistent with f and G. Any play that is consistent with f and $g^{\alpha i}$ after a position that ends in a vertex outside $F_i(X_{<\alpha})$ stays out of $F_i(X_{<\alpha})$. As the play continues from v, each pair associated with a vertex in the play must be lexicographically greater than or equal to the pair associated with the next vertex.

Consequently, there must be some point of the play after which the pairs associated with vertices do not change. Assume that from some point on in the play, the pair associated with each vertex is $\langle\alpha, i\rangle$. From this point on all vertices reached are in $X_{<\alpha}$, and G at these vertices, as a refinement of $g^{iX_{<\alpha}}$, not only stays out of $\mathrm{Dom}(X_{<\alpha} \cap B_i, \delta, f/g)$, but actually stays out of B_i. After this point the same winning strategy for Mr. $(1-\delta)$ for Γ^i is played. Thus, the play after this point is consistent with f and $g^{\alpha i}$ and wins Γ^i for Mr. $(1-\delta)$. This means that the play is in the complement of U_i and in the complement of $[B_i]$, and therefore wins Γ for Mr. $(1-\delta)$. □

Lemma 7.1.40 (Strategy for Mr. δ). *Mr. δ has a forgetful strategy F refining f that wins against g from all vertices $v \in P$.*

Proof. By assumption, there exists for each i a forgetful strategy f^i refining f that wins Γ^i against g^{iP} from all vertices in $\mathrm{Win}(\Gamma^i, \delta, f/g^{iP})$. Outside of P, let $F(v) = f(v)$. For $v \in P$, the strategy F for Mr. δ is defined as follows.

First we choose a *goal* at v which is an integer in the interval $[1,\ldots,m]$. Recall that each B_k is the union of some collection β_k of colors C^s. Suppose that $\text{LAR}(v) = s_1 s_2 \cdots s_n$. Replace each s_j by the values of k in increasing order such that $C^{s_j} \in \beta_k$; the result is some sequence $\sigma_1(v)$. Prune $\sigma_1(v)$ by discarding all but the rightmost occurrence of each value k in the sequence; let $\sigma_2(v)$ be the result. The goal is the leftmost value in $\sigma_2(v)$.

Before we proceed, let us make a few remarks about the goals that will be used later. If $v \in B_j - B_k$ then j appears to the right of k in $\sigma_2(v)$. Assume that, in a given play, there is a point w such that no later point v belongs to B_k. Then the part of $\sigma_2(v)$ after k cannot decrease in length. In fact, the goal chosen along this play eventually does not change and equals the index j (not necessarily equal to k) of a set B_j that is never reached after some point on this play.

Now we are ready to define F at vertex $v \in P$:

$$F(v) = \begin{cases} f^i(v) & \text{if } v \notin \text{Dom}^+(P \cap B_i, \delta, f/g) \\ \textit{decrease}(P \cap B_i, \delta, f/g)(v) & \text{otherwise,} \end{cases}$$

where i is the goal picked at v for Mr. δ .

Once any play consistent with F and g reaches a vertex in P, it remains in P. Indeed, any move made by Mr. $(1-\delta)$ according to g from a vertex in $F_i(P)$ must be to a vertex in $F_i(P)$ or P; just examine the two cases corresponding to the two summands of $F_i(P)$. Further, any move made by Mr. δ according to F from a vertex in P at which the goal is i must also be to a vertex in $F_i(P)$ or P; just examine the two cases in the definition of F. Finally notice that each $F_i(P) \subseteq P$. For, suppose $v \in F_i(P) - P$. Since $v \notin P$, Mr. $(1-\delta)$ has a strategy that forces the play to stay out of P and wins Γ from v. But since $v \in F_i(P)$, then – starting from v – it is possible for Mr. δ either to reach $P \cap B_i$ (and therefore to reach P) or else to win Γ^i and hence to win Γ f since $U_i \subseteq W_\delta$; this gives the desired contradiction. Thus the play stays in P, since $F_i(P) \subseteq P$ for each i.

We now show that F wins Γ for Mr. δ against g from any vertex $v \in P$. Consider an element of $PLAY(A)$ that is consistent with F and g after a position p that ends at vertex v in P as it continues after v.

Case 1: There is a point in the play after which the goal i does not change, and $\text{Dom}(P \cap B_i, \delta, f/g)$ is not reached. Then play after this point is that of f^i against g^{iP} and wins Γ^i (and hence Γ) for Mr. δ.

Case 2: There is no point after which the goal i remains the same. This means that the play keeps hitting all of the sets B_i for $1 \le i \le m$, and this constitutes a win for Mr. δ.

Case 3: There is a point after which the goal i does not change, but there is no point after which $\text{Dom}(P \cap B_i, \delta, f/g)$ is not reached. Then it must still be the case (see the remarks about the goals just before the definition of F) that the play keeps hitting all of the sets B_i for $1 \leq i \leq m$, and this constitutes a win for Mr. δ. □

We thus have exhibited a forgetful strategy for each player that is a refinement of his initial strategy, and that wins against the opponent's initial strategy at any of his winning vertices. Also $P = \mathrm{Win}(\Gamma, \delta, f/g)$, and any vertex not in P is in $\mathrm{Win}(\Gamma, 1-\delta, g/f)$. This proves Lemma 7.1.36 and thus the Forgetful Determinacy Theorem.

7.2 The Monadic Second-Order Theory of One Unary Function

In this section we prove that the monadic theory (and the weak monadic theory) of a unary function with a countable domain is decidable. It immediately follows that the satisfiability and finite satisfiability problems for the class $[all, (\omega), (1)]_=$ are also decidable.

We will then show that this theory is not elementary recursive, i.e. it cannot be decided in k-fold exponential time, for any fixed k. In fact we prove a stronger result, giving an explicit non elementary recursive lower bound even for the *first order theory of one unary function.*

Rabin's Tree Theorem easily generalizes to SnS, the monadic theory of the infinite n-ary tree. As we will show next, it also implies the decidability of SωS, the monadic theory of the countably branching tree. To avoid an infinite vocabulary one usually works with with the structure

$$T^\omega := (\omega^*, <, \prec)$$

where $<$ is the prefix relation and $\prec$ the lexicographic order on ω^*.

Note that (the graph of) each successor function $succ_i$ ($i \in \omega$) is first-order definable on T^ω, by a formula saying that $x \prec y$ and there are precisely i elements z satisfying $x \prec z \wedge z \prec y$.

Example 7.2.1. For future reference we exhibit some definable relations on T^ω.

1. The formula

$$\alpha(Z) := Z\lambda \wedge \forall x(Zx \rightarrow Zx0) \wedge \forall x \forall y(Zy \wedge x < y \rightarrow x0 \leq y)$$

   expresses that $Z = \{0\}^*$.
2. The class of all sets $C \subseteq \{0^n10^m : n, m < \omega\}$ is definable on T^ω by

$$\begin{aligned}\beta(Z) \quad := \quad & \exists Y(\alpha(Y) \wedge \forall x \exists z(Zx \rightarrow Yz \wedge z1 \leq x \wedge \\ & \forall y(z1 \leq y < x \rightarrow y0 \leq x))).\end{aligned}$$

3. Finally the class $\mathcal{S}$ of all sets $C \subseteq \{0^n10^m : n, m < \omega\}$ that contain for every n at most one word of the form 0^n10^m, is defined by

$$\gamma(Z) := \beta(Z) \wedge \forall x \forall y \forall z (Zx \wedge Zy \wedge z1 < x \wedge z1 < y \rightarrow x = y).$$

Theorem 7.2.2. SωS, *the monadic theory of* T^ω, *is decidable.*

Proof. Let $A := \{\lambda\} \cup \{1^{n_1}01^{n_2}0 \cdots 01^{n_k} : 1 \leq k,\ 1 \leq n_i\} \subseteq \{0,1\}^*$. Note that $T^\omega \simeq (A, <|_A, \prec|_A)$ where $<|_A$ and $\prec|_A$ are the restrictions of $<$ and $\prec$ to A. Further A is first-order definable on T^2, e.g. by the formula

$$\delta(Z) := Z\lambda \wedge \neg Z0 \wedge \forall x(\neg Zx00 \wedge (Zx \rightarrow (Zx1 \wedge Zx10))).$$

Exercise 7.2.3. Prove that $\delta(Z)$ indeed defines A.

Thus, every monadic formula ψ in the language of T^ω can be translated into an S2S-formula φ such that

$$T^\omega \models \psi \iff T^2 \models \varphi.$$

Hence, the decidability of SωS follows from the decidability of S2S. □

7.2.1 Decidability Results for One Unary Function

We now prove that Rabin's Tree Theorem implies the decidability of the following theories:

(i) The monadic theory of one unary function over a countable domain.
(ii) The weak monadic theory of one unary function.

The decidability of the satisfiability and the finite satisfiability problems for $[all, (\omega), (1)]_=$ is a simple corollary of these results.

Let $\mathcal{K}_f^\omega$ be the class of structures $\mathfrak{A} = (A, f)$ with one unary function over a countable (i.e. finite or countably infinite) domain. As in Sect. 6.4.2 we refer to a structure $\mathfrak{A} = (A, f)$ as an *algebra*. Further, if $B \subseteq A$ is closed under f, we write (B, f) for the subalgebra with universe B, i.e. we do not distinguish notationally between f and its restriction to B.

We want to prove that $\mathrm{Th}_{\mathrm{mon}}(\mathcal{K}_f^\omega)$, the monadic theory of one unary function with a countable domain, is can be interpreted in SωS. To do this we make some general observations on algebras.

Elements a, b of an algebra $\mathfrak{A} = (A, f)$ are *connected* if $f^n(a) = f^m(b)$ for some $n, m \in \mathbb{N}$. Connectedness is an equivalence relation and every algebra can be written as the disjoint union of its connected components. We call a countable disjoint union $\mathfrak{A} = \bigcup_{n<\omega} \mathfrak{A}_n$ of algebras an *ω-sum.*

The algebras $(\mathbb{Z}_n, succ)$ and $(\omega, succ)$ are called the *basic algebras*.

Lemma 7.2.4. *For every connected algebra* $\mathfrak{A}$ *one of the following two alternatives holds:*

(i) *$\mathfrak{A}$ contains precisely one basic subalgebra and this subalgebra is finite.*
(ii) *$\mathfrak{A}$ contains (infinitely many) copies of $(\omega, succ)$, but no finite basic subalgebra.*

Proof. Let a be an element of an algebra $\mathfrak{A}$ and consider the set $T(a) = \{f^n(a) : n \in \mathbb{N}\}$. Obviously $(T(a), f)$ is a subalgebra of $\mathfrak{A}$. If $f^n(a) \neq f^m(a)$ for all $n \neq m$ then $(T(a), f) \simeq (\omega, succ)$. Otherwise, let (m, n) be the lexicographically minimal pair of natural numbers such that $f^m(a) = f^{m+n}(a)$, and let $S(a) = \{f^m(a), f^{m+1}(a), \ldots, f^{m+n-1}(a)\}$. Now, $(S(a), f)$ is a subalgebra of $\mathfrak{A}$ which is isomorphic to $(\mathbb{Z}_n, succ)$. This proves that every algebra contains a basic subalgebra.

Now suppose that $\mathfrak{A}$ contains two distinct basic subalgebras $\mathfrak{B}, \mathfrak{C}$. The two subalgebras have a common element a; otherwise they would lie in different connected components of A which is impossible. If $\mathfrak{B}$ is finite, then it consists of elements $f^k(a)$ and therefore $\mathfrak{C}$ includes $\mathfrak{B}$. But then $\mathfrak{C}$ properly includes $\mathfrak{B}$ and therefore cannot be basic. By symmetry, $\mathfrak{C}$ cannot be finite either. Thus both subalgebras are of the type $(\omega, succ)$, but an algebra of that type has infinitely many isomorphic subalgebras. □

Definition 7.2.5. The *enveloping algebra* $(\mathfrak{B}, g)$ of a basic algebra $\mathfrak{A} = (A, f)$ is defined as follows. Let $N = \omega - \{0\}$, $B = AN^*$ and

$$\begin{aligned} g(an_1 \cdots n_k n_{k+1}) &:= an_1 \cdots n_k \\ g(a) &:= f(a). \end{aligned}$$

Lemma 7.2.6. *Let $\mathfrak{A}, \mathfrak{A}'$ be basic algebras of the same type, let $\mathfrak{B}$ be the enveloping algebra of $\mathfrak{A}$, and $\mathfrak{C}$ be any countable connected extension of $\mathfrak{A}'$. Then every isomorphism $\pi : \mathfrak{A}' \to \mathfrak{A}$ can be extended to an embedding, i.e. an injective homomorphism $\hat{\pi} : \mathfrak{C} \hookrightarrow \mathfrak{B}$. Thus, every countable algebra can be embedded into an ω-sum of enveloping algebras.*

Proof. Let $\mathfrak{C} = (C, f)$ and $\mathfrak{B} = (B, g)$. For $x, y \in C - A'$, define $x < y$ if $x = f^k(y)$ for some $k > 0$. It is easy to see that this is a partial order where each y is preceded by only finitely many x's. It follows that one can choose an enumeration $c_0, c_1, \ldots,$ of C such that $j < i$ whenever $f(c_i) = c_j$ and c_i is not contained in $\mathfrak{A}'$. Suppose that $\hat{\pi}(c_j)$ is already defined for all $j < i$. If c_i belongs to $\mathfrak{A}'$, let $\hat{\pi}(c_i) = \pi(c_i)$. If c_i does not belong to $\mathfrak{A}'$, then by assumption $\hat{\pi}$ is already defined on $f(c_i)$; let $\hat{\pi}(f(c_i)) = w \in AN^*$. Take the minimal element $k \in N$ such that $wk \notin \{\hat{\pi}(c_j) : j < i\}$ and set $\hat{\pi}(c_i) := wk$. By construction $\hat{\pi}$ is one-one and by the definition of an enveloping algebra $g(wk) = w$ so $\hat{\pi}$ is a homomorphism. □

Lemma 7.2.7. *There exists a monadic formula $\chi(x, y, Z)$ in the language of T^ω with the following properties: For all $C \subseteq \omega^*$, the binary relation*

$$F_C := \{(a, b) : T^\omega \models \chi[a, b, C]\} \subseteq \omega^* \times \omega^*$$

is either the empty set or the graph of a unary function f_C on a domain $B_C \subseteq \omega^*$ such that (B_C, f_C) is an ω-sum of enveloping algebras. Further, every ω-sum of enveloping algebras, is isomorphic to some (B_C, f_C), for $C \subseteq \omega^*$.

Proof. Let $\mathcal{S}$ be the class of all $C \subseteq \omega^*$ such that

(i) $C \subseteq \{0^n 10^m : n, m < \omega\}$.
(ii) For every n, C contains at most one word of the form $0^n 10^m$.

According to Example 7.2.1, $\mathcal{S}$ is definable on T^ω.

Given $C \in \mathcal{S}$, let $A_{n,C} := \{0^n 10^i : i \leq m\}$ if $0^n 10^m \in C$ and $A_{n,C} := \{0^n 10^i : i \leq \omega\}$ if C contains no word of the form $0^n 10^m$. Further, let

$$B_C := \bigcup_{n<\omega} A_{n,C} N^*$$

where, as above, $N = \omega - \{0\}$. Note that B_C is definable in terms of C on T^ω, i.e. there exists a monadic formula $\delta(y, Z)$ such that for all $C \in \mathcal{S}$

$$B_C = \{b \in \omega^* : T^\omega \models \delta[b, C]\}.$$

Exercise 7.2.8. Construct such a formula $\delta(y, Z)$.

The function $f_C : B_C \to B_C$ is defined as follows: If $a = 0^n 10^i w$ for some $w \in N^* - \{\varepsilon\}$ then $f(a)$ is the predecessor of a; if $a = 0^n 10^i \in C$ then $f(a) = 0^n 1$; finally, if $a = 0^n 10^i \notin C$ then $f(a) = a0 = 0^n 10^{i+1}$.

It is easy to see that the set of triples (a, b, C) such that $C \in \mathcal{S}$, $a \in B_C$ and $b = f_C(a)$ is definable by an SωS-formula $\chi(x, y, Z)$.

Exercise 7.2.9. Construct $\chi(x, y, Z)$ explicitly.

Note that $(A_{n,C}, f_C) \simeq (Z/m\mathbb{Z}, succ)$ if $0^n 10^m \in C$, and $(A_{n,C}, f_C) \simeq (\omega, succ)$ if C contains no word $0^n 10^m$. As a consequence (B_C, f_C) is an ω-sum of enveloping algebras. Conversely for every ω-sum $\mathfrak{A} = \bigcup_{n<\omega} \mathfrak{B}_n$ of enveloping algebras, let

$$C = \{0^n 10^m : \mathfrak{B}_n \text{ is an enveloping algebra for } (\mathbb{Z}_m, succ)\}.$$

Then $(B_C, f_C) \simeq \mathfrak{A}$. □

Theorem 7.2.10 (Rabin). $\mathrm{Th}_{\mathrm{mon}}(\mathcal{K}_f^\omega)$, *the monadic theory of a unary function on a countable domain, is decidable.*

Proof. Let $\chi(x, y, Z)$ be the SωS-formula constructed in the previous lemma, and

$$\alpha(X, Z) := \forall x \exists y (Xx \to Xy \wedge \chi(x, y, Z)).$$

For all non-empty $A, C \subseteq \omega^*$ we have that $T^\omega \models \alpha[A, C]$ if and only if (A, f_C) is a subalgebra of (B_C, f_C).

Since, by Lemma 7.2.6, every countable algebra can be embedded in an ω-sum of enveloping algebras, it follows that for every countable algebra $\mathfrak{A}$, there exist $A, C \subseteq \omega^*$ such that $T^\omega \models \alpha[A, C]$ and $\mathfrak{A} \simeq (A, f_C)$.

Now, let ψ be a sentence of the monadic logic of one unary function. Without loss of generality we can assume that ψ is in term-reduced form, i.e., the function f appears only in atoms of form $fx = y$ where x, y are variables. To translate ψ into an SωS-formula $\varphi(X, Z)$, replace all atoms $fx = y$ by $\chi(x, y, Z)$, relativize all individual quantifiers to X and all set quantifiers to subsets of X. Then, ψ is a theorem of $\mathrm{Th}_{\mathrm{mon}}(\mathcal{K}_f^\omega)$ if and only if

$$T^\omega \models \forall X \forall Z(\alpha(X, Z) \rightarrow \varphi(X, Z)).$$

This proves that $\mathrm{Th}_{\mathrm{mon}}(\mathcal{K}_f^\omega)$ is decidable. □

Corollary 7.2.11. *The weak monadic theory of a unary function on a countable domain is decidable.*

Proof. Finiteness is definable in SωS (see Example 7.1.4). Hence we can modify the translation from ψ to $\varphi(X, Z)$ by relativizing all set quantifiers to *finite* subsets of X. In this way also the weak monadic theory of one unary function is interpreted in SωS. □

Notice that the Löwenheim-Skolem Theorem generalizes to monadic second-order logic; if a monadic sentence has model then it has a (at most) countable model. It follows that the weak monadic theory of unary function is decidable.

Corollary 7.2.12. *The satisfiability and the finite satisfiability problems for the standard class* $[\mathit{all}, (\omega), (1)]_=$ *are decidable.*

Proof. Let ψ be a first-order sentence with monadic predicates $Z_1, \ldots, Z_m$ and one unary function. If ψ is satisfiable then, by the Löwenheim-Skolem Theorem, ψ has a countable model. Thus, ψ is unsatisfiable if and only if $\forall Z_1 \cdots \forall Z_m \neg\psi$ is a theorem of the $\mathrm{Th}_{\mathrm{mon}}(\mathcal{K}_f^\omega)$.

Further, ψ has no finite model if and only

$$\exists Y \forall z Yz \rightarrow \forall Z_1 \cdots \forall Z_m \neg\psi$$

is a a theorem of the weak monadic theory of one unary function. (The premise tells us that Y is the whole universe.) By Theorem 7.2.10 and Corollary 7.2.11, these problems are decidable. □

7.2.2 The Theory of One Unary Function is not Elementary Recursive

We denote by $Sat(\mathcal{K}_f)$ the set of satisfiable formulae in the first order theory of one unary function, i.e. in our usual notation $Sat[\mathit{all}, (0), (1)]_=$.

Definition 7.2.13. Let $\exp_k(n)$ be the k-fold iterated exponential function with base two:

$$\exp_0(n) = n; \qquad \exp_{k+1}(n) = 2^{\exp_k(n)}$$

and furthermore, let

$$\exp_\infty(n) := \exp_n(1) = 2^{2^{\cdot^{\cdot^{2}}}} \Big\}n$$

be the 'tower of twos' function. A problem is said to be *elementary recursive* if and only if it it decidable in time $O(\exp_k(n))$ for some fixed $k \in \mathbb{N}$.

We will show that $Sat(\mathcal{K}_f)$ is not elementary recursive. We first prove a technical lemma that we need in the proof, to the effect that, in first-order logic with equality, the number of occurrences of any predicate in a given formula can be reduced to one:

Lemma 7.2.14. *There exists a polynomial time algorithm which, given a prenex first-order sentence ψ and a predicate P, transforms ψ into a new sentence ψ' of the same vocabulary with the following properties*

(i) ψ' contains at most one occurrence of P.
(ii) $|\psi'| = O(|\psi|)$.
(iii) ψ and ψ' are equivalent on every structure of cardinality at least two.

Proof. Let $\psi := Q_1x_1 \cdots Q_kx_k\varphi$ and $P\bar{y}_1, \dots P\bar{y}_m$ be the list of atoms in φ containing the predicate P; every $\bar{y}_i$ is a r-tuple of variables from $x_1, \dots, x_k$. The idea of the proof is to replace every atom $P\bar{y}_i$ by an equality $z = z_i$ where $z, z_1, \dots, z_m$ do not occur in φ. The equivalence of these equalities with the original atoms is asserted by the formula

$$\alpha := \forall u \forall v_1 \cdots \forall v_r([\bigvee_{i=1}^{m} (u = z_i \wedge \bigwedge_{j=1}^{r} v_j = y_{ij})] \to (z = u \leftrightarrow P\bar{v})).$$

This formula has just one occurrence of P and its length is linear in the length of ψ; thus the formula

$$\psi' := Q_1x_1 \cdots Q_kx_k \exists z \exists z_1 \cdots \exists z_m (\alpha \wedge \varphi[P\bar{y}_i/(z = z_i)])$$

has the required properties. Here $\varphi[P\bar{y}_i/(z = z_i)]$ is the formula obtained by replacing $P\bar{y}_i$ by $z = z_i$. □

Theorem 7.2.15. *There exists a constant $c > 0$ such that*

$$Sat(\mathcal{K}_f) \notin \text{NTIME}(\exp_\infty(cn/\log n)).$$

Proof. As in most of our lower bound proofs we use a reduction from an appropriate domino problem: We will show that for every domino system $\mathcal{D}$

$$\text{DOMINO}(\mathcal{D}, \exp_\infty(n)) \leq_{n \log n} Sat(\mathcal{K}_f).$$

Suppose that $\mathcal{D}$ contains the tiles $d_1, \ldots, d_r$. First we express the tiling conditions in a language with $r+1$ binary predicates $P_0, \ldots, P_r$ together with equality. Given an initial condition $w = w_0, \ldots, w_{n-1} \in D^*$, there exists a formula ψ of this language which has a model of cardinality t if and only if $\mathcal{D}$ tiles $Z(t)$ with initial condition w (where t is an arbitrary number not smaller than n).

Intuitively, the interpretation of P_0 is the graph of the successor function and for $i > 0$, P_i contains the points that are tiled by d_i. We abstain from writing down ψ in full detail; it asserts that each model $\mathfrak{B} = (B, P_0, \ldots, P_r)$ looks as follows:

(i) P_0 is the graph of a permutation of B with only one cycle;
(ii) the predicates $(P_i)_{1 \leq i \leq r}$ partition $B \times B$;
(iii) the adjacency conditions imposed by H and V are satisfied;
(iv) there is an element $0 \in B$ such that the points $(0,0), \ldots, (0, n-1)$ are tiled by $w_0, \ldots, w_{n-1}$. Here $0, \ldots, n-1$ are the first n points with respect to the order defined by P_0 and 0.

Only *(iv)* depends on w and on n. The formula ψ can be constructed in such a way that it has length $O(n \log n)$ and (in view of Lemma 7.2.14 above) that every relation symbol P_i occurs only once in ψ. In fact by using the tree representation of the numbers up to n (see Lemma 6.1.10) we could do even with length $O(n)$ at the expense of two new relation symbols.

Exercise 7.2.16. Construct ψ explicitly.

On the second step of the proof we construct a sequence of first-order formulae of length $O(n \log n)$ to interpret structures $\mathfrak{B} = (B, P_0, \ldots, P_r)$ of cardinality at most $\exp_\infty(n)$ in algebras $\mathfrak{A} = (A, f)$:

Let $\mathfrak{A} = (A, f)$ be a model of the formula

$$\alpha :- \exists x \forall y (fx = x \wedge f^{n+1}y = x).$$

Then $\mathfrak{A}$ is a tree of height at most $n+1$ with f mapping every node to its parent. Each node $a \in A$ defines the subtree T_a containing all descendants of a. The height of a is meant to be the height of T_a. For $0 \leq m \leq n$ we inductively define equivalence relations E_m on the nodes of height at most m: All leaves are E_0-equivalent. Two nodes are E_m-equivalent, if for every E_{m-1}-equivalence class K they either both have no child in K or both have at least one child in K. Note that for each m there may exist up to $\exp_\infty(m)$ E_m-equivalence classes.

Exercise 7.2.17. Prove the last claim. Notice that there is exactly one E_0 equivalence class. There are at most two E_1 equivalence classes: one comprises nodes of height zero and the other comprises nodes of height one.

We construct formulae $\beta_m(x, y)$ expressing that x and y have height at most m and that they are E_m-equivalent:

$$\begin{aligned}
\beta_0(x,y) &:= \forall z(fz \neq x \wedge fz \neq y) \\
\beta_m(x,y) &:= \forall z(f^{m+1}z \neq x \wedge f^{m+1}z \neq y) \wedge \\
& \forall u \exists v((fu = x \to fv = y) \wedge (fu = y \to fv = x) \wedge \\
& \qquad \beta_{m-1}(u,v)).
\end{aligned}$$

Translating β_n into prenex normal form gives a formula of length $O(n \log n)$; if we wouldn't insist on prenex normal form we could do with length $O(n)$ by using a fixed stock of variables (independent of n).

Now take the formulae

$$\begin{aligned}
\delta(x) &:= (fx \neq x) \wedge \\
& (f^2x = fx) \wedge (\forall y.\ fy = x)(\forall z.\ fz = x)(\beta_{n-1}(y,z) \to y = z) \\
\gamma &:= (\forall x.\ \delta(x))(\forall y.\ \delta(y))(\beta_n(x,y) \to (x = y)).
\end{aligned}$$

Here $(\forall x.\ \varphi)\psi$ is used as an abbreviation for $\forall x(\varphi \to \psi)$ (the quantifier $\forall x$ is relativized to φ). The formula $\delta(x)$ expresses that x is a child of the root and that no two children of x are E_{n-1}-equivalent. If α and γ are true in $\mathfrak{A}$ then the set

$$B := \{a \in A : \mathfrak{A} \models \delta[a]\}$$

has cardinality at most $\exp_\infty(n)$.

Next we encode $r+1$ binary relations on B using the following idea: For $i = 0, \ldots, r$ define functions $g_i : \{0,1\} \times \{0,1\} \to \mathbb{N}$ by

$$g_i(j,k) = 2 + 4i + 2j + k.$$

The images of different g_i are disjoint. It is not difficult (but a bit lengthy) to write down formulae $\pi_i(x, y)$ which state:

> There exists a z which is a child of the root such that for every E_{n-1}-equivalence class K and for all j, $k \in \{0,1\}$, the following holds: If x has j and y has k children in K then z has precisely $g_i(j,k)$ children in K.

Let $P_i := \{(a,b) \in B \times B \mid \mathfrak{A} \models \pi_i[a,b]\}$. We thus have defined a first-order interpretation I, that associates with every algebra $\mathfrak{A}$ a structure $I(\mathfrak{A}) = (B, P_0, \ldots, P_r)$ with $r+1$ binary relations. For every structure $\mathfrak{C}$ with $r+1$ binary relations, there exists a algebra $\mathfrak{A}$ such that $\mathfrak{A} \models \alpha \wedge \gamma$ and $I(\mathfrak{A}) \simeq \mathfrak{C}$. Moreover the formulae $\pi_i(x, y)$ can be constructed in such a way that they contain just one occurrence of β_{n-1} (see Lemma 7.2.14); they therefore have length $O(n \log n)$.

Given an arbitrary formula η with binary relations $P_0, \ldots, P_r$, let η' be the formula obtained by relativizing every quantifier to a new unary predicate D and by transforming the formula in such a way that D occurs only once. Then translate η into the formula

$$\varphi := \alpha \wedge \gamma \wedge \eta'[Dx/\delta(x), P_ixy/\pi_i(x,y)]$$

where, as usual, $\eta'[Dx/\delta(x), P_ixy/\pi_i(x,y)]$ is the formula obtained by replacing atoms Dx by $\delta(x)$ and atoms P_ixy by $\pi_i(x,y)$ for arbitrary variables x, y. The resulting formula φ contains one unary function and no relation symbols; it is satisfiable if and only if η has a model of size at most $\exp_\infty(n)$.

If we apply this translation to the formula ψ constructed above, the resulting formula φ has length $O(n \log n)$ because it contains at most one occurrence of α, γ, δ and $\pi_0, \ldots, \pi_r$ and because the lengths of these formulae are bounded by $O(n \log n)$. Furthermore φ is satisfiable iff ψ has a model $\mathfrak{B}$ with at most $\exp_\infty(n)$ elements which is true if and only if $w \in \text{DOMINO}(\mathcal{D}, \exp_\infty(n))$. □

7.3 The Shelah Class

The *Shelah class* is the standard fragment $[\exists^*\forall\exists^*, all, (1)]_=$ of first-order logic with equality. Thus, a *Shelah sentence* is a prenex sentence φ of first-order logic with equality such that:

- The prefix of φ is dominated by $\exists^*\forall\exists^*$.
- The vocabulary of φ contains no function symbols of arity ≥ 2 and at most one unary function symbol. There is no restriction on the number of predicates or their arities.

Theorem 7.3.1. *Both the satisfiability and the finite satisfiability problems for Shelah sentences are decidable.*

The (finite) satisfiability problem is reduced to the (finite) satisfiability problem for the class $[\exists^*\forall\exists^*, (\omega), (1)]_=$ whose decidability follows from the decidability of the (finite) satisfiability problem for the class $[all, (\omega), (1)]_=$ proved above (Corollary 7.2.12). The Shelah class has infinity axioms (see the beginning of this chapter), but the two reductions are very similar. To avoid unnecessary repetitions, we prove only that the satisfiability problem for Shelah class is decidable. However, the proof remains valid if the term satisfiability is interpreted as finite satisfiability.

7.3.1 Algebras with One Unary Operation

Throughout this section, we restrict attention to first-order formulae (that may use the equality sign) whose only function symbol of positive arity is a unary function symbol Par (an allusion to "parent"). For future reference, let FO(Par) be the corresponding fragment of first-order logic with equality.

Accordingly a *(total) algebra* is a nonempty set with a unary operation Par. A *partial algebra* is a nonempty set with a partial unary operation Par. We introduce some terminology and prove simple properties of partial algebras.

Suppose that $\mathfrak{A}$ is a partial algebra and x, y, z, with or without subscripts, are elements of A. If $y = \mathrm{Par}(x)$, then y is the *parent* of x and x is a *child* of y. Define $\mathrm{Par}^0(x) = x$ and $\mathrm{Par}^{i+1}(x) = \mathrm{Par}(\mathrm{Par}^i(x))$. If $y = \mathrm{Par}^i(x)$ for some $i \geq 0$, then x is *younger* than y and a *descendent* of y, and y is *older* then x and an *ancestor* of x.

We write $x \leq y$ if x is younger than y. We write $x < y$ if $y = \mathrm{Par}^i(x)$ for some $i > 0$. Clearly, $<$ is transitive. However it is not necessarily irreflexive. An element x *cyclic* if $x < x$ and *acyclic* otherwise. Clearly, $<$ is a partial order on acyclic elements.

Elements x, y are *comparable* if $x \leq y$ or $y \leq x$. A sequence of elements $x_0, \ldots, x_k$ is a *path* from x_1 to x_k if, for every $i < k$, either $x_{i+1} = \mathrm{Par}(x_i)$ or $x_i = \mathrm{Par}(x_{i+1})$. Here k may be zero. Elements x, y are *connected* if there is a path from x to y. Connectivity is an equivalence relation. The equivalence classes are the *components* of $\mathfrak{A}$. $\mathfrak{A}$ is *connected* if it has only one component.

Lemma 7.3.2. *If $x_0, x_1, \ldots, x_l$ is a shortest path from $x = x_0$ to $y = x_l$, then there is $k \leq l$, such that*
(i) $x_{i+1} = \mathrm{Par}(x_i)$ for $i < k$, and (ii) $x_i = \mathrm{Par}(x_{i+1})$ for $i \geq k$.

Proof. Suppose that $(x_0, x_1, \ldots, x_l)$ is a shortest path from x to y. Call a pair $e_i = (x_i, x_{i+1})$ *positive* (resp. *negative*) if $x_{i+1} = \mathrm{Par}(x_i)$ (resp. $x_i = \mathrm{Par}(x_{i+1})$). If e_i is negative and e_{i+1} is positive then $x_{i-1} = \mathrm{Par}(x_i) = x_{i+1}$ which allows us to shorten the given path. Thus, no negative e_i is followed by a positive e_{i+1}. If all e_i are positive then y is an ancestor of x and the desired $k = l$. Otherwise the desired k is the least number such that e_k is negative. □

A pair e_i may be both positive and negative. Consider for example the path $0, 1, 2, 3$ in the case when $\mathrm{Par}(0) = 1$, $\mathrm{Par}(1) = 2$, $\mathrm{Par}(2) = 1$, and $\mathrm{Par}(3) = 2$.

Corollary 7.3.3. *If x, y are connected then they have a common ancestor.*

Lemma 7.3.4. *Every connected partial algebra $\mathfrak{A}$ has the following properties.*

(i) *If x, y are cyclic then $x < y$.*
(ii) *The cyclic elements together with the inherited operation Par form a connected total algebra (called the* cycle *of $\mathfrak{A}$).*
(iii) *If x is any cyclic element and n is the minimal number such that $\mathrm{Par}^n(x) = x$ then the cycle consists of the n elements $\mathrm{Par}^i(x)$, $i < n$.*

(iv) Every cyclic element has exactly one cyclic child.
(v) Every element is a descendent of any cyclic element.
(vi) If there is a cyclic element then $\mathfrak{A}$ is total.

Proof. *(i):* Let z be a common ancestor of x and y, so that $\mathrm{Par}^k(x) = \mathrm{Par}^l(y) = z$ for some k and l. Since y is cyclic, there is $n > 0$ such that $\mathrm{Par}^n(y) = y$. Clearly, $\mathrm{Par}^{in}(y) = y$ for every i. Choose i so that $in \geq l$. Then

$$\mathrm{Par}^{k+(in-l)}(x) = \mathrm{Par}^{in-l}(z) = \mathrm{Par}^{in-l}(\mathrm{Par}^l(y)) = \mathrm{Par}^{in}(y) = y.$$

(ii): If $x = \mathrm{Par}^n(x)$ for some $n > 0$, then x has a parent $y = \mathrm{Par}(x)$. Furthermore, y is cyclic because

$$\mathrm{Par}^n(y) = \mathrm{Par}^{n+1}(x) = \mathrm{Par}(\mathrm{Par}^n(x)) = \mathrm{Par}(x) = y.$$

(iii): Let x and n be as in *(iii)*. By *(i)*, every cyclic element is an ancestor of x and therefore equals to some $\mathrm{Par}^i(x)$.

(iv): Let x and n be as in *(iii)* and y be any cyclic child of x. By *(iii)*, $y = \mathrm{Par}^i(x)$ for some $i < n$. By the minimality of n, $i = n - 1$.

(v): Let x be a cyclic element and y any element. By *(ii)*, the common ancestor z of x and y is cyclic. By *(i)*, $z \leq x$. By transitivity, $y \leq x$.

(vi): By *(ii)*, every cyclic element has a parent. By *(iv)*, every acyclic element has a parent. □

In addition to the family terminology, we have the following geometric picture in mind. If x, y are acyclic and $x < y$ then y is higher than x; if a component has cyclic elements, they form a horizontal plateau above all acyclic elements.

Lemma 7.3.5. *There are at most two shortest paths from x to y. If $x_0, \dots, x_l$ and $y_0, \dots, y_l$ are distinct shortest paths from x to y then there are $m < n \leq l$ such that:*

(i) $x_i = y_i = \mathrm{Par}^i(x)$ if $i \leq m$, and $x_i = y_i = \mathrm{Par}^{l-i}(y)$ if $i \geq n$,
(ii) $x_m = \mathrm{Par}^{n-m}(x_n)$ and $x_n = \mathrm{Par}^{n-m}(x_m)$.

In other words, both paths rise to a cycle of even length, traverse the cycle in different directions to the opposite element and then descend together.

Proof. By Lemma 7.3.2, there exist k_1, k_2 such that $x_{i+1} = \mathrm{Par}(x_i)$ for $i < k_1$, and $x_i = \mathrm{Par}(x_{i+1})$ for $i \geq k_1$, and $y_{i+1} = \mathrm{Par}(y_i)$ for $i < k_2$, and $y_i = \mathrm{Par}(y_{i+1})$ for $i \geq k_2$. Set $m = \min\{k_1, k_2\}$ and $n = \max\{k_1, k_2\}$.

(i): Check by induction on i that $x_i = y_i$ for $i \leq m$. Similarly $x_i = y_i$ for $i \geq n$.

(ii): If $m = n$ then the two paths coincide. Thus $m < n$. Without loss of generality, $k_1 = m$ and $k_2 = n$. By the definition of k_1, k_2, we have $x_m = \mathrm{Par}^{n-m}(x_n)$, $y_n = \mathrm{Par}^{n-m}(y_m)$. □

Lemma 7.3.6. *If x, y are connected then, for each l, there are at most three ancestors of y on distance l from x.*

Proof. It suffices to prove the lemma in the special case when $x \geq y$. Indeed, let x be arbitrary and P be a be a shortest path $x = x_0, x_1, \ldots, x_k$ such that $x_k \geq y$. Then $x_i \not\geq x_k$ for all $i < k$. By Lemma 7.3.2, $x_i = \text{Par}^i(x)$ for $i \leq k$. By Claim *(v)* of Lemma 7.3.4, every x_i is acyclic. It easy to see that P is an initial segment of any shortest path from x to an ancestor of y. Thus it suffices to prove that there are at most three ancestors of y on distance $l - k$ from x_k.

In the rest of the proof, we suppose that $x \geq y$. For brevity, call ancestors of y red. Red elements form a line or a cycle with a handle. The handle is the red line rising to one of the red cyclic elements; call that red cyclic element *central*.

By Lemma 7.3.2, a shortest path from x to a red element consists of red elements. If red elements form a cycle without a handle, then every red element has at most one red child. If the cycle is small then there are no red elements on distance l from x. If the cycle is sufficiently large, then there are at most two red elements on distance l from x. One is given by a positive path (that is a path $x_0, \ldots, x_l$ where each $x_{i+1} = \text{Par}(x_i)$) from x and the other by a negative path from x. If red elements form a line then there are at most two red elements on distance l from x; one is given by a positive path and the other, if it exists, is given by a negative path. Similarly, there are at most two red elements of distance l from x if red elements form a cycle with a handle and x belongs to the handle and the distance from x to the central red element is $\leq l$. The remaining case is this: red elements form a cycle with a handle and x is either cyclic or acyclic but close to the central red element. In either scenario, there are at most two cyclic red elements on distance l from x and at most one acyclic red element on distance l from x. □

Finally we will say that z is *between* elements x and y if the following conditions are satisfied:

- $x < z < y$.
- If x is cyclic and $l = \min\{i : \text{Par}^i(x) = y\}$ then $z = \text{Par}^j(x)$ for some positive $j < l$.

Let $(x, y]$ be the set of elements z such that either $z = y$ or z is between x and y. Similarly, let $[x, y]$ be the set of elements z such that $z = x$, $z = y$ or z is between x and y.

7.3.2 Canonic Sentences

Sentences as Sets of Clauses. As usual, the *depth* of a term is defined by induction: The depth of an individual constant or variable is zero, and $\text{Depth}(\text{Par}(t)) = 1 + \text{Depth}(t)$. A *literal* is an atomic formula or the negation of such. If α is an atomic formula then $+\alpha$ is α and $-\alpha$ is the negation of α. The *constituent terms* of an atomic formula $P(t_1, \dots, t_r)$ are the terms $t_1, \dots, t_r$, and the *constituent terms* of a non-atomic formula φ are the constituent terms of the atomic subformulae of φ. A predicate (that is a relation name) is *proper* if it is different from the equality sign. A literal $\pm P(t_1, \dots, t_r)$ is *proper* if the predicate P is so.

Definition 7.3.7. A literal α is *admissible* if it has one of the three following forms:

- a proper literal with all constituent terms of depth 0,
- an inequality with both constituent terms of depth 0,
- an equality $t_1 = \mathrm{Par}(t_2)$ where each t_i is of depth 0.

We reserve the variable u to be used with the universal quantifier; it is the *universal variable*. All other variables are *existential*. (We will not consider formulae with more than one universal quantifier in this section.)

Strings of any ordered alphabet are ordered lexicographically (if s_1 is a proper prefix of s_2 then s_1 precedes s_2 in the lexicographical order). The lexicographical order is total. Without loss of generality, we suppose that constants and variables are strings in some finite alphabet, that every constants lexicographically precedes every variable and that the universal variable u is the lexicographically first variable.

Definition 7.3.8. A *clause* K is a conjunction of admissible literals (called the *constituent literals* of K) satisfying the following conditions:

- If distinct variables v_1, v_2 occur in K and v_1 lexicographically precedes v_2 then K contains the inequality $v_1 \neq v_2$.
- If an existential variable v and a constant c occur in K then K contains the literal $c \neq v$.
- If a constant c occurs in K then K contains either the equality $c = u$ or the inequality $c \neq u$.

If a variable v occurs in a clause K and a constant d does not then $K(v/d)$ is the clause obtained from K by replacing v with d and making the obvious additional changes to ensure that the result is a clause. Let $\mathrm{EV}(K)$ be the collection of existential variables of a clause K.

Definition 7.3.9. The E-*closure* $\bar{K}$ of a clause K with existential variables $v_1, \dots, v_m$ is the formula $(\exists v_1 \dots \exists v_m)K$.

Definition 7.3.10. A first-order sentence φ is *canonic* if

- φ has the form $(\forall u)[\bar{K}_1 \vee \cdots \vee \bar{K}_m]$ where $K_1, \dots, K_m$ are clauses (the *constituent clauses* of φ), and
- the constants (resp. variables) of φ form an initial segment in the lexicographical order of constants (resp. variables).

Theorem 7.3.11. *The satisfiability problem for Shelah sentences reduces to the satisfiability problem for canonic sentences.*

Proof. Let φ_0 be a Shelah sentence. Without loss of generality, φ_0 contains a universal quantifier, so that the prefix of φ consists of two batches of existential quantifiers separated by a universal quantifier. Remove the first batch and replace the corresponding variables with constants in the quantifier-free part of the sentence. Let φ be the resulting sentence. Clearly, φ is satisfiable if and only if φ_0 is so.

In the remainder of the proof, we perform several transformation on φ until we get a canonic sentence. In each case, the output is equivalent to the input and both will be called φ. Think about φ as the current formula.

Transformation 0: Start. Transform the quantifier free part $\mathrm{QF}(\varphi)$ of φ to an equivalent quantifier free formula that is built from literals by means of conjunctions and disjunctions; those literals will be called the *constituent literals* of φ. For each literal α, let $\mathrm{Depth}(\alpha)$ be the sum of the depths of the constituent terms of α.

Transformation 1: Term-depth reduction. Let I be the collection of inadmissible constituent literals of φ and $d = \sum_{\alpha \in I} \mathrm{Depth}(\alpha)$. If $d > 0$, choose a subterm $\mathrm{Par}(t)$ of depth 1 of a term in I and replace $\mathrm{QF}(\varphi)$ with

$$(\exists v)[v = \mathrm{Par}(t) \wedge \varphi']$$

where v is a fresh existential variable and φ' is the result of replacing $\mathrm{Par}(t)$ with v in $\mathrm{QF}(\varphi)$. Repeat this procedure until $d = 0$.

Transformation 2: Elimination of inadmissible constituent literals. Reduce $\mathrm{QF}(\varphi)$ to disjunctive normal form and let I be the collection of inadmissible constituent literals of φ. It is easy to see that every $\alpha \in I$ is an equality. If $I \neq \varnothing$, choose a literal $t = t'$ in I. Without loss of generality, t precedes t' in the lexicographical order. Do the following for every disjunct of $\mathrm{QF}(\varphi)$ that has the chosen literal as a conjunct: delete the chosen conjunct and replace t' with t in the remaining conjuncts. Repeat this procedure until $I = \varnothing$.

Transformation 3: Finale. Use the equivalences

$$(\exists v)(\alpha \vee \beta) \equiv (\exists v)\alpha \vee (\exists v)\beta \quad \text{and} \quad (\exists v)(\gamma) \equiv \gamma,$$

where γ does not contain v, to transform φ into a canonical sentence. □

In the rest of this section we prove the decidability of the satisfiability problem for canonic sentences. For brevity, we make the following conventions:

- A clause is identified with the set of its constituent literals.
- A canonic sentence is identified with the set of constituent clauses.
- Canonic sentences are called simply sentences.

7.3.3 Terminology and Notation

Generalized Structures. We generalize the notion of structures in two ways. First, we allow a structure to include a variable assignment. Second, we allow first-order statements to be undefined.

Vocabularies. We extend the notion of *vocabulary* by allowing vocabularies to contain individual variables. In addition to variables, a vocabulary may contain the function name Par, various predicates and individual constants.

A model or structure $\mathfrak{A}$ of vocabulary Υ interprets every variable $v \in \Upsilon$ as a an element $\mathfrak{A}(v)$ of A. The vocabulary of a structure $\mathfrak{A}$ will be denoted $\mathrm{Voc}(\mathfrak{A})$. If $v \notin \mathrm{Voc}(\mathfrak{A})$ and a belongs to A, then $\mathfrak{A}(v/a)$ is the expansion of $\mathfrak{A}$ to v (or, more exactly to the vocabulary $\mathrm{Voc}(\mathfrak{A}) \cup \{v\}$.

The vocabulary $\mathrm{Voc}(\psi)$ of a formula ψ contains the predicates, constants and free variables of ψ and the function name Par if it occurs in ψ. We say that a structure $\mathfrak{A}$ is a structure *for* a formula ψ if $\mathrm{Voc}(\mathfrak{A})$ includes $\mathrm{Voc}(\psi)$ but does not contain any bound variables of ψ.

The satisfaction relation is defined in the obvious way. We require, however, that $\mathfrak{A}$ is a structure for ψ if $\mathfrak{A}$ satisfies ψ. Instead of giving a detailed definition of satisfaction, we give the following useful criterion.

Lemma 7.3.12. *Let $\mathfrak{A}$ be a structure for a sentence φ. $\mathfrak{A}$ satisfies φ if and only if, for every $x \in A$, some $K \in \varphi$ is satisfied in $\mathfrak{A}(u/x)$.*

Proof. Obvious. □

A structure $\mathfrak{A}$ is a model *of* ψ if and only if it satisfies ψ.

Truth Values. An *atomic statement* about a structure $\mathfrak{A}$ is a statement of the form $P(a_1, \dots, a_r)$ where P is an r-ary predicate in the vocabulary of $\mathfrak{A}$ and $a_1, \dots, a_r$ are elements of A. (An atomic statement $P(a_1, \dots, a_r)$ can be represented by the pair $(P, (a_1, \dots, a_r))$.) Usually each such statement is either true or false in $\mathfrak{A}$. We allow proper atomic statements to be undefined. In other words, atomic statements will take values in the set {*true*, *false*, *undef*}, but each equality statement will evaluate to either *true* or *false*.

It follows that if $\mathfrak{A}$ is a structure for an atomic formula α then the truth value $\mathrm{TV}_{\mathfrak{A}}(\alpha)$ of α at $\mathfrak{A}$ belongs to the set {*true*, *false*, *undef*}. To compute truth values of arbitrary first-order formulae at $\mathfrak{A}$, we order the three truth values as follows: *false* < *undef* < *true*. Then conjunction and universal quantification are minimization operators while disjunction and existential quantification are maximization operators. The negation is consistent with the usual negation and leaves *undef* intact. A generalized structure $\mathfrak{A}$ *satisfies* a formula ψ is ψ evaluates to *true* in $\mathfrak{A}$. It is easy to see that a sentence is satisfiable if and only if it is satisfiable in a generalized structure.

In the rest of this section, structures are the two-way generalized structures.

Literal Statements, Edges and Amalgams. An *literal statement* about a structure $\mathfrak{A}$ is an atomic statement about $\mathfrak{A}$ or the negation of such. A literal statement $\pm P(a_1, \ldots, a_r)$ is *proper* if P is so.

An *edge* of structure $\mathfrak{A}$ is a true literal statement about $\mathfrak{A}$. A structure $\mathfrak{A}$ is a *substructure* of a structure $\mathfrak{B}$ of the same vocabulary (and $\mathfrak{B}$ is a *superstructure* of $\mathfrak{A}$) if every element of A is an element of B and every edge of $\mathfrak{A}$ is an edge of $\mathfrak{B}$. If $\mathfrak{A}$ is a substructure of $\mathfrak{B}$ then, for every $a \in A$, $\mathrm{Par}_{\mathfrak{A}}(a) = \mathrm{Par}_{\mathfrak{B}}(a)$; this is so because equalities take only boolean values *true* and *false*. However, $\mathfrak{B}$ may have an edge $\pm P(\bar{a})$ that is not an edge of $\mathfrak{A}$ even though all elements of $\bar{a}$ are elements of A; in this case $P(\bar{a})$ evaluates to *undef* in $\mathfrak{A}$. If every edge of $\mathfrak{B}$ with elements from A is also an edge of $\mathfrak{A}$, then $\mathfrak{A}$ is an *induced substructure* of $\mathfrak{B}$. Notice that each nonempty subset X of a structure $\mathfrak{A}$ that is closed under Par gives rise to an induced substructure of $\mathfrak{A}$ with universe X; this substructure is called the *restriction* of $\mathfrak{A}$ to X and denoted $\mathfrak{A}|X$.

Structures $\mathfrak{A}, \mathfrak{B}$ of the same vocabulary *agree* on the common part X of their universes (or simply agree) if X is empty or else X is closed under Par in both structures and $\mathfrak{A}|X = \mathfrak{B}|X$.

The *amalgam* of pairwise agreeing structures $\mathfrak{A}_1, \ldots, \mathfrak{A}_k$ is the unique structure $\mathfrak{A}$ such that (i) A is the union of $A_1, \ldots, A_k$ and (ii) the collection of edges of $\mathfrak{A}$ is the union of the collection of edges of $\mathfrak{A}_1, \ldots, \mathfrak{A}_k$. Structures $\mathfrak{A}_i$ are *summands* of $\mathfrak{A}$.

$\mathfrak{A}$ is the *amalgam* of $\mathfrak{A}_1, \ldots, \mathfrak{A}_k$ *over* a set X if, for all $i < j$, X is the common part of A_i and A_j. If $\mathfrak{A}$ is the *amalgam* of $\mathfrak{A}_1, \ldots, \mathfrak{A}_k$ over a set X, we say also that $\mathfrak{A}$ is the amalgam of $\mathfrak{A}_1, \ldots, \mathfrak{A}_k$ over the structure $\mathfrak{A}_1|X$.

Lemma 7.3.13. *If structures $\mathfrak{A}_1, \ldots, \mathfrak{A}_k$ pairwise agree and each of them satisfies a sentence φ, then their amalgam satisfies φ as well.*

Proof. We use the satisfiability criterion of Lemma 7.3.12. An arbitrary element x of the amalgam belongs to the universe A_i of some summand $\mathfrak{A}_i$. Since $\mathfrak{A}_i \models \varphi$, there is a clause $K \in \varphi$ such that some expansion of $\mathfrak{A}_i(u/x)$ satisfies K. The corresponding expansion of the amalgam satisfies K as well. □

Nobles and Plebeians. Distinguished elements are *princes*; they also are *royal*. Ancestors of princes are *noble*, and the other elements are *plebeian*. (Notice that children of a prince may be plebeian.) We use this terminology extensively. The collection of all noble elements of a structure $\mathfrak{A}$ is denoted by Noble($\mathfrak{A}$). The restriction of $\mathfrak{A}$ to Noble($\mathfrak{A}$) is the *noble substructure* of $\mathfrak{A}$. A component of $\mathfrak{A}$ is *royal* if it contains a prince; otherwise it is *plebeian*. An expansion $\mathfrak{A}^+$ of $\mathfrak{A}$ to a collection V of variables if *plebeian* if all elements $\mathfrak{A}^+(v)$, $v \in V$, are plebeian.

Let κ be a positive integer. A noble element x is a κ-*baron* if the distance from x to the nearest prince is at most κ.

Lemma 7.3.14. *In a structure with n princes, the number of κ-barons is $\leq 3\kappa n^2$.*

Proof. For each κ-baron x, there are princes y_1, y_2 and a number $l \leq \kappa$ such that $y_1 \leq x$ and Distance$(x, y_2) = l$. By Lemma 7.3.6, there are at most three descendents of y_1 on distance l from y_2. □

The Adjunct Logic. Recall FO(Par), the first-order logic of unary function Par, introduced in Sect. 7.3.1. The *adjunct logic* of Par is the extension of FO(Par) with the strict younger-than relation $x < y$ defined in Sect. 7.3.1. Notice that the adjunct logic is a fragment of second-order logic.

The non-strict younger-than relation $x \leq y$ is definable in the obvious way: $x < y \vee x = y$. A priori, one may expect that $x < y$ is equivalent to $x \leq y \wedge x \neq y$, but this is not necessarily true. The condition $x \leq y \wedge x \neq y$ is sufficient for $x < y$ but it is not necessary in general. If x is cyclic then $x < x$ but the condition $x \leq y \wedge x \neq y$ fails.

We are interested in formulae of the adjunct logic where Par occurs only in atomic formulae of the form $t_1 = \text{Par}(t_2)$ with both terms t_1 and t_2 of depth zero. Only such formulae will be called *adjunct formulae.*

Lemma 7.3.15. *In a given algebra $\mathfrak{A}$, each κ-baron x is definable by an existential adjunct formula with $< \kappa$ quantifiers.*

Proof. Since x is a κ-baron, there are princes y_1, y_2 and a number $l \leq \kappa$ such that $y_1 \leq x$ and Distance$(x, y_2) = l$. It takes no quantifiers to express that $y_1 \leq x$. By Lemma 7.3.6, there are at most three descendants of y_1 on distance l from y_2. In the proof of Lemma 7.3.6, we have described the possible paths of length l from y_1 to descendants of y_1. It is easy to see that $l - 1$ quantifiers suffice to specify the path leading to x. □

Capitals, Towns and Villages.

Definition 7.3.16. The *syntactical partial algebra* of a clause K is the following partial algebra $\mathfrak{P}$: Elements of $\mathfrak{P}$ are the constants and variables of K, and $\text{Par}_{\mathfrak{P}}(t) = t'$ if the clause $t' = \text{Par}(t)$ belongs to K. A subclause K' of K is a *component* of K if there exists a component P' of $\mathfrak{P}$ such that a literal $\alpha \in K$ belongs to K' if and only if all constants and variables of α belong to P'

Notice that a clause may contain inter-component literals.

Definition 7.3.17. The component of a clause K containing u is the *capital* of K, any non-capital component containing a constant is a *town* of K, and any non-capital component without constants is a *village* of K. A variable that belongs to the capital (resp. a town, a village) is a *capital* (resp. *town, village*) variable.

7.3.4 1-Satisfiability

Definition 7.3.18. Let $\mathfrak{A}$ be a structure.

- A subset X of A is *1-regular* if the plebeian elements of X are pairwise comparable.
- Every equality or inequality statement α about $\mathfrak{A}$ is *1-relevant* to $\mathfrak{A}$. An equality or inequality formula α is *1-relevant* to $\mathfrak{A}$ if all variables of α belong to $\mathrm{Voc}(\mathfrak{A})$. In either case, $\mathfrak{A}$ *1-satisfies* α if $\mathfrak{A} \models \alpha$.
- A proper literal statement $\alpha = \alpha(x_1, \ldots, x_r)$ about $\mathfrak{A}$ is *1-relevant* to $\mathfrak{A}$ if the set $\{x_1, \ldots, x_r\}$ is 1-regular. $\mathfrak{A}$ *1-satisfies* α if either α is 1-irrelevant to $\mathfrak{A}$ or $\mathfrak{A} \models \alpha$. $\mathfrak{A}$ is *1-regular* if all edges of $\mathfrak{A}$ are 1-relevant to $\mathfrak{A}$.
- A proper literal $\alpha = \alpha(v_1, \ldots, v_r)$ is *1-relevant* to $\mathfrak{A}$ if $\mathfrak{A}$ is a structure for α and the literal statement $\alpha(\mathfrak{A}(v_1), \ldots, \mathfrak{A}(v_r))$ is 1-relevant to $\mathfrak{A}$. $\mathfrak{A}$ *1-satisfies* α if $\mathfrak{A}$ is a structure for α and either α is 1-irrelevant to $\mathfrak{A}$ or $\mathfrak{A} \models \alpha$.
- $\mathfrak{A}$ *1-satisfies* a clause K if it 1-satisfies all literals of K. $\mathfrak{A}$ *1-satisfies* $\bar{K}$ if some expansion of $\mathfrak{A}$ 1-satisfies K.
- $\mathfrak{A}$ *1-satisfies* a sentence φ if $\mathfrak{A}$ is a structure for φ and, for every $x \in A$, there exists $K \in \varphi$ such that $\mathfrak{A}(u/x)$ 1-satisfies $\bar{K}$.

If $\mathfrak{A}$ 1-satisfies a formula ψ, we write $\mathfrak{A} \models_1 \psi$ and say that $\mathfrak{A}$ is a 1-model of ψ.

Lemma 7.3.19. *(i) Suppose that a structure $\mathfrak{A}$ 1-satisfies a sentence φ and $\mathfrak{B}$ is the substructure of $\mathfrak{A}$ obtained by removing all 1-irregular edges. Then $\mathfrak{B} \models_1 \varphi$.*
(ii) An amalgam of 1-regular structures is 1-regular.
(iii) An amalgam of structures $\mathfrak{A}_i$ 1-satisfies a sentence φ if each $\mathfrak{A}_i$ 1-satisfies φ.

Proof. *(i):* 1-irregular edges cannot witness the 1-satisfiability of literals.

(ii): Any edge of the amalgam belongs to one of the summands.

(iii): If u takes value in A_i, then $\mathfrak{A}_i$ provides the necessary witnessing instantiation of existential variables. □

Theorem 7.3.20. *A sentence φ is satisfiable if it is 1-satisfiable. Moreover, if φ is 1-satisfiable in a structure $\mathfrak{A}$ then it is satisfiable in a structure whose noble substructure is that of $\mathfrak{A}$.*

Proof. Assume that a sentence φ is 1-satisfiable and let $\mathfrak{A}$ be a 1-model of φ. Let $c = \mathrm{Card}(\mathfrak{A})$. We may suppose without loss of generality that $c \leq \aleph_0$ because the proof of the Löwenheim-Skolem Theorem (see [76]) remains valid in the case of 1-satisfiability. On the ground of Lemma 7.3.19, suppose without loss of generality that $\mathfrak{A}$ is 1-regular.

Fix a choice function that, given a plebeian acyclic component X of $\mathfrak{A}$, produces an element of X. Call the produced elements and their ancestors *pseudo noble*.

Stratify $\mathfrak{A}$ by assigning a numerical depth $\delta(x)$ to each element. The depth of any noble, cyclic and pseudo noble element is zero. The depth of any other element x is the distance from x to the closest ancestor of depth zero. The depth $d = \delta(\mathfrak{A})$ of $\mathfrak{A}$ is the supremum of the depths of its elements.

Let κ be the number of existential variables in φ, n be the maximal number of proper literals in the clauses of φ and $\varepsilon = 2^{-n}$. If $c = \aleph_0$, set $s = \aleph_0$. Otherwise choose s to be a positive integer so large that $c(s\kappa+1)^{d+1}(1-\varepsilon)^s < 1$.

Construct an increasing series of structures $\mathfrak{A}_i$, $i \le d$, as follows. $\mathfrak{A}_0$ is the amalgam of $\mathfrak{A}$ and $s\kappa$ isomorphic copies of $\mathfrak{A}$ over Noble($\mathfrak{A}$). $\mathfrak{A}_{i+1}$ is the amalgam of $\mathfrak{A}_i$ and $s\kappa$ additional isomorphic copies of $\mathfrak{A}_i$ over $\{x \in A_i : \delta(x) \le i\}$. If d is infinite, then $\mathfrak{A}_d$ is the amalgam of all structures $\mathfrak{A}_i$.

Fix witnessing isomorphisms from $\mathfrak{A}$ to other summands of $\mathfrak{A}_0$ and from each $\mathfrak{A}_i$, $i < d$, to other summands of $\mathfrak{A}_{i+1}$. These isomorphisms give rise to *canonic automorphisms* of $\mathfrak{A}_d$. For example, for each summand $\mathfrak{A}_2'$ of $\mathfrak{A}_3$, there exists a canonic automorphism θ of $\mathfrak{A}_d$ that fixes $\mathfrak{A}_1$ and maps $\mathfrak{A}_2$ onto $\mathfrak{A}_2'$. In addition the trivial automorphism of $\mathfrak{A}_d$ will also be called *canonic*.

Check by induction on i that, for each $x \in A_i$, there is a composition of canonic automorphism that maps element of A to x. Check by induction on i that, in the case of finite $\mathfrak{A}$, $\text{Card}(\mathfrak{A}_i) \le c(s\kappa+1)^{i+1}$ and thus $\text{Card}(\mathfrak{A}_d) \le c(s\kappa+1)^{d+1}$.

For each undefined proper atomic statement $P(x_1, \ldots, x_r)$ about $\mathfrak{A}_d$, toss a fair coin and make the statement true (resp. false) if the coin comes up heads (resp. tails). The result is a random structure $\mathfrak{B}$. It suffices to prove that the probability $\Pr[\mathfrak{B} \not\models \varphi] < 1$.

Given an arbitrary $x \in B$, we estimate the probability that $\mathfrak{B}(u/x)$ 1-satisfies the existential closure of some $K \in \varphi$. Since there is an automorphism of $\mathfrak{A}_d$ that maps an element of $\mathfrak{A}$ to x, we may suppose without loss of generality that $x \in A$. Since $\mathfrak{A} \models_1 \varphi$, there exists $K \in \varphi$ such that some expansion $\mathfrak{A}^+$ of $\mathfrak{A}$ to EV(K) 1-satisfies K. Fix appropriate K and $\mathfrak{A}^+$.

We construct an expansion $\mathfrak{C}$ of $\mathfrak{B}(u/x)$ to EV(K) in stages.

Preliminary stage. If $\mathfrak{A}^+(v)$ is noble or $\mathfrak{A}^+(v) \ge x$, set $\mathfrak{C}(v) = \mathfrak{A}^+(v)$.

Let V' be the collection of the remaining existential variables of K. The rest of the construction proceeds by induction on $\delta(\mathfrak{A}^+(v))$. For each finite $i \le d$, let $V_i = \{v \in V' : \delta(\mathfrak{A}^+(v)) \le i\}$. Let f be an injective map from the collection of plebeian components of $\mathfrak{A}$ that do not contain x to the collection of summands of $\mathfrak{A}_0$ different from $\mathfrak{A}$.

Stage 0. Suppose that $v \in V_0$. Then $\mathfrak{A}^+(v)$ lies in a component $\mathfrak{A}' \in$ Domain(f). Set $\mathfrak{C}(v) = \theta(\mathfrak{A}^+(v)$ where θ is the canonic automorphism that takes $\mathfrak{A}$ to $\mathfrak{A}'$.

Stage (i+1). We suppose that $\mathfrak{C}$ is defined on V_i and, for each $v \in V_i$, there is a composition of canonic automorphisms that takes $\mathfrak{A}^+(v)$ to $\mathfrak{C}(v)$. Let $X = \{\mathfrak{A}^+(w) : w \in V_i\}$, $j = i+1$ and

$$Y = \{y : \delta(y) = 0\} \cup \{y \in A : x \leq y\} \cup X.$$

Partition $V_j - V_i$ into blocks with the same closest ancestor in Y. Let $y \in Y$ and $v_1, \ldots, v_r$ be all variables $v' \in V_j - V_i$ such that y is the closest ancestor of $\mathfrak{A}^+(v')$ in Y. Clearly, $\mathfrak{A}^+(v_1), \ldots, \mathfrak{A}^+(v_r)$ are incomparable in $\mathfrak{A}$.

We define an auxiliary automorphism θ_y of $\mathfrak{A}_d$. If y is noble or $x \leq y$, θ_y is the trivial automorphism. If Component$(y) \in$ Domain(f) but $y \notin X$, let θ_y be the canonic automorphism that moves $\mathfrak{A}$ to the summand $f(\text{Component}(y))$. In the remaining case, $y = \mathfrak{A}^+(w)$ for some $w \in V_i$; let $theta_y$ be a composition of canonic automorphisms such that $\theta(\mathfrak{A}^+(w)) = y$.

Choose canonic automorphisms $\eta_1, \ldots, \eta_r$ of $\mathfrak{A}_d$ that fix $\mathfrak{A}_i$ and move $\mathfrak{A}_j$ to different summands of $\mathfrak{A}_{j+1}$. Set $\mathfrak{C}(v_q) = \eta_q(\theta_y(\mathfrak{A}^+(v_q)))$.

Lemma 7.3.21. *(i)* $\delta(\mathfrak{C}(v)) = \delta(\mathfrak{A}^+(v))$.
(ii) $\mathfrak{C}(v_1) = \mathfrak{C}(v_2)$ *if and only if* $\mathfrak{A}^+(v_1) = \mathfrak{A}^+(v_2)$.
(iii) $\mathfrak{C}(v_1) < \mathfrak{C}(v_2)$ *if and only if* $\mathfrak{A}^+(v_1) < \mathfrak{A}^+(v_2)$.

Proof. *(i)*. Obvious.

(ii): The *only if* direction is obvious. Prove the *if* direction by contradiction. Suppose that $\mathfrak{C}(v_1) = \mathfrak{C}(v_2)$ and consider all cases when $\mathfrak{C}$ has been defined at v_1 and v_2.

(iii): Induction on $e = \delta(\mathfrak{A}^+(v_2))$. □

Let $\alpha = \alpha(v_1, \ldots, v_r) \in K$ and $x_q = \mathfrak{A}^+(v_q)$. First we suppose that α is 1-relevant to $\mathfrak{A}^+$ and check that $\mathfrak{C} \models \alpha$. If α is an equality or inequality, use claim 2 of the previous lemma. Suppose that α is proper. If every x_q is noble then, according to the preliminary stage of the construction of $\mathfrak{C}$, $\mathfrak{C} \models \alpha$. So suppose that some of elements x_q are plebeian and all plebeian elements x_q are pairwise comparable. We illustrate the proof on an example where $r = 3$, all elements x_q, and plebeian, $d_q = \delta(x_1)$ and $d_1 < d_2 < d_3$. According to the stage d_1 of the construction of $\mathfrak{C}$, some canonic automorphism θ_1 takes x_1 to $\mathfrak{C}(v_1)$. According the stage d_2 of the construction of $\mathfrak{C}$, some automorphism θ_2 fixes $\mathfrak{A}_{d_2}$ and moves x_2 to $\mathfrak{C}(v_2$. According to the stage d_3 of the construction of $\mathfrak{C}$, some automorphism θ_3 fixes $\mathfrak{A}_{d_3}$ and moves x_3 to $\mathfrak{C}(v_2)$. Let θ be the composition of automorphisms $\theta_1, \theta_2, \theta_3$. Since $\mathfrak{C} \models \alpha(x_1, x_2, x_3)$, $\mathfrak{C} \models \alpha(\theta(x_1), \theta(x_2), \theta(x_3)) = \alpha(\mathfrak{C}(v_1), \mathfrak{C}(v_2), \mathfrak{C}(v_3))$.

Second we suppose that α is 1-irrelevant to $\mathfrak{A}^+$ and check that the truth value of α is undefined in $\mathfrak{A}_d$. To simplify notation let x_1 and x_2 be incomparable plebeians. First suppose that $\mathfrak{A}^1$ of x_1 differs from the component $\mathfrak{A}^2$ of x_2. If $\mathfrak{A}^k \in$ Domain(f) for some k then the component of $\mathfrak{A}_d$ that includes $f(\mathfrak{A}^k)$

contains $\mathfrak{C}(x_k)$ but not $\mathfrak{C}(x_{3-k})$. If neither $\mathfrak{A}^k$ belongs to Domain(f) and $\mathfrak{A}_d^k$ is the component of $\mathfrak{A}_d$ that includes $\mathfrak{A}^k$, then $\mathfrak{C}(v_1) \in \mathfrak{A}_d^1 \neq \mathfrak{A}_d^2 \in \mathfrak{C}(v_2)$.

Second suppose that $\mathfrak{A}^1 = \mathfrak{A}^2$. Without loss of generality, $\delta(x_1) \leq \delta(x_2) = e$. According to stage e of the construction of $\mathfrak{C}$, $\mathfrak{C}(x_1)$ and $\mathfrak{C}(x_2)$ belong to different summands of $\mathfrak{A}_{e+1}$ and thus are incomparable.

Thus, for every α 1-irrelevant to $\mathfrak{A}^+$, probability $\Pr[\mathfrak{C} \models \alpha] = 1/2$. By the definition of n, $\Pr[\mathfrak{C} \models K] \geq 2^n = \varepsilon$. Hence $\Pr[\mathfrak{C} \not\models K] \leq 1 - \varepsilon$.

Since each $\mathfrak{A}_i$ is composed of $s\kappa$ summands, there exist versions $\mathfrak{C}_1, \dots, \mathfrak{C}_s$ of $\mathfrak{C}$ such that, for every $v \in \text{Var}(K)$, if $\mathfrak{A}^+(v)$ is plebeian and $\mathfrak{A}^+(v) \not\geq x$ then all elements s elements $\mathfrak{C}_1(v), \dots, \mathfrak{C}_s(v)$ are distinct. It follows that, for every $\alpha \in K$ 1-irrelevant to $\mathfrak{A}^+$, the events $\mathfrak{C}_1 \models \alpha, \dots, \mathfrak{C}_s \models \alpha$ are independent. The probability that none of expansions $\mathfrak{C}_1, \dots, \mathfrak{C}_s$ models K is $\leq (1-\varepsilon)^s$.

Thus the probability p that of the event $\mathfrak{B}(u/x) \not\models K$ is $\leq (1-\varepsilon)^s$. In the case of infinite $\mathfrak{A}$, $p = 0$. Since B is countable, the event $\mathfrak{B} \not\models \varphi$ is the intersection of countably many events of probability zero and thus has probability zero.

In the case of finite $\mathfrak{A}$, by the choice of s, we have:

$$\Pr[\mathfrak{B} \not\models \varphi] \leq p \cdot \text{Card}(\mathfrak{B}) \leq c(s\kappa + 1)^{d+1}(1-\varepsilon)^s < 1.$$

□

Corollary 7.3.22. *A sentence is satisfiable if and only it is 1-satisfiable.*

7.3.5 2-Satisfiability

Definition 7.3.23. Suppose that $\mathfrak{A}$ is a structure for a clause K.

- A proper literal $\alpha \in K$ is *2-relevant* to $\mathfrak{A}$ with respect to K if it is 1-relevant to $\mathfrak{A}$ and some component of K contains every variable $v \in \text{Var}(\alpha)$ such that $\mathfrak{A}(v)$ is plebeian.
- An equality or inequality literal $\alpha \in K$ is *2-relevant* to $\mathfrak{A}$ with respect to K.
- $\mathfrak{A}$ *2-satisfies* K if, for every $\alpha \in K$, either α is 2-irrelevant to $\mathfrak{A}$ wrt K or $\mathfrak{A} \models \alpha$.

Definition 7.3.24. – A structure $\mathfrak{A}$ *2-satisfies* the existential closure $\bar{K}$ of a clause K if $\mathfrak{A}$ is a structure for $\bar{K}$ and some expansion of $\mathfrak{A}$ satisfies K.
- $\mathfrak{A}$ *2-satisfies* a sentence φ if $\mathfrak{A}$ is a structure for φ and, for every $x \in A$, there exists $K \in \varphi$ such that $\mathfrak{A}(u/x)$ 2-satisfies $\bar{K}$.

Lemma 7.3.25. *The amalgam of structures $\mathfrak{A}_i$ 2-satisfies a sentence φ if each $\mathfrak{A}_i$ 2-satisfies φ.*

Proof. Obvious. □

Theorem 7.3.26. *A sentence φ is 1-satisfiable if it is 2-satisfiable. Moreover, if φ is 2-satisfiable in a structure $\mathfrak{A}$ then it is 1-satisfiable in a structure whose noble substructure is that of $\mathfrak{A}$.*

Proof. Assume that $\mathfrak{A}$ 2-satisfies φ and let $c = \mathrm{Card}(\mathfrak{A})$. Without loss of generality, $c \le \aleph_0$. If $\mathfrak{A}$ is infinite, let $s = \aleph_0$. Otherwise let s be a positive integer so big that $s(1-2^{-n})^s < 1$ where n is the maximal number of proper literals in clauses of φ.

Let m be the maximal number of non-capital components in the clauses of φ and construct isomorphisms $\eta_r : \mathfrak{A} \to \mathfrak{A}_r$, $0 \le i < ms$, such that η_0 is the identity map and, for all $q < r$, $A_q \cap A_r = \mathrm{Noble}(\mathfrak{A})$. Let $\mathfrak{A}^*$ be the amalgam of structures $\mathfrak{A}_i$ and construct a random superstructure $\mathfrak{B}$ of the amalgam as follows. For every undefined proper 1-regular atomic statement α about $\mathfrak{A}^*$, toss a fair coin and make α true (resp. false) if the coin comes up heads (resp. tails). It suffices to prove that $\Pr[\mathfrak{B} \models_1 \varphi] < 1$.

Let x be an arbitrary element of B. By symmetry, we may suppose that $x \in A$. Fix $K \in \varphi$ and an expansion $\mathfrak{A}^+$ of $\mathfrak{A}(u/x)$ to $\mathrm{EV}(K)$ such that $\mathfrak{A}^+ \models_3 K$.

Let K_0 be the capital component of K and $K_1, \dots, K_l$ be the other components of K. Call an expansion $\mathfrak{C}$ of $\mathfrak{B}(u/x)$ to $\mathrm{EV}(K)$ *admissible* if there is an injective function

$$f = f_{\mathfrak{C}} : \{0, \dots, l\} \to \{r : r < ms\}$$

such that $f(0) = 0$ and $\mathfrak{C}(v) = \eta_{f(i)}(\mathfrak{A}^+(v))$ for all $v \in \mathrm{Var}(K_i)$. Let $\mathfrak{C}$ range over the admissible expansions.

Estimate the probability $\Pr[\mathfrak{C} \not\models_1 K]$. Let $\alpha \in K$. If α is 2-relevant to $\mathfrak{C}$ wrt K then some K_i contains all variables of α with plebeian interpretations in $\mathfrak{C}$. Since η_i is an isomorphism and $\mathfrak{A} \models_2 K$, $\mathfrak{C} \models \alpha$. Assume that α is 1-relevant but 2-irrelevant to $\mathfrak{C}$ with respect to K. There are variables $v_1, v_2 \in \mathrm{Var}(\alpha)$ with plebeian interpretations in $\mathfrak{C}$ which belong to different components of K. $\mathfrak{C}(v_1)$ and $\mathfrak{C}(v_2)$ belong to different summands of $\mathfrak{A}^*$ and $\mathrm{TV}_{\mathfrak{A}^*}(\alpha) =$ *undef.* The probability $\Pr[\mathfrak{C} \not\models \alpha] = 1/2$. Thus $\Pr[\mathfrak{C} \not\models_1 K] \ge 2^{-n}$.

There exist s admissible expansions $\mathfrak{C}$ such that functions $f_{\mathfrak{C}}$ have disjoint ranges. The events $\mathfrak{C} \models_1 K$ are independent. Hence the probability p that $\mathfrak{B}(u/x)$ fails to 1-satisfy $\bar{K}$ is $\le (1-2^{-n})^s$. If $s = \aleph_0$ then $p = 0$. The event $\mathfrak{B} \not\models_1 \varphi$ is the intersection of $\aleph_0$ events of probability zero and thus has probability zero.

Suppose that s is finite. Then $\Pr[\mathfrak{B} \not\models_1 \varphi]$ is bounded by cp which is < 1 by the choice of s. □

Corollary 7.3.27. *A sentence is 2-satisfiable if and only if it is 1-satisfiable.*

7.3.6 Refinements

We start with constructing a computable function $\rho(\Upsilon, \kappa)$ where Υ is a vocabulary and κ a positive integer. The we use ρ to define a refinement relation on sentences. The particular choice of ρ will not be exploited in this section, but it will be exploited later.

Definition 7.3.28. Noble elements a, b of a structure $\mathfrak{A}$ are *κ-similar* if, for every existential adjunct formula $\psi(v)$ with $\leq \kappa$ quantifiers, $\mathfrak{A}(v/a) \models \psi$ if and only if $\mathfrak{A}(v/b) \models \psi$.

κ-similarity is an equivalence relation; its equivalence classes will be called κ-similarity classes.

Definition 7.3.29. $\rho_0(\Upsilon, \kappa)$ is a computable bound on the number of κ-similarity classes in Υ-structures. It is supposed that the function $\rho_0(\Upsilon, \kappa)$ is monotone in both arguments.

The existence of a computable bound on the number of κ-similarity classes in Υ-structures is obvious.

Definition 7.3.30. $\rho_1(\Upsilon, \kappa) = 4(\kappa + 1)\ell\rho_0(\Upsilon, \kappa)$ where ℓ is the number of individual constants in Υ. Further, $\rho(\Upsilon, \kappa) = \kappa \cdot (\rho_1(\Upsilon, \kappa))^\kappa$.

Definition 7.3.31. A clause L is an *immediate refinement* of a clause K if it satisfies the following conditions.

- L implies $\bar{K}$.
- L contains $\leq \rho(\kappa)$ new constants.
- L contains less existential variables than K.

Definition 7.3.32. A sentence φ' is an *immediate refinement* of a sentence φ if φ' is obtained from φ by replacing a clause $K \in \varphi$ with an arbitrary collection of immediate refinements of K.

Lemma 7.3.33. *A sentence is satisfiable if it has a satisfiable refinement.*

Proof. Obvious. □

Definition 7.3.34. The *refinement* relation on clauses (resp. sentences) is the transitive closure of the immediate refinement relation.

Lemma 7.3.35. *The refinement relation on sentences is well ordered.*

Proof. Assign the ordinal $\sum_{k<\kappa} m_k \omega^k$ to a given sentence φ; here κ is the number of existential variables in φ and m_k is number of clauses in φ with k existential variables. Refinement decreases the ordinal. □

Lemma 7.3.36. *The number of refinements of a given sentence φ is finite and computable from φ.*

Proof. The number of refinements is $\leq 2^N$ where N is the number N of clauses that appear in φ and its refinements is computable from φ. Thus it suffices to prove that N is finite and computable from φ.

Let $\kappa = \text{Card}(\text{EV}(\varphi))$. Refinements of a given clause $K \in \varphi$ form a tree of height κ where K is the root and a clause L_2 is a child of a clause L_1 if and only if L_2 is an immediate refinement of L_1.

□

Lemma 7.3.37. *The poset of all refinements of a given sentence φ is computable from φ.*

Proof. Obvious. □

Definition 7.3.38. A sentence is *modest* if it has no 3-satisfiable refinements.

Theorem 7.3.39 (Refinement Theorem). *The satisfiability problem reduces to that for modest sentences.*

A decision algorithm for the satisfiability of modest sentences is not supposed to recognize modesty. It just happens to work correctly on modest sentences.

Proof. Let F be a decision algorithms for satisfiability of modest sentences. Construct the poset of refinements of the given sentence φ and then, using F, traverse the poset bottom up and check the satisfiability of refinements of φ. If any of the refinements is 3-satisfiable then φ is so. Otherwise, φ is modest and $F(\varphi)$ tells us if φ is satisfiable. □

Lemma 7.3.40. *Degenerated clauses of a satisfiable modest sentence have no existential variables.*

Proof. By contradiction suppose that a degenerate clause K of is satisfiable modest sentence φ contains an existential variable v. K contains an equality $u = c$. Let $\mathfrak{A}$ be a model of φ and $x = \mathfrak{A}(c)$. Then the structure $\mathfrak{A}(u/x)$ has an expansion $\mathfrak{A}^+$ that satisfies some clause $L \in \varphi$. If $L \neq K$, then $\varphi - \{K\}$ is satisfied in $\mathfrak{A}$ which contradicts the modesty of φ. Assign a new constant c' to the element $\mathfrak{A}^+(v)$ and let $\mathfrak{A}^*$ be the resulting expansion of $\mathfrak{A}$. Refine φ by replacing K with $K(v/c')$ and let φ^* be the resulting expansion. Clearly, $\mathfrak{A}^*$ satisfies φ^*. □

7.3.7 Villages

In this section we prove the clauses of a satisfiable modest sentence have no villages. Fix a positive integer κ and restrict attention to formulae that use only the first κ existential variables.

Matching Nobles. Recall that noble elements a, b of a structure $\mathfrak{A}$ are κ-similar if and only if

$$\mathfrak{A}(v/a) \models \psi \iff \mathfrak{A}(v/b) \models \psi$$

for every existential adjunct formula $\psi(v)$ with $\leq \kappa$ quantifiers.

Definition 7.3.41. Similar nobles $a < b$ are *κ-matching* if they are κ-similar and $\text{Distance}(a, b) > 2\kappa$.

Since κ is fixed, terms "κ-similar" and "κ-matching" are abbreviated to "similar" and "matching" respectively. Recall the definition of ρ_1 (Definition 7.3.30).

Lemma 7.3.42. *In any structure of vocabulary Υ, if N is a set of noble elements of cardinality $> \rho_1(\Upsilon, \kappa)$ then there exists a pair $a < b$ of matching noble elements such that $(a, b]$ includes the κ-vicinity of some element of N.*

Proof. Let $\mathfrak{A}$ be an arbitrary Υ-structure and N a subset of A of cardinality $> \rho_1(\Upsilon, \kappa)$. There is a prince p of $\mathfrak{A}$ with $> 4(\kappa + 1)\rho_0(\Upsilon, \kappa)$ ancestors in N. Further, there exists a finite $X \subseteq N$ such that every $x \geq p$, all elements of X are E-equivalent and $\text{Card}(X) > 4(\kappa + 1)$. Either all elements of X are acyclic or they all are cyclic. Call elements of X red.

Case 1: Red elements are acyclic. Choose a to be the minimal red element and b the maximal red element.

Case 2: Red elements are cyclic. Let δ be the maximal distance between red elements. Choose a and b such that $\text{Distance}(a, b) = \delta$ and $\text{Card}((X \cap (a, b]) \geq \text{Card}((X \cap (b, a])$. Then $\text{Card}((X \cap (a, b]) \geq 2(\kappa + 1)$ and therefore there exists $\geq 2(\kappa + 1)$ red elements between a and b.

It is easy to see that in either case there is a red element x such that $(a, b]$ includes the κ-vicinity $\{y : \text{Distance}(x, y) \leq \kappa\}$ of x. □

Folds. Let $a < b$ be matching elements of a structure $\mathfrak{A}$ and $a' = \text{Par}(a)$. We extend $\mathfrak{A}$ to a structure $\mathfrak{B} = \mathfrak{A} + \text{Fold}(a, b)$ which consists of $\mathfrak{A}$ and an additional plebeian component $\text{Fold}(a, b)$.

Let η be a one-to-one correspondence between the interval $(a, b]$ of $\mathfrak{A}$ and a set X disjoint from A. If $x \in (a, b]$ then $\eta(x) \in X$, and if $x \in X$ then $\eta(x) \in (a, b]$. In either case $\eta^2(x) = x$.

We construct the desired plebeian component $\text{Fold}(a, b)$ of $\mathfrak{B}$ on the elements of X so that $B = A \cup X$. If $x \in X$ and $x \neq \eta(b)$, set $\text{Par}_{\mathfrak{B}}(x) = \eta(\text{Par}_{\mathfrak{A}}(\eta(x)))$. If $x = \eta(b)$, set $\text{Par}_{\mathfrak{B}}(x) = \eta(a')$.

Next we define the truth value $t = \text{TV}_{\mathfrak{B}}(\alpha)$ of a proper atomic statement $\alpha = P(x_1, \ldots, x_r)$ about $\mathfrak{B}$ such that the set $\{x_1, \ldots, x_r\}$ intersects X. Set $t = \textit{undef}$ if at least one of the following conditions is satisfied:

- There exists q such that $x_q \in A$ but x_q is not a baron (that is not a κ-baron).

– There exist p, q such that $x_p \in X$ and $x_q \in X$ but $\text{Distance}(x_p, x_q) \geq \kappa$.
– The set $\{x_1, \dots, x_r\}$ is not 1-regular.

Suppose that none of these condition is satisfied. Then there exists $x \in \{x_1, \dots, x_q\}$ such that every plebeian $x_q = \text{Par}^{i_q}_{\mathfrak{B}}(x)$ for some $i_q < \kappa$. Set $t = \text{TV}_{\mathfrak{A}}(P(y_1, \dots, y_r))$ where elements y_q are defined as follows. If x_q is a baron then $y_q = x_q$, and if x_q is plebeian then $y_q = \text{Par}^{i_q}_{\mathfrak{A}}(\eta(x))$.

Deterministic Construction. Let φ be a satisfiable modest sentence with κ existential variables and let n be the maximal number proper literals in clauses of φ. Fix a model $\mathfrak{A}$ of φ of cardinality $c \leq \aleph_0$. If $c = \aleph_0$, let $s = \aleph_0$. Otherwise let s be a positive integer so large that $c(1-2^n)^s < 1$. Construct the amalgam $\mathfrak{A}^*$ of the following structures over the union of royal components of $\mathfrak{A}$:

– $\mathfrak{A}$ itself and s additional copies of $\mathfrak{A}$,
– ns isomorphic copies of $\mathfrak{A} + \text{Fold}(a, b)$ for every pair of matching elements $a < b$ in $\mathfrak{A}$.

Lemma 7.3.43. *$\mathfrak{A}^*$ 2-satisfies φ. Moreover, for every summand $\mathfrak{B}$ of $\mathfrak{A}^*$ of the form $\mathfrak{A}$ + Fold(a, b) and every element $x \in B - A$, there exists $K \in \varphi$ such that some plebeian expansion of $\mathfrak{A}^*(u/x)$ 2-satisfies K.*

Proof. Let $\mathfrak{B}$ and x be as the lemma and a', η be as in the definition of folds. Since $\mathfrak{A} \models \varphi$, there exists $K \in \varphi$ such that some expansion $\mathfrak{A}^+$ of $\mathfrak{A}(u/\eta(x))$ to $\text{EV}(K)$ 2-satisfies K. Fix appropriate K and $\mathfrak{A}^+$. We construct an expansion $\mathfrak{C}$ of $\mathfrak{A}^*(u/x)$ to $\text{EV}(K)$.

First we deal with capital existential variables v. Clearly, $\eta(x) \in (a, b]$. Since a, b are similar and different, there is no prince in the κ-vicinity of $\eta(x)$. Hence $\mathfrak{A}^+$ maps the capital component of K onto a connected subset Y_0 of $\mathfrak{A}$ of cardinality $\leq \kappa$. Y_0 is not necessarily included in the interval $(a, b]$. It may overspill over the a-side of the interval (in which case it contains a) or over the b-side of interval (in which case it contains $b' = \text{Par}_{\mathfrak{A}}(b)$) but not both because $\text{Distance}(a, b) > 2\kappa$. If $\mathfrak{A}^+(v)$ is between a and b, set $\mathfrak{C}(v) = \eta(\mathfrak{A}^+(v))$. Let $Y = Y_0 - (a, b]$.

Case 1: Y contains a. Let Z be the superset of Y that contains all elements $\text{Par}^i_{\mathfrak{A}}(a)$ with $i < \kappa$. Let Δ be the quantifier-free diagram of Z in the adjunct language. Since a, b are matching, there exists a one-to-one mapping θ of Z to a subset of A such that $\theta(a) = b$ and θ preserves Δ. (Notice that θ maps nobles to nobles; because the nobility is expressible in a quantifier free way in the adjunct language.) If $\mathfrak{A}^+(v) \in Y$, set $\mathfrak{C}(v) = \eta(\theta(\mathfrak{A}^+(v)))$. It is easy to see that $\mathfrak{C}$ 2-satisfies the capital of K. Indeed, if α is a literal in the capital of K that is 2-relevant to $\mathfrak{C}$ and α contains some v with $\mathfrak{A}^+(v) \in Y$ then Z contains the $\mathfrak{A}^+$ interpretations of all variables in α. By the definition of θ, it preserves the truth value of α. It remains to use the definition of η.

Case 2: Y contains b'. This case is similar to the previous one. Let Z be the superset of Y that contains all elements $\text{Par}^i_{\mathfrak{A}}(b)$ with $i < \kappa$. Let Δ be the quantifier-free diagram of Z in the adjunct language. Since a, b are matching, there exists a one-to-one mapping θ of Z to a subset of A such that $\theta(b) = a$ and θ preserves Δ. If $\mathfrak{A}^+(v) \in Y$, set $\mathfrak{C}(v) = \eta(\theta(\mathfrak{A}^+(v)))$. It is easy to see that $\mathfrak{C}$ 2-satisfies the capital of K.

Second we deal with town variables. Let T be the union of all towns of K. Choose $\mathfrak{C}$ such that $\mathfrak{C} \models T$ and every $\mathfrak{C}(v)$, $v \in \mathrm{Var}(T)$, is plebeian. The possibility of such a choice follows from the following claim.

Claim. *Some plebeian expansion of* $\mathfrak{A}$ *satisfies* T.

Proof. Assume the contrary and assign a new constant to every non-royal baron of $\mathfrak{A}$; let $\mathfrak{A}'$ be the resulting expansion of $\mathfrak{A}$. Construct a refinement φ' of φ by replacing K with all clauses $K(v/\xi)$ where v is a town variable of K and ξ is a new constant. We check that $\mathfrak{A}' \models \varphi'$. That gives the desired contradiction because φ is modest.

Let $x \in A'$. Since $\mathfrak{A} \models \varphi$, there exist $L \in \varphi$ and an expansion $\mathfrak{X}$ of $\mathfrak{A}(u/x)$ to $\mathrm{EV}(L)$ that satisfies L. Let $\mathfrak{X}'$ be the expansion of $\mathfrak{A}'$ to $\mathrm{EV}(L)$ consistent with $\mathfrak{X}$. Clearly $\mathfrak{X}' \models L$ if $L \neq K$. Suppose $L = K$. By the assumption, some $\mathfrak{X}(v)$, $v \in \mathrm{Var}(T)$, is noble. Since every town of K contains a constant, $\mathfrak{X}(v)$ is a baron and therefore interprets a new constant ξ in $\mathfrak{A}'$. Then $\mathfrak{A}' \models K(v/\xi)$. □

Third we deal with village variables. For every village V of K, choose $\mathfrak{C}$ such that $\mathfrak{C} \models V$ and every $\mathfrak{C}(v)$, $v \in \mathrm{Var}(V)$, is plebeian. The possibility of such a choice follows from the following claim.

Claim. *Some plebeian expansion of* $\mathfrak{A}^*$ *satisfies* V.

Proof. Assume the contrary. For every expansion X of $\mathfrak{A}$ to $\mathrm{Var}(V)$ that satisfies V, there exists a variable $v_X \in \mathrm{Var}(V)$ such that $X(v_X)$ is noble.

Case 1: There are $\leq \rho(\mathrm{Voc}(\mathfrak{A}), \kappa)$ elements $X(v_X)$. Assign a new constant to each of these elements and refine φ by replacing K with clauses $K(v/\xi)$ where $v \in \mathrm{Var}(V)$ and ξ is a new constant. It is easy to see that the expansion of $\mathfrak{A}$ to the new constants satisfies the refinement of φ which contradicts the modesty of φ.

Case 2: There are $> \rho(\mathrm{Voc}(\mathfrak{A}), \kappa)$ elements $X(v_X)$. By the definition of ρ, there exist matching elements $a' < b'$ of $\mathfrak{A}$ such that $(a', b']$ includes the κ-vicinity of some $X(v_X)$. Let $\mathfrak{A}'$ be a summand of $\mathfrak{A}^*$ of the form $\mathfrak{A}+\mathrm{Fold}(a', b')$ different from $\mathfrak{A}_0$ and let η be as in the definition of folds (with a', b' playing the role of a, b). The plebeian extension X of $\mathfrak{A}^*$ to $\mathrm{Var}(V)$ that sends every v to $\eta'(\mathfrak{A}^+(v))$. Obviously $X \models V$. This gives us the desired contradiction. □

It remains to check that $\mathfrak{C} \models_2 K$. Obviously, $\mathfrak{C}$ satisfies all equalities and inequalities of K. Suppose that $\alpha \in K$ is 2-relevant to $\mathfrak{C}$ wrt K, so that some component K' of K contains all variables of α with plebeian interpretations in $\mathfrak{C}$. Since $\mathfrak{C}$ is plebeian, K' contains all variables of α. By the construction, $\mathfrak{C} \models \alpha$. □

Randomization. Let $\mathfrak{A}$ and $\mathfrak{A}^*$ be as above. The construction of $\mathfrak{A}^*$ involved a cardinal s but the particular choice of s has not been exploited in the previous subsection; we will do that in this subsection. Construct a random superstructure $\mathfrak{B}$ of $\mathfrak{A}^*$ by assigning, randomly and independently, a proper truth value to each undefined atomic statement about $\mathfrak{A}^*$.

Lemma 7.3.44. *With a positive probability, $\mathfrak{B}$ 2-satisfies φ. Moreover, the probability of the following event is positive: For every $x \in B$, there exists a clause $K \in \varphi$ and an expansion $\mathfrak{C}$ of $\mathfrak{B}(u/x)$ to $EV(K)$ such that $\mathfrak{C} \models_2 (K)$ and $\mathfrak{C}(v)$ is plebeian for every village variable v of K.*

Proof. By Lemma 7.3.43, it suffices to restrict attention to elements x that belong to summands of $\mathfrak{A}^*$ that are isomorphic to $\mathfrak{A}$. By symmetry, it suffices to restrict attention to elements $x \in A$. Consider such an element x. Fix a clause $K \in \varphi$ and an expansion $\mathfrak{A}^+$ of $\mathfrak{A}(u/x)$ such that $\mathfrak{A}^+ \models K$. We construct the desired village-plebeian expansion $\mathfrak{C}$ of $\mathfrak{A}^*(u/x)$. Let $\mathfrak{C}_0$ be the reduct of $\mathfrak{A}^+$ obtained by disinterpreting the village variables of K.

Claim. *For every village V of K, there exists an expansion $\mathfrak{X}$ of $\mathfrak{A}(u/x)$ to Var(V) such that (i) $\mathfrak{X}$ satisfies V and (ii) there exist matching elements $a < b$ such that $(a, b]$ contains all elements $X(v)$, $v \in$ Var(V).*

Proof. Suppose the contrary and let V be a counter-example village. Let Y be the set of all elements $\mathfrak{X}(v)$ where $\mathfrak{X}$ ranges over expansions of $\mathfrak{A}(u/x)$ to Var(V) that satisfy V and v ranges over Var(V). Recall the definition of ρ_1 (Definition 7.3.30).

Case 1: Card(Y) $\leq \rho_1(\text{Voc}(\mathfrak{A}), \kappa)$. Introduce a new constant for each element of Y and check that be the corresponding expansion of $\mathfrak{A}$ satisfies the refinement of φ obtained by replacing K with all clauses $K(v/\xi)$ where $v \in$ Var(V) and ξ is a new constant. This contradicts the modesty of φ.

Case 2: Card(Y) $> \rho(\text{Voc}(\mathfrak{A}), \kappa)$. Use the definition of ρ_1. □

It follows that, for every village V, there exist an expansion $\mathfrak{X}$ of $\mathfrak{A}(u/x)$ to Var(V) such that (i) $\mathfrak{X}$ satisfies V and (ii) some folding component of $\mathfrak{A}^*$ contains all elements $\mathfrak{X}(v)$, $v \in$ Var(V). By the construction of $\mathfrak{A}^*$, there exists an expansion $\mathfrak{X}$ of $\mathfrak{C}_0$ such that (i) $\mathfrak{X}$ satisfies all villages of K and (ii) $\mathfrak{X}$ maps the variables of different villages to different folding components of $\mathfrak{A}^*$; call such expansions of $\mathfrak{C}_0$ *admissible*.

Obviously, an admissible expansion $\mathfrak{X}$ satisfies all equalities and inequalities of K. Suppose that $\alpha \in K$ is 2-relevant to $\mathfrak{X}$ wrt K. Obviously, $\mathfrak{X} \models \alpha$ if α has no village variables or α has only village variables. In the remaining case, some village V contains all variables of α that have plebeian interpretations in $\mathfrak{X}$, and some other variables of α have noble interpretations. By the construction of $\mathfrak{B}$, $\Pr[\mathfrak{X} \models \alpha] = 1/2$. Hence $\Pr[\mathfrak{X} \not\models_2 K] \leq 1 - 2^{-n}$ where n is the maximal number of proper literals in clauses of φ.

By the construction of $\mathfrak{B}$, there exist admissible expansions $\mathfrak{X}_1, \ldots, \mathfrak{X}_s$ of $\mathfrak{C}_0$ such that no two of them use the same folding component. The events

$\mathfrak{X}_r \models_2 K$ are independent. Hence the probability p that $\mathfrak{B}(u/x)$ fails to 2-satisfy $\bar{K}$ in a village-plebeian way is $\leq (1-2^{-n})^s$. If $s = \aleph_0$ then $p = 0$. The event $\mathfrak{B} \not\models_1 \varphi$ is the intersection of $\aleph_0$ events of probability zero and thus has probability zero.

Suppose that s is finite. Then $\Pr[\mathfrak{B} \not\models_1 \varphi]$ is bounded by cp which is < 1 by the choice of s. □

Theorem 7.3.45. *No clause of a satisfiable modest sentence φ has any villages.*

Proof. By contradiction suppose that a clause $K \in \varphi$ has a village V. By the previous lemma, some $\mathfrak{B} \models_2 \varphi$. Let ℓ be the number of different components of K and an amalgam $\mathfrak{B}'$ of $\ell + 1$ copies $\mathfrak{B}_0, \ldots, \mathfrak{B}_{\ell-1}$ of $\mathfrak{B}$ over Noble($\mathfrak{B}$). For each $j \leq \ell$, let η_j be an isomorphism from $\mathfrak{B}$ onto $\mathfrak{B}_j$ that is the identity map on Noble($\mathfrak{B}$). We suppose that $\mathfrak{B}_0 = \mathfrak{B}$ and η_0 is the identity map.

Since φ is modest, $\varphi - \{K\}$ is not 2-satisfiable and thus there is $x \in B$ such that K is the only clause of φ 2-satisfied by some expansion of $\mathfrak{B}(u/x)$. By the previous lemma, there exists an expansion $\mathfrak{C}$ of $\mathfrak{B}(u/x)$ that maps all variables of V to plebeians. Introduce a new constant for each element $\eta_j(\mathfrak{C}(v))$ where $j \leq \ell$ and $v \in \text{Var}(V)$ and let $\mathfrak{B}^*$ be the resulting expansion of $\mathfrak{B}'$. Let K_j be the result of replacing every $v \in \text{Var}(V)$ by the constant with value $\eta_j(\mathfrak{C}(v))$. Construct a refinement φ^* of φ by replacing K with clauses K_j. We check that $\mathfrak{B}^* \models_2 \varphi^*$.

Let y be an arbitrary element of B'. We need to find a clause $L \in \varphi^*$ such that some expansion $\mathfrak{X}$ of $\mathfrak{B}^*(u/y)$ 2-satisfies L. Since $\mathfrak{B}' \models \varphi$, there exists a clause M such that some expansion $\mathfrak{D}$ of $\mathfrak{B}'(u/y)$ 2-satisfies M. If $M \neq K$, let $L = M$; the desired $\mathfrak{X}$ is the expansion of $\mathfrak{B}^*(u/y)$ to EV(L) consistent with $\mathfrak{D}$. Suppose that $M = K$. Since K has only ℓ components, some summand $\mathfrak{B}_j$ is disjoint from $\mathfrak{D}(K)$. Let $L = K_j$; again the desired $\mathfrak{X}$ is the expansion of $\mathfrak{B}^*(u/y)$ to EV(L) consistent with $\mathfrak{D}$ on EV(L) $- V$.

Thus φ^* is 2-satisfiable which contradicts the modesty of φ.

Theorem 7.3.45 is proved □

7.3.8 Contraction

Without loss of generality, we restrict attention to sentences with at least one constant.

Lemma 7.3.46. *Every satisfiable modest sentence φ has a model where each component has a prince.*

Proof. Let $\mathfrak{A}$ be model of φ and $\mathfrak{B}$ the substructure of $\mathfrak{A}$ obtained by removing all plebeian components. We check that $\mathfrak{B} \models \varphi$. Let x be an arbitrary element of B. Since $\mathfrak{A} \models \varphi$, there exists an expansion $\mathfrak{A}^*$ of $\mathfrak{A}(u/x)$ to EV(φ) that satisfies some clause $K \in \varphi$. Clearly, B contains the $\mathfrak{A}^*$-interpretations of all capital and town variables of K. By Theorem 7.3.45, K has no villages.

Hence, the expansion of $\mathfrak{B}(u/x)$ to $\mathrm{EV}(\varphi)$ that is consistent with $\mathfrak{A}^*$ satisfies K. □

In the sequel we restrict attention to structures without plebeian components. Fix a positive integer κ.

Definition 7.3.47. Noble(x) is the closest noble ancestor of x. Define $x \preceq y$ if there is $i \leq \kappa$ such that $\mathrm{Par}^i(\mathrm{Noble}(x)) = \mathrm{Noble}(y)$ and $\mathrm{Distance}(x, y) \leq \kappa$. As expected, $x \prec y$ if $x \preceq y$ and $x \neq y$.

The condition "there is $i \leq \kappa$ such that $\mathrm{Par}^i(\mathrm{Noble}(x)) = \mathrm{Noble}(y)$" cannot be replaced by a weaker condition $\mathrm{Noble}(x) \leq \mathrm{Noble}(y)$. For example, consider the case when x is a noble element of a cycle of length $> 2\kappa$ and $y = \mathrm{Par}^\kappa(x)$, so that $x \prec y$ but $y \not\prec x$. The weaker condition would give $y \prec x$.

Definition 7.3.48. If there is a unique noble y with $\mathrm{Par}^i(y) = x$ then $y = \mathrm{Par}^{-i}(x)$.

Collapsing a Noble Interval. Assume that $\mathfrak{A}$ is a structure, $a < b$ are matching nobles of $\mathfrak{A}$, $a' = \mathrm{Par}(a)$ and $b' = \mathrm{Par}(b)$. We define a new structure $\mathfrak{B}$ on elements $A - (a, b]$. Set $\mathrm{Par}_{\mathfrak{B}}(x) = \mathrm{Par}_{\mathfrak{A}}(x)$ unless $x = a$ in which case set $\mathrm{Par}_{\mathfrak{B}}(x) = b'$. Given a proper atomic statement $\alpha = P(y_1, \ldots, y_r)$ about $\mathfrak{B}$, we define the truth value $\mathrm{TV}_{\mathfrak{B}}(\alpha)$ of α in $\mathfrak{B}$. Let $\mathrm{Trouble}_a(\alpha)$ be the set of elements $x_q \preceq a$, and let $\mathrm{Trouble}_b(\alpha)$ be the set of elements $x_q succ b$. If at least one of the trouble sets is empty, set $\mathrm{TV}_{\mathfrak{B}}(\alpha) = \mathrm{TV}_{\mathfrak{A}}(\alpha)$. Suppose that both trouble sets are nonempty. If every element of $\mathrm{Trouble}_b(\alpha)$ is noble, set

$$\mathrm{TV}_{\mathfrak{B}}(P(y_1, \ldots, y_r) = \mathrm{TV}_{\mathfrak{A}}(x_1, \ldots, x_r)$$

where $x_q = y_q$ unless there is $i < \kappa$ such that $y_q = \mathrm{Par}^{i+1}(b)$ in which case $x_q = \mathrm{Par}^{i+1}(a)$. If every element of $\mathrm{Trouble}_a(\alpha)$ is noble, set

$$\mathrm{TV}_{\mathfrak{B}}(P(y_1, \ldots, y_r) = \mathrm{TV}_{\mathfrak{A}}(x_1, \ldots, x_r)$$

where $x_q = y_q$ unless there is $i \leq \kappa$ such that $y_q = \mathrm{Par}^{-i}(a)$ in which case $x_q = \mathrm{Par}^{-i}(b)$. Otherwise set $\mathrm{TV}_{\mathfrak{B}}(\alpha) = undef$.

Lemma 7.3.49. *Let φ be a modest sentence with κ existential variables that is satisfied in $\mathfrak{A}$. Then $\mathfrak{B}$ 2-satisfies φ.*

Proof. Let x be an arbitrary element of B. Fix a clause $K \in \varphi$ and and expansion $\mathfrak{A}^+$ of $\mathfrak{A}(u/x)$ that satisfies K. We construct an expansion $\mathfrak{C}$ of $\mathfrak{B}(u/x)$ that satisfies K. Let $X = \{\mathfrak{A}^+(v) : v \in \mathrm{Var}(\mathrm{Capital}(K))$ and X^* be the union of X and the set of barons.

If X is disjoint from $(a, b]$, set $\mathfrak{C}(v) = \mathfrak{A}^+(v)$ for all $v \in \mathrm{Var}(K)$; it is clear that $\mathfrak{C} \models K$. Suppose that X intersects $(a, b]$. Then either $x \preceq a$ or $b \prec \mathrm{Noble}(x)$.

Case 1: $x \preceq a$. Since a and b are matching, there is a partial isomorphism η from $\mathfrak{A}$ to $\mathfrak{A}$ such that Domain(η) $= X^*$, η fixes the barons, η takes nobles to nobles and $\eta(a) = b$. Set

$$\mathfrak{C}(v) = \begin{cases} \eta(\mathfrak{A}^+(v)) & (\mathfrak{A}^+(v) \in (a, b], \\ \mathfrak{A}^+(v) & \text{otherwise.} \end{cases}$$

Obviously $\mathfrak{C}$ satisfies the equalities and inequalities of K. Suppose that $\alpha(v_1, \dots, v_r)$ is a proper literal of K that is 2-relevant to $\mathfrak{C}$ wrt K, $x_q = \mathfrak{A}^+(v_q)$ and $y_q = \mathfrak{C}(v_q)$.

Subcase 1: Every $x_q succ a$ is noble. For every such x_q there is $i \leq \kappa$ such that $x_q = \mathrm{Par}^{i+1}(a)$ and $y_q = \mathrm{Par}^{i+1}(b)$, so that $\mathrm{Trouble}_b(\alpha(y_1, \dots, y_r))$ consists of nobles. By the definition of $\mathfrak{B}$,

$$\mathrm{TV}(\alpha(y_1, \dots, y_r) = \mathrm{TV}_{\mathfrak{A}}(\alpha(x_1, \dots, x_r)).$$

Subcase 2: Some plebeian $x_q succ a$. Then every town variable v_p has a baronial interpretation in $\mathfrak{C}$ and every $x_p \preceq a$ is noble. Thus every $y_p = \eta(x_p)$. Hence

$$\mathrm{TV}(\alpha(y_1, \dots, y_r) = \mathrm{TV}_{\mathfrak{A}}(\alpha(\eta(x_1), \dots, \eta(x_r))) = \mathrm{TV}_{\mathfrak{A}}(\alpha(x_1, \dots, x_r)).$$

Case 2: $x succ b$. Since a and b are matching, there is a partial isomorphism η from $\mathfrak{A}$ to $\mathfrak{A}$ such that Domain(η) $= X^*$, η fixes the barons, η takes nobles to nobles and $\eta(b) = a$. Set

$$\mathfrak{C}(v) = \begin{cases} \eta(\mathfrak{A}^+(v)) & (\mathfrak{A}^+(v) \in (a, b], \\ \mathfrak{A}^+(v) & \text{otherwise.} \end{cases}$$

Obviously $\mathfrak{C}$ satisfies the equalities and inequalities of K. Suppose that $\alpha(v_1, \dots, v_r)$ is a proper literal of K that is 2-relevant to $\mathfrak{C}$ wrt K, $x_q = \mathfrak{A}^+(v_q)$ and $y_q = \mathfrak{C}(v_q)$.

Subcase 1: Every $x_q \preceq a$ is noble. For every such x_q there is $i \leq \kappa$ such that $x_q = \mathrm{Par}^{-i}(a)$ and $y_q = \mathrm{Par}^{-i}(a)$, so that $\mathrm{Trouble}_a(\alpha(y_1, \dots, y_r))$ consists of nobles. By the definition of $\mathfrak{B}$,

$$\mathrm{TV}(\alpha(y_1, \dots, y_r) = \mathrm{TV}_{\mathfrak{A}}(\alpha(x_1, \dots, x_r)).$$

Subcase 2: Some plebeian $x_q \preceq a$. Then every town variable v_p has a baronial interpretation in $\mathfrak{C}$ and every $x_p succ b$ is noble. Thus every $y_p = \eta(x_p)$. Hence

$$\mathrm{TV}(\alpha(y_1, \dots, y_r) = \mathrm{TV}_{\mathfrak{A}}(\alpha(\eta(x_1), \dots, \eta(x_r))) = \mathrm{TV}_{\mathfrak{A}}(\alpha(x_1, \dots, x_r)).$$

□

Creating a Cycle. Again, assume that $\mathfrak{A}$ is a structure, $a < b$ are matching nobles of $\mathfrak{A}$ and $a' = \text{Par}(a)$. Suppose that the component of $\mathfrak{A}$ that contains a has no cycle. We define a new structure $\mathfrak{B}$ on elements $A - \{x : \text{Noble}(x) > b\}$. Set $\text{Par}_{\mathfrak{B}}(x) = \text{Par}_{\mathfrak{A}}(x)$ unless $x = b$ in which case set $\text{Par}_{\mathfrak{B}}(b) = a'$. Let $\alpha = P(x_1, \ldots, x_r)$ be a proper atomic statement about $\mathfrak{B}$ that is 1-relevant to $\mathfrak{B}$, Trouble(α) be the set of elements $x_q \prec b$, and $\text{Trouble}_a(\alpha)$ be the set of elements $x_q succ a'$.

If at least one of the trouble set is empty, set $\text{TV}_{\mathfrak{B}}(\alpha) = \text{TV}_{\mathfrak{A}}(\alpha)$.

If $\text{Trouble}_b(\alpha)$ is not empty and $\text{Trouble}_a(\alpha)$ contains only nobles, set $\text{TV}_{\mathfrak{B}}(\alpha) = \text{TV}_{\mathfrak{A}}(\alpha')$ where α' is obtained from α by replacing $\text{Par}_{\mathfrak{A}}^{i+1}(a)$ with $\text{Par}_{\mathfrak{A}}^{i+1}(b)$ for $i < \kappa$.

If $\text{Trouble}_a(\alpha) \neq \varnothing$ and $\text{Trouble}_b(\alpha)$ contains only nobles, set $\text{TV}_{\mathfrak{B}}(\alpha) = \text{TV}_{\mathfrak{A}}(\alpha')$ where α' is obtained from α by replacing $\text{Par}_{\mathfrak{A}}^{-i}(b)$ with $\text{Par}_{\mathfrak{A}}^{i}(a)$ for $i < \kappa$.

Otherwise set $\text{TV}_{\mathfrak{B}}(\alpha) = undef.$

Lemma 7.3.50. *Let φ be a modest sentence with κ existential variables that is satisfied in $\mathfrak{A}$. Then $\mathfrak{B}$ 2-satisfies φ*

Proof. Let x be an arbitrary element of B. Fix a clause $K \in \varphi$ and and expansion $\mathfrak{A}^+$ of $\mathfrak{A}(u/x)$ that satisfies K. We construct an expansion $\mathfrak{C}$ of $\mathfrak{B}(u/x)$ that satisfies K. Let $X = \{\mathfrak{A}^+(v) : v \in \text{Var}(\text{Capital}(K))$ and Y be the union of X and the set of barons.

If X is disjoint from $\{y : b \prec \text{Noble}(y)$, set $\mathfrak{C}(v) = \mathfrak{A}^+(v)$ for all $v \in \text{Var}(K)$; it is clear that $\mathfrak{C} \models K$. Suppose that X intersects $\{y : b \prec \text{Noble}(y)$. Then $x \prec b$.

Since a and b are matching, there is a partial isomorphism η from the restriction $\mathfrak{A}|Y$ to $\mathfrak{A}$ such that η fixes the barons, takes nobles to nobles and $\eta(b) = a$. Set $\mathfrak{C}(v) = \eta(\mathfrak{A}^+(v))$ if v is a capital variable and $Noble(\mathfrak{A}^+(v)) > b$. Otherwise set $\mathfrak{C}(v) = \mathfrak{A}^+(v)$.

Obviously $\mathfrak{C}$ satisfies the equalities and inequalities of K. Suppose that $\alpha = \alpha(v_1, \ldots, v_r)$ is a proper literal of K that is 2-relevant to $\mathfrak{C}$ wrt K. Let $x_q = \mathfrak{A}^+(v_q)$ and $y_q = \mathfrak{C}(v_q)$.

Subcase 1: For every v_q, if $\text{Noble}(x_q) > b$ then x_q is noble. Thus $x_q \leq b$ or $x_q > b$. In the first case, $y_q = x_q$. In the second case, there is $i < \kappa$ such that $x_q = \text{Par}_{\mathfrak{A}}^{i+1}(b)$ and $y_q = \text{Par}_{\mathfrak{A}}^{i+1}(a)$. By the definition of $\mathfrak{B}$, $\text{TV}_{\mathfrak{B}}(\alpha(y_1, \ldots, y_r)) = \text{TV}_{\mathfrak{A}}(\alpha(x_1, \ldots, x_r) = true.$

Subcase 2: Some plebeian $x_q > b$. For every town variable v_p, x_p is a baron and therefore $y_p = x_p = \eta(x_p)$. For every capital variable v_p with $x_p \prec b$, there is $i < \kappa$ such $\text{Par}_{\mathfrak{A}}^{-i}(b)$ and therefore $y_p = \text{Par}_{\mathfrak{A}}^{-i}(a) = \eta(x_q)$. Thus

$$\text{TV}_{\mathfrak{C}}(\alpha(v_1, \ldots, v_r)) = \text{TV}_{\mathfrak{B}}(\alpha(y_1, \ldots, y_r)) =$$
$$\text{TV}_{\mathfrak{A}}(\alpha(\eta(x_1), \ldots, \eta(x_r))) = \text{TV}_{\mathfrak{A}}(\alpha(x_1, \ldots, x_r) = true.$$

□

Limiting the Number of Nobles. Recall the definitions of $\rho_0(\Upsilon, \kappa)$ and $\rho_1(\Upsilon, \kappa)$ in Section 7.3.6.

Theorem 7.3.51. *Every modest satisfiable sentence φ of vocabulary Υ with κ existential quantifiers has a model with $< \rho_1(\Upsilon, \kappa)$ nobles.*

Proof. By Lemma 7.3.50, there exists a structure $\mathfrak{A}_2$ that 2-satisfies φ and has only finitely many noble elements. By the proof of Theorem 7.3.26, there is a structure $\mathfrak{A}_1$ that 1-satisfies φ and has exactly the same noble elements as $\mathfrak{A}_2$. By Theorem 7.3.20, there exists a structure $\mathfrak{A}_0$ that satisfies φ and has exactly the same noble elements as $\mathfrak{A}_1$.

Let ℓ be the number of individual constants in Υ and $\mathfrak{A}$ be a structure for φ with the minimal number of nobles. By Lemma 7.3.49, $\mathfrak{A}$ has no matching noble elements. It follows that, for every prince p and every similarity class X, p has $< 2\kappa+2$ ancestors in X. Thus, the number of nobles in $\mathfrak{A}$ is bounded by $2(\kappa + 2)\ell\rho_0(\Upsilon, \kappa)$ which is $< \rho_1(\Upsilon, \kappa)$. □

7.3.9 Towns

Theorem 7.3.52. *The clauses of a modest satisfiable sentence have no town variables.*

Proof. By contradiction suppose that a clause K of a modest satisfiable sentence φ has town variables. Let $\Upsilon = \text{Voc}(\varphi)$ and $\kappa = \text{Card}(\text{EV}(\varphi))$. By Theorem 7.3.51, φ has a model $\mathfrak{A}$ where the number of nobles is bounded by $\rho_1(\Upsilon, \kappa)$. Since φ is modest, $\varphi - \{K\}$ is unsatisfiable and therefore there exist elements $x \in A$ such that $\mathfrak{A}(u/x)$ satisfies $\bar{K}$ but does not satisfies any $\bar{L}$ with $L \in \varphi - \{K\}$; call such elements x *red*.

For each red x, fix an expansion $\mathfrak{A}_x$ of $\mathfrak{A}$ to $\text{EV}(K)$ that 2-satisfies K. Call red elements x, y equivalent if, for every variable v of K, both $\mathfrak{A}_x(v)$ and $\mathfrak{A}_y(v)$ are plebeian or they both are noble and equal. The number of red equivalence classes is bounded by $\rho_2(\Upsilon, \kappa) = (\rho_1(\Upsilon, \kappa))^\kappa$.

Without loss of generality, we may suppose that the expansions $\mathfrak{A}_x$ are chosen in such a way that if x, y are equivalent red elements then $\mathfrak{A}_x$ and $\mathfrak{A}_y$ coincide on the town variables of K. It suffices to check that the expansion $\mathfrak{A}'_y$ of $\mathfrak{A}(u/y)$ obtained from $\mathfrak{A}_y$ by replacing $\mathfrak{A}_y(v)$ with $\mathfrak{A}_x(v)$ for all town variables of K 2-satisfies K. But this is obvious.

Call an element y of $\mathfrak{A}$ pseudo-noble if there exists a red element x such that $y = \mathfrak{A}_x(v)$ for some town variable v of K. The number of pseudo-nobles is bounded by $\kappa\rho_2(\Upsilon, \kappa)$ which is equal to $\rho(\Upsilon, \kappa)$. A pseudo-noble element may be noble or plebeian.

Assign new constants to all non-royal pseudo-noble elements of $\mathfrak{A}$; let $\mathfrak{A}^*$ be the resulting expansion of $\mathfrak{A}$. If X is a red equivalence class and $x \in X$, let K_X be the result of replacing each town variable v of K with a constant for $\mathfrak{A}_x(v)$. Refine φ by replacing K with all clauses K_X; let φ^* be the refined sentences. It is easy to see that $\mathfrak{A}^*$ 2-satisfies φ^*, which contradicts the modesty of φ. □

7.3.10 The Final Reduction

In this section, we reduce the satisfiability problem for modest sentences to the satisfiability problem for $[\exists^*\forall\exists^*, (\omega), (1)]_=$. Of course this means also that the finite satisfiability problem for modest sentences reduces to the finite satisfiability problem for $[\exists^*\forall\exists^*, (\omega), (1)]_=$. Both the satisfiability and finite satisfiability problems for $[all, (\omega), (1)]_=$ are decidable (see Sect. 7.2). This concludes the proof of Theorem 7.3.1.

Remark. In fact, we reduce the satisfiability problem for modest sentences to the satisfiability problem for a special subfragment of $[\exists^*\forall\exists^*, (\omega), (1)]_=$. Instead or referring to Sect. 7.2, one can prove directly that the satisfiability problem for the subfragment is decidable. We will not do that here.

We will use the fact that the clauses of modest sentences have no town or village variables (Theorems 7.3.45 and 7.3.52). In this section, the term "sentence" is further restricted to mean sentences whose clauses have no town or village variables. All variables of such a clause belong to the same component, namely to the capital of the clause. This trivializes the distinction between 1-satisfiability and 2-satisfiability. However, the 1-satisfiability of a sentence remains a nontrivial necessary and sufficient condition of the satisfiability of the sentence; we will take advantage of this condition. Recall that we deal only with vocabularies that contain constants.

Definition 7.3.53. A *kingdom* is a structure $\mathfrak{Q}$ such that the vocabulary of $\mathfrak{Q}$ contains an initial segment of individual constants and every element of $\mathfrak{Q}$ is royal. If $\mathfrak{Q}$ is a kingdom, then $\text{Mod}(\mathfrak{Q})$ is the class of structures $\mathfrak{A}$ such that the noble substructure of $\mathfrak{A}$ is isomorphic to $\mathfrak{Q}$ and $\mathfrak{A}$ has no plebeian components.

Notice that in every structure $\mathfrak{A} \in \text{Mod}(\mathfrak{Q})$, all plebeian element are acyclic.

Definition 7.3.54. The *kingdom-constraint 1-satisfiability problem for modest sentences* is the following decision problem:

Instance: A kingdom $\mathfrak{Q}$ and a sentence φ such that $\text{Voc}(\varphi) = \text{Voc}(\mathfrak{Q})$.
Question: Is φ 1-satisfiable in $\text{Mod}(\mathfrak{Q})$?

Lemma 7.3.55. *The satisfiability problem for modest sentences reduces to the kingdom-constraint 1-satisfiability problem for modest sentences.*

Proof. By Corollary 7.3.22, the satisfiability problem for modest sentences reduces to the 1-satisfiability problem for modest sentences. It suffices to prove that there exists an algorithm F that, given a modest sentence φ, computes a collection $F(\varphi)$ of kingdoms such that φ is 1-satisfiable only if there exists $\mathfrak{Q} \in F(\varphi)$ such that φ is 1-satisfiable in $\mathrm{Mod}(\mathfrak{Q})$. The existence of such F follows from Theorem 7.3.51. For example, $F(\varphi)$ can be the collection of all kingdoms of cardinality $< \rho_1(\Upsilon, \kappa)$ where $\Upsilon = \mathrm{Voc}(\varphi)$ and κ is the number of existential variables in φ. One can be more parsimonious though and take advantage of the modesty of φ. If a clause $K \in \varphi$ contains an equality $d = \mathrm{Par}(c)$ where c, d are constants then we can ignore kingdoms where $d \neq \mathrm{Par}(c)$. Indeed, every model $\mathfrak{A}$ of φ contains an element x such that $\mathfrak{A}(u/x)$ satisfies K (otherwise $\mathfrak{A}$ satisfies $\varphi - \{K\}$ which contradicts the modesty of φ) and thus $d = \mathrm{Par}(c)$ in $\mathfrak{A}$. □

Definition 7.3.56. A sentence φ of vocabulary Υ is *plebeian* if every clause $K \in \varphi$ satisfies the following conditions:

1. Either there is a constant $c \in \Upsilon$ such that K contains the equality $c = u$ (in which case K is *royal*), or else K contains the inequality $c \neq u$ for every constant $c \in \Upsilon$ (in which case, K is *plebeian*).
2. K contains the inequality $c \neq v$ for every constant $c \in \Upsilon$ and every existential variable v of K.

Corollary 7.3.57. *Suppose that $\mathfrak{Q}$ is a kingdom, $\mathfrak{A} \in Mod(\mathfrak{Q})$ and $\mathfrak{A}$ satisfies a clause K of a plebeian sentence φ. Then:*

- *$\mathfrak{A}(v)$ is plebeian for every existential variable of K.*
- *$\mathfrak{A}(u)$ is plebeian if K is plebeian.*

Lemma 7.3.58. *The kingdom-constraint 1-satisfiability problem for modest sentences reduces to the kingdom-constraint 1-satisfiability problem for plebeian sentences.*

Proof. Let $\mathfrak{Q}$ be a kingdom of vocabulary Υ and φ a modest sentence with $\mathrm{Voc}(\varphi) = \Upsilon$. We construct a pseudo-plebeian sentence φ^* of vocabulary Υ that is satisfiable in $\mathrm{Mod}(\mathfrak{Q})$ if and only if φ is 1-satisfiable in $\mathrm{Mod}(\mathfrak{Q})$. Call a clause $K \in \varphi$ *bad* if it does not contain all constants of Υ. If φ has no bad clauses, set $\varphi^* = \varphi$. Suppose φ has bad clauses. We illustrate the transformation of φ to the desired φ^* on an example where φ has only one bad clause K and there are exactly two constants c, d in Υ that do not occur in K. The transformation is done in two stages.

First replace K with clauses

$$K(u/c), K(u/d), K \cup \{c \neq u, d \neq u\}$$

to ensure that condition 1 is satisfied. Clearly, an arbitrary $\mathfrak{A} \in \mathrm{Mod}(\mathfrak{Q})$ 1-satisfies φ if and only if it 1-satisfies the modified sentence φ'. Let L ranges over the three new clauses.

Second replace each L a number of new clauses to ensure that condition 2 is satisfied. We illustrate the second stage on the example where L is the plebeian clause $K \cup \{c \neq u, d \neq u\}$ with exactly two existential variables v and w. The new classes are

- $L(v/c)(w/d)$, $L(v/d)(w/c)$,
- $L(v/c) \cup \{c \neq w, d \neq w\}$,
- $L(v/d) \cup \{c \neq w, d \neq w\}$,
- $L(w/c) \cup \{c \neq v, d \neq v\}$,
- $L(w/d) \cup \{c \neq v, d \neq v\}$,
- $L \cup \{c \neq v, c \neq w, d \neq v, d \neq w\}$.

Clearly an arbitrary $\mathfrak{A} \in \text{Mod}(\mathfrak{Q})$ 1-satisfies φ'' if and only if it 1-satisfies the resulting φ^*. □

Recall the syntactical partial algebra of a clause K defined in the paragraph 'Capitals, Towns and Villages' of Sect. 7.3.3.

Definition 7.3.59. Suppose that φ is a sentence and K ranges over the clauses of φ.

- A proper literal $\alpha \in K$ is *regular* in K if it satisfies the following condition.
  - Let v_1, v_2 be arbitrary distinct variables in α. In case K is royal, suppose additionally that v_1, v_2 are existential. Then $v_1 < v_2$ or $v_2 < v_1$ in the syntactical partial algebra of K.
- $\text{Reg}(K)$ is the result of deleting all irregular proper literals from K. $\text{Reg}(\varphi) = \{\text{Reg}(K) : K \in \varphi\}$.
- φ is *regular* if it is plebeian and $\text{Reg}(\varphi) = \varphi$.

Lemma 7.3.60. *The kingdom-constraint 1-satisfiability problem for plebeian sentences reduces to the kingdom-constraint 1-satisfiability problem for regular sentences.*

Proof. Let $\mathfrak{Q}$ be a kingdom of vocabulary Υ and φ a constant-saturated sentence with $\text{Voc}(\varphi) = \Upsilon$. We check that if $\mathfrak{A} \in \text{Mod}(\mathfrak{Q})$ and $\mathfrak{A} \models_1 \text{Reg}(\varphi)$ then $\mathfrak{A} \models_1 \varphi$.

Let $x \in B$. Since $\mathfrak{A} \models_1 \text{Reg}(\varphi)$, there exists $K \in \varphi$ such that some expansion $\mathfrak{B}$ of $\mathfrak{A}(u/x)$ 1-satisfies $\text{Reg}(K)$. Let $\alpha \in K - \text{Reg}(K)$ and let v_1, v_2 witness the irregularity of α. Clearly elements $x_1 = \mathfrak{B}(v_1)$ and $x_2 = \mathfrak{B}(v_2)$ are plebeians. Let $\mathfrak{P}$ be the syntactical partial algebra of K. Recall that we consider only sentences without town or village variables in this section. Hence v_1, v_2 are both capitals variables and therefore are connected in $\mathfrak{P}$. Let w be the youngest common ancestor of v_1, v_2 in $\mathfrak{P}$. Since v_1, v_2 are incomparable in $\mathfrak{P}$, there exists distinct children w_1, w_2 of w such that $v_1 < w_1$ and $v_2 < w_2$ in $\mathfrak{P}$. It follows that x_1, x_2 are incomparable and thus the statement α about $\mathfrak{B}$ is 1-irrelevant to $\mathfrak{B}$. Hence $\mathfrak{B}$ 1-satisfies α. □

Call a kingdom $\mathfrak{Q}$ *perfect* if different individual constants have different interpretations in $\mathfrak{Q}$.

Lemma 7.3.61. *The kingdom-constraint 1-satisfiability problem for regular plebeian sentences reduces to the perfect-kingdom-constraint 1-satisfiability problem for regular sentences.*

Proof. Obvious. □

Recall that a sentence is monadic if its vocabulary contains no predicates (or function names) of arity > 1. In the case of monadic sentences, there is no need to distinguish between satisfiability and 1-satisfiability.

Lemma 7.3.62. *The perfect-kingdom-constraint 1-satisfiability problem for regular sentences reduces to the perfect-kingdom-constraint satisfiability problem for monadic regular sentences.*

Proof. We demonstrate the proof on an example. Suppose that, in addition to the function symbol Par and individual constants, the vocabulary Υ of the given instance $(\mathfrak{Q}, \varphi)$ of the perfect-kingdom-constraint 1-satisfiability problem for regular sentences contains only a ternary predicate P. Let κ be the number of existential variables in φ.

Let Υ' be a monadic vocabulary obtained from Υ be replacing P with unary predicates (P, e_1, e_2, e_3) where each e_i is an individual constant of Υ or a natural number $\leq \kappa$ and exactly one of expressions e_i is 0. We illustrate the intended meaning of the new predicates on examples. $(P, c, 0, 2)(x)$ codes the fact that $P(c, v, \mathrm{Par}^2(v))$ holds and $\mathrm{Par}^2(v)$ is plebeian. $(P, 0, 3, d)(v)$ codes the fact $P(v, \mathrm{Par}^3(v), d)$ holds and $\mathrm{Par}^3(v)$ is plebeian.

We construct an instance $(\mathfrak{Q}', \varphi')$ of the perfect-kingdom-constraint 1-satisfiability problem for monadic regular sentences such that the vocabulary of $(\mathfrak{Q}', \varphi')$ equals Υ'. The desired φ' is constructed by replacing the atomic P-formulae in φ. Suppose that $\alpha = P(t_1, t_2, t_3)$ is an atomic formula in a clause K of φ.

Case 1: At least two of terms t_i are constants. If t_2, t_3 are constants, replace α with $(P, 0, t_2, t_3)(t_1)$. If t_3, t_1 are constants and t_2 is a variable, replace α with $(P, t_1, 0, t_3)(t_2)$. If t_1, t_2 are constants and t_3 is a variable, replace α with $(P, t_1, t_2, 0)(t_3)$.

Case 2: Exactly one of the terms t_i is a constant. Because of symmetry, we consider only the subcase when t_1 is a constant. There exist a unique variable $v \in \{t_2, t_3\}$ and natural numbers $k_2, k_3 \leq \kappa$ such that $t_2 = \mathrm{Par}^{k_2}(v)$ and $t_3 = \mathrm{Par}^{k_3}(v)$. Replace α with $(P, t_1, k_1, k_2)(v)$.

Case 3: None of terms t_i is a constant. Since φ is plebeian, there exist a unique variable $v \in \{t_1, t_2, t_3\}$ and natural numbers k_1, k_2, k_3 such that K implies that each $t_i = \mathrm{Par}^{k_i}(v)$. Replace α with $(P, k_1, k_2, k_3)(v)$.

$\mathfrak{Q}'$ is obtained from $\mathfrak{Q}$ as follows: Replace every edge $\pm P(t_1, t_2, t_3)$ with a new edge $(P(0, t_2, t_3)(t_a)$. It is easy to see that φ' is satisfied in $\mathrm{Mod}(\mathfrak{Q}')$ if and only if φ is satisfied in $\mathrm{Mod}(\mathfrak{Q})$.

Indeed, suppose that $\mathfrak{A} \models \varphi$. Call an edge $\pm P(x_1, x_2, x_3)$ of $\mathfrak{A}$ relevant if either x_1, x_2, x_3 are royal or there exist a plebeian $y \in \{x_1, x_2, x_3\}$ and natural numbers $k_1, k_2, k_3 \leq \kappa$ such that each plebeian $x_i = \mathrm{Par}^{k_i}(y)$. Remove all irrelevant edges from $\mathfrak{A}$. Delete irrelevant edges and replace each relevant edge $\pm P(x_1, x_2, x_3)$ with a new unary edge in the obvious way. For example, if x interprets a constant c and $y = \mathrm{Par}^i(z)$ then the new unary edge is $(P, c, k, 0)(z)$. The resulting structure $\mathfrak{A}'$ belongs to $\mathrm{Mod}(\mathfrak{Q}')$ and satisfies φ.

Conversely, suppose that a structure $\mathfrak{B} \in \mathrm{Mod}(\mathfrak{Q}')$ satisfies φ'. There exists a structure $\mathfrak{A} \in \mathrm{Mod}(\mathfrak{Q})$ such that $\mathfrak{B} = \mathfrak{A}'$. Clearly, $\mathfrak{A}$ satisfies φ. □

Lemma 7.3.63. *The perfect-kingdom-constraint 1-satisfiability problem for monadic regular sentences reduces to the satisfiability problem for the class* $[all, (\omega), (1)]_=$.

Proof. The reduction is obvious. □

We have proved

Theorem 7.3.64. *The satisfiability problem for modest sentences reduces to the satisfiability problem for* $[\exists^*\forall\exists^*, (\omega), (1)]_=$.

The satisfiability problem for $[all, (\omega), (1)]_=$ is decidable; see Sect. 7.2 in this connection. Theorem 7.3.1 is proved.

7.4 Historical Remarks

The early history of finite automata on infinite objects involves Church, Büchi and Trakhtenbrot [83, 63, 512]. It culminated with Büchi's decision procedure for the monadic theory of the free algebra with one constant and one unary function, in other words, the monadic theory of one successor, or S1S. Recall that here monadic means monadic second-order. Büchi's decision procedure uses finite automata in an essential way. Every finite automaton gives rise to a Σ_1^1 formula, and every formula φ is reducible to such an automaton formula which is thus a normal form for φ.

Later Shelah gave a simpler and more direct proof of the decidability of S1S that does not use automata; see [467, 231]. But the automaton approach has its own advantages and far-reaching generalizations. The interplay of automata and logic has proven to be very powerful. Here we concentrate on results relevant to our main subject.

In [430], Rabin used automata on infinite trees to prove the decidability of the monadic theory of the infinite binary tree, in other words, the monadic theory of two successors, or S2S. That is the famous Rabin's Tree Theorem. S2S is quite rich. Many interesting mathematical theories have been proved decidable by interpreting them in S2S. We return to this subject below.

The general idea of Rabin's proof of the Tree Theorem is simple and clear but some lemmas are very hard. The hardest is the Complementation Lemma for tree automata which is based on a transfinite induction on countable ordinals ordinals (up to ω_1, the first uncountable ordinal). In [433], Rackoff gave a simpler algorithm (with a simpler correctness proof) to check whether a given automaton accept any tree at all. He also gave a simpler construction of the complementation automaton. "The proof that our construction works, however, is difficult and very similar in complexity to Rabin's proof (...) that his (more difficult) construction works", wrote Rackoff.

The game approach was pioneered by Büchi and Landweber [68, 66]. In [67], Büchi used games to give an alternative proof of Rabin's Complementation Theorem. His proof is still very hard and retains the induction over countable ordinals. In the meantime, Gurevich and Harrington proved the Forgetful Determinacy Theorem (FDT) which is of independent interest and showed that it implies Rabin's Tree Theorem rather easily [236]; our derivation of Rabin's Tree Theorem from FDT follows [236]. Independently, Muchnik has found an alternative game-theoretic proof of Rabin's Tree Theorem [400]. In [540], Alexander and Vladimir Yakhnis filled in all the details missing in the sketchy extended abstract [236] and strengthened Gurevich-Harrington's results in several ways; in particular they provided explicit winning strategies for the players. Zeitman adapted their proof to the case of graph games [546]. Our exposition follows [546] (though, for the sake of brevity, we prove a slightly less general result sufficient for our purposes). Several other proofs of Rabin's Tree Theorem have been published recently, notably [401] and [548]. For more information on automata on infinite objects, see [507].

As we said above, many mathematical theories have been proved decidable by interpretation into S2S. In particular, Rabin [430] established the decidability of the monadic theory of one unary function over a countable domain; our exposition follows Rabin's proof. Earlier, Ehrenfeucht [143] had announced (without proof) that the first-order theory of one unary function is decidable. Gurevich [227] observed that Rabin's results imply that the satisfiability problem and the finite satisfiability problems for the class $[all, (\omega), (1)]_=$ are decidable. In 1974, Meyer [390] announced that the first-order theory of one unary function has non-elementary complexity. This result was strengthened by Compton and Henson [89] to the particular lower complexity bound proved in Sect. 7.2.2.

The decision problem for the Shelah class was left open in [225] where it was conjectured that the satisfiability and finite satisfiability problems for the Shelah class were decidable. Modulo this conjecture, the decision problem for prefix-vocabulary classes with equality and at least one function symbol of positive arity had been settled. Shelah found a beautiful and intricate proof of the conjecture [468], which explains the name "Shelah class". Unfortunately Shelah's paper is far too sketchy. The detailed proof above is published for the first time. We acknowledge Shelah's help.

8. Other Decidable Cases

8.1 First-Order Logic with Two Variables

We denote by L_k the restriction of first-order logic to formulae of relational vocabulary (i.e. without function symbols) that contain only the variables $x_1, \dots, x_k$.

Logics with only a bounded number of variables are important in many branches of mathematical logic and its applications, including finite model theory, model checking, database query languages and knowledge representation. Of course, interesting sentences in L_k are *not* in prenex normal form. Quite to the contrary, one extensively uses the possibility to re-use variables.

Example. To express that a graph $G = (V, E)$ contains a path of length 37, a sentence in prenex normal form needs 38 variables. By re-using variables, the same property is expressible in L_2, by a sentence of the form

$$\exists x \exists y (Exy \wedge \exists x (Eyx \wedge \exists y (Eyx \wedge \cdots) \cdots))).$$

The decision problem for L_k is unsolvable (even for formulae without equality) for all $k \geq 3$ since L_3 extends the prefix class $[\forall\exists\forall]$.

We prove in this section that L_2 has the finite model property, a result due to Mortimer [396] (see the historical remarks in Sect. 8.4).

Theorem 8.1.1 (Mortimer). *L_2 has the finite model property. As a consequence, Sat(L_2) is decidable.*

The bound on the model size implied by Mortimer's proof is doubly exponential. Recently Grädel, Kolaitis and Vardi [208] strengthened Mortimer's Theorem by proving an (essentially optimal) small model property for L_2 with a single exponential bound on the model size. In addition this new proof is much simpler than Mortimer's proof.

The first step towards this result is a reduction to a normal form for L_2, essentially due to Scott [459]. A similar reduction appears in Mortimer's paper.

Lemma 8.1.2. *For each sentence $\psi \in L_2$ one can construct in polynomial time a sentence $\varphi \in L_2$ of the form*

$$\varphi := \forall x \forall y \alpha \wedge \bigwedge_{i=1}^{m} \forall x \exists y \beta_i$$

where α *and* β_i *are quantifier-free such that*

(i) $\varphi \models \psi$.
(ii) For every model $\mathfrak{A}$ *of* ψ *there exists a unique expansion* $\mathfrak{B}$ *of* $\mathfrak{A}$ *such that* $\mathfrak{B} \models \varphi$.
(iii) If n *is the length of* ψ, *then* φ *contains at most* n *relation symbols and has length* $O(n \log n)$.

Proof. If ψ has not the required form then choose (an occurrence of) a subformula of form $Qy\eta$ of ψ where Q is $\exists$ or $\forall$ and η is quantifier-free. Select a unary predicate R not contained in ψ and let ψ' be the result of replacing $Qy\eta$ in ψ by Rx. If $Qy\eta$ occurs in ψ inside the scope of a quantifier $\exists x$ or $\forall x$ (which is always the case if x appears in η) then ψ' is a sentence; otherwise it may contain a free occurrence of x. The formula

$$\psi' \wedge \forall x(Rx \leftrightarrow Qy\eta)$$

satisfies the properties *(i)* and *(ii)*. For $Q = \exists$ this formula is equivalent to

$$\psi' \wedge \forall x \forall y(\eta \rightarrow Rx) \wedge \forall x \exists y(Rx \rightarrow \eta)$$

and for $Q = \forall$, it is equivalent to

$$\psi' \wedge \forall x \forall y(Rx \rightarrow \eta) \wedge \forall x \exists y(\eta \rightarrow Rx).$$

Let φ be obtained by iterating this reduction step until the formula has the desired form. If at the end, φ contains a free variable, replace it by its universal closure. Finally, use that a conjunction of $\forall\forall$-formulae can be combined to a single $\forall\forall$-formula. □

Note that the prenex normal form of φ is in the $\forall^2\exists^*$-class.

Recall that a k-table of vocabulary σ is a σ-structure with universe $\{1, \ldots, k\}$. Further, for any σ-structure $\mathfrak{A}$ and distinct elements $a_1, \ldots, a_k$ of $\mathfrak{A}$ we denote by $T_{\mathfrak{A}}[a_1, \ldots, a_k]$ the unique k-table $\mathfrak{B}$ such that the function $a_1 \mapsto 1, \ldots, a_k \mapsto k$ is an isomorphism from $\mathfrak{A}|_{\{a_1,\ldots,a_k\}}$ (i.e., the substructure of $\mathfrak{A}$ induced by $a_1, \ldots, a_k$) to $\mathfrak{B}$.

Definition 8.1.3. An element a of $\mathfrak{A}$ is a *king* if no other element realizes the same 1-table, i.e. if $T_{\mathfrak{A}}[b] \neq T_{\mathfrak{A}}[a]$ for all $b \neq a$.

By Lemma 8.1.2 we can restrict attention to L_2-sentences

$$\varphi := \forall x \forall y \alpha \wedge \bigwedge_{i=1}^{m} \forall x \exists y \beta_i$$

where α and β_i are quantifier-free. Moreover we assume that $\beta_i(x,y) \models x \neq y$, for all $i \leq m$. This is no loss of generality since the equivalence

$$\forall x \exists y \eta(x,y) \equiv \forall x \exists y (x \neq y \wedge (\eta(x,x) \vee \eta(x,y)))$$

holds on all structures with at least two elements.

Fix such a sentence φ of vocabulary σ. Obviously we can assume that σ consists of unary and binary predicate symbols only.

Proposition 8.1.4. *Suppose that $\mathfrak{A} \models \varphi$. Let K be the set of kings in $\mathfrak{A}$ and $P = \{T_{\mathfrak{A}}[a] : a \in A\}$ the set of 1-tables realized in $\mathfrak{A}$. Then φ has a model with at most $(m+1)|K| + 3m(|P| - |K|) = O(n2^r)$ elements (where $n = |\varphi|$ and r is the number of predicate symbols in σ).*

Proof. Since $\mathfrak{A} \models \forall x \exists y \beta_i$ for $i = 1, \ldots, m$ there exist Skolem functions $f_i : A \to A$ such that $\mathfrak{A} \models \beta_i[a, f_i(a)]$ for all $a \in A$ and $i = 1, \ldots, m$. The *court* of $\mathfrak{A}$ is the substructure $\mathfrak{C} \subseteq \mathfrak{A}$ induced by $C := K \cup \{f_i(k) : k \in K, i = 1, \ldots m\}$, i.e. by the kings and their immediate dependents (to account for the case that $\mathfrak{A}$ is a republic we allow here that $C = \varnothing$). Further, let $Q \subseteq P$ be the set of 1-tables realized by the kings.

We extend $\mathfrak{C}$ to a model $\mathfrak{D} \models \varphi$ with universe

$$D := C \cup ((P - Q) \times \{1, \ldots, m\} \times \{0, 1, 2\}).$$

The construction is in four steps:

1. $\mathfrak{D}$ is an extension of $\mathfrak{C}$, i.e. $\mathfrak{D}|_C = \mathfrak{C}$.
2. We specify the 1-tables of the remaining elements: for $b = (\mathfrak{B}, i, j) \in D - C$, set $T_{\mathfrak{D}}[b] := \mathfrak{B}$.
3. Fix a 'choice function' $h : P \to A$ assigning to every 1-table $\mathfrak{B} \in P$ an element $h(\mathfrak{B})$ of $\mathfrak{A}$ with $T_{\mathfrak{A}}[h(\mathfrak{B})] = \mathfrak{B}$. To guarantee for all $d \in D$ and $i = 1, \ldots, m$ that $\mathfrak{D} \models \exists y \beta_i[d]$ we provide appropriate witnesses as follows:
   a) If d is a king, this is trivial since already $\mathfrak{C} \models \exists y \beta_i[d]$.
   b) Let $d \in C - K$ be a non-royal member of the court. If $f_i(d)$ also belongs to the court then again already $\mathfrak{C} \models \exists y \beta_i[d]$. Otherwise $T_{\mathfrak{A}}[f_i(d)] = \mathfrak{B} \in P - Q$. Let $e = (\mathfrak{B}, i, 0)$ and set $T_{\mathfrak{D}}[d, e] := T_{\mathfrak{A}}[d, f_i(d)]$. Thus $\mathfrak{D} \models \exists y \beta_i[d]$.
   c) Let $d = (\mathfrak{B}, j, \ell) \in D - C$ and $a = h(\mathfrak{B})$. Consider the element $b = f_i(a)$ witnessing that $\mathfrak{A} \models \exists y \beta_i[a]$.
   If b is a king then set $T_{\mathfrak{D}}[d, b] := T_{\mathfrak{A}}[a, b]$. (Note that there arises no conflict, since all witnesses for kings belong to the court and d is outside the court; thus $T_{\mathfrak{D}}[d, b]$ has not been defined yet.)
   Otherwise $T_{\mathfrak{A}}[b] = \mathfrak{B}' \in P - Q$. Let $e = (\mathfrak{B}', i, \ell + 1 (\mathrm{mod}\ 3))$ and set $T_{\mathfrak{D}}[d, e] := T_{\mathfrak{A}}[a, b]$. Since $\mathfrak{A} \models \beta_i[a, b]$ it follows that $\mathfrak{D} \models \beta_i[d, e]$ and hence $\mathfrak{D} \models \exists y \beta_i[d]$. Note that no conflicts arise since no 2-table $T_{\mathfrak{D}}[d, e]$ is defined twice (see Fig. 8.1).

4. To complete the construction choose, for every pair of $d, d' \in D$ (with $d \neq d'$) for which $T_{\mathfrak{D}}[d, d']$ is not defined yet, two distinct elements a, a' of $\mathfrak{A}$ such that $T_{\mathfrak{A}}[a] = T_{\mathfrak{D}}[d]$, $T_{\mathfrak{A}}[a'] = T_{\mathfrak{D}}[d']$ and set $T_{\mathfrak{D}}[d, d'] := T_{\mathfrak{A}}[a, a']$.

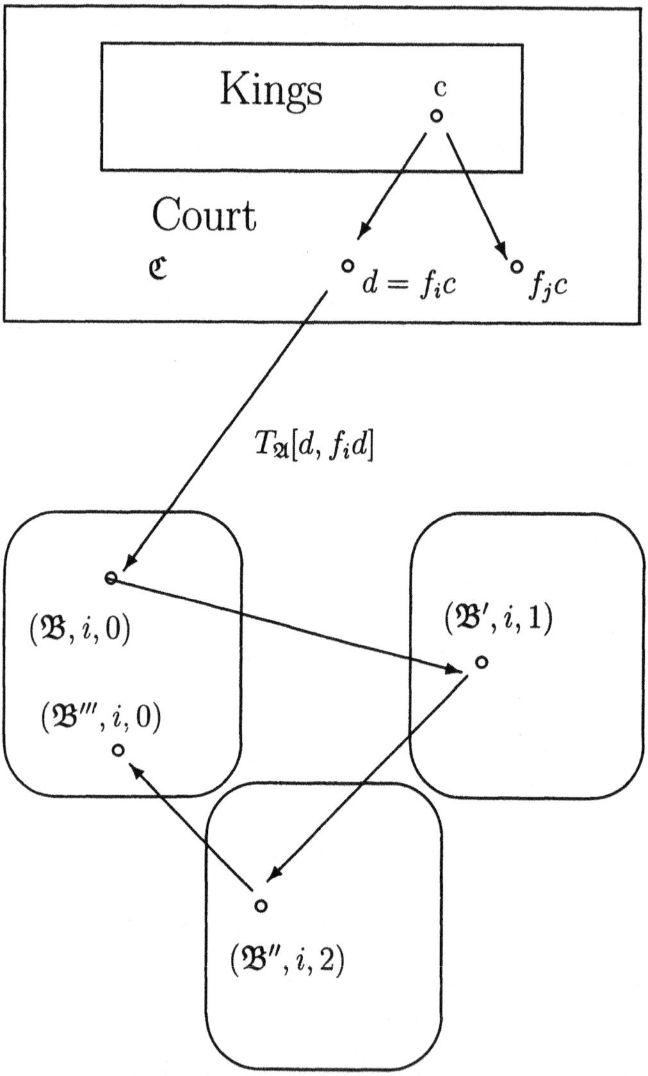

Figure 8.1. Providing witnesses for $\forall x \exists y \beta_i$ in $\mathfrak{D}$

Since σ contains only unary and binary predicates this completes the definition of $\mathfrak{D}$. It is easy to verify that $\mathfrak{D} \models \varphi$. Since a signature with r predicate symbols admits 2^r 1-tables the bound on the model size is established. □

This gives an almost optimal upper bound for the complexity of $Sat(L_2)$. The normal form φ associated with a given L_2-sentence ψ of length n can be constructed in polynomial time and has at most n predicate symbols. Thus only models up to size $2^{O(n)}$ need to be checked which by Lemma 6.0.4 can be done in nondeterministic exponential time.

Corollary 8.1.5 (Grädel, Kolaitis, Vardi). $Sat(L_2) \in \text{NTIME}(2^{O(n)})$.

Note that for formulae that are already in the normal form provided by Lemma 8.1.2, the complexity is $\text{NTIME}(2^{O(n/\log n)})$ since a formula of length n (when written with a fixed number of symbols) contains at most $O(n/\log n)$ distinct predicates. By Corollary 6.2.14 this bound is optimal, even for formulae in the class $[\forall\forall \wedge \forall\exists]$.

However, an L_2-sentence of length n may well have $\Omega(n)$ nested quantifiers, so it is not clear whether the upper bound $\text{NTIME}(2^{O(n/\log n)})$ can be achieved for arbitrary L_2-sentences.

Exercise 8.1.6 (L_2 with constants). Let L_k^+ be the extension of L_k to formulae that may contain constant symbols. Show that the small model property of Proposition 8.1.4 and the complexity bound of Corollary 8.1.5 also hold for L_2^+. Hint: Constants are kings. Use the same model construction as in the proof of Proposition 8.1.4 but instead of 2-tables, define $r+2$-tables $T_{\mathfrak{D}}[c_1, \ldots, c_r, d, e]$ where $c_1, \ldots, c_r$ are the interpretations of the constant symbols.

Note that this result cannot be extended to function symbols of positive arity. Indeed, since $[\forall, (0), (2)]_=$ and $[\forall^2, (0,1), (1)]$ are reduction classes, it follows that even the extension of L_1 by unary functions or the extension of L_2 (without equality) by a single unary function are undecidable for satisfiability.

Also other rather modest extensions of L_2 lead to undecidable satisfiability problems (see Sect. 5.3, and, for a broader treatment [211]).

Remark. An interesting two-variable logic is C_2, the extension of L_2 by counting quantifiers $\exists^{\geq m}$ and $\exists^{\leq m}$, for arbitrary $m \in N$. Note that C_2 does not have the finite model property; indeed the sentence

$$\forall x \exists^{=1} y Exy \wedge \forall x \exists^{\leq 1} y Eyx \wedge \exists x \forall y \neg Eyx,$$

asserts that E is the graph of an injective but not surjective function and is thus satisfiable only over infinite domains. However it was proved by Grädel, Otto and Rosen that the satisfiability problem for C_2 is decidable [210].

Remark. In applications of logic to computer science and linguistics (related e.g. to knowledge representation systems, automatic verification, concurrent systems) a number of logic problems arose that are closely related to the classical decision problem, but are not covered by the standard framework. Indeed, many logics used in computer science applications can be seen as (parts of) fragments of first-order logic. In many cases, however, the relevant fragments are *not* those determined by quantifier prefix and vocabulary that were traditionally studied by logicians. Among important fragments are those determined by the number of variables and, in fact, a number of logics used in computer science can be embedded into L_2 or C_2. In particular this is the case for *propositional modal logics* and for a number of *knowledge representation logics* (or *concept logics*) belonging to the so-called KL-ONE family [27]. Thus the results on decidability, upper complexity bounds and finite model property for L_2 and C_2 immediately imply corresponding results for those logics.

We note in this context that the standard propositional modal logics do not have the full expressive power of L_2 (see e.g. [21]); they also have lower decision complexity (see [242]). Andréka, van Benthem and Németi [22] study another interesting fragment of first-order logic, the so-called *guarded fragment.* It is characterized by the restriction that quantifiers can be used only in the form

$$\exists \bar{y}(\alpha(\bar{x},\bar{y}) \wedge \varphi(\bar{x},\bar{y}))$$

where $\alpha(\bar{x},\bar{y})$ is atomic and $\varphi(\bar{x},\bar{y})$ is itself a guarded formula whose free variables all have to occur in α. It is easy to see that propositional modal logic can be embedded into the guarded fragment of first-order logic. Andréka, van Benthem and Németi prove that the guarded fragment is decidable and that it has nice model-theoretic properties. A slightly weaker result was proved in [19].

8.2 Unification and Applications to the Decision Problem

8.2.1 Unification

Let X be a countable set of variables and Ω a vocabulary of function symbols. The set of Ω-terms with variables in X is denoted by $T(\Omega, X)$. A *substitution* is a function $\pi : X \to T(\Omega, X)$ such that $\pi(x) = x$ for all but finitely many $x \in X$. Since $T(\Omega, X)$ is the free Ω-algebra generated by X, a substitution uniquely extends to a homomorphism $\pi : T(\Omega, X) \to T(\omega, X)$; the term $\pi(t)$ is obtained by simultaneous replacement of all occurrences of variables x by $\pi(x)$.

A *unifier* of two terms $s, t \in T(\Omega, X)$ is a substitution π such that $\pi(s) = \pi(t)$. A substitution π is *more general* than a substitution τ if $\tau = \sigma\pi$ for some substitution σ.

An instance of the *unification problem* is a finite list $(s_1, t_1), \ldots, (s_n, t_n)$ of pairs of terms. A solution for such an instance is a unifier π (preferably a most general unifier) such that $\pi(s_1) = \pi(t_1), \ldots, \pi(s_n) = \pi(t_n)$.

Representing Terms by Labeled Dag's. Since a term may contain multiple occurrences of a particular subterm, it is often more efficient to represent terms by labeled directed acyclic graphs (dag's) instead of strings. A *labeled dag* for $T(\Omega, X)$ is a finite directed acyclic graph $G = (V, E)$ such that

- every node $v \in V$ has a unique label $\ell(v) \in \Omega \cup X$;
- for each variable $x \in X$, there is at most one node with label x;
- if v has label $f \in \Omega$, and f has arity k, then there are precisely k arcs leaving v; they have the labels $1, \dots, k$, respectively.

A root of a dag is a node of in-degree 0. The leaves are the nodes of out-degree 0. Note that leaves are labeled by a constant or a variable.

Every node v in a dag of this form represents a term $t_v \in T(\Omega, X)$:

- If v is a leaf, then $t_v := \ell(v)$.
- If the label of v is a function symbol f of arity k, and the arcs labeled $1, \dots, k$ lead to the vertices $v_1, \dots, v_k$, then $t_v = f t_{v_1} \cdots t_{v_k}$.

If G is a dag with single root w we also write t_G for the term t_w represented by the root of G. Note that a term, written as string of symbols, may be exponentially longer than its dag representation. Indeed, consider the dag G consisting of nodes $v_0, \dots, v_{n+1}$ labeled by a binary function symbol f, a node v_n with label x, and with two arcs labeled 1,2 from v_i to v_{i+1} for each $i \le n$ (see Fig. 8.2). Obviously $|t_G| \ge 2^n$.

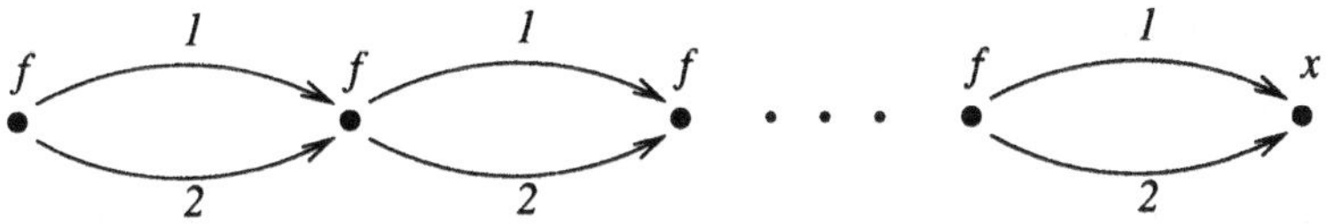

Figure 8.2. Compact dag-representation of a term

Theorem 8.2.1. *The unification problem is solvable in polynomial time.*

Proof. A given finite list I of pairs of terms (s, t) is represented by a dag G such that

- every term in I is represented by some node of G;
- distinct nodes of G represent distinct terms (i.e. the dag representation is concise);
- for each pair (s, t) the dag contains an undirected edge, labeled with the equality sign, that connects the nodes representing s and t.

Given the list I as a string of symbols, a corresponding dag representation can be constructed in polynomial time.

If I contains a pair (s, t) where s and t start with a different function symbol, then we can immediately stop and declare the terms to be non-unifiable *(clash failure)*. If s and t are identical terms then we can eliminate the pair (s,t) from I (in the dag representation this means that we eliminate self-loops). A substitution unifies two terms $s = fs_1 \dots s_k$ and $t = ft_1 \dots t_k$ if and only if it is a unifier of $(s_1,t_1),\dots,(s_k,t_k)$. We can thus remove the edge connecting the roots of s and t and replace it by k edges connecting the corresponding children of them.

This process is repeated until at least one term of each pair in the list is a variable. This takes only polynomial time and results in no significant increase in the size of the dag representation since we just manipulate edges.

Let (x,t) be a pair in the list where x is variable and $t \neq x$. We check whether x occurs in t; if yes, then we stop and declare the list to be non-unifiable *(occur check failure)*. Otherwise we replace the occurrences of x in all other terms of the list by t (i.e. we replace all arcs leading to x by arcs leading to the root of t). By repeated application of this procedure we either find that the list is non-unifiable or transform it into a finite collection of pairs $(x_1,s_1),\dots,(x_n,s_n)$ where $x_1,\dots,x_n$ are distinct variables and $s_1,\dots,s_n$ are terms that do not contain $x_1,\dots,x_n$. The substitution $\pi : x_i \mapsto s_i$ (for $i = 1,\dots,n$) is a unifier of the original list. Indeed, π obviously unifies $(x_1,s_1),\dots,(x_n,s_n)$ and all operations performed by the algorithm on the given list of pairs preserve the set of unifiers of the list. Obviously the algorithm runs in polynomial time. □

Exercise 8.2.2. Prove that the unifier constructed in the proof of Theorem 8.2.1 is in fact a most general unifier.

A dag is called *simple* if the only nodes with in-degree greater than 1 are leaves. Given a term t (as a string of symbols) a simple dag representing t can be constructed in linear time with logarithmic workspace. Conversely, given a simple dag G with a single root, the term represented by G can also be written out in linear time and logarithmic space.

We will show now, that the unification problem is in fact *complete* for polynomial time under log-space reductions. It is therefore unlikely that the unification problem is in NC, i.e. admits a parallel algorithm running in polylogarithmic time with polynomial hardware. Indeed, P-complete problems have efficient parallel algorithms only if NC = P. It is widely conjectured that NC is a proper subclass of P. We refer to [215, 416] for background on P-completeness and parallel complexity.

Theorem 8.2.3 (Dwork, Kanellakis, Mitchell). *The unification problem is log-space complete for* P*, even for simple dags.*

Proof. We reduce a variant of the circuit value problem (CVP) to the unification problem. Here a circuit is a directed acyclic graph with four types of nodes. *Input nodes* have no incoming edges and precisely one outgoing edge, called an *input edge.* NAND *nodes* have have two incoming edges and one outgoing edge. *Branching nodes* have one incoming edge and at least one outgoing edge. Finally the circuit has a unique *output node* with one incoming edge and no outgoing edge. Each input edge is labeled with either 0 or 1.

The labeling of input edges extends in the usual way to an assignment of Boolean values to all other edges of the circuit. The circuit value problem CVP is the problem of deciding, for a given a circuit of this form, whether the value assigned to the output edge is one. It is well known, that this problem is P-complete (see [215, Sect.6.2]).

Given a circuit C, we will construct a simple dag G representing two terms t_1, t_2 with roots r_1, r_2, respectively. For every edge e of C, we include in G two nodes $a(e,1)$ and $b(e,1)$ belonging to t_1 and two nodes $a(e,2)$ and $b(e,2)$ belonging to t_2. Given nodes u, v of G we write $u \sim v$ to indicate that the terms t_u and t_v are unifiable. The idea of the construction is to encode the Boolean value $f(e)$ of edge e by the possibilities to unify the terms $t_{a(e,i)}$ and $t_{b(e,j)}$: If $f(e) = 0$ then $a(e,1) \sim b(e,2)$ and $b(e,1) \sim a(e,2)$; if $f(e) = 1$ then $a(e,1) \sim a(e,2)$ and $b(e,1) \sim b(e,2)$.

For input edges, this property is ensured by making the nodes corresponding children of the roots. If e is the jth input edge and $f(e) = 1$, then (for $i = 1, 2$) there exist arcs labeled $2j - 1$ from r_i to $a(e,i)$ and arcs labeled $2j$ from r_i to $b(e,i)$. If $f(e) = 0$ for the jth input edge e, then we twist the arcs from r_2, i.e. the arc labeled $2j-1$ from r_2 leads to $b(e,2)$ and the arc labeled $2j$ from r_2 leads to $a(e,2)$.

For branching nodes and NAND nodes we add gadgets to the dag in order to propagate the unification properties correctly. Suppose that e is an incoming edge to a branching node with outgoing edges $e_1, \ldots, e_k$. Then, for each $j = 1, \ldots, k$ and $i = 1, 2$ we put arcs from $a(e,i)$ to $a(e_j,i)$ and from $b(e,i)$ to $b(e_j,i)$.

For NAND nodes with incoming edges e', e'' and outgoing edge e we add the subgraph shown in Fig. 8.3 (for simplicity we write $a'(1)$ for $a(e',1)$ etc.).

To complete the description of G we have to define the labeling of the nodes by function symbols and variables. Every node with $k > 0$ outgoing arcs is labeled with the k-ary function symbol f_k. If e is the output edge of G, then the nodes $a(e,1)$ and $a(e,2)$ are labeled with the constant symbol c and the nodes $b(e,1)$ and $b(e,2)$ with a different constant d. All other leaves are labeled by distinct variables.

It remains to show that the output edge of C has value 1 if and only if t_1 and t_2 are unifiable. This follows if we can prove that the encodings of Boolean values by unification properties is correctly propagated through the representations of the branching nodes and NAND gates of C in G. For branching nodes this is obvious.

For NAND-nodes the verification is split into two steps.

Claim 1. *The four possible assignments of Boolean values to e' and e'' force unifications among the terms represented by $u', v', w', u'', v'', w''$ as follows:*

- *if $f(e') = f(e'') = 0$ then $u' \sim u'', v' \sim v''$ and $w' \sim w''$;*
- *if $f(e') \neq f(e'')$ then $u' \sim v'', v' \sim w''$ and $w' \sim u''$;*
- *if $f(e') = f(e'') = 1$ then $u' \sim w'', v' \sim u'', w' \sim v''$.*

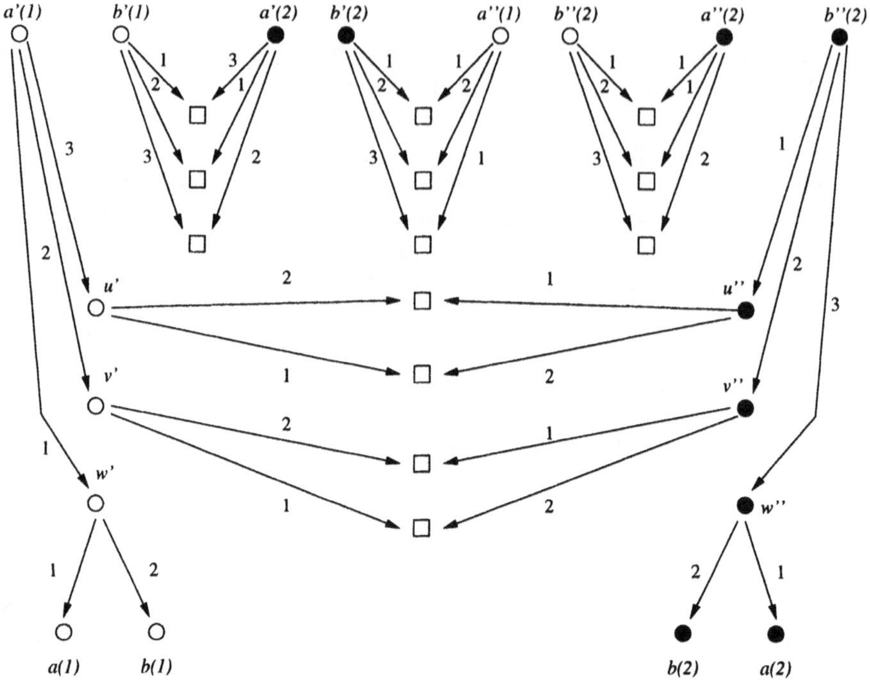

Figure 8.3. NAND gates

Claim 2. *The three possible patterns of unifications among the terms represented by* $u', v', w', u'', v'', w''$ *force unifications among the terms represented by* $a(e,1), b(e,1), a(e,2), b(e,2)$ *as follows:*

– *if* $u' \sim u'', v' \sim v''$ *and* $w' \sim w''$ *then* $a(e,1) \sim a(e,2)$ *and* $b(e,1) \sim b(e,2)$*;*
– *if* $u' \sim v'', v' \sim w''$ *and* $w' \sim u''$ *then* $a(e,1) \sim a(e,2)$ *and* $b(e,1) \sim b(e,2)$*;*
– *if* $u' \sim w'', v' \sim u'', w' \sim v''$ *then* $a(e,1) \sim b(e,2)$ *and* $b(e,1) \sim a(e,2)$*.*

Exercise 8.2.4. Verify these two claims.

Suppose that $f(e) = 0$ for the output edge e of C. If t_1 and t_2 were unifiable, then the unification properties would propagate through C in such a way to enforce that $a(e,1) \sim b(e,2)$. However, this is impossible since $a(e,1)$ and $a(e,2)$ are labeled with different constants. On the other side, the two roots can be unified if the output edge has value 1. □

Corollary 8.2.5 (Dwork, Kanellakis, Stockmeyer). *Unification is P-complete, even if both terms are represented by trees and have no common variable, each variable appears at most twice in a term and no function symbol has arity greater than two.*

Proof. Note that the dag G constructed in the proof of Theorem 8.2.3 consists of two trees that share only variable nodes. Let $x_1, \ldots, x_n$ be the variables appearing in both terms t_1, t_2. Replace the single occurrence of x_j in t_i by a new variable $x_{i,j}$. We force $x_{1,j}$ and $x_{2,j}$ to be equal by increasing the arity of the roots by n, and adding new arcs labeled j from the root r_i to a new leaf labeled $x_{i,j}$, for every $j = 1, \ldots, n, i = 1, 2$. The terms transformed in this way are unifiable if and only if the original terms are. Finally we reduce the maximal arity to two by replacing each subterm $f_k(s_1, \ldots, s_k)$ with a function symbol of arity $k > 2$ by $g(s_1, g(s_2, \ldots, g(s_{k-1}, s_k)) \cdots)$ where g_k is a new binary function symbol (see Fig. 8.4). □

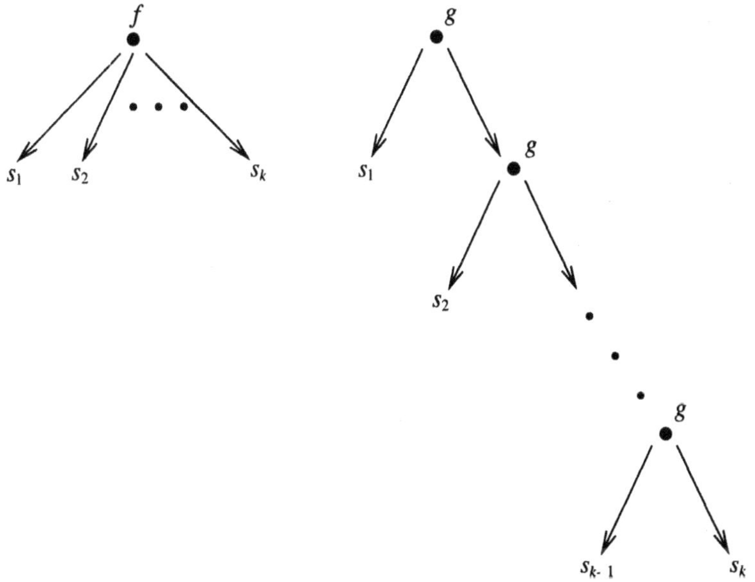

Figure 8.4. Elimination of function symbols with arity > 2

8.2.2 Herbrand Formulae.

A Herbrand formula is a first-order formula in prenex normal whose quantifier-free part is a conjunction of atomic and negated atomic formulae.

Theorem 8.2.6. *The satisfiability problem for Herbrand formulae without equality is P-complete.*

Proof. To decide whether a given Herbrand formula is satisfiable we first transform it to Skolem normal form. We then rename variables so that no variable occurs in more than one atom.

This preserves the propositional structure, so we obtain a sentence of the form

$$\psi := \forall x_1 \cdots \forall x_n \bigwedge_i \alpha_i \wedge \bigwedge_j \neg\beta_j$$

where α_i and β_j are atomic formulae $Pt_1 \cdots t_k$ that do not share variables. Due to Theorem 2.1.12 and the equivalences

$$\forall \bar{x}(\psi \wedge \varphi) \equiv \forall x\psi \wedge \forall \bar{x}\varphi \equiv \forall \bar{x}\psi \wedge \forall \bar{y}\varphi[\bar{x}/\bar{y}] \equiv \forall \bar{x}\forall \bar{y}(\psi \wedge \varphi[\bar{x}/\bar{y}])$$

the sentence obtained in this way has a Herbrand model if and only if the original sentence was satisfiable.

Clearly, ψ has a Herbrand model if and only if it does *not* contain two complimentary literals $Ps_1 \cdots s_k$ and $\neg Pt_1 \cdots t_k$ such that $(s_1, t_1), \ldots, (s_k, t_k)$ are unifiable. By Theorem 8.2.1 this is decidable in polynomial time.

A trivial reduction from the unification problem implies that the satisfiability problem for Herbrand formulae is P-complete. Indeed, a pair of terms (s, t) that do not share variables is unifiable if and only if the universal closure of $Ps \wedge \neg Pt$ (for a monadic predicate P) is unsatisfiable. □

Remark. For Herbrand formulae with functions and equality, satisfiability is undecidable. As we have proved in Chap. 4.1 this applies already to the class $[\forall, (0), (2)]_= \cap$ HERBRAND of Herbrand sentences with one universally quantified variable and two unary function symbols. In contrast, Wirsing [534] proved that the class $[all, all, (1)]_= \cap$ HERBRAND is decidable by reducing it to S2S, the monadic second-order theory of the infinite binary tree (see Chap. 7.1).

8.2.3 Positive First-Order Logic

The *positive fragment of first-order logic*, denoted FO^+, is the set of first-order formulae (of arbitrary vocabulary) in which the negation sign $\neg$ does not appear.

Note that $Sat(\mathrm{FO}^+)$ is trivial: every positive formula is satisfiable over a domain with only one element. However, the validity problem is more difficult.

Theorem 8.2.7 (Kozen). *The validity problem for* FO^+ *is* NP-*complete.*

Proof. To see that $Val(\mathrm{FO}^+) \in \mathrm{NP}$ we first show that we can restrict attention to sentences of purely functional vocabulary (i.e. without relation symbols besides equality). Given a σ-structure $\mathfrak{A} = (A, R_1, \ldots, R_s, f_1, \ldots, f_t)$, let $\mathfrak{A}_0 = (A, \varnothing, \ldots, \varnothing, f_1, \ldots f_t)$ be the structure with the same universe and the same functions as $\mathfrak{A}$, but where all relations are empty. We call the structures $\mathfrak{A}_0$ reduced.

Obviously, positive formulae are preserved under augmenting relations. Thus, if $\psi \in \mathrm{FO}^+$ and $\mathfrak{A}_0 \models \psi$ then $\mathfrak{A} \models \psi$. It therefore suffices to check whether ψ holds in all reduced structures. Let ψ_0 be the formula obtained from ψ by substituting *false* for all atomic formulae $R_i t_1 \cdots t_m$. The vocabulary of ψ_0 is functional and, since ψ and ψ_0 are equivalent on reduced structures, we have that

$$\psi_0 \in Val(\mathrm{FO}^+) \iff \psi \in Val(\mathrm{FO}^+).$$

We can thus assume that ψ is built from equalities $(s = t)$ of terms by means of conjunction, disjunction, existential and universal quantifiers. Since ψ is now purely functional, its Herbrand structure $\mathfrak{H}$ is uniquely determined.

We next reduce the validity problem for FO^+ to the problem of evaluating an *existential* sentence in the Herbrand structure $\mathfrak{H}$. Indeed, ψ is valid if and only if $\neg\psi$ is unsatisfiable. Let $\forall x_1 \cdots \forall x_k \eta$ (with η quantifier-free) be the Skolem normal form of $\neg\psi$. Note than the Skolem normal form can be computed in polynomial time and it preserves the propositional structure. Thus η is the negation of a positive formula and ψ is valid if and only if $\forall x_1 \cdots \forall x_k \eta$ is unsatisfiable. By Theorem 2.1.12, this is true if and only if $\mathfrak{H} \models \exists x_1 \cdots \exists x_k \neg\eta$.

We are now in a position to reduce $Val(\mathrm{FO}^+)$ *nondeterministically* to the unification problem. We just have to eliminate disjunctions.

Suppose that the given existential formula has the form

$$\exists x_1 \cdots \exists x_k \bigwedge_{i=1}^{n} \bigvee_{j=1}^{m_i} \varphi_{i,j}.$$

We nondeterministically choose in each conjunct $\bigvee_{j=1}^{m_i} \varphi_{i,j}$ one $\varphi_{i,j(i)}$ among the $\varphi_{i,j}$ and consider the simplified formula

$$\exists x_1 \cdots \exists x_k \bigwedge_{i=1}^{n} \varphi_{i,j(i)}.$$

Clearly, the original sentence is a logical consequence of the simplified one. Conversely, if $\mathfrak{H} \models \exists x_1 \cdots \exists x_k \bigwedge_{i=1}^{n} \bigvee_{j=1}^{m_i} \varphi_{i,j}$ then the right choice of the $\varphi_{i,j(i)}$ yields a sentence that holds in $\mathfrak{H}$.

This nondeterministic reduction step is repeated until all disjunctions are eliminated. The resulting formula has the form

$$\exists x_1 \cdots \exists x_k \bigwedge_{i=1}^{m} s_i = t_i.$$

Such a sentence holds in $\mathfrak{H}$ if and only if the terms $(s_1, t_1), \ldots, (s_m, t_m)$ are unifiable, which, by Theorem 8.2.1 can be determined in polynomial time.

This proves that $Val(\mathrm{FO}^+) \in \mathrm{NP}$.

To see that $Val(\mathrm{FO}^+)$ is NP-hard, we present a reduction from SAT. Given a Boolean formula ψ in conjunctive normal form with propositional variables $X_1, \ldots, X_n$, let

$$\exists 0 \exists 1 \exists x_1 \cdots \exists x_n \psi[X_i/(x_i = 1), \neg X_i/(x_i = 0)]$$

be the existential closure of the positive first-order formula that is obtained from ψ by replacing every positive literal X_i by the equality $(x_i = 1)$ and every negative literal $\neg X_i$ by $(x_i = 0)$. Clearly, this formula is valid if and only if ψ is satisfiable. □

Remark. Kozen also considered the more general *entailment problem* whether $\Sigma \models \psi$, where $\psi \in \mathrm{FO}^+$ and Σ is a finite set of atomic sentences and negations of atomic sentences of form $t_1 = t_2$ and $Rt_1 \cdots t_k$ where $t_1, \ldots, t_k$ are terms without variables. Theorem 8.2.7 implies that this problem is NP-hard even for $\Sigma = \varnothing$. Kozen [322] proved that it is NP-complete for arbitrary Σ.

8.3 Decidable Classes of Krom Formulae

Recall that a Krom formula is a first-order formula whose quantifier-free part is a conjunction of Krom clauses, i.e. of subformulae $(\alpha \vee \beta)$ where α and β are atoms or negated atoms.

In Chap. 5.1.1 we proved that the following prefix classes of Krom sentences without functions and equality are reduction classes (even when restricted to Horn sentences):

- $[\forall\exists^*\forall]$ (Krom 1970)
- $[\exists\forall\exists\forall], [\forall\exists^2\forall]$ (Aanderaa and Börger 1971, Orevkov 1973)
- $[\forall^2\exists\forall], [\forall\exists\forall^2]$ (Lewis 1972)

In this section we show that the satisfiability problem is decidable for the Aanderaa-Lewis class $[\forall\exists\forall] \cap \mathrm{KROM}$ and for the Maslov class $[\exists^*\forall^*\exists^*] \cap \mathrm{KROM}$.

The remaining Krom classes in pure predicate logic are $[\forall\exists\forall\exists^k] \cap \mathrm{KROM}$ (with $k > 0$) and $[\forall\exists\forall\exists^*] \cap \mathrm{KROM}$. It is open whether these classes are decidable for satisfiability.

8.3.1 The Chain Criterion

We describe a convenient graph-theoretic criterion for the unsatisfiability of Krom formulae. We start with the propositional case.

Let Φ be a set of propositional Krom clauses $(Y \vee Z)$ where Y, Z are literals, i.e. propositional variables or their negations. (To simplify notation we always identify $\neg\neg X$ with X.)

With Φ we associate a directed graph $G(\Phi)$ whose vertices are the literals of Φ, i.e. the propositional variables appearing in Φ and their negations. There is an arc in $G(\Phi)$ from Y to Z if and only if some clause of Φ is equivalent to the implication $(Y \to Z)$. (Note that a clause $(Y \vee Z)$ gives two arcs $(\neg Y \to Z)$ and $(\neg Z \to Y)$ and a clause consisting of a single literal Y is equivalent to the implication $\neg Y \to Y$).

Lemma 8.3.1 (Chain Criterion). *A set Φ of propositional Krom clauses is unsatisfiable if and only if there exists a variable X and a cycle in $G(\Phi)$ that contains both X and $\neg X$.*

Proof. We write $Y \xrightarrow{\Phi} Z$ if there is a path in $G(\Phi)$ from Y to Z. Let $V(\Phi)$ be the set of propositional variables of Φ. Suppose that there exists an assignment $\varepsilon : V(\Psi) \to \{0,1\}$ that makes Φ true. Clearly, if $\varepsilon(Y) = 1$ and $Y \xrightarrow{\Phi} Z$, then also $\varepsilon(Z) = 1$. It is therefore impossible that $X \xrightarrow{\Phi} \neg X \xrightarrow{\Phi} X$, i.e. there cannot be cycle in $G(\psi)$ containing both X and $\neg X$.

Conversely, suppose that there is no variable X with $X \xrightarrow{\Phi} \neg X \xrightarrow{\Phi} X$. An assignment ε that satisfies Φ can be constructed as follows.

Initially, let S be the set of all literals of Φ. Pick any literal $Y \in S$ such that not $Y \xrightarrow{\Phi} \neg Y$. Set $\varepsilon(Y) = 1$ and remove Y and $\neg Y$ from S; further, for all Z such that $Y \xrightarrow{\Phi} Z$, set $\varepsilon(Z) = 1$ and remove $Z, \neg Z$ from S. Repeat this procedure until S is empty.

No matter how the literals Y are chosen, there never arises a conflict, in the sense that for some Z, both $Y \xrightarrow{\Phi} Z$ and $Y \xrightarrow{\Phi} \neg Z$. Indeed it would the follow that also $Z \xrightarrow{\Phi} \neg Y$ and $\neg Z \xrightarrow{\Phi} Y$ and therefore $Y \xrightarrow{\Phi} \neg Y \xrightarrow{\Phi} Y$, contradicting our assumption.

Moreover, any assignment ε constructed in this way satisfies Φ. Otherwise, Φ would contain a clause $(U \vee V)$ such that $\varepsilon(U) = \varepsilon(V) = 0$. This means that in the course of the procedure the literals $\neg U, \neg V$ have been set *true*. Suppose that this happened first with $\neg U$. But since there is an arc $\neg U \to V$ in $G(\Phi)$, the procedure then would have set $\varepsilon(V) = 1$. □

Corollary 8.3.2. 2-SAT, *the satisfiability problem for propositional Krom-formulae, is in* NLOGSPACE.

Exercise 8.3.3. Prove this corollary (use that NLOGSPACE is closed under complementation). Further, prove that 2-SAT is in fact complete for NLOGSPACE via log-space reduction.

Remark. Using more general variants of the chain criterion, Aspvall, Plass and Tarjan [25] present a linear time algorithm for evaluating Krom sentences of quantified propositional logic and Grädel [207] proved that this problem is also complete for nondeterministic logarithmic space.

For future reference we make the following simple observation.

Lemma 8.3.4. *Suppose that there is a cycle of $G(\Phi)$ containing X and $\neg X$. Then for any literal Y on that cycle, there also exists a cycle containing Y and $\neg Y$.*

Proof. This is an immediate consequence of the fact that with every path from X to Y, $G(\Phi)$ also contains a path from $\neg Y$ to $\neg X$. □

Although we have formulated the chain criterion for sets of propositional Krom clauses, it is applicable also in the context of first-order Krom formulae.

Indeed, let ψ be a first-order Krom sentence and $\forall x_1 \cdots \forall x_k \varphi$ its functional form. The *Herbrand expansion* $E(\psi)$ of ψ is the (in general infinite) conjunction over all formulae $\varphi[t_1, \ldots, t_k]$ obtained by substituting terms from the Herbrand universe H of ψ for the variables.

A *literal* of $E(\psi)$ is an atomic statement $Pt_1 \cdots t_k$ of $E(\psi)$ or the negation of such. In the same way as above we construct a (possibly infinite) directed graph $G(\psi)$ whose nodes are the literals of $E(\psi)$ and apply the chain criterion.

Lemma 8.3.5. *$E(\psi)$ is inconsistent (and thus ψ unsatisfiable) if and only if there is a cycle in $G(\psi)$ containing an atomic statement and its negation.*

A simple application of this proves the PSPACE-completeness for the restriction of the Bernays-Schönfinkel class to Krom formulae.

Recall that we already proved in Sect. 2.2.4 that satisfiability of relational Krom formulae with prefix $\exists^2\forall^*$ is a PSPACE-hard problem (see Theorem 2.2.50). The chain criterion gives a corresponding upper bound.

Theorem 8.3.6. *The satisfiability problem for $[\exists^*\forall^*] \cap$ KROM is PSPACE-complete.*

Proof. A relational sentence of form $\exists x_1 \ldots \exists x_n \forall y_1 \cdots \forall y_m \varphi$ (where φ is quantifier-free) has Herbrand universe $H = \{c_1, \ldots, c_n\}$. Thus the Herbrand expansion $E(\psi)$ is a finite *propositional* Krom formula

$$E(\psi) := \bigwedge_{u_1 \in H} \cdots \bigwedge_{u_m \in H} \varphi[x_1/c_1, \ldots, x_n/c_n, y_1/u_1, \ldots, y_m/u_m]$$

where we consider the atomic statements $Pt_1 \cdots t_k$ as propositional variables.

It thus suffices to show, that it can be checked with polynomial workspace (with respect to the length of the original formula) whether there is a cycle in the graph $G(\psi)$ that contains some atom $P\bar{t}$ and also its negation. Note that $G(\psi)$ may have exponentially many vertices. However it is not necessary to construct ψ and $G(\psi)$ explicitly.

Instead, we use a *nondeterministic* procedure to guess a literal $P\bar{t}$ and an appropriate cycle through $G(\psi)$ containing both $P\bar{t}$ and $\neg P\bar{t}$; at each moment the only data stored on the work-tape are $P\bar{t}$, the current arc $R\bar{u} \to R'\bar{v}$ of the cycle, and the single bit, whether $\neg P\bar{t}$ has already been reached.

Suppose that the algorithm has already established that there is a path from $P\bar{t}$ to $R\bar{u}$. In the next step, a literal $R'\bar{v}$ is guessed such that $R\bar{u} \to R'\bar{v}$ is an arc of $G(\psi)$; this is the case iff there exist a clause $(\neg R\bar{z} \vee R'\bar{z}')$ of φ and a substitution taking $\bar{z}$ to $\bar{u}$ and $\bar{z}'$ to $\bar{v}$. Then $R\bar{u}$ is replaced by $R'\bar{v}$. This is repeated until $\neg P\bar{t}$ is reached; then proceeds in the same way to find a path from $\neg P\bar{t}$ to $P\bar{t}$. If the algorithm finds a cycle, it accepts. If after 2^n iterations, no cycle has been found the algorithm rejects.

Thus, we have described a nondeterministic algorithm requiring polynomial space, which determines (un)satisfiability of φ. Since nondeterministic algorithms can be simulated by deterministic ones with only quadratic increase of space, the result follows. □

Exercise 8.3.7. Prove that the satisfiability problem for $[\exists^*\forall^k] \cap$ KROM is complete for NLOGSPACE, for any fixed k.

Exercise 8.3.8. [106] Prove that the satisfiability of monadic Krom sentences, i.e. of the class $[all, (\omega)] \cap$KROM is decidable in polynomial time. Hint: Let ψ be a monadic Krom sentences in prenex normal form with quantifier-free part φ. Transform ψ to a new formula be replacing φ with $\varphi \wedge \alpha$ where α is the conjunction of

(i) all Krom clauses that are tautological consequences of clauses of φ, and

(ii) all Krom clauses C that can be obtained from a clause of φ by a substitution of universally quantified variables by arbitrary variables of φ, such that C contains either two universal variables, or an existential variable x and a universal variable y dominated by x.

Repeat this process till a fixed point ψ' is reached. Prove that

1. ψ is satisfiable if and only if ψ' is.
2. ψ' is satisfiable if and only if its quantifier-free part is propositionally consistent.
3. The length of ψ' is polynomially bounded with respect to the length of ψ.

In fact, the satisfiability problem for monadic Krom sentences is complete for P [106].

8.3.2 The Aanderaa-Lewis Class

In this section and the next one we consider the two maximal decidable relational Krom prefix classes.

Theorem 8.3.9 (Aanderaa, Lewis). *The satisfiability problem for* $[\forall\exists\forall] \cap$ KROM *is decidable.*

The proof proceeds as follows. First it is shown that the Aanderaa-Lewis class can be reduced to a subclass of sentences containing only binary atoms with certain restrictions of the pattern of variables that occur inside the clauses.

We then consider the chain criterion explained in the previous section. Note that the Herbrand universe for an $\forall\exists\forall$-sentence ψ can be identified with ω, and the Skolem function for the existential variable corresponds to the successor function. Thus the vertex set of $G(\psi)$ consists of the literals Pmn and $\neg Pmn$ where $m, n \in \omega$ and P is a predicate of ψ, Given a natural number k, we write $G(\psi, k)$ for the subgraph of $G(\psi)$ containing only literals Smn with $m, n \leq k$.

The proof of Theorem 8.3.9 reduces the unsatisfiability of the given $\forall\exists\forall$-Krom sentence ψ to arithmetic considerations on linear Diophantine equations. This will imply that in fact the chain criterion for a small subgraph $G(\psi, k)$ of $G(\psi)$, with k polynomially bounded in $|\psi|$, is necessary and sufficient for the unsatisfiability of ψ.

In fact these considerations will imply that $Sat([\forall\exists\forall] \cap \text{KROM})$ is complete for NLOGSPACE.

Exercise 8.3.10. Show that the satisfiability problem for $\forall\exists\forall$-Krom sentences is log-space reducible to that for $\forall\exists\forall$-Krom sentences with binary predicates only. Hint: $\forall\exists\forall$-sentences have only two independent variables. For details, see [133].

Definition 8.3.11. A *signed predicate* is a predicate symbol either alone or preceded by a negation sign. The functional form of an $\forall\exists\forall$-Krom sentence is $\forall x \forall y \varphi(x, x', y)$. The clauses of φ have the form $(Suv \vee S'wz)$ where S, S' are signed predicates and $u, v, w, z \in \{x, x'y\}$. A clause is *monadic* if it is of form $(Sxx \vee S'xx)$ and it is *elementary* it is of one of the forms

$$(Sxy \vee Sxy), \qquad (Syx \vee S'yx), \qquad (Sxy \vee S'x'y), \qquad (Syx \vee S'yx').$$

Exercise 8.3.12. Show that the satisfiability problem for binary $\forall\exists\forall$-Krom sentences is log-space reducible to that for binary $\forall\exists\forall$-Krom sentences with monadic and elementary clauses only [133, pp. 235–237].

In the sequel we assume that $\psi := \forall x \exists u \forall y \varphi$ is an $\forall\exists\forall$-Krom sentence with only binary predicates and only monadic and elementary clauses. By the chain criterion, ψ is unsatisfiable if and only if there exists a predicate P and numbers $m, n \in \omega$ such that both Pmn and $\neg Pmn$ lie on a cycle of $G(\psi)$. We show next that we can in fact assume $m = n$.

Indeed, let C be the cycle containing Pmn and $\neg Pmn$. If C contains a literal Srr (for some signed predicate S and $r \in \omega$) then by Lemma 8.3.4 both Srr and its negation lie on a cycle of $G(\psi)$. Otherwise ψ remains unsatisfiable even if we omit all its monadic clauses.

An *elementary chain* is a path in $G(\psi)$ defined by elementary clauses only. Clearly, if $G(\psi)$ contains an elementary chain from Sij to $S'k\ell$ then it also contains elementary chains from $S\, i+p\; j+q$ to $S\, k+p\; \ell+q$, for all natural numbers p, q.

Let $p = \max(m, n) - m$, $q = \max(m, n) - n$ and $r = m + p = n + q = \max(m, n)$. Given that $G(\psi)$ contains elementary chains from Pmn to $\neg Pmn$ and back, it follows that both Prr and $\neg Prr$ lie on an elementary cycle of $G(\psi)$. We have proved:

Lemma 8.3.13. *ψ is unsatisfiable if and only there exists an atomic statement Prr such that both Prr and $\neg Prr$ lie on a cycle of $G(\psi)$.*

We now show how this problem is related to a problem on linear Diophantine equations. In the sequel Γ is the set of signed predicates of ψ and S, S' always stand for signed predicates.

Definition 8.3.14. A chain from S to S' is a path in $G(\psi)$ from a literal Smn to a literal $S'pq$. The *yield* $Y(C)$ of a chain from Smn to $S'pq$ is the pair $(p-n, q-m) \in \mathbb{Z} \times \mathbb{Z}$.

Let E be the set of *elementary* chains in $G(\psi)$ and let K be the set of chains that begin and end with monadic atomic statements. Obviously chains in K can be decomposed into subchains that are either elementary or consist of a single arc coming from a monadic clause.

For any $S, S' \in \Gamma$ let $\Xi(S, S') \subseteq \mathbb{Z} \times \mathbb{Z}$ be the set of yields of elementary chains from S to S'. Further, let $\Theta(S, S')$ be the set of yields (p,p) of chains in K from S to S'.

We thus have the following modified criterion for the unsatisfiability of ψ.

Lemma 8.3.15. *ψ is unsatisfiable if and only there exists a predicate P such that*

$$(0,0) \in \Theta(P, \neg P) \cap \Theta(\neg P, P).$$

Semilinear Sets. Consider equations of the form $\boldsymbol{a} \cdot \boldsymbol{x} = b$ where $\boldsymbol{a} \in \mathbb{Z}^n$, $b \in \mathbb{Z}$ and $\boldsymbol{u} \cdot \boldsymbol{v} := \sum_{i=1}^{n} u_i v_i$. We are interested in solutions over the natural numbers, i.e. in $\boldsymbol{v} \in \mathbb{N}^n$ such that $\boldsymbol{a} \cdot \boldsymbol{v} = b$. Such a solution $\boldsymbol{v}$ is *minimal* if $\boldsymbol{v} \neq \boldsymbol{0}$ and there is no solution $\boldsymbol{u} \neq \boldsymbol{v}$ such that $u_i \leq v_i$ for all $i \leq n$.

Lemma 8.3.16. *If $\max(|a_1|, \ldots, |a_n|, |b|) \leq c$ and $\boldsymbol{v}$ is a minimal solution of $\boldsymbol{a} \cdot \boldsymbol{x} = b$, then $\max(v_1, \ldots, v_n) \leq nc^2$.*

Proof. Clearly $v_i = 0$ if $a_i = 0$. Thus we can eliminate these components and assume that $a_i \neq 0$ for all i. If all $v_i < c$ then we are done. Otherwise, we can assume that $v_i \geq c$ for $i = 1, \ldots, k$ and $v_i < c$ for $i = k+1, \ldots, n$ (where $k \geq 1$).

Suppose that among $a_1, \ldots, a_k$ there are both positive and negative components, e.g. $a_1 > 0$ and $a_2 < 0$. Then $\boldsymbol{u} = \boldsymbol{v} + (a_2, -a_1, 0, \ldots, 0)$ is a smaller solution of the equation than $\boldsymbol{v}$ which contradicts the minimality of $\boldsymbol{v}$. Hence $a_1, \ldots, a_k$ are all positive or all negative. It follows that

$$|\sum_{i=1}^{k} a_i v_i| = |b - \sum_{i=k+1}^{n} a_i v_i| \leq nc^2$$

and therefore $v_i \leq nc^2$ for all i. □

Definition 8.3.17. For $\boldsymbol{u} \in \mathbb{Z}^n$ and a finite set $V \subset \mathbb{Z}^n$, let

$$L(\boldsymbol{u}, V) := \{\boldsymbol{u} + \boldsymbol{v}_1 + \cdots + \boldsymbol{v}_m : m \in \mathbb{N}, \boldsymbol{v}_i \in V\}.$$

A *semilinear set* is a subset of $\mathbb{Z}^n$ that can be written as a finite union of sets of form $L(\boldsymbol{u}, V)$. A *presentation* of an element of a semilinear set is a decomposition as a sum of $\boldsymbol{u}$ and $\boldsymbol{v}_i$ and the size of the presentation is the number of summands in the decomposition.

Exercise 8.3.18. Let $U, V \subseteq \mathbb{N}^n$ be the sets of minimal solutions of, respectively, $\boldsymbol{a} \cdot \boldsymbol{x} = b$ and $\boldsymbol{a} \cdot \boldsymbol{x} = \boldsymbol{0}$. Prove that the set of all solutions of $\boldsymbol{a} \cdot \boldsymbol{x} = b$ is the semilinear set $\bigcup_{\boldsymbol{u} \in U} L(\boldsymbol{u}, V)$.

Let $s = |\Gamma|$ denote the number of signed predicates in ψ.

Note that $\Xi(S, S')$ and $\Theta(S, S')$ can be seen as functions F that associate with every sentence ψ (on the special form considered here) a set $F(\psi) \subseteq \mathbb{Z} \times \mathbb{Z}$, We call such functions F *tractable* if there exist polynomials p and q such that for all ψ

(i) $F(\psi)$ is semilinear and admits a presentation in which every integer has magnitude bounded by $p(s)$.
(ii) Any element of $F(\psi)$ which has a presentation of size N is the yield of some chain in $G(\psi, Nq(s))$.

We will show that $\Xi(S, S')$ and $\Theta(S, S')$ are tractable for all $S, S' \in \Gamma$, and that the polynomial bounds p and q are independent of S and S'.

An (S, S')-word is a word $W \in \Gamma^*$ that begins with S and ends with S'. Suppose that for all signed predicates $S, S' \in \Gamma$, we have a set $X(S, S') \in \mathbb{Z} \times \mathbb{Z}$. For each sequence $W = S_0 \cdots S_n \in \Gamma^*$, let

$$\Sigma_X(W) := \{\boldsymbol{u}_0 + \cdots + \boldsymbol{u}_{n-1} : \boldsymbol{u}_i \in X(S_i, S_{i+1})\}.$$

We will use the following simple combinatorial fact on sets $\Sigma_X(W)$.

Lemma 8.3.19. *For all $S, S' \in \Gamma$ and $\boldsymbol{w} \in \mathbb{Z} \times \mathbb{Z}$ there exists an (S, S')-word W with $\boldsymbol{w} \in \Sigma_X(W)$ if and only if there words $U, Z_1, \ldots, Z_n \in \Gamma^*$ such that*

(i) U is an $(S'S')$-word.
(ii) Z_i is an (S_i, S_i) – word for some S_i occurring in U.
(iii) Each of $U, Z_1, \ldots, Z_n$ has length $\leq s^2 + 1$.
(iv) $\boldsymbol{w} = \boldsymbol{u} + \sum_{i=1}^{n} \boldsymbol{z}_i$ where $\boldsymbol{u} \in \Sigma_X(U)$ and $\boldsymbol{z}_i \in \Sigma_X(Z_i)$.

Exercise 8.3.20. Prove this Lemma. Hint: In one direction, apply the pigeonhole principle to excise sequences $Z_1, \ldots, Z_n$ from the given sequence W until it is short. In the other direction construct W by interpolating Z_i at an occurrence of S_i in U (see [133, pp. 238–239]).

We will apply Lemma 8.3.19 for several different definitions of $X(S,S')$. First, let $E(S,S') \in \mathbb{Z}\times\mathbb{Z}$ be the set of pairs (p,q) such that for all $m,n \geq 1$, the implication $(Smn \to S'm+pn+q)$ is equivalent to an instance of an elementary clause of ψ. Clearly $E(S,S') \subseteq \{(0,0),(0,1),(1,0),(0,-1),(-1,0)\}$.

Lemma 8.3.21. *For all $S,S' \in \Gamma$, $\Xi(S,S')$ is semilinear. In fact $\Xi(S,S')$ is the union of sets $L(\boldsymbol{u},V)$ where every integer occurring in $\boldsymbol{u}$ or an element of S has absolute value bounded by s^2. Thus, $\Xi(S,S')$ is tractable.*

Proof. If W is an (S,S')-word of length k such that $(p,q) \in \Sigma_E(S,S')$ then there are elementary chains in $G(\psi)$ from Smn to $S'm+pn+q$ for all $m,n \geq k-1$. Conversely, the yield of any elementary chain from S to S' clearly belongs to $\Sigma_E(W)$ for some W. Thus $\Xi(S,S')$ is the union over all sets $\Sigma_E(W)$ for (S,S')-words W. By the previous Lemma, each $\Sigma_E(W)$ can be represented as

$$L(\boldsymbol{u},\Sigma_E(Z_1)\cup\cdots\cup\Sigma_E(Z_n))$$

where $\boldsymbol{u} \in \Sigma_E(U)$ for some (S,S')-word U of length $\leq s^2+1$ and where $\{Z_1,\ldots,Z_n\}$ is the set of all words in Γ^* that and have length $\leq s^2+1$ and begin and end with the same symbol (which must occur also in U).

Finally, note that for words W of length m, $\Sigma_E(W)$ contains only pairs (p,q) with $|p|,|q| \leq m-1$. □

The *diagonal* of a set $V \in \mathbb{Z}\times\mathbb{Z}$ is $D(V) := V \cap \{(m,m) : m \in \mathbb{Z}\}$.

Lemma 8.3.22. *For all S,S', $D(\Xi(S,S'))$ is tractable.*

Proof. Let $\boldsymbol{a} \in \mathbb{Z}\times\mathbb{Z}$, $B = \{\boldsymbol{b}_1,\ldots,\boldsymbol{b}_m\} \subseteq \mathbb{Z}\times\mathbb{Z}$, and let (p,p) be in the diagonal of $L(\boldsymbol{a},B)$. Thus, there exists $\boldsymbol{x} = (x_1,\ldots x_m) \in \mathbb{N}^m$ such that

$$p = a_1 + \sum_{i=1}^{m} b_{i1}x_i = a_2 + \sum_{i=1}^{m} b_{i2}w_i$$

i.e., $\boldsymbol{w}$ is a solution of

$$(*) \qquad \sum_{i-1}^{m}(b_{i1}-b_{i2})x_i = a_2 - a_1.$$

As noted in Exercise 8.3.18 this implies that $\boldsymbol{w} \in L(\boldsymbol{u},V)$ where $\boldsymbol{u}$ is a minimal solution of $(*)$ and V is the set of minimal solutions for the associated homogeneous equation. Let $V = \{\boldsymbol{v}_1,\ldots,\boldsymbol{v}_n\}$ and $\boldsymbol{w} = \boldsymbol{u}+\sum_{j=1}^{n} e_j\boldsymbol{v}_j$ where $e_1,\ldots,e_n \geq 0$. Thus

$$p = a_1 + \textstyle\sum_{i=1}^{m}(u_i + \sum_{j=1}^{n} e_j v_{ji})b_{i1} = a_1 + \sum_{i=1}^{m} u_i b_{i1} + \sum_{j=1}^{n} e_j \sum_{i=1}^{m} v_{ji} b_{i1}.$$

Thus $D(L(\boldsymbol{a},B))$ is semilinear: For can be written as the union, taken over all minimal solutions $\boldsymbol{u}$ of (*), of the sets of pairs (p,p) where

$$p \in L(a_1 + \textstyle\sum_{i=1}^{m} u_i b_{i1}, \{\sum_{i=1}^{m} v_{ji} b_{i1} : j = 1,\ldots,k\}).$$

Taking the union of this expression over all pairs $(\boldsymbol{a}, B)$ in a presentation of $\Xi(S,S')$ we obtain a presentation for $D(\Xi(S,S')$. The numbers in the presentation of $\Xi(S,S')$ are bounded in magnitude by s^2, so $m < (2s^2+1)^2$ and with Lemma 8.3.16 it follows that the absolute values of the components of $\boldsymbol{u}$ and $\boldsymbol{v}_i$ are bounded by $(2s^2+1)^2 s^4$. Applying Lemma 8.3.16 once more we infer that all integers in the presentation of $D(\Xi(S,S'))$ are polynomially bounded in s. Finally, suppose that we have a presentation of size N of $(p,p) \in D(\Xi(S,S'))$. Thus

$$p = a_1 + \sum i = 1^m u_i b_{i1} + \sum_{j=1}^N \sum_{i=1}^m v_{ji} b_{i1}.$$

This is a sum of a_1 and the b_{i1} with at most $m \max\{u_i\} + Nm \max\{v_j i\} N$ terms. Thus, we obtain a presentation of (p,p) as a member of $\Xi(S,S')$ of size bounded by N times a polynomial in s. Thus, since $\Xi(S,S')$ is tractable, so is its diagonal. □

Let $Y(S,S') = \Xi(S,S') \cup \{(0,0)\}$ if some clause of ψ is equivalent to $(Sxx \to S'xx)$ and $Y(S,S') = \Xi(S,S')$ otherwise. Clearly, $Y(S,S')$ is tractable.

If $(p,p) \in Y(S,S')$ then for all sufficiently large m, there exist chains in K from Smm to $S'\,m+p\;m+p$; either the chain is elementary or $p=0$ and the chain consist of a single arc from Smm to $S'mm$. Hence, for all (S,S')-words W such that $(p,p) \in \Sigma_Y(W)$ there are chains in $G(\psi)$ from Smm to $S'\,m+p\;m+p$ for all large enough m. On the other side every chain from Smm to $S'\,m+p\;m+p$ can be decomposed into subchains that are either elementary or consist of a single arc coming from a monadic clause. Thus for every $(p,p) \in \Theta(S,S')$ there exists an (S,S')-word W such that $(p,p) \in \Sigma_Y(W)$.

Hence $\Theta(S,S')$ is the union of the sets $\Sigma_Y(W)$ for (S,S')-words W. Again we apply Lemma 8.3.19 to decompose W into words of length $\leq s^2+1$.

Lemma 8.3.23. *$\Theta(S,S')$ is tractable, for all $S,S' \in \Gamma$.*

Proof. Let $W = S_1 \cdots S_k \in \Gamma^*$, with $k \leq s^2+1$ and let each set $Y(S_i,S_{i+1})$ be presented as in the previous lemma. Then $\Sigma_Y(W)$ is the union of all sets $L(\boldsymbol{u}_1+\cdots+\boldsymbol{u}_{k-1}, V_1 \cup \cdots \cup V_{k-1})$ such that $(\boldsymbol{u}_i, V_i)$ is in the given presentation of $Y(S_i,S_{i+1})$. Clearly, $\Sigma_Y(W)$ is tractable. Let q be the polynomial such that, for all such W, every element of $\Sigma_Y(W)$ with a presentation of size N is the yield of some chain in $G(\psi, Nq(s))$ and let $q'(s)$ bound the numbers in the presentation of $\Sigma_Y(W)$.

For any word $U \in \Gamma^*$ of length $\leq s^2+1$, let $\mathfrak{S}_U$ be the set of $\boldsymbol{a}_0 + \boldsymbol{a}_1 + \cdots + \boldsymbol{a}_m$ where $m \geq 0$, $\boldsymbol{a}_0 \in \Sigma_Y(U)$ and $\boldsymbol{a}_i \in \Sigma_Y(Z_i)$ for some word $Z_i \in \Gamma$ of length $\leq s^2+1$ that begins and ends with s_i for some S_i occurring in U. By Lemma 8.3.19 and the arguments given above, $\Theta(S,S')$ is the union of of the sets $\mathfrak{S}_U$ taken over all words $U \in \Gamma$ that have length $\leq s^2+1$, begin with S and end with S'. For such a $\mathfrak{S}_U$, a presentation is given by the union of the sets

$$L(b_0 + b_1 + \cdots + b_r, V_0 \cup V_1 \cup \cdots \cup V_r \cup \{b_1, \ldots, b_r\})$$

where (b_0, V_0) is in the presentation of $\Sigma_Y(U)$ and, for each i, (b_i, V_i) is in the presentation of $\Sigma_Y(Z_i)$ for some word of length $\leq s^2 + 1$ that begins and ends with the same symbol of U; moreover we can assume that all (b_i, V_i) are distinct (since we have included the set $\{b_1, \ldots, b_r\}$ with the union of the V_i).

To find a bound on r, note that each b_i and each element of V_i is bounded in magnitude by $q'(s)$. The union of all the V_i has therefore at most $2q'(s)+1$ elements. If $r > q'(s)(2q'(s)+1)$ then there is a j such that $V_j \subseteq \bigcup_{i \neq j} V_i$ and $b_j = b_i$ for some $i \neq j$ so we can simplify the given expression. It follows that r can be bounded by a polynomial $q''(s)$ and therefore also the magnitude of $b_0 + b_1 + \cdots + b_r$ is polynomially bounded in s.

The tractability of $\Theta(S, S')$ now follows easily. □

We can now summarize the proof of Theorem 8.3.9 and, in fact, determine precisely the complexity of $Sat([\forall\exists\forall] \cap \text{KROM})$.

Given an $\forall\exists\forall$-Krom sentence we first reduce it (using only logarithmic work space) to an $\forall\exists\forall$-Krom sentence ψ with only binary predicates and only monadic and elementary clauses. By Lemma 8.3.15 ψ is unsatisfiable if and only there exists a predicate P such that

$$(0,0) \in \Theta(P, \neg P) \cap \Theta(\neg P, P).$$

Since the sets $\Theta(P, \neg P)$ and $\Theta(\neg P, P)$ are tractable, the chain criterion for unsatisfiability need to be applied only to $G(\psi, p(s))$ where p is a polynomial and s is the number of signed predicates of ψ. This can be done with nondeterministic logarithmic space.

Clearly the problem is also hard for NLOGSPACE (since already the satisfiability problem for propositional Krom formulae is).

Corollary 8.3.24 (Denenberg, Lewis). *The satisfiability problem for $\forall\exists\forall$-Krom sentences is complete for* NLOGSPACE.

Exercise 8.3.25. [41] Show that $[\forall^*, (\omega, \omega)] \cap \text{KROM}$ is decidable by an effective reduction to the Aanderaa-Lewis class.

Exercise 8.3.26. [42] Show that $[\forall^*, (\omega, \omega), (1)] \cap \text{KROM}$ is decidable. Hint: Use the preceding exercise.

Exercise 8.3.27. [133] Show that $[\forall^*, (\omega, \omega), (1)]_= \cap \text{KROM}$ is decidable. Hint: For any formula ψ in the latter class, let ψ' be the result of replacing equality with a new predicate letter I and conjoining to the quantifier-free part the formula

$$Ixx \wedge Pxfx \wedge Qfxx \wedge (Qxy \to Qfxy) \wedge (Pfxy \to Pxy) \wedge$$
$$\wedge (Qxy \to \neg Ixy) \wedge (Pxy \to \neg Ixy)$$

where P and Q are new dyadic predicate letters, f is the function sign, and x and y are universally quantified variables.

Now we have: ψ is satisfiable iff (1) ψ has a finite model, or (2) ψ has a model over the natural numbers, with f interpreted as the successor function. But (2) holds iff ψ' is satisfiable. Hence the set of sentences ψ for which (2) holds is recursive, hence the set of satisfiable ψ is r.e., hence the class is decidable.

8.3.3 The Maslov Class

The Maslov class is the class of $\exists^*\forall^*\exists^*$-Krom sentences in pure predicate logic. Its satisfiability problem was proved to be decidable by Maslov [378] using his 'inverse method'. We present here a different decidability proof due to Fermüller [160] based on the resolution method (see [286, 439]).

We first recall some basic definitions and facts on resolution. Here, a clause is a set of literals, i.e. atomic formulae and negations of atomic formulae. It is read as the disjunction over its elements. Given a clause C and a substitution π we write $C\pi$ for the clause obtained by applying the substitution π to all variables appearing in C.

A clause set is read as the universal closure of the conjunction over all its elements. Thus, every clause set corresponds to a formula $\psi = \forall y_1 \cdots \forall y_r \varphi$ where φ is a conjunction of clauses and every formula of this form gives rise to a clause set S.

Definition 8.3.28 (Resolution). Let C, D and E be clauses. E is an $\mathcal{R}$-resolvent – or Robinson-resolvent — of C and D if

(i) There exist substitutions σ, τ such that $C\sigma$ and $D\tau$ have no common variables.
(ii) There exist $A \subseteq C\sigma$ and $B \subseteq D\tau$ such that $(\overline{A} \cup B)$ can be unified. Here $\overline{A}$ denotes the set of complementary literals to the literals in A. Let θ be the most general unifier of $(\overline{A} \cup B)$.
(iii) $E = ((C\sigma - A) \cup (D\tau - B))\theta$.

We write $\mathcal{R}(S)$ for the set of $\mathcal{R}$-resolvents of S. It is well-known that $\mathcal{R}$-resolution is complete, i.e. a clause set S is unsatisfiable if and only if the empty clause can be derived from S by $\mathcal{R}$-resolution.

An important tool to prove the completeness of certain resolution strategies is the concept of *semantic trees.*

Definition 8.3.29. A semantic tree based on a set K of atoms is a binary tree whose edges are labeled by the atoms of K and their negations in the following way. Fix an enumeration $\alpha_0, \alpha_1, \ldots$ of K. The two edges leaving the root are labeled $\alpha_0, \neg\alpha_0$; if α_i labels the edges entering a node, then α_{i+1} and $\neg\alpha_{i+1}$ label the outgoing edges from that node.

For any clause set S the *Herbrand base* of S is the set of all atomic statements $Pt_1 \cdots t_r$ where P is a predicate symbol occurring in S and $t_1, \ldots, t_r$ are elements of the Herbrand universe of S. Let now T be a semantic tree that is based on a subset of the Herbrand base of S.

A clause $C \in S$ *fails* at a node v of T if there exists a ground instance $C[t_1, \ldots, t_r]$ of C such that the complement of every literal in $C[t_1, \ldots, t_r]$ appears as a label on the path from the root to v. A node v is a *failure node* for S if there exists a clause of S that fails at v and no clause of S fails at a node $u < v$ (i.e. at a node u that precedes v on the path from the root to v). A node whose two successors are both failure nodes is called an *inference node*. We say that T is *closed* for S if every branch of T contains a failure node.

The following fact is a simple consequence of Herbrand's Theorem.

Lemma 8.3.30. *If S is unsatisfiable, then there exists a finite subset K of the Herbrand base of S such that every semantic tree based on K is closed for S.*

Exercise 8.3.31. Derive Lemma 8.3.30 from Herbrand's Theorem.

A *resolution strategy* F is a mapping that assigns to any clause set S a set $F(S) \supseteq S$ of Robinson-resolvents of clauses in S or, as in the case of M-resolution (to be introduced below), a set of *instances* of $\mathcal{R}$-resolvents of S.

Definition 8.3.32. A resolution strategy F is *complete with respect to semantic trees* for a class $\mathcal{C}$ of clause sets if for every unsatisfiable $S \in \mathcal{C}$ there exists a semantic tree T which is based on a finite subset of the Herbrand base for S such that a) T is closed for S, and b) for any two clauses C, D of S that fail immediately below, but not at, an inference node v of T there exists a resolvent $E \in F(S)$ of C and D such that E fails at v.

Lemma 8.3.33. *Let Fbe a resolution strategy which is complete via semantic trees for some class $\mathcal{C}$ of clause sets. Then F is a complete refutation calculus for $\mathcal{C}$, i.e. allows to derive the empty clause from every unsatisfiable clause set $S \in \mathcal{C}$.*

Proof. Let $S \in \mathcal{C}$ be unsatisfiable and T be a finite semantic tree satisfying the conditions of Definition 8.3.32. Since $S \subseteq F(S)$, T is closed for $F(S)$ and any failure node for S is either a failure node for $F(S)$ or some predecessor of this node is a failure node for $F(S)$. Either the empty clause belongs to S (and then nothing must be proved) or there exists at least one inference node v for S in T (otherwise we could find a branch of T without a failure node, which is impossible since T is closed for S). Let $E \in F(S)$ be the resolvent of the two clauses that fail at the successor nodes of v. Clearly, E makes v a failure node for $F(S)$. This means that T contains strictly less failure nodes for $F(S)$ than for S. Thus every application of the resolution strategy F strictly reduces the number of failure nodes until we reach a set containing the empty clause. □

Exercise 8.3.34. Prove that $\mathcal{R}$-resolution is complete with respect to semantic trees for all $\mathcal{C}$.

We now introduce a special resolution strategy, called M-resolution, which will establish the decidability of the Maslov class.

We recall some basic definitions. The depth of a term is defined as usual: variables and constants have depth 0 and the depth of $ft_1 \cdots t_r$ is the maximum of the depths of its arguments $t_1, \ldots, t_r$ increased by one. The depth of a formula or of a clause is the maximal depth of the terms occurring in it. A term, clause or formula is called *functional* if it has depth > 0.

Definition 8.3.35. Two functional terms $fs_1 \cdots s_k$ and $gt_1 \cdots t_k$ are *congruent* if $(t_1, \ldots, t_k)$ is a permutation of $(s_1, \ldots, s_k)$. A clause C is *uniform* if either C is function-free or there exists a functional argument t of some literal of C such that each argument of any literal of C is either a constant, an argument of t, or congruent to t.

Consider a formula $\psi := \exists x_1 \cdots \exists x_p \forall y_1 \cdots \forall y_q \exists z_1 \cdots \exists z_r \varphi$ in the Maslov class. Its functional form is $\forall \bar{y} \varphi'$ where φ' is obtained from φ by substituting constants $c_1, \ldots, c_p$ for $x_1, \ldots, x_p$ and Skolem functions $f_1\bar{y}, \ldots, f_r\bar{y}$ for $z_1, \ldots, z_r$. The set $S(\varphi')$ of Krom clauses of φ' satisfies the following conditions:

(i) All terms have depth ≤ 1 (i.e. there is no nesting of function symbols).
(ii) Every clause is uniform.

Let M be the set of Krom clauses satisfying *(i)* and *(ii)*.

Definition 8.3.36 (M-Resolution). The set $\mathcal{R}_M(S)$ of M-resolvents of a clause set $S \subseteq M$ is the the set of all clauses $E\pi \in M$, where E is an $\mathcal{R}$-resolvent of clauses in S and π is a substitution (based on the functions and variables of S).

Lemma 8.3.37. *For every $S \subseteq M$ there exists a semantic tree T for S such that any two clauses which fail immediately below an inference node v of T yield an M-resolvent that fails at v.*

Proof. Given a clause set S, let $\alpha_0, \alpha_1, \ldots$ be an enumeration of the Herbrand base such that deeper atoms succeed less deep ones and within atoms of the same depth uniform atoms precede those that are not uniform.

Let $C, D \in S$ be two clauses that fail immediately below an inference node v of the semantic tree T defined by this enumeration. By the completeness of $\mathcal{R}$-resolution with respect to semantic trees there exists an $\mathcal{R}$-resolvent E of C and D which fails at v. We show there also exists an M-resolvent $E\pi$ of C and D that fails at v.

We distinguish three cases:

1. If E is function-free then E is itself an M-resolvent.
2. Suppose that E has depth 1. Since $C, D \in M$ there exists a functional term t such that all functional terms of E, are congruent to t. If E is uniform, then $E \in M$ and we are done. Otherwise E contains variables that do not occur in t. Since E fails at v there exists an instance $E\sigma$ of E such that all literals in $E\sigma$ appear as labels on the path from the root to v. Note that, due to the chosen enumeration of the Herbrand base, $E\sigma$ is uniform on $\sigma(t)$. Let y be any variable of E that does not appear in t; y can only occur as the argument of a predicate symbol Therefore either $\sigma(y)$ is a constant, or $\sigma(y) = \sigma(s)$ where s is an argument of t, or $\sigma(y) = \sigma(t)$. We define a substitution π as follows: For every variable y that occurs in E but not in t, let $\pi(y) := \sigma(y)$ if $\sigma(y)$ is a constant and $\pi(y) := s$ if $\sigma(y) = \sigma(s)$ where $s = t$ or s is an argument of t. For all other variables z, let $\pi(z) := z$. Obviously, $E\pi$ is uniform on t and therefore $E\pi \in M$. But we also have that $(E\pi)\sigma = E\sigma$, so $E\pi$ indeed fails at v.
3. Otherwise C and D have smaller depth than E. But this is impossible: If θ is a most general unifier of two uniform functional atoms of depth 1, then no $\theta(x)$ is a functional term. Therefore, the greater depth of the resolvent could arise only if there is some function-free atom α resolved upon and there is a functional atom β, not being resolved upon, in the same clause as α. By uniformity, β contains all variables of α. Since β has greater depth than α, it follows that $\beta\theta$ has greater depth than $\alpha\theta$, for all substitutions θ. In particular this holds for θ being the most general unifier used to generate E. Note that $\alpha\theta$ is the resolved atom. Let now σ be the substitution such that all literals in $E\sigma$ appear on the path from the root to v. It then follows that $E\sigma$ has greater depth than $\alpha\theta\sigma$.
But the enumeration of the atoms underlying the semantic tree has been chosen in such a way that the depths of the atoms on the path to v do not exceed the depth of the resolved atom, whence we have a contradiction.

□

By Lemma 8.3.33 and 8.3.37, it follows that M-resolution is complete for clause sets in M. For a given finite vocabulary of constants, predicate and function symbols the set of clauses in M is finite.

Theorem 8.3.38 (Maslov). *Sat*($[\exists^*\forall^*\exists^*] \cap$ KROM) *is decidable.*

In fact, since the number of clauses in M (up to renaming of variables) is exponentially bounded in the size of the underlying vocabulary, it follows that the satisfiability problem for the Maslov class is solvable in deterministic exponential time, a result that was first proved by Denenberg and Lewis [106] based on another resolution-based decidability proof for the Maslov class due to Joyner [286]. They also a established a matching lower bound.

Theorem 8.3.39 (Denenberg, Lewis). *The satisfiability problem for the Maslov class is complete for* $\mathrm{DTIME}(2^{O(n)})$ *via log-lin reductions.*

8.4 Historical Remarks

The finite model property of L_2 was established by Mortimer [396]. An immediate consequence of this is the decidability of $Sat(L_2)$ a result that is sometimes attributed to Scott [459]. However, what Scott actually proved is that L_2-sentences can be effectively transformed into prenex sentences in the $\forall^2\exists^*$-class such that satisfiability is preserved. At that time it had not been detected yet that, contrary to Gödel's claim [188, p. 326], his decidability proof for the $\forall^2\exists^*$-class can *not* be extended to formulae with equality (see Chapter 4). Thus, Scott's reduction appeared to give a proof for the decidability of $Sat(L_2)$, but in fact, it applies only to L_2-sentences without equality.

Mortimer's proof for the finite model property of L_2 is much more complicated than the one presented here which is due to Grädel, Kolaitis and Vardi [208]; moreover the bound on the model size that can be derived from Mortimer's proof is doubly exponential, whereas the new proof gives a single exponential bound. One of the reasons for this improvement and simplification is that the new proof exploits the full power of the normal form for L_2 given by Lemma 8.1.2 whereas Mortimer only makes use of the weaker fact that L_2 can be reduced to sentences with quantifier rank 2. The corresponding lower bound for L_2 is an immediate consequence of a result by Fürer [177].

The first unification algorithm was published by Herbrand [253]; he introduced three properties of first-order formulae, called A, B and C. While properties B and C are the basis for the celebrated Herbrand Theorem, the concept A was more or less forgotten. Herbrand described a unification algorithm which forms part of the test whether a formula satisfies property A.

In his seminal paper on the resolution method J. Robinson [439] presents a unification algorithm and proves that this algorithm computes a most general unifier. This basic unification algorithm was later rediscovered by Knuth. In its naive form, Robinson's unification algorithm is of exponential time and space complexity. Later Robinson proposed a more succinct representation of terms which improved the space complexity of his algorithm, but the time complexity remained exponential. The first published polynomial-time unification algorithm is due to Venturini-Zilli [524]; her algorithm has quadratic time complexity. The asymptotically best unification algorithm known today is the linear time algorithm by Paterson and Wegman [417]. The P-completeness of the unification problem was proved by Dwork, Mitchell and

Kanellakis [137]. For a number of special cases that can be solved efficiently in parallel, we refer to the paper by Dwork, Kanellakis and Stockmeyer [138] from where we have taken the proof of Theorem 8.2.3. For more information on unification theory we refer to the survey [28] by Baader and Siekmann.

The observation that the P-completeness of unification implies the P-completeness of the satisfiability problem for Herbrand formulae was made by Denenberg and Lewis [106]. The results of Sect. 8.2.3 on positive first-order logic are taken from Kozen's paper [322].

The properties of Krom formulae were studied by Krom [329, 330, 331, 333]. The chain criterion for unsatisfiability of Krom sentences was proved in [331]. More general variants, applying to quantified propositional Krom formulae, were formulated by Aspvall, Plass and Tarjan [25] and by Grädel [207].

The decidability of the Aanderaa-Lewis class $[\forall\exists\forall]\cap$KROM was proved in [2, 12]. The proof presented here is taken from [106] and [133]. At roughly the same time as Robinson [439] introduced the resolution method, Maslov [378] proposed a related technique, the so-called 'inverse method' for establishing deducibility in first-order logic. Using the inverse method Maslov proved that the satisfiability problem for $\exists^*\forall^*\exists^*$-Krom sentences is decidable. In a later paper [381], Maslov proved the decidability of his class K which generalizes a number of known decidable classes in pure predicate logic, such as the monadic class and the Gödel-Kalmàr Schütte class. We refer to [544] for a modern exposition on the inverse method and the class K. Aanderaa and Goldfarb [10] proved that the Maslov class has the finite model property. The decidability proof presented here is due to Fermüller [160] (see also [163, 286] for more background and applications of the resolution method to the decision problem). The complexity results for decidable Krom classes are due to Denenberg and Lewis [105, 106, 107] For further remarks on Krom classes, in particular concerning undecidability results, we refer to Sect. 5.5.

A. Appendix: Tiling Problems

Cyril Allauzen[1] and **Bruno Durand**[2]

A.1 Introduction

In this appendix, we prove the undecidability of the following problems:

- the origin constrained domino problem (Wang in [531]): given a set of tiles and a tile as input, ask whether it is possible to form a tiling of the plane which contains the given tile;
- the unconstrained domino problem (Berger's Theorem in [33]): the input is a tile set and the question is whether one can tile the plane with it;
- the periodic domino problem (Berger and Gurevich-Koryakov in [237]): the input is also a tile set, but the question is whether it can be used to form a periodic tiling of the plane.

The last construction provides a direct proof of the recursive inseparability result of Berger and Gurevich-Koryakov (Theorem 3.1.7 in this book and reference [237]). Its intuitive meaning is that it is not possible to separate, with any computing device, tile sets that cannot tile the plane from tile sets that can tile the plane periodically.

In order to study these problems, we present recursive transformations of Turing machines into tile sets. These constructions are not independent of each other, thus the reader will probably not understand the last one if he did not understand the first one.

We do not present in this Appendix the original proofs of these theorems: they were based on Berger's construction (see [33]). We present a simplified proof inspired by R. Robinson's ideas in [440]. Both proofs are based on the construction of an *aperiodic* tile set (*i.e.* a tile set that can tile the plane, but not periodically) but Berger's aperiodic tile set contains more that 30 000 tiles whereas Robinson's contains 56 tiles. The reason why these aperiodic tile sets are fundamental in these proofs is the following: imagine that any tile set that can tile the plane can also tile it periodically. Then, one could solve the unconstrained domino problem in the following way: form all possible

[1] Elève du Département de Mathématiques et Informatique, ENS-Lyon, 46 Allée d'Italie, 69364 Lyon Cedex 07, France.

[2] Laboratoire de l'Informatique du Parallélisme, ENS-Lyon CNRS, 46 Allée d'Italie, 69364 Lyon Cedex 07, France.

$n \times n$ squares of tiles. If no correct tiling is formed then stop and answer "no". If one of these squares induces a periodic tiling of the plane, then stop and answer "yes". Else increment n and iterate the process. If no aperiodic tile set exists then the domino problem is decidable by the previous algorithm (and this was conjectured by Wang in [532]).

In the sequel, we first study the origin constrained domino problem; independantly, we construct a particular aperiodic tile set with *ad hoc* properties, and then we prove the undecidability of the domino problem. In the end, we adapt the proof in order to get the undecidability of the periodic domino problem and the inseparability theorem.

A.2 The Origin Constrained Domino Problem

We now consider the origin constrained domino problem, for which a tile set and a particular tile are given as input. The problem is to form a tiling of the plane containing this chosen tile. Wang proved that this problem is undecidable in [532]. Another version of this problem consists of forming a nontrivial tiling of the plane which is blank almost everywhere, using a tile set and a blank tile (see [440]).

The proof of this theorem consists of a reduction from the halting problem for Turing machines on an empty input. Let us consider such a machine with a bi-infinite tape. Let $Q = \{q_o, q_1, \ldots, q_k\}$ be the set of states, $q_0 \in Q$ be the initial state, and $Q_f \subset Q$ be the set of halting states. Let $S = \{s_0, s_1, \ldots, s_l\}$ be the set of symbols, and $s_0 \in S$ the blank symbol. Let $M = \{L, R\}$ be the possible movements for the head of the machine. The action of the machine is determined by the transition function $\gamma : (Q - Q_f) \times S \rightarrow S \times M \times Q$. In the sequel, we transform such a machine into a tile set.

The idea of the transformation is to force rows of the tilings to represent the tape of the machine while columns will represent the evolutions of cells of the tape during the computation. In other words, the tilings represent the space × time diagram of the (possibly infinite) computation of the machine on an empty tape.

More precisely, a configuration of the machine (tape, position of the head, and state) is represented on the upper and lower edges of our tile set. This tile set is described in Figures A.1 to A.3.

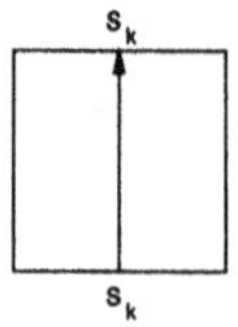

Figure A.1. Alphabet tile

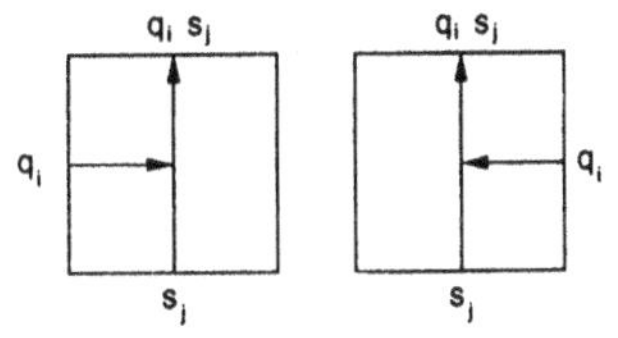

Figure A.2. Merging tiles

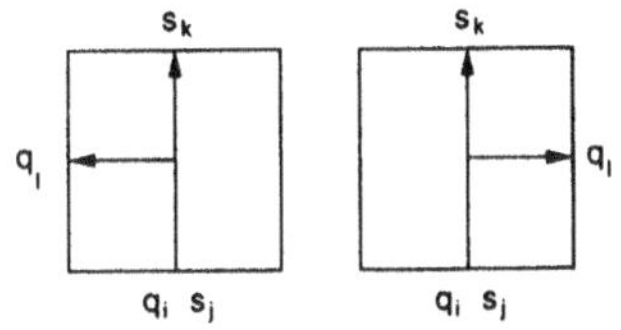

Figure A.3. Action tiles

Note that we use labeling and arrows to represent colours. It is an easy exercise to prove that these representations are equivalent (*i.e.* it is possible to transform a tile set with arrows into a tile set with colours without modifying the set of valid tilings).

An "alphabet" (Fig. A.1) tile transmits without modification the symbol s_k, and is constructed for all k. Merging tiles (Fig. A.2) combine a state q_i with a symbol s_j. For the sake of simplicity, we construct all tiles corresponding to all combinations of states and symbols. However not all of them will be able to take part of our tilings. The first tile (resp. the second tile) of Fig. A.3 is constructed if and only if $\gamma(q_i, s_j) = (s_k, L, q_l)$ (resp. $\gamma(q_i, s_j) = (s_k, R, q_l)$).

Assume that we have a row of tiles whose upper edges represent the tape of the machine at time t. Assume in addition that the machine is not in a halting state. Then, one of these tiles contains an up arrow labeled $q_i s_j$, and all others contain an up arrow with a symbol label (such as s_k). Then there is only one possibility to tile the next row: it must represent the configuration of the machine at time $t + 1$.

If we assume that the machine starts on a blank tape, then we can use tiles of Fig. A.4 in order to represent its initial configuration.

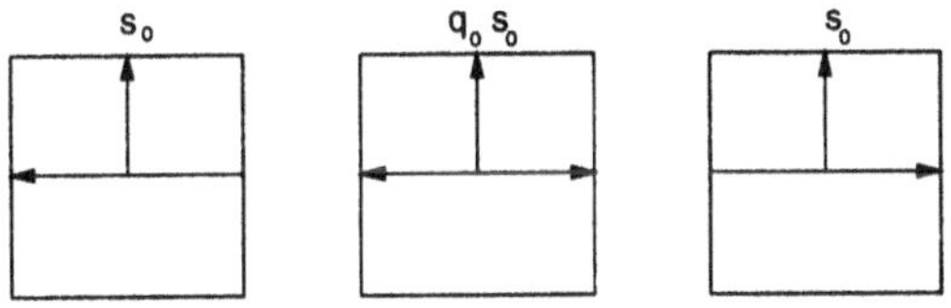

Figure A.4. Tiles involved in an initial configuration

Let us add now a blank tile to the tile set defined in Figures A.1 to A.4. We obtain a tile set associated to the considered Turing machine. Let us select the second tile of Fig. A.4 as the imposed tile. Then in order to tile the plane, this tile must have, to its left, the first tile of Fig. A.4, and, to its right, the third tile of the same Fig.. This implies that the only way to tile the bottom half-plane is to use the blank tile. Thus, the upper sides of the row on which the imposed tile appears represent the initial configuration. Hence, one can tile the plane with the previous tile set if and only if the considered Turing machine does not halt. The undecidability of the origin constrained domino problem is proved.

A.3 Robinson's Aperiodic Tile Set

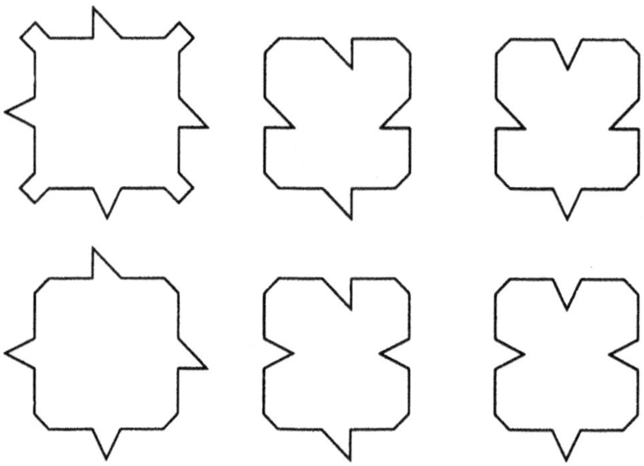

Figure A.5. Robinson's aperiodic tile set

Robinson's aperiodic tile set is formed by the 6 tiles of Fig. A.5 and by their images by all possible rotations and symmetries. We prove in the sequel that this tile set can only form aperiodic tilings of the plane. These tiles are not exactly Wang tiles since they have bumps and humps instead of colours. Nevertheless, it is an easy exercise to transform this tile set into a set of coloured Wang tiles.

Let us first consider Robinson's tile set without the particular shape of the corners. Then the two tiles of the first column are the same. We represent bumps and humps by arrows: symmetrical bumps are represented by a centered arrow, non-symmetrical ones are represented by two arrows, one centered and the other one slightly shifted. We obtain the tiles of Fig. A.6 and we add to them all the tiles obtained by symmetries and rotations. Then, we transform arrows into colours and obtain a set of Wang tiles. For the sake of simplicity we prefer to work with tiles of Fig. A.6 rather than with the Wang tiles obtained by the above transformation.

We call the first tile of Fig. A.6 "a cross" and other tiles "arms". The direction of the pictured cross is "up-right". Arms point on the direction of their main arrow. We represent these tiles with the help of the abbreviated

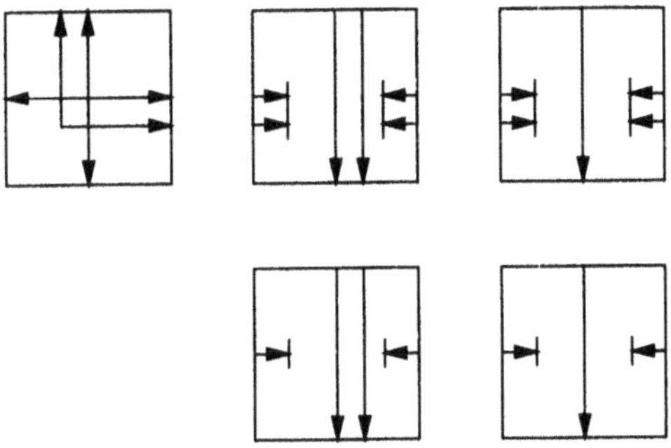

Figure A.6. The five basic tiles

symbols of Fig. A.7. The cross represents a cross-tile of any orientation while the other one represents any arm, whose main arrow is oriented as pictured.

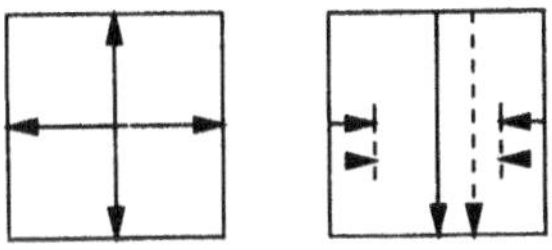

Figure A.7. Abbreviated notations

On another hand, we can suppress the bumps on the sides of Robinson's tiles and we obtain two polygons represented in Fig. A.8, which we call respectively bumpy and humpy.

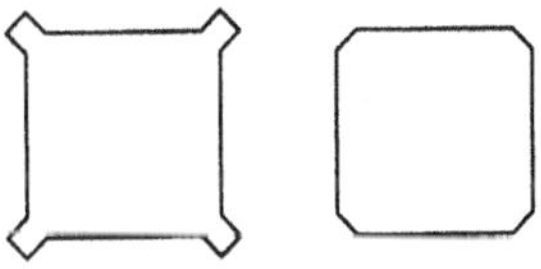

Figure A.8. A bumpy and a humpy tile

Assume now that the whole plane is tiled by these two polygons. Then in the neighborhood of each corner, one bumpy tile and three humpy ones must be found. This implies that either the bumpy tile appears on every other row, and on on every other cell on that row, or it appears on every other column, and on every other cell of that column.

We can thus obtain the periodic pattern of Figure A.9: bumpy tiles appear on cells whose coordinates are both odd. There are other possible tilings with these tiles, one of them is depicted in Figure A.10. More precisely, either every even row contains only humpy tiles and on every odd row humpy and bumpy tiles alternate; or the same situation holds for columns instead of rows. The important property is that given a bumpy tile, there exist a line (vertical or horizontal) centered on that tile on which there are bumpy tiles every two cells.

Let us now consider a tiling of the plane with the basic tiles. Assume that a cross appears somewhere. To the right of this cross, we find a sequence

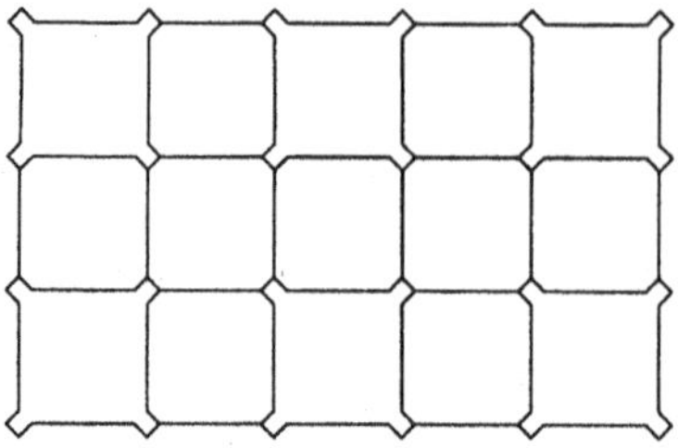

Figure A.9. First solution

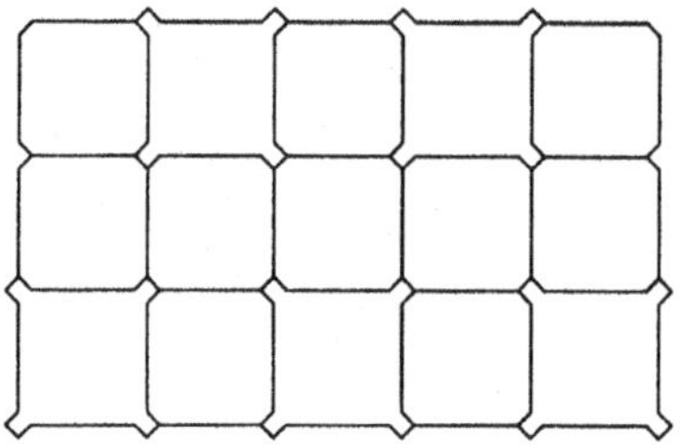

Figure A.10. Second solution

of right arms (possibly empty or infinite), and then there is another cross. Two consecutive crosses on a row are face-to-face or back-to-back. If they are face-to-face, then the vertical arm between these crosses points downwards (see Fig. A.11). If they are back-to-back, then the vertical arm between these crosses may point either upwards or downwards (see Fig. A.12).

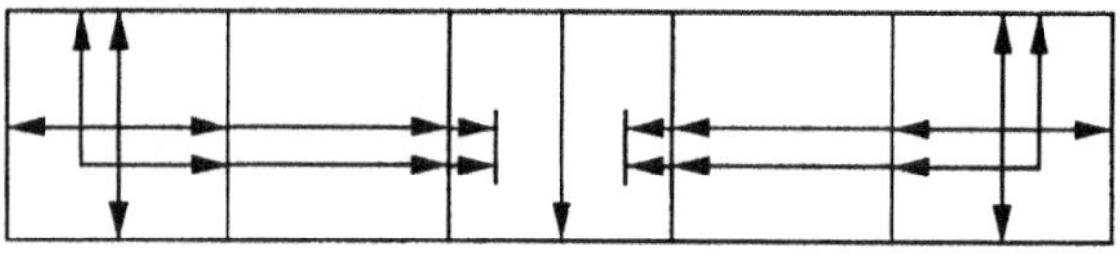

Figure A.11. Face-to-face crosses

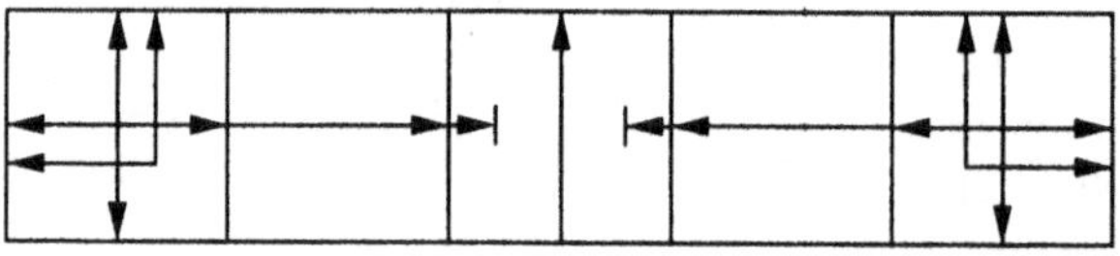

Figure A.12. Back-to-back crosses

The distance between two consecutive crosses is odd: let us assume for instance that both of them point upwards (see Fig. A.13). Then all tiles between them have tails of arrows on their upper side – they are either vertical or horizontal arms. The tile above the left cross (resp. the right cross) is necessarily a left arm (resp. a right arm) because this tile has a left tail (resp. right tail) on its bottom side. Thus the upper-right tile of the cross is also a cross and by iteration of this argument, one proves that there is one cross every two cells on this row. Hence the number of tiles between two crosses is odd.

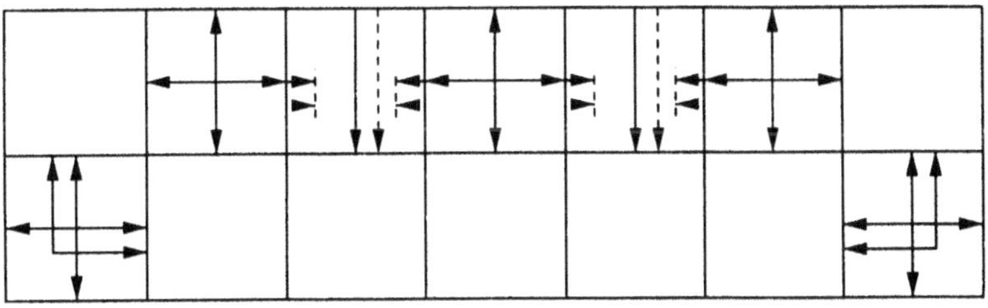

Figure A.13. This row is forced by two face-to-face crosses

Let us now consider a tiling of the plane with Robinson's tiles. We then have the constraint that bumpy tiles must appear. In the sequel, we call these tiles "1-squares"; note that all of them are crosses.

Let us consider a 1-square. It faces another 1-square located two tiles farther in a direction (vertical or horizontal) because it is bumpy. Thus we obtain the construction of Fig. A.14 or a rotated version. We are sure that we have arrows on the 3 neighbor cells the direction of which is imposed. Hence we are sure that the "center" cell is a cross (see Fig. A.15). Observe now that two crosses cannot be found side-by-side because arrows diverge from the center of any cross. Thus there is only one possible position of the bumpy tiles that must be found on this row: it must face the two bumpy tiles we started with. We are in the case of Fig. A.9. Thus we must obtain a figure similar to Fig. A.16 that we call a "3-square". A cross must be present at the center of this square but its orientation is not imposed: we obtain exactly 4 possible 3-squares with 1-squares as corners.

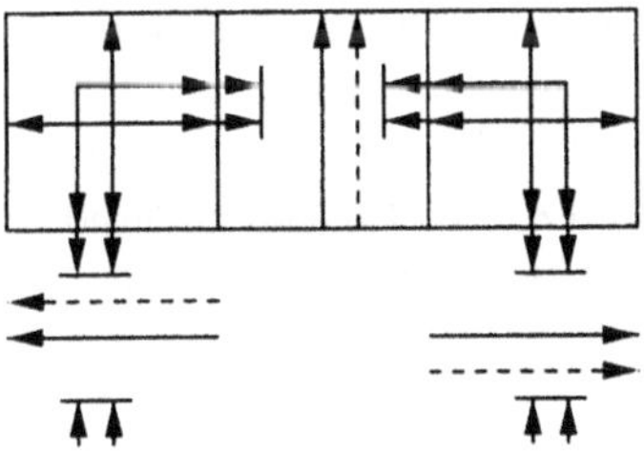

Figure A.14. The first step of the construction of a 3-square

Let us now consider the central cross of a 3-square. It must face another cross. Then this cross must be at the center of another 3-square, otherwise double arms would not coincide. We thus obtain a 7-square as in Fig. A.17. By iteration of the same construction, we can construct a $(2^n - 1)$-square for all $n \in \mathbb{N}$. By König's Lemma, it is possible to tile the plane with Robinson's tiles. Note that we have just proved a stronger result: all tilings of the plane with Robinson's tiles consist of nested $(2^n - 1)$-squares. Note also that we do not obtain directly a tiling of the whole plane, but we obtain at least a tiling of a quarter of the plane (that is why we have invoked König's Lemma).

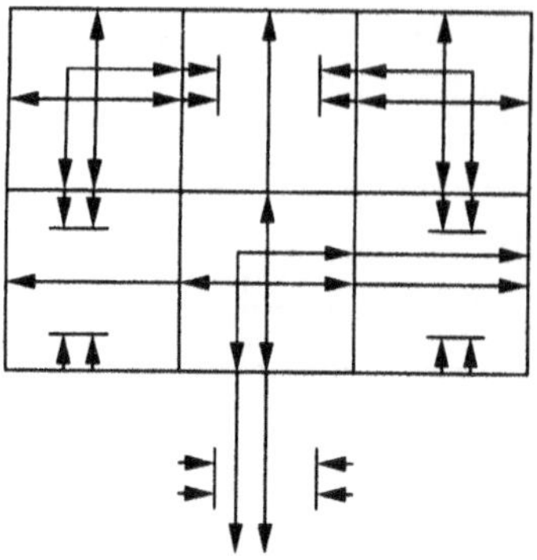

Figure A.15. The second step of the construction of a 3-square

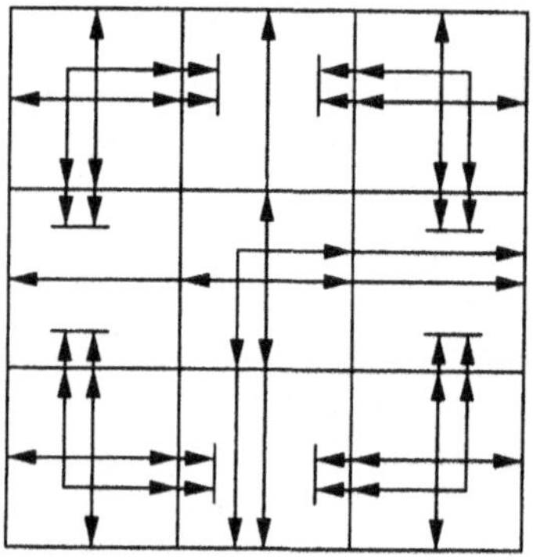

Figure A.16. A 3-square

Let us now prove that all tilings by Robinson's tile are not periodic. We have proved above that 1-squares must appear in all tilings. Each 1-square uniquely determines the position of the 3-square in which it is included; there are only 4 possible positions, according to the orientation of the cross. Then, the 3-square uniquely determines the position of the 7-square in which it is included (4 possible positions). By iteration, for all $n \in \mathbb{N}$, the tiling contains a $(2^n - 1)$-square. These $(2^n - 1)$-squares are not periodic structures hence all tilings by Robinson's tile are not periodic.

A.4 The Unconstrained Domino Problem

The goal of this section is to transform a Turing machine into a tile set which can tile the plane if and only if the machine does not halt on a blank tape. The basic idea is that the computation of the machine should be represented in a uniform manner in any tiling of the plane. More precisely, we construct a tile set such that, given a time t, there exists a size n such that the t first steps of the computation of the machine are represented in any $n \times n$ square correctly tiled. Thus, we just have to force a tiling error when the halting state appears, and the theorem will be proved. Note that this notion of uniform

Figure A.17. A 7-square

distribution of computations is analogous to the notion of quasiperiodicity: a tile set is quasi periodic if any finite pattern that appears in the tiling appears in all sufficiently large squares.

Let us consider again Robinson's tiles presented in Section A.3. We proved that in the center of all $(2^{n+1}-1)$-squares one must find a cross, and that a $(2^{n+1}-1)$-square is formed by four (2^n-1)-squares. The centers of these (2^n-1)-squares are four crosses that delimit a square of size 2^n that we call a 2^n-frame. Furthermore, all crosses that are face-to-face delimit a 2^k-frame.

Observe now that two of these frames intersect if and only if a corner of one of them is the center of the other one. Equivalently, a 2^n-frame intersects only one 2^{n+1}-frame and four 2^{n-1}-frames (see Fig. A.18).

We now modify the five basic tiles (Fig. A.6) by giving two different colours for the slightly shifted arrow. This arrow can be either red or green with the following constraint: on the arms, the colour of the horizontal arrows (resp. of vertical arrows) must be the same. For the central tile (with two double arms), if the vertical arrow is green, then the horizontal one must be red (resp. red and green). All tiles excepted the last one of Fig. A.6 are duplicated once, hence we obtain 9 tiles. We give the humpy shape to the green cross and we obtain a new tile set.

Thus, if we do not consider colours, the tilings that can be formed with this tile set are exactly the same ones as with Robinson's tiles. The only difference is that frames are coloured, and if 2^n-frames are red then 2^{n+1}-frames are green and conversely. The "humpy" constraint on the green cross imposes that 2-frames are green and thus that a 2^n-frame is green if n is odd and red if n is even (see Fig. A.18 and A.19).

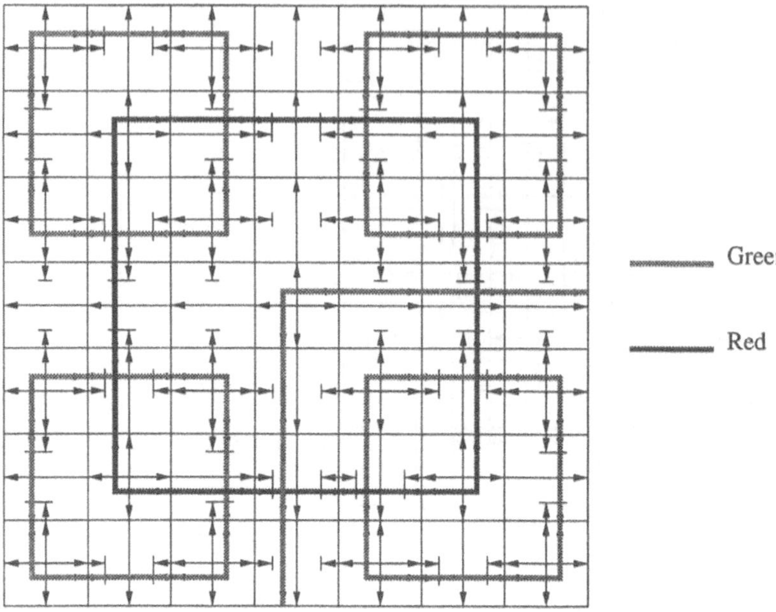

Figure A.18. A 4-frame

Let us now consider only red squares (that are 4^n-frames). These squares do not intersect, and any tiling of the plane contains a sequence of frames of size $4, 4^2, \ldots, 4^n, \ldots$

In the sequel, we call 4^n-zone the surface delimited by a 4^n-frame. In this zone, we are interested in "free" rows. These rows are those which do not intersect a 4^k-frame where $k < n$. In a 4^n-zone, if we suppress columns and rows that intersect a 4^{n-1}-zone, we obtain four 4^{n-1}-zones side-by-side. Thus if the number of free rows is denoted by F_n in a 4^n-zone, then it satisfies $F_n = 2F_{n-1}$. As $F_1 = 3$, $F_n = 2^n + 1$. The same argument proves that there are $F_n = 2^n+1$ free columns in a 4^n-zone. Hence there are $(2^n+1)^2$ free cells in a 4^n-zone (located on a free row and on a free column). It is approximatively the square root of the number of all cells of the zone.

It is easy to modify our tile set so that cells that are not free are marked with a different colour. In order to do that, we impose that red squares send vertically and horizontally an "obstruction" colour outside their borders (see Fig. A.20). This transformation is very easy in its principle but tedious to prove. The reader can do this transformation as an exercise.

Let us now consider a Turing machine. On each free tile, we superimpose the tiles representing the space-time local behavior of the machine as represented in Figures A.1 to A.3. The tiles that are obstructed in only one direction (vertically or horizontally) will transmit the state of the tape in the other direction (horizontally or vertically). We do not modify tiles that are obstructed in both directions. Furthermore, we impose that the lowest free

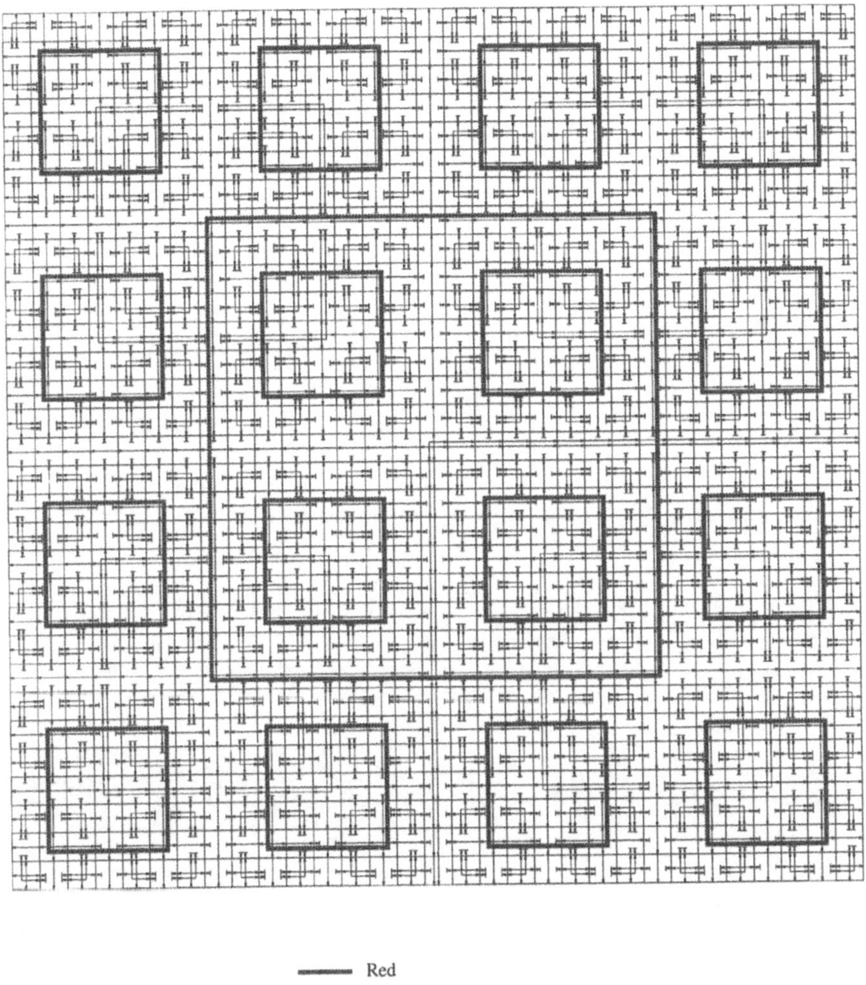

Figure A.19. A 16-frame

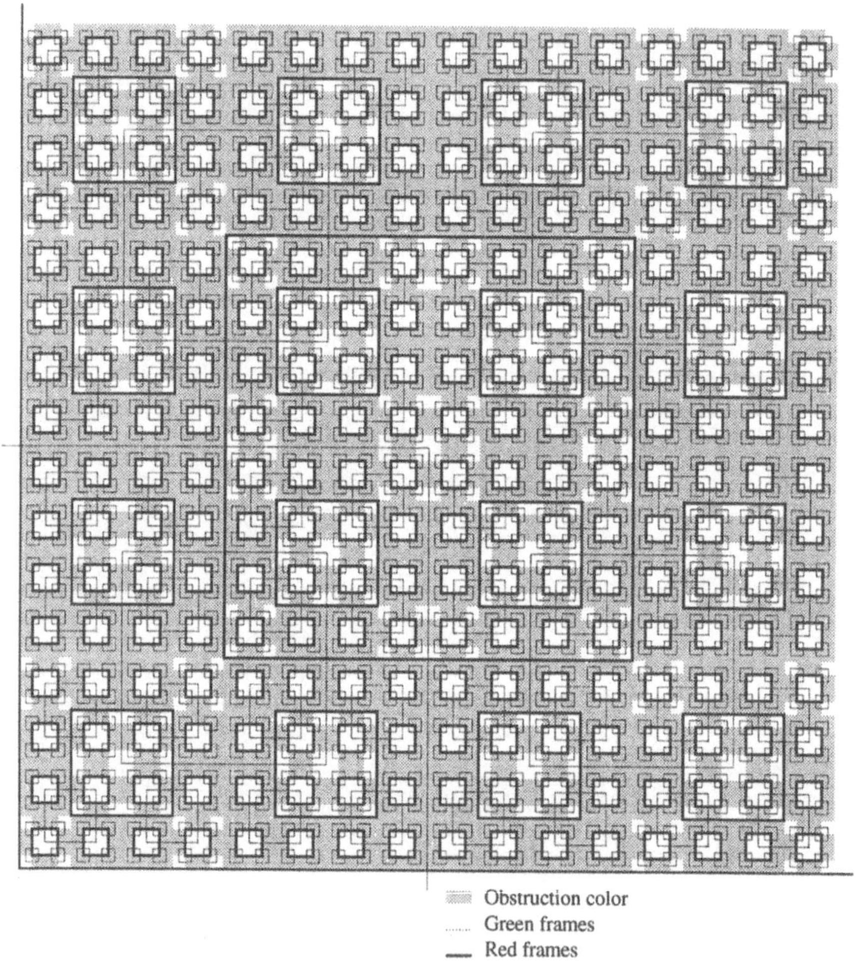

Figure A.20. Computation zones

cells of a zone are associated to the blank colour of the tape (*i.e.* s_0) excepted the center one which is associated with the head in an initial state (*i.e.* $q_0 s_0$). We do not give full details for this construction but its justification is that a copy of a down-most tile in a red zone may only appear as a down-most tile in another red zone. The center one is recognized because it is the only free down-most cell which also belongs to a green frame.

Thus with this tile set, we can tile arbitrarily large squares of the plane if and only if the Turing machine does not halt. Berger's theorem is proved. Note that the computation at time step t can be found in any $4t^2 \times 4t^2$ square of the tiling, since these squares contain a 4^n-zone with $4^n > t^2$.

A.5 The Periodic Problem and the Inseparability Result

With the previous construction, we can also prove the undecidability of the periodic domino problem: is it possible, with a given tile set, to tile the plane periodically? We modify the previous tile set so that it can tile the plane periodically if and only if the associated Turing machine halts. Otherwise it can tile the plane as in the previous proof (not periodically).

First, we transform the halting state of the machine in such a way that if the machine enter this state, then it stays in this state forever and does nothing on the tape. When a colour $q_f s_i$ (where q_f is a halting state) arrives on a red frame, then this frame is transformed into a purple frame. This means that we duplicate tiles for red frames into tiles for purple frames and that if a halting state arrives on the frame, it is only possible to put these purple tiles. We impose another difference between red and purple frames: outside the zone, the edges of purple tiles are blank, except for corners where the outside edges are blue.

In other words, if the machine halts, then we have cut out sufficiently large squares on which the halting state appears. These squares are blank on their edges and blue on their corners. Then, the neighbors of such a square might be identical squares. Thus if the machine halts, then it is possible to form a periodic tiling on the plane with the tile set. The periodic pattern is a 4^n-zone with n large enough so that the Turing machine halts.

Conversely, if one can construct a periodic tiling of the plane, then a purple tile (hence a purple square) must appear in it – otherwise we have a tiling of the plane with the tiles of the previous section and we proved that it cannot be periodic. As there is a purple frame in the tiling, we can find inside it a halting computation of the Turing machine.

The undecidability result is proved but as an exercise, the reader can prove that the tile set can form only periodic tilings if the machine halts, and can form only aperiodic tilings if the machine does not halt.

We can strengthen this result and obtain the inseparability theorem (Theorem 3.1.7 in this book). Let us consider Turing machines with two halting states (1 and 2). Then we can do the previous construction with the difference that if the halting state 2 appears in the tiling, then a tiling error is forced. For instance we impose that if a tile represents this state on its upper edge, then we cannot put another tile as north neighbor. If the halting state 1 appears, then the construction is as presented before for the undecidability of the periodic tiling.

Thus, if we denote by τ_x the tile set associated to the Turing machine φ_x, as the sets $H_1 = \{x,\ \varphi_x(0) = 1\}$, $H_2 = \{x,\ \varphi_x(0) = 0\}$, and $H = \{x,\ \varphi_x(0) \text{ diverges}\}$ are recursively inseparable (see Chapt. 2), we have the following property:

- if $x \in H_1$, then τ_x can tile the plane periodically,
- if $x \in H_2$, then τ_x admits no tiling,
- if $x \in H$, then τ_x admits a tiling but not a periodic one.

Hence we have proved the recursive inseparability of the tile sets that cannot tile the plane and those that can be used to tile the plane periodically.

Annotated Bibliography

1. S. Aanderaa. *A New Undecidable Problem with Applications in Logic.* PhD thesis, Harvard University, 1966.
   Proves the undecidability of the class of all formulae $\forall x\exists u\forall y\alpha \in [\forall\exists\forall, (\omega, \omega)]$ where in α only atomic formulae $P_i xy, P_i yu, R_j x$ appear. The proof introduces a novel kind of automaton which is shown to be computation universal. The result is simplified and extended in [13] to formulae α with only one binary predicate and in addition monadic atomic formulae $R_j x, R_j y, R_j u$. This solves the classification of the $[\forall\exists\forall, (\omega, \omega)]$ subclasses with respect to the occurring atomic formulae (see [134]); namely any class in whose formulae the combination of at least three atomic formulae including the pair Pxy, Pyu or the pair Pyx, Puy is allowed to occur, yields a conservative reduction class.
2. S. Aanderaa. On the decision problem for formulas in which all disjunctions are binary. In *Proc. 2nd Scandinavian Logic Symp.*, pages 1–18, 1971. Proves that the following Krom class of formulae with only binary predicates and without functions or equality is a conservative reduction class: $[\exists\forall \wedge \forall\exists\forall]$. The same result and the same proof have been obtained independently in [39] and are explained in Chap. 2 of the book. Sketches the decidability of $\forall\exists\forall \cap$ KROM, the proof sketch has been elaborated in [12]. Proves also that slight extensions of Krom formulae yield simple satisfiable formulae without recursive models. This construction has been simplified and extended in [45], see Sect. 2.1.3 of this book.
3. S. Aanderaa. On the solvability of the extended $\forall\exists \wedge \exists\forall^*$–Ackermann class with identity. In E. Börger, G. Hasenjäger, and D. Rödding, editors, *Logic and Machines: Desision Problems and Complexity*, Lecture Notes in Computer Science No. 171, pages 270–284. Springer, 1984.
   The paper sketches a proof for the decidability of the class of sentences of form $\forall y\exists x Pyx \wedge \exists x_0 \forall y_1 \ldots \forall y_n \alpha$ where P is a binary predicate and α a quantifier-free formula containing only monadic and binary predicates and equality. See the related weaker result in [17]. The proof uses Ramsey's Theorem [435].
4. S. Aanderaa and E. Börger. The Horn complexity of Boolean functions and Cook's problem. In B. Mayoh and F. Jensen, editors, *Proc. 5th. Scandinavian Logic Symposium*, pages 231–256. Aalborg University Press, 1979.
   Introduces the notion of Horn complexity for Boolean functions, measuring the minimal length of a defining propositional formula which is Horn in all except its input variables. Shows that Horn complexity and network (and therefore Turing) complexity for Boolean functions are polynomially equivalent. A shorter proof appears in [5].
5. S. Aanderaa and E. Börger. The equivalence of Horn and network complexity for Boolean functions. *Acta Informatica*, 15:303–307, 1981.
   See comment to [4].

6. S. Aanderaa, E. Börger, and Y. Gurevich. Prefix classes of Krom formulae with identity. *Archiv math. Logik u. Grundlagenforschung*, 22:43–49, 1982.
The Krom class $[\forall\exists\forall\exists]_=$ is shown to be a conservative reduction class, see Exercise 2.1.19 in this book. The Krom class $[\forall^3\exists]_=$ is shown to be a reduction class with respect to satisfiability. The method in [7] can be applied to infer that this reduction class is also conservative. The decidability established for the Krom class $\forall\exists\forall$ in [2] is extended to formulae with equality. This together with the reduction in [41] yield the decidability of the Krom class $[\forall\exists\forall^*, (\omega,\omega)]_=$.
7. S. Aanderaa, E. Börger, and H. Lewis. Conservative reduction classes of Krom formulas. *Journal of Symbolic Logic*, 47:110–129, 1982.
Exhibits a method by which the Krom reduction class $\forall\exists\forall\forall$ and similar undecidable classes can be shown to be conservative reduction classes.
8. S. Aanderaa and D. Cohen. Modular machines and the Higman-Clapham-Valier embedding theorem. In *Word Problems II; the Oxford Book*, pages 17–28. North-Holland, 1980.
This and [9] provide a very simple proof of the unsolvability of the word problem for finitely presented groups which uses modular machines. In [86] these machines are used to give a very simple proof of undecidability and incompleteness theorems for predicate logic and elementary number theory respectively starting from the existance of a modular machine with unsolvable halting problems. Cohen [85] uses [50] to prove degree representation theorems for halting, word and confluence problems of modular machines.
9. S. Aanderaa and D. Cohen. Modular machines, the word problem for finitely presented groups and Collin's Theorem. In *Word Problems II; the Oxford Book*, pages 1–16. North-Holland, 1980.
10. S. Aanderaa and W. Goldfarb. The finite controllability of the Maslov case. *Journal of Symbolic Logic*, 39:509–518, 1974.
Generalizing two lemmas devised by Gödel [187] in the proof for the finite model property of the Gödel-Kalmar-Schütte class, it is shown that the Maslov class, proved to be decidable in [378], has the finite model property.
11. S. Aanderaa and F. Jensen. On the existence of recursive models for Krom formulas. Preprint Series 33, Aarhus University, 1972.
It is shown that every consistent Krom formula without functions or equality has a recursive model. The proof uses a priority argument. See also [147] .
12. S. Aanderaa and H. Lewis. Prefix classes of Krom formulas. *Journal of Symbolic Logic*, 38:628–642, 1973.
Proves Lewis' result that the two Krom classes $\forall\exists\forall^2, \forall^2\exists\forall$ are reduction classes and Aanderaa's result that the Krom class $\forall\exists\forall$ is decidable, correcting the incorrectly stated Lemma 4 in [2]. See Chapters 5 and 8 in this book.
13. S. Aanderaa and H. Lewis. Linear sampling and the $\forall\exists\forall$ case of the decision problem. *Journal of Symbolic Logic*, 39:519–548, 1974.
See comment to [1].
14. S. Abiteboul, R. Hull, and V. Vianu. *Foundations of Databases*. Addison-Wesley, 1995.
15. M. Abramsky. The classical decision problem and partial functions. *Archiv math. Logik u. Grundlagenforschung*, 20:3–12, 1980.
The classical decision problem for prefix-vocabulary classes with at least one function of positive arity is generalized to the case when functions are in general partial. The classification survives.
16. W. Ackermann. Über die Erfüllbarkeit gewisser Zählausdrücke. *Math. Annalen*, 100:638–649, 1928.
Generalizes the decidability result for the class $[\forall\exists, all]$ of [35] to the class

$[\exists^*\forall\exists^*, all]$ of relational prenex sentences with only one universal quantifier by reducing the latter to the former. The proof is in terms of validity and shows finite controllability for the class. Other decision procedures for this can be obtained from [254, 478].

17. W. Ackermann. Beiträge zum Entscheidungsproblem der mathematischen Logik. *Math. Annalen*, 112:419–432, 1936.
The paper starts with a proof for the reduction class $[\exists\forall\exists\forall^*, all]$, reducing Kalmar's class $[\exists^*\forall^2\exists\forall^*, (0,0,1)]$ (see [291]) or Skolem's class $[\forall^*\exists^*, all]$ (see [477]) to it. The reduction formulae can really be made of form $\forall x \exists u Sxu \wedge \exists v \forall y_1 \cdots \forall y_n \alpha$ where S is a binary predicate and α is quantifier-free. This motivates the decision procedure provided for finite satisfiability, and for the satisfiability in infinite and only infinite domains for formulae of form $\forall x \exists Sxy \wedge \forall y_1 \cdots \forall y_n \alpha$ with $n \leq 4$ (consisting in a reduction to Presburger arithmetic for $n = 3$, see [423]). See [297] for an improvement of the reduction class to $[\exists\forall\exists\forall^*, (0,1)]$ and [547] for a simplification of part of the decidability proof. See [318].
18. W. Ackermann. *Solvable Cases of the Decision Problem.* North Holland, Amsterdam, 1954.
A comprehensive treatment of solvable cases of the Entscheidungsproblem known at that time. Many decision procedures then known are extended to the case with equality or to second-order logic. The formulation is semantic and considers validity, determining the cardinality of domains for which a given formula is valid or not valid.
19. N. Alechina. On a decidable generalized quantifier logic corresponding to a decidable fragment of predicate logic. *Journal of Logic, Language and Information*, 4:177–189, 1995.
20. H. Andréka and I. Németi. The generalised completeness of Horn predicate-logic as a programming language. DAI Ressearch Report 21, University of Edinburgh, 1976.
See the comment to Example 2.1.21 to the Aanderaa-Börger Theorem.
21. H. Andréka, J. van Benthem, and I. Németi. Back and forth between modal logic and predicate logic. *Bulletin of the Interest Group in Pure and Applied Logic*, 3:685–720, 1995.
22. H. Andréka, J. van Benthem, and I. Németi. Modal languages and bounded fragments of predicate logic. ILLC Research Report and Technical Notes Series ML-96-03, University of Amsterdam, 1996.
23. V. Arvind and S. Biswas. On bandwith restricted versions of the satisfiability problem of propositional CNF formulas. *Theor. Computer Science*, 68:123–134, 1989.
Uses the method of economical description of Turing machines (see Sect. 2.1.1) for completeness results for restricted variants of the propositional satisfiability problem.
24. C. Ash. Sentences with finite models. *Zeitschr. f. math. Logik u. Grundlagen d. Math.*, 21:401–404, 1975.
Shows (using Vaught's test [519] and Ramsey's Theorem [435]) the finite model property for the following first-order cases: The class of all formulae with either only monadic predicates and equality or only monadic predicates and functions but without equality; the class of all universal sentences with only monadic predicates, one monadic function and equality. Examples are given which show that these results are sharp. See Sect. 6.5.
25. B. Aspvall, M. Plass, and R. Tarjan. A linear time algorithm for testing the truth of certain quantified Boolean formulas. *Inform. Process. Letters*, 8(3):121–123, 1979.

Presents an algorithm for evaluating Krom sentences of quantified propositional logic which runs in linear time on a random access machine. See [207] for an NLOGSPACE-algorithm for the same problem.

26. G. Asser. Das Repräsentantenproblem im Prädikatenkalkül der ersten Stufe mit Identität. *Zeitschr. f. math. Logik u. Grundlagen d. Math.*, 1:252–263, 1955.
Provides an arithmetical characterisation of first-order spectra which implies that the latter are Kalmár-elementary. It is shown that not every Kalmár-elementary set is a first-order spectrum. The problem of an inductive characterisation of first-order spectra is stated together with what became known as (the still open) Asser's problem, namely whether the complement of every spectrum is also a spectrum. See also [399] for some extensions.
27. F. Baader, H. Bürckert, B. Hollunder, A. Laux, and W. Nutt. Terminologische Logiken. *KI*, 3/92:23–33, 1992.
28. F. Baader and J. Siekmann. Unification theory. In D.M. Gabbay, C.J. Hogger, and J.A. Robinson, editors, *Handbook of Logic in Artificial Intelligence and Logic Programming*. Oxford University Press, Oxford, UK, 1994.
29. J. Balcázar, J. Díaz, and J. Gabarró. *Structural Complexity*, volume I. Springer-Verlag, 1988.
30. J. Balcázar, J. Díaz, and J. Gabarró. *Structural Complexity*, volume II. Springer-Verlag, 1990.
31. H. Behmann. Beiträge zur Algebra der Logik, insbesondere zum Entscheidungsproblem. *Math. Annalen*, 86:136–229, 1922.
This paper contains an anti-prenex normal-form for formulae of monadic predicate logic without functions and equality and a solution of their decision problem (in terms of validity). Behmann's decidability proof is also given in modified form in [82, §46] (using [426] and [253, Ch. 2.9.2]) and in [267, pp. 193–195]. The decidability result is extended in the paper to (a) formulae with equality, b) second-order formulae all whose predicates (free or bound) are monadic. (See [476, 365]).
32. J. Bennett. *On spectra.* PhD thesis, Princeton University, 1962.
33. R. Berger. The undecidability of the domino problem. *Mem. AMS*, 66, 1966.
The unconstrained domino problem, introduced in [531], is proved to be undecidable. Robinson considerably simplified the proof [440]. Gurevich and Koryakov showed that instances with no solutions and instances with periodic solution are recursively inseparable [237]; see Appendix A of this book for a new proof.
34. L. Berman. The complexity of logical theories. *Theor. Computer Science*, 11:71–77, 1980.
Alternating Turing machines (as defined in [73]) are formalized by formulae of Presburger arithmetic or the theory of real numbers with addition, relating machine alternatious to quantifier changes in the describing formulae. Together with the decision procedure developed by [166] using Ehrenfeucht-Fraïssé games [170, 144] this yields a precise characterization of the complexity of Presburger arithmetic and the theory of real addition in terms of time, space, and alternation depth of alternating Turing machines. See also [89, 201, 206].
35. P. Bernays and M. Schönfinkel. Zum Entscheidungsproblem der mathematischen Logik. *Math. Annalen*, 99:342–372, 1928.
This is the first paper that gives a decision procedure for a class of formulae of predicate logic (without fuctions or equality) containing predicates of arity > 1, namely the Bernays-Schönfinkel prefix class $[\exists^*\forall^*, all]$ (formulated in terms of satisfiability). The proof shows that each satisfiable formula

$\exists x_1 \ldots \exists x_n \forall y_1 \ldots y_m \in [\exists^*\forall^*, all]$ has a model over a domain of n elements. The decideability result for $[\exists^*\forall^*, all]$ was extended by Ramsey [435] to the $[\exists^*\forall^*]$-class with equality. The paper contains also a decision procedure for monadic predicate logic showing that any monadic formula α is valid if and only if it is valid in a domain of cardinality 2^n where n is the number of predicates occuring in α (see [365]). Furthermore it contains a reduction of the prefix class $[\forall\exists, all]$ to propositional logic; see the generalization in [16].

36. A. Blass and Y. Gurevich. Henkin quantifiers and complete problems. *Annals of Pure and Applied Logic*, 32:1–16, 1986.
37. A. Blass and Y. Gurevich. Randomizing reductions of search problems. *SIAM Journal on Computing*, 22:949–975, 1993.
See the comment to [344].
38. C. Böhm and G. Jacopini. Flow diagrams, Turing machines and languages with only two formation rules. *Communications of the ACM*, 9:366–371, 1966.
39. E. Börger. *Reduktionstypen in Krom- und Hornformeln.* PhD thesis, Universität Münster, Institut für Grundlagenforschung, 1971.
Proves the conservative reduction class property for the Krom classes $[\exists\forall\exists\forall, (\omega, k)]$, $[\forall\exists\exists\forall, (\omega, k)]$, $[\exists^*\forall\exists\forall, (0, k)]$, $[\exists^*\forall\exists\forall, (0, 0, 1)]$, found independently in [2]. Continuing the investigations in [444], for the Krom classes with prefix $\forall\exists^2\forall$ or $\forall\exists^*\forall$ the reduction class property is shown to hold also if the interpretations of the binary predicates are restricted to antisymmetric and antitransitive or to disjoint or to reflexive relations. Contains also some undecidable Krom classes where the number of binary predicates is restricted to a small number. Some of the proofs are reported in [43]. See Chap. 5.1.
40. E. Börger. Reduktionstypen der klassischen Prädikatenlogik, Teil 1: Der Satz von Trachtenbrot und das Prädikatenproblem für Kromklassen. Technical report, Lecture Notes, Institut für math. Logik und Grundlagenforschung, Universität Münster, 1972.
Contains the new proof for Trakhtenbrot's Theorem which is used in Sect. 2.1.2 and undecidability proofs for Krom classes; proves in particular the conservativity of Lewis' class $[\forall^2\exists\forall, (0, \omega, k)] \cap \text{KROM} \cap \text{HORN}$ for some k of the size of a universal 2-register machine, see Exercise 5.1.5.
41. E. Börger. Eine entscheidbare Klasse von Kromformeln. *Zeitschr. f. math. Logik u. Grundlagen d. Math.*, 19:117–120, 1973.
$[\forall^*, (\omega, \omega)] \cap \text{KROM}$ is shown to be decidable by an effective reduction to Aanderaa's decidable class $\forall\exists\forall \cap \text{KROM}$ [2, 12]. See the extension in [42].
42. E. Börger. Reduction classes of Krom formulae with only one predicate symbol and function symbols. Report n.1973.1, pp. 20., Dept. of Computer Science, University of Salerno, 1973.
$[\forall^*, (\omega, \omega), (1)] \cap \text{KROM}$ is shown to be decidable, extending the reduction in [41]. $[\forall^2, (0, 1), (2)] \cap \text{KROM} \cap \text{HORN}$ is shown to be a reduction class. Other reductions classes are established which have been improved in [48].
43. E. Börger. Beitrag zur Reduktion des Entscheidungsproblems auf Klassen von Hornformeln mit kurzen Alternationen. *Archiv math. Logik u. Grundlagenforschung*, 16:67–84, 1974.
Contains some results of [39], in particular for restrictions of Krom and Horn reduction classes to standard mathematical theories.
44. E. Börger. La Σ_3-complétude de l'ensemble des types de réduction. *Logique et Analyse*, 65/66:89–94, 1974.
Shows that the reduction class property is complete for the third level of the Kleene-Mostowski arithmetical hierarchy. See [58].
45. E. Börger. On the construction of simple first-order fomulae without recursive models. In *Proc. Coloquio sobra logica simbolica, Madrid*, pages 9–24, 1975.

Simplifies and extends Aanderaa's construction [2] of slight extensions of satisfiable Krom formulae without recursive models to various complexity classes. See Sect. 2.1.3 of this book.

46. E. Börger. Recursively unsolvable algorithmic problems and related questions reexamined. In K. Potthoff G.H. Müller, A. Oberschelp, editor, *Logic Colloquium Kiel*, Lecture Notes in Mathematics No. 499, pages 10–24. Springer, 1975.
47. E. Börger. Zwei Reduktionstypen aus Kromformeln mit Prädikaten- und Funktionssymbolen. Unpublished, pp. 22, Manuscript, 1976.
Contains the full proofs for the results in [48].
48. E. Börger. Two new reduction classes in Krom formulae with predicate and function symbols. *Journal of Symbolic Logic*, 42:442, 1977.
Improving results in [42] it is shown that $[\forall^2, (1), (0,1)] \cap \mathrm{KROM} \cap \mathrm{HORN}$ is a reduction class; the results of [42] are restated.
49. E. Börger. Bemerkung zu Gurevich's Arbeit über das Entscheidungsproblem für Standardklassen. *Archiv math. Logik u. Grundlagenforschung*, 19:111–114, 1978.
The theorem in [227] that $[\forall, (0), (\omega)]_=$ is a conservative reduction class is shown to hold also for Horn formulae with only ternary alternations. Gurevich's proof in [227] reduces recursively inseparable unconstrained domino problems (see [237]) and produces reduction formulae which are not Horn and whose alternation length depends on the number of colours of the underlying domino problem. The proof in this paper uses a simple reduction of 2-register machines which has been improved by Löwen to Krom formulae. See Theorem 4.1.1.
50. E. Börger. A new general approach to the theory of the many-one equivalence of decision problems for algorithmic systems. *Zeitschr. f. math. Logik u. Grundlagen d. Math.*, 25:135–162, 1979.
51. E. Börger. Logical description of computation processes. In F. Gecseg, editor, *Fundamentals of Computation Theory*, Lecture Notes in Computer Science No. 117, pages 410–424. Springer, 1981.
Survey superseded by [53, 56]. Contains also a full proof of the theorem appearing in [442], see Theorem 5.1.10. in this book.
52. E. Börger. Undecidability versus degre complexity of decision problems for formal grammars. Report on the first GTI-Workshop, Dept. of Math. and Computer Science, University of Paderborn, 1983. pp. 44–55.
Proves that the usual reduction scheme of Post correspondence problems to formal grammars implies a degree representation theorem for formal grammar decision problems.
53. E. Börger. Decision problems in predicate logic. In *Logic Colloquium 82*, pages 263–301. Elsevier (North Holland), 1984.
Surveys the 1982 status of the classical decision problem for first-order logic, including also new completeness results for logical descriptions of computational problems with respect to various recursive complexity classes. An extension can be found in [56].
54. E. Börger. Spectralproblem and completeness of logical decision problems. In D. Rödding E. Börger, G. Hasenjäger, editor, *Logic and Machines: Decision Problems and Complexity*, Lecture Notes in Computer Science No. 171, pages 333–356. Springer, 1984.
Surveys the history of the spectrum problem, defines the economical description of Turing machines appearing in Chapter 2 of this book and applies it to give a new proof of the Spectrum Hierarchy Theorem (Theorem 2.2.26 in this book).

55. E. Börger. Unsolvable decision problems for Prolog programs. In E. Börger, editor, *Computation Theory and Logic*, Lecture Notes in Computer Science No. 270, pages 37–48. Springer, 1987.
Refines the Aanderaa–Börger description of register machines in [2, 39] to a general scheme by which various decision problems for Prolog programs can easily shown to be recursively unsolvable. In particular Flannagan's conjecture is proved that the floundering property for queries with respect to MU-PROLOG is undecidable.
56. E. Börger. Logic as machine: Complexity relations between programs and formulae. In *Trends in Theoretical Computer Sciene*, pages 59–94. Computer Science Press, 1988.
Unifying treatment of the results in [54, 45, 4, 5, 58], applied to an investigation of the relations between complexity properties for programs and for their first-order formalizations. In particular PROLOG interpretations of bounded and unbounded computations of various machines are shown. Much of the material is covered in chapter 2 of this book.
57. E. Börger. Computability, Complexity, Logic. *Studies in Logic and the Foundations of Math.*, 128, 1989. North-Holland.
58. E. Börger and K. Heidler. Die m-Grade logischer Entscheidungsprobleme. *Archiv math. Logik u. Grundlagenforschung*, 17:105–112, 1976.
Using the reduction method in [46] and the degree representation theorems in [50] it is shown that the initial halting problems of 2-register machines and the validity problems for recursive classes of formulae are the same with respect to m-quivalence; under a natural closure condition on the considered classes of formulae, the equivalence is shown to hold with respect to recursive isomorphism. The same is shown for immmortality problems of 2-register machines and satisfiability problems of recursive classes of formulae. As a corollary the exact complexity of various meta-decision problems (decidability, reduction class property, etc.) is established in the Kleene-Mostowski arithmetical hierarchy, see also [44] and [222].
59. E. Börger and H. Kleine Büning. The r.e. complexity of decision problems for commutative semi-Thue systems with recursive rule set. *Zeitschr. f. math. Logik u. Grundlagen d. Math.*, 26:459–469, 1980.
60. E. Börger and H. Kleine Büning. The reachability problem for Petri nets and decision problems for Skolem arithmetic. *Theor. Computer Science*, 11:123–143, 1980.
Studies the decision problem for the class C_0 of closed universal Horn formulae in prenex conjunctive normal form of extended Skolem arithmetic without equality (i.e. first-order formulae built up from the multiplication sign, constants for the natural numbers and free occurring predicates). It is shown that the decision problem for C_0 is exponentially time bounded equivalent to the reachability problem for Petri nets if restricted to the class of a) Krom formulae with b) only monadic predicates and c) without terms containing a variable more than once. It is shown that leaving out one of the restrictions a) to c) yields classes of formulae whose decision problem can assume any prescribed r.e. many-one degree.
61. E. Börger and U. Löwen. Logical decision problems and complexity of logic programs. *Fundamenta Informaticae*, 10:1–34, 1987.
Surveys old and proves some new results on characterizations of complexity classes by logical decision problems. Contains in particular a classification of decision problems for classes of Bernays–Schönfinkel formulae with and without equality and with restrictions on the arity of predicates.

62. J. Büchi. Weak second-order arithmetic and finite automata. *Zeitschr. f. math. Logik u. Grundlagen d. Math.*, 6:66–92, 1960.
Review and corrigenda by McNaughton in Journal of Symbolic Logic 28, 1963, pages 100-102. Proves, among other results, that a language is regular if and only if it is definable in monadic second-order logic. See also [511].
63. J. Büchi. On a decision method in restricted second-order arithmetic. In E. Nagel et el., editor, *Proc. Internat. congr. on Logic, Methodology and Philosophy of Science*, pages 1–11. Stanford University Press, 1962.
The seminal paper where Büchi introduces his automata and uses them to prove the decidability of the monadic second-order theory of the natural numbers with zero and the successor function.
64. J. Büchi. Turing machines and the Entscheidungsproblem. *Math. Annalen*, 148:201–213, 1962.
In this paper, for the first time since 1937, Turing's reduction method (see [513]) is taken and simplified dramatically by using Skolemization of prenex normal forms and by restricting the attention to their interpretations over the corresponding canonical domain of terms (Herbrand domains). This allows Büchi to completely eliminate that part of Turing's proof for the unsolvability of the Entscheidungsproblem where the data structures of Turing machines (sequences of words and formulae) are formalized. Büchi also observes that this refined reduction can be easily trimmed to a conservative one. Thus the conservative reduction class $[\exists \wedge \forall\exists\forall, (\omega, 3)]$ is established answering some of the open problems stated in [498, p. 177] and the ground is prepared for answering the remaining open finite case $\forall\exists\forall$ (see [288]), as well as the infinite prefix cases, see Chap. 3 of this book. For a further refinement of the method which allows to reduce also the propositional structure of reduction formulae, see [2, 39].
65. J. Büchi. The monadic theory of ω_1. In *Decidable Theories II*, Lecture Notes in Mathematics No. 328, pages 1–127. Springer, 1973.
A well-written account on the results by the author and others on the decidable monadic and weak monadic (second-order) theories of ordinals. In particular, the monadic theory of ω_1 is proven decidable. Later Büchi and Zaiontz extended this result to any ordinal $< \omega_2$ [369] which is the best possible result in a sense [238].
66. J. Büchi. Using determinacy to eliminate quantifiers. In *Fundamentals of Computation Theory*, Lecture Notes in Computer Science No. 56, pages 367–378. Springer, 1977.
A sketch of interesting ideas to extend the results of [68]. See [67].
67. J. Büchi. State-strategies for games in $F_{\sigma\delta} \cap G_{\delta\sigma}$. *Journal of Symbolic Logic*, 48:1171–1198, 1983. Realizing ideas sketched in [66], Büchi uses games to give an alternative proof of Rabin's Complementation Lemma [430]. The proof retains the induction over countable ordinals. See [236].
68. J. Büchi and L. Landweber. Solving sequential conditions by finite-state strategies. *Trans. Amer. Math. Soc.*, 138:295–311, 1969.
A pioneering paper on constructive determinacy. The authors prove a constructive determinacy of a game $G(W)$ where the winning set W is the set of infinite strings accepted (in the appropriate sense) by a finite automaton A. The proof yields an algorithms that, given A, decides which player has a winning strategy and constructs a finite automaton that executes the winning strategy. See [369].
69. A. Bullock and H. Schneider. On generating the finitely satisfiable formulas. *Notre Dame Journal of Formal Logic*, XIV:373–379, 1973.

A set of four rules is given which is shown to generate precisely the finitely satisfiable equality-free first-order formulae. See also [241].

70. J. Burch and D. Dill. Automatic verification of pipelined microprocessor control. In *Proceed. of Conf. on Computer-Aided Verification*, 1994.
71. R. Cafera and N. Peltier. Decision procedures using model building techniques. In H. Kleine Büning, editor, *Computer Science Logic, CSL'95. Selected papers*, Lecture Notes in Computer Science No. 1092, pages 130 – 144. Springer, 1996.
72. H. Carstens. *Über die Kompliziertheit numerischer Modelle.* PhD thesis, Institut für math. Logik und Grundlagenforschung der Universität Münster i.W., 1972. See also: Reducing hyperarithmetic sequences, in: Fundamenta Mathematicae 89, 1975, 5-11.
73. A. Chandra, D. Kozen, and L. Stockmeyer. Alternation. *Journal of the ACM*, 28:114–133, 1981.
74. A. Chandra, H. Lewis, and J. Makowsky. Embedded implicational dependencies and their inference problem. In *Procedings of 13th Annual ACM Symposium on Theory of Computing*, pages 342–354, 1981.
75. C. Chang and H. Keisler. An improved prenex normal form. *Journal of Symbolic Logic*, 27:317–326, 1962.
Proves that in a logic with equality each formula α is logically equivalent to a prenex conjunctive normal form whose alternations have length $\max(2, p)$ where p is the number of predicate symbols occuring in α. For the case without equality the same normal form is shown with alternation length $\max(3, p)$.
76. C. Chang and J. Keisler. *Model theory.* North-Holland, 1990.
77. B. Chlebus. On the computational complexity of satisfiability in propositional logics of programs. *Theor. Computer Science*, 21:179–212, 1982.
78. B. Chlebus. Domino-tiling games. *Journal of Computer and System Sciences*, 32:374–392, 1986.
The paper investigates the computational complexity of strategy problems for games in which two players build bounded domino tilings. In particular it is shown that the square tiling game is complete for PSPACE, that the rectangle tiling game is complete for EXPTIME and that the high tiling game is complete for double exponential time. These results are shown to provide simple proofs for the hardness part of the following completeness results for propositional logic satisfiability problems: PSPACE-completeness of quantified propositional logic [493, 491], EXPTIME-completeness of propositional dynamic logic [167], PSPACE-completeness for propositional modal logic [337], double exponential time completeness of propositional dynamic logic with double-star programs [77]. See also [203, 206].
79. C. Christen. *Spektren und Klassen elementarer Funktionen.* PhD thesis, ETH Zürich, 1974.
The results of this thesis are presented partly in [486]. See also the survey [53].
80. A. Church. A note on the Entscheidungsproblem. *Journal of Symbolic Logic*, 1:40–41, 1936. Correction ibid 101–102.
The first proof in the literature for the unsolvability of the Entscheidungsproblem (in terms of provability). The proof goes as follows: Every partial recursive function can be represented in a theory which is a finite extension of predicate logic (namely using finitely many number theoretic axioms which describe how to compute those functions for given arguments). Therefore this theory is undecidable. As a consequence provability in predicate logic is undecidable. See [513] for an independent proof which follows a different line of arguments.

81. A. Church. Special cases of the decision problem. *Revue philosophique de Louvain*, 49:203–221, 1951. A correction, ibid. 50, 1952, pp. 270–272. See [306].
82. A. Church. *Introduction to mathematical logic.* Princeton University Press, 1956. pp X+378
The book contains two sections on the decision problem, one on solutions in special cases (§46) and one on reductions (§47), together with numerous historical references.
83. A. Church. Logic, arithmetic and automata. In *Proc. Intern. Congress. Math.*, pages 23–35., 1963.
84. A. Church and W. Quine. Some theorems on definability and decidability. *Journal of Symbolic Logic*, 17:179–187, 1952.
Strengthens Kalmár's reduction in [295] to one symmetric binary relation. The proof is given by proving first that every binary relation R of natural numbers is first-order definable in terms of a symmetric binary relation R'. This result is strengthened in [93] to the first-order definability in terms of a single binary relation, of any class of predicates over the natural numbers.
85. D. Cohen. Degree problems for modular machines. *Journal of Symbolic Logic*, 45:510–528, 1980.
86. D. Cohen. Modular machines, undecidability and incompleteness. In E. Börger, G. Hasenjäger, and D. Rödding, editors, *Logic and Machines: Desision Problems and Complexity*, Lecture Notes in Computer Science No. 171, pages 237–247. Springer, 1984.
See the comment to [8].
87. A. Colmerauer and P. Roussel. *The birth of Prolog.* PrologIA, 1992.
88. K. Compton. 0-1 laws in logic and combinatorics. In I. Rival, editor, *NATO Adv. Study Inst. on Algorithms and Order*, pages 353–383. D. Reidel, 1988.
An excellent survey on 0-1 laws.
89. K. Compton and C. Henson. A uniform method for proving lower bounds on the computational complexity of logical theories. *Annals of Pure and Applied Logic*, 48:1–79, 1990.
The method proposed in this paper starts from appropriate inseparability results for certain logical problems which are obtained by standard first-order formalizations of Turing machines. Then interpretations of one theory in another are used to transfer lower complexity bounds. These interpretations use well-known techniques for proving the undecidability of theories by interpreting models of a theory already known to be undecidable (see for explanations the Sect. 3.2 on existential interpretation in this book, and [148, 429, 432, 506]. They avoid the need to code machine computations into the models of the theory being studied and replace such machine codings by often much simpler definability considerations. As a result the authors obtain new proofs of essentially all previously known lower bounds for logical theories.
90. J. Conway. Unpredictable iterations. In *Proc. 1972 Number theory Conference*, pages 49–52. University of Colorado, 1972.
91. S. Cook. The complexity of theorem-proving procedures. In *Proceed. of the 3rd Annual ACM Symposium on Theory of Computing*, pages 151–158, 1971.
The seminal paper which defines NP (see also [343]) and NP-completeness. In particular the satisfiability problem for propositional formulae in conjunctive normal form (with ternary disjunctions) is shown to be NP-complete. For an extension of this completeness to the Σ-levels of the polynomial time hierarchy see [392, 493, 491]. For a good introduction see [486] and for a comprehensive treatment [178]. The paper contains also the proof that the satisfiability problem for Boolean Krom formulae is in P. (For a proof showing

that the (provability by unit resolution of the) unsatifiability of Boolean Krom formulae is complete for non-deterministic logarithmic space see [284], see [150] for a linear time algorithm.) See [23] for reductions to restricted versions of propositional satisfiability problems which use the method of economical description of Turing machines of [56] (see Chap. 2 of this book).

92. W. Craig. Incompletability, with respect to validity in every finite nonempty domain, of first-order functional calculus. In *Proceedings of the International Congress of Mathematicians*, volume 1, page 721, Cambridge/Mass., 1950.
Announces a proof for the undecidability of the class of all finitely satisfiable first-order formulae. The proof is said to be using Turing's method and to have been found independently from Trakhtenbrot's different proof in [509].
93. W. Craig and W. Quine. On reduction to a symmetric relation. *Journal of Symbolic Logic*, 17:188, 1952.
See [84].
94. E. Dahlhaus, A. Israeli, and J. Makowsky. On the existence of polynomial time algorithms for interpolation problems in propositional logic. *Notre Dame Journal of Formal Logic*, 29:497–509, 1988.
95. M. Dauchet. Simulation of Turing machines by a regular rewrite rule. *Theor. Computer Science*, 103:409–420, 1992.
It is proved that each Turing machine can be simulated by just one, a left linear, rewrite rule. This shows that the termination problem of rewrite rule systems with only one, a left linear, rule is undecidable. See the analogous result for Krom and Horn implications established in [129].
96. M. Davis. *Leibniz's dream.* To appear.
97. M. Davis. Unsolvable problems. In J. Barwise, editor, *Handbook of Mathematical Loghic*, pages 567–594. North-Holland, 1977.
98. M. Davis, Y. Matijasevich, and J. Robinson. Hilbert's tenth problem. Diophantine equations: Positive aspects of negative solutions. In *AMS-Proceedings of Symposia in Pure Mathematics*, pages 323–378, 1976.
99. M. Davis, H. Putnam, and J. Robinson. The decision problem for exponential Diophantine equations. *Annals of Math.*, 74:425–436, 1961.
100. A. Dawar. *Feasible Computation through Model Theory.* PhD thesis, University of Pennsylvania, 1993.
101. A. Degtyarev, Y. Gurevich, and A. Voronkov. Herbrand's Theorem and Equational Reasoning: Problems and Solutions. *Bulletin of EATCS*, 60, 1996.
The article (written in a popular form) explains that a number of different algorithmic problems related to Herbrand's Theorem happen to be equivalent. Among these problems are the intuitionistic provability problem for the existential fragment of first-order logic with equality, the intuitionistic provability problem for the prenex fragment of first-order with equality, and the simultaneous rigid E unification problem (SREU). The article explains an undecidability proof of SREU and decidability proofs for special cases. It contains an extensive bibliography on SREU.
102. A. Degtyarev, Yu. Matiyasevich, and A. Voronkov. Simultaneous rigid *E*-unification and related algorithmic problems. In *11th Annual IEEE Symposium on Logic in Computer Science*, page 11, 1996.
103. A. Degtyarev and A. Voronkov. Decidability problems for the prenex fragment of intuitionistic logic. In *11th Annual IEEE Symposium on Logic in Computer Science*, page 10, 1996.
A proof-theoretic technique for investigating provability in intuitionistic logic with and without equality is introduced. Authors give an analogue of skolemization for intuitionistic logic with equality. It is proved that the prefix fragment of intuitionistic logic with equality without function symbols is

PSPACE-complete. Using a result of [102] it is proved that the prefix fragment of intuitionistic logic with equality in the signature with one unary function symbol and any number of constants is decidable.

104. A. Degtyarev and A. Voronkov. The undecidability of simultaneous rigid E-unification. *Theor. Computer Science*, 166:10, 1996.
It is proved that simultaneous rigid E-unification is undecidable. The result has a number of consequences. One consequence is that the $\exists^*$-fragment of intuitionistic logic with equality and function symbols is undecidable.

105. L. Denenberg. *Computational Complexity of Logical Problems: Formulas, Dependencies, and Circuits.* PhD thesis, Harvard University, Division of Applied Sciences, 1984.
Proves results on Krom and Horn classes. See [107].

106. L. Denenberg and H. Lewis. The complexity of the satisfiability problem for Krom formulas. *Theor. Computer Science*, 30:319–341, 1984.

107. L. Denenberg and H. Lewis. Logical syntax and computational complexity. In *Computation and Proof Theory. Proceedings of the Logic Colloquium 83, Aachen, Part II*, Lecture Notes in Mathematics No. 1104, pages 101–115. Springer, 1984.
Survey of results on the computational complexity of classes of first-order formulae with restricted prefix-vocabulary and truth-functional structure. In particular the following classes are dicussed: monadic predicate logic, the Bernays-Schönfinkel class, the Ackermann class and the Gödel-Kalmár-Schütte class, also when restricted to Krom and Horn formulae. For proofs of the results surveyed see [105, 106, 137, 175, 352, 353, 421].

108. J. Denton. *Applications of the Herbrand Theorem.* PhD thesis, Harvard University, 1963.
Contains a proof that $[\forall\exists\forall^*, (0,1)]$ is a reduction class.

109. M. Deutsch. *Normalformen aufzählbarer Prädikate.* PhD thesis, Universität Münster, Institut für Grundlagenforschung, 1968.
See comment to [124].

110. M. Deutsch. Zur Darstellung koaufzählbarer Prädikate bei Verwendung eines einzigen unbeschränkten Quantors. *Zeitschr. f. math. Logik u. Grundlagen d. Math.*, 21:443–454, 1975.
Refines Trakhtenbrot's [509] spectral representation of recursively enumerable predicates to formulae $\alpha_f \in [\forall\exists^*\forall, (\omega, 1)]$ in which the equality occurs only once. The improvement is based on the Davis-Putman-Robinson exponential Diophantine normal form for recursively enumerable predicates, see [99]. See comment to [124].

111. M. Deutsch. Zur Theorie der spektralen Darstellung von Prädikaten durch Ausdrücke der Prädikatenlogik I. Stufe. *Archiv math. Logik u. Grundlagenforschung*, 17:9–16, 1975.
See comment to [124].

112. M. Deutsch. Zur Reduktionstheorie des Entscheidungsproblems. *Zeitschr. f. math. Logik u. Grundlagen d. Math.*, 27:113–117, 1981.
See comment to [124].

113. M. Deutsch. Reductions for the satisfiability with a simple interpretation of the predicate variable. In E. Börger, G. Hasenjäger, and D. Rödding, editors, *Logic and Machines: Desision Problems and Complexity*, Lecture Notes in Computer Science No. 171, pages 285–311. Springer, 1984.

114. M. Deutsch. Ein neuer Beweis und eine Verschärfung für den konservativen Reduktionstyp $\forall\exists\forall^\infty(0,1)$. *Zeitschr. f. math. Logik u. Grundlagen d. Math.*, 32:551–574, 1986.
See comment to [124].

115. M. Deutsch. Eine Bemerkung zum Reduktionstyp $\forall^3\exists^\infty(0,1)$. *Zeitschr. f. math. Logik u. Grundlagen d. Math.*, 33:179–186, 1987.
See comment to [124].
116. M. Deutsch. Eine Verschärfung eines Satzes von Kostyrko zur Reduktionstheorie mit einer Anwendung auf die spektrale Darstellung von Prädikaten. *Zeitschr. f. math. Logik u. Grundlagen d. Math.*, 33:347–358, 1987.
See comment to [124].
117. M. Deutsch. Eine Bemerkung zur spektralen Darstellung aufzählbarer und koaufzählbarer Prädikate durch Ausdrücke aus $\forall\exists\forall\exists^\infty(\rho,1)$. *Zeitschr. f. math. Logik u. Grundlagen d. Math.*, 34:67–77, 1988.
See comment to [124].
118. M. Deutsch. Eine Bemerkung zur spektralen Darstellung von ρ-stelligen aufzählbaren und koaufzählbaren Prädikaten durch Ausdrücke aus $\exists^\infty\forall^3\exists(\rho,1), \forall^\infty\exists(\rho,1)$ und $\forall^3\exists(\infty,1)$. *Zeitschr. f. math. Logik u. Grundlagen d. Math.*, 34:67–77, 1988.
See comment to [124].
119. M. Deutsch. Eine weitere Verschärfung zum konservativen Reduktionstyp $\forall\exists\forall\exists^\infty(0,1)$ mit einer Anwendung auf die spektrale Darstellung von Prädikaten. *Zeitschr. f. math. Logik u. Grundlagen d. Math.*, 35:137–153, 1989.
See comment to [124].
120. M. Deutsch. Zum Reduktionstyp $\exists^\infty\forall\exists\forall(\rho,1)$ und zur spektralen Darstellung ρ-stellinger aufzählbarer und koaufzählbarer Prädikate durch Ausdrücke aus $\exists^\infty\forall\exists\forall(\rho,1)$. *Zeitschr. f. math. Logik u. Grundlagen d. Math.*, 35:517–529, 1989.
See comment to [124].
121. M. Deutsch. Eine Bemerkung zu spektralen Darstellungen von ρ-stellingen aufzählbaren und koaufzählbaren Prädikaten durch Ausdrücke aus $\exists\forall\exists\forall^\infty(\rho,1)$ und $\forall\exists^2\forall(\infty,1)$. *Zeitschr. f. math. Logik u. Grundlagen d. Math.*, 36:163–184, 1990.
See comment to [124].
122. M. Deutsch. Weitere Verschärfungen zu den Reduktionstypen $\forall\exists^\infty\forall(0,1)$, $\exists^\infty\forall^3\exists(0,1)$, $\forall^3\exists(\infty,1)$. *Zeitschr. f. math. Logik u. Grundlagen d. Math.*, 36:339–355, 1990.
See comment to [124].
123. M. Deutsch. Reduktionstyp und spektrale Darstellung mit dem Präfix $\exists\forall\exists\forall$. *Zeitschr. f. math. Logik u. Grundlagen d. Math.*, 37:273–5288, 1991.
See comment to [124].
124. M. Deutsch. Ein neuer Beweis und eine Verschärfung für den Reduktionstyp $\forall\exists\forall^\infty(0,1)$ mit einer Anwendung auf die spektrale Darstellung von Prädikaten. *Zeitschr. f. math. Logik u. Grundlagen d. Math.*, 38:559–564, 1992.
Here and in [109]–[123] Deutsch shows that the conservative reduction classes in question can be sharpened by restricting the interpretation of the unique binary predicate to an ε-like relation over transitive sets. It is also shown that the corresponding reduction formulae yield a spectral representation (in a slightly stronger sense than the one introduced by [509]) of arbitrary pairs of a recursively enumerable predicate P and the complement of a recursively enumerable predicate Q satisfying $Px_1 \ldots x_n \to \neg Qx_1 \ldots x_n$.
125. P. Devienne. Weighted graphs: A tool for studying the halting problem and time complexity in term rewriting systems and logic programming. *Theor. Computer Science*, 75:157–215, 1990. See comment to [454].

126. P. Devienne, P. Lebegue, A. Parrain, J. Routier, and J. Würtz. Smallest Horn clause programs. *Journal of Logic Programming*, 19:1–41, 1994.
Surveys the results of [129, 128, 127, 244], see Theorem 5.2.14.

127. P. Devienne, P. Lebegue, and J. Routier. Halting problem of one binary Horn clause is undecidable. In *Proceedings of 10th Symposium on Theoretical Aspects of Computer Science STACS'93*, Lecture Notes in Computer Science No. 665, pages 48–57. Springer, 1993.
See the comment to [129].

128. P. Devienne, P. Lebegue, and J. Routier. The emptiness problem of one binary recursive Horn clause is undecidable. In *International Logic Programming Symposium ILPS'93*, pages 250–265. MIT Press, Vancouver 1993.
See the comment to [129].

129. P. Devienne, P. Lebegue, J. Routier, and J. Würtz. One binary Horn clause is enough. In Springer Verlag, editor, *Proceedings of 11th Symposium on Theoretical Aspects of Computer Science STACS'94*, Lecture Notes in Computer Science No. 775, pages 21–33. Springer, 1995.
The undecidability of the class of formulae $\overline{\forall}(\pi \wedge (\rho \rightarrow \sigma) \wedge \neg\tau)$ with atomic formulae π, ρ, σ, τ containing functions, proved independently and by different methods in [127, 128, 244] and surveyed in [126], is presented once more, reformulated as computational universality (reduction class) property for $\{\overline{\forall}(\pi_n \wedge (\rho \rightarrow \sigma) \wedge \neg\tau) : n \geq 0, \pi_n, \rho, \sigma, \tau \text{ atomic}\}$ and going through Conway's [90] encoding of 2-register machines by periodically linear functions. See Theorem 5.2.14.

130. J. Dopp. *Logique construites par une méthode de déduction naturelle.* Nauwelaerts, Louvain and Gauthier-Villars, Paris, 1962.

131. P. Downey. Undecidability of Presburger arithmetic with a single monadic predicate letter. Technical report, Center for Research in Computer Technology, Harvard University, Cambridge, MA, 1972.

132. B. Dreben. Solvable Surányi subclasses: An introduction to the Herbrand theory. In *Annals of the Computation Laboratory of Harvard University vol.31*, pages 32–47. Harvard University Press, Cambridge, Massachsetts, 1961.
Proves that every formula $\forall x \forall y \exists u \forall z \alpha$ in Surányi's class $[\forall^2\exists\forall, (\omega, \omega)]$ [498] in which no atomic formula of form Pzx does occur is satisfiable if and only if it is satisfiable in a finite domain whose cardinality is computable. The proof uses the technique of Herbrand expansions. See [189].

133. B. Dreben and W. Goldfarb. *The decision problem: solvable cases of quantificational formulas.* Addison-Wesley, 1979.
The book is based on [190] and investigates the relations between the syntactic form of a first-order formula and structural properties of its Herbrand expansions with a particular emphasis on the ways these structural properties can permit decision procedures. A detailed list of the classes which are shown to be solvable by this method is given on pages 263–265.

134. B. Dreben, A. Kahr, and H. Wang. Classification of $\forall\exists\forall$ formulas by letter atoms. *Bulletin of the American Math. Soc.*, 68:528–532, 1962.
The paper studies the decision problem of solvable subclasses of the Kahr-Moore-Wang reduction class $[\forall\exists\forall, (0, \omega)]$ which are determined by the forms of variable combinations appearing in the atomic formulae. These forms are for formulae $\forall x \exists u \forall y \alpha \in [\forall\exists\forall, (0, \omega)]$ the following: The diagonal forms xx, xu, ux, uu, yy and xy, yx, uy, yu. The four classes of formulae where any diagonal form can occur together with xy and either uy or yx (or with yu and either yx and uy) are shown to be solvable. By [288] the classes where three of the forms out of xy, yx, uy, yu are permitted to occur form reduction classes (see Exercise 3.1.10). See comment to [1].

135. B. Durand. The surjectivity problem for 2D cellular automata. *Journal of Computer and Systems Science*, 49(3):718–725, 1994.
A drastic simplification of Kari's construction in [304]: the proof uses only the constrained tiling problem and no more techniques developped by Robinson [440] and Berger [33].
136. B. Durand. A Random NP-complete problem for inversion of 2D cellular automata. *Theoretical Computer Science*, 148(1):19–32, 1995.
The reduction presented is based on tiling problems.
137. C. Dwork, P. Kanellakis, and J. Mitchell. On the sequential nature of unification. *Journal of Logic Programming*, 1:35–50, 1984.
The P-completeness of the unification problem for first-order terms is proved. See also [138] and Sect. 8.2.
138. C. Dwork, P. Kanellakis, and L. Stockmeyer. Parallel algorithms for term matching. *SIAM Journal of Computing*, 17:711–731, 1988.
See the comment to [137].
139. R. Dyckhoff. Contraction-free sequent calculi for intuitionistic logic. *Journal of Symbolic Logic*, 57:795–807, 1992.
Provides a simple decision procedure for the intuitionistic propositional calculus, based upon a variant of Gentzen's sequent calculus LJ. Provides also the history of the method which has been discovered independently by multiple authors.
140. H.-D. Ebbinghaus. Extended logics: The general framework. In J. Barwise and S. Feferman, editors, *Model Theoretic Logics*, pages 25–76. Springer, 1985.
141. H.-D. Ebbinghaus and J. Flum. *Finite Model Theory.* Springer, 1995.
142. H.-D. Ebbinghaus, J. Flum, and W. Thomas. *Mathematical Logic.* Springer, 1984.
143. A. Ehrenfeucht. Decidability of the theory of one function. *Notices of the AMS*, 6:268, 1959.
Announces (without proof) that the first-order theory of one unary function is decidable. A proof of the much stronger result that the monadic second-order theory of one unary function is decidable is due to Rabin [430]. See Sect. 7.2.
144. A. Ehrenfeucht. An application of games to the completeness problem for formalized theories. *Fund. Math.*, 49:129–141, 1961.
145. T. Eichholz. Semantische Untersuchungen zur Entscheidbarkeit im Prädikatenkalkül mit Funktionsvariablen. *Archiv math. Logik u. Grundlagenforschung*, pages 19–28, 1957.
Shows the decidability for the class of first-order formulae with functions but without equality in which each term and each atomic formula contains exactly one variable. (Extension of the decidability of monadic predicate logic.).
146. H. Enderton. *A Mathematical Introduction to Logic.* Academic Press, 1972.
147. Y. Ershov. Skolem functions and constructive models. *Algebra i Logika*, 12:644–654, 1973.
English translation 1975. Exhibits a sufficient condition for the existence of recursive models for certain axiomatizations of first-order theories. This condition applies in particular to Krom formulae and therefore implies the result of [11].
148. Y. Ershov, I. Lavrov, A. Taimanov, and M. Taitslin. Elementary theories. *Russian Math. Surveys*, 20:35–105, 1965. English translation.
149. Y. Ershov and M. Taitslin. Undecidability of certain theories. *Algebra i Logika*, 2:37–42, 1963.
150. S. Evan, A. Itai, and A. Shamir. On the complexity of timetable and multicommodity flow problems. *SIAM Journal of Computing*, 5(4):691–703, 1976.
See [91].

151. R. Fagin. *Contributions to the model theory of finite structures.* PhD thesis, University of California, Berkley, 1973.
152. R. Fagin. Generalised first order spectra and polynomial time recognizable sets. In R. Karp, editor, *Complexity of Computation. SIAM-AMS Proc.7*, pages 43–73, 1974.
Introduces the notion of a generalized spectrum (i.e. the class of finite models of an existential second-order sentence). Proves what is now called Fagin's Theorem, namely that generalized spectra coincide with NP. This paper has been very influential for finite model theory and descriptive complexity. See Sect. 2.2.3.
153. R. Fagin. Monadic generalized spectra. *Zeitschr. f. math. Logik u. Grundlagen d. Math.*, 21:89–96, 1975.
154. R. Fagin. A spectrum hierarchy. *Zeitschr. f. math. Logik u. Grundlagen d. Math.*, 21:123–134, 1975.
155. R. Fagin. A two-cardinal characterization of double spectra. *Zeitschr. f. math. Logik u. Grundlagen d. Math.*, 21:121–122, 1975.
156. R. Fagin. Probabilities on finite models. *Journal of Symbolic Logic*, 41:50–58, 1976.
Proves the 0-1 law for first-order logic. The same result was proved by Glebskiĭ et al. [183]. See [88] for a survey on 0-1 laws.
157. S. Feferman. *Lectures on proof theory.* Lecture Notes in Computer Science No. 70. Springer, 1968.
158. C. Fermüller. Deciding some Horn clause sets by resolution. In Kurt Gödel Society, editor, *Jahrbuch der Kurt Gödel Gesellschaft 1989*, pages 60 – 73, 1990.
Three classes of clause sets are proved to be decidable via termination of proof search procedures based on different resolution refinements. The classes do not correspond to prenex classes but rather are motivated by formats of logical databases and are characterized in terms of the structure of occurring terms.
159. C. Fermüller. Deciding classes of clause sets by resolution. Doctoral dissertation, Technische Universität Wien, 1991.
This dissertation summarizes some results on resolution as decision procedure. In particular extensions of the monadic class, as well as the Ackermann, Gödel, Skolem, and Maslov classes are shown to be decidable via terminating proof search procedures based on different variants and refinements of resolution. The decidability proofs are mostly accompanied by undecidability results for slight syntactical generalisations of the decidable classes.
160. C. Fermüller. A resolution variant deciding some classes of clause sets. In E. Börger, H. Kleine Büning, and M.M. Richter, editors, *Computer Science Logic, CSL'90*, Lecture Notes in Computer Science No. 533, pages 128 – 144. Springer, 1991.
Using a particular variant of resolution, the decidability is proved for a class of clause sets which contains clausal versions of the Gödel-Kalmar-Schütte class, the so-called extended Skolem class, Maslov's class K [381], etc. Also an extension of the monadic class is proved to be decidable via resolution. These results extend some results of [286].
161. C. Fermüller. A simple resolution based decision procedure for the Maslov class. In *Developments in Theoretical Computer Science, Proceedings of the 7th IMYCS, Smolenice, November 1992*, Topics in Computer Mathematics, pages 197 – 204. Gordon & Breach, 1994.
A simplified decidability proof of Maslov's class (prefix $\exists^*\forall^*\exists^*$ and Krom matrix) is presented. In fact it is shown that disallowing nesting of function

symbols in ordinary resolution is refutationally complete and terminating for a class of clause sets containing a clausal version of the Maslov class. The resulting decision procedure is optimal with respect to worst case complexity.

162. C. Fermüller and A. Leitsch. Hyperresolution and automated model building. *Journal of Logic and Computation*, to appear.
Extends hyperresolution as decision procedure for clausal classes (see [158], [341], [342] and [163]) to a method of automated model building. The method works for all decision classes of hyperresolution defined in the papers above. The following finite model property is shown: If C is a finite set of clauses which is stable under the operator of hyperresolution and all positive clauses in C are linear (i.e. every variable occurs at most once) then C has a finite model.

163. C. Fermüller, A. Leitsch, T. Tammet, and N. Zamov. *Resolution Methods for the Decision Problem*. Lecture Notes in Computer Science No. 679. Springer, 1993.
This monograph extends Joyner's work [286] in several aspects: **1.** all decision classes are generalized (particularly the Ackermann class, the monadic class, the Gödel class and the Maslov class) to nonprenex types (function symbols need not be Skolem symbols). **2.** Other resolution refinements than A-orderings are investigated as decision procedures (e.g. hyperresolution and π-orderings). It is shown that Maslov's K-class [381] can be decided by a π-ordering refinement. Hyperresolution is shown to decide the classes PVD (see [342]) and OCC1N (a generalization of classes introduced in [158]); these classes are not extensions of prenex classes. OCC1N plays a major role in finite model building (see [162]). **3.** Extension of resolution decision procedures to methods of automated model building: A backtracking–free method is defined to extract finite models from inference–stable sets of clauses obtained from the Ackermann class and the monadic class; the method is a variant of the method in [501].

164. C. Fermüller and G. Salzer. Ordered paramodulation and resolution as decision procedure. In A. Voronkov, editor, *Logic Programming and Automated Reasoning, LPAR'93, St. Petersburg, July 1993*, Lecture Notes in Artificial Intelligence No. 698, pages 122 – 133. Springer, 1993.
This paper extends results on resolution as decision procedure by considering also clauses with equality literals. A version of ordered paramodulation and resolution is used to decide a class of clause sets that corresponds to an extension of the Ackermann class with equality. By encoding Turing machines it is also shown that slight modifications of the defining conditions for this class lead to undecidability.

165. J. Ferrante and C. Rackoff. A decision procedure for the first order theory of real addition with order. *SIAM Journal of Computing*, 4(1):69–76, 1975.

166. J. Ferrante and C. Rackoff. *The Computation Complexity of Logical Theories*. Lecture Notes in Mathematics No. 718. Springer, 1979.
A detailed (althogh not comprehensive) treatment of upper and lower complexity bounds for various first-order logical theories. In particular the following theories are considered: One injective function or one successor (with and without an additional monadic predicate), two successors with the length identity predicate, well-order, lexicographical order, natural numbers with order, functions over the natural numbers which are almost everywhere 0 with addition, integers or reals with $0, +, \leq$, finite abelian groups, pairing functions. Most of the upper bounds are obtained by the technique of Ehrenfeucht-Fraïssé games [144, 170]. The lower bounds are obtained by direct logical (or arithmetical) description of Turing machines which run in bounded time or

space; these formalizations are in the line of the reduction techniques used in this book. For a different technique using interpretations to transfer lower bounds from one theory to another and a more complete treatment of the known lower bounds for logical theories see [89]. For yet another method see [169].

167. M. Fischer and R. Ladner. Propositional dynamic logic of regular programs. *Journal of Computer and System Sciences*, pages 194–211, 1979. See also [78].

168. M. Fischer and M. Rabin. Super-exponential complexity of Presburger arithmetic. In R.M. Karp, editor, *Complexity of Computation. Proc. of SIAM-AMS Symposium in Applied Mathematics, vol.7*, pages 27–44, 1974.
This paper starts a long series of investigations into the computational complexity of decision procedures for Presburger arithmetic. Similar results are proved for the theories of finite abelian (or cyclic) groups, of integer multiplication, of real (or complex) addition. For the further development of the theory see [89, 166, 199, 201, 206].

169. K. Fleischmann, B. Mahr, and D. Siefkes. Bounded concatenation theory as a uniform method for proving lower comlexity bounds. In R. Gandy and M. Hyland, editors, *Logic Colloquium '76*, pages 471–490. North-Holland, Amsterdam, 1977.

170. R. Fraïssé. Sur quelques classifications des systèmes de relations. *Publ. Sci. Univ. Alger. Sér. A*, 1:35–182, 1954.

171. G. Frege. *Begriffsschrift, eine der arithmetischen nachgebildete Formelsprache des reinen Denkens.* 1879.

172. J. Friedman. A computer program for a solvable case of the decision problem. *Journal of the ACM*, 10:348–356, 1963.
The decision procedure from [173] is implemented by a FORTRAN program which decides the validity of formulae in $[\exists^2\forall^*, (0,\omega)]$.

173. J. Friedman. A semi-decision procedure for the functional calculus. *Journal of the ACM*, 10:1–24, 1963.
Proposes decision tables as method for a systematic treatment of solvable cases of the Entscheidungsproblem. The method is shown to provide a decision procedure for the validity of formulae in Skolem normal form $\exists y_1 \ldots \exists y_m \forall z_1 \ldots \forall z_n \alpha$ where the quantifier-free part α satisfies one of the following conditions: a) every atomic formula contains at least one of the individual varialbes z_i or contains only one individual variable or contains both y_1 and y_2 and no other individual variables, b) every atomic formula contains at least one of the individual variables z_i or contains only one individual variable or contains all of y_j. For a decision procedure for $[\forall^2\exists, all]$ using this method, see [289]. See comment to [172].

174. J. Friedman. Extensions of two solvable cases of the decision problem. *Journal of Symbolic Logic*, 22:108, 1975.

175. M. Fürer. Alternation and the Ackermann case of the decision problem. *L'Enseignement Mathématique*, II(XXVII):137–162, 1982.
The paper establishes a $\mathrm{DTIME}(c^{n/\log n})$ lower bound for the $\forall\exists^2$-subcase of the Ackermann's class $[\exists^*\forall\exists^*]$. It is proved by formalizing the acceptance problem for linear space bounded alternating Turing machines. Lewis [352] independently proves the same lower boung by using alternating pushdown automata. Fürer also proves that $[\exists^*\forall\exists^*, (\omega)] \in \mathrm{DTIME}(c^{n/\log n})$ of [352] by using an alternating Turing machine working in space $O(n/\log n)$. Fürer's machine enters no universal states for formulae in $[\exists^*\forall\exists, (\omega)]$, so this class is in $\mathrm{NSPACE}(n/\log n)$. As a corollary it follows that $[\exists^*\forall\exists^*] \in \mathrm{DTIME}(2^{O((n/\log n)^2)})$ and $[\exists^*\forall\exists] \in \mathrm{NSPACE}((n/\log n)^2)$. See Chap. 6 of this book.

176. M. Fürer. The complexity of Presburger arithmetic with bounded quantifier alternation. *Theor. Computer Science*, 18:105–111, 1982.
It is shown that the decision problem for Presburger arithmetic with only bounded quantifier alternations has nondeterministic exponential complexity. The entire theory has double exponential complexity [89, 166, 168]. It is claimed that it is typical for first-order theories that bounding quantifier alternations reduces the decision complexity by one exponential step. For counterexamples to this claim, see [198, 206].

177. M. Fürer. The computational complexity of the unconstrained limited domino problem (with implications for logical decision problems). In E. Börger, G. Hasenjäger, and D. Rödding, editors, *Logic and Machines: Desision Problems and Complexity*, Lecture Notes in Computer Science No. 171, pages 312–319. Springer, 1984.
Refutes a conjecture of Lewis [349] by showing that for numbers k given in binary the class of k-limited domino problems has an $\Omega(c^n)$ nondeterministic time lower bound (upper bound $O(d^n)$). It follows by an argument in [351] that the class $\forall\exists \wedge \forall^2$ has a nondeterministic time complexity between $\Omega(c^{n/\log n})$ and $O(d^{n/\log n})$ for some constants $c, d > 1$, see Chap. 6.2. This improves the lower bound derived in [349] for $\exists \wedge \forall\exists \wedge \forall^2$ and gives an essentially optimal lower complexity bound for L_2 (see Sect. 8.1).

178. M. Garey and D. Johnson. *Computers and Intractability: A Guide to the Theory of NP-Completeness.* H. Freeman, San Francisco, 1979.

179. J. Genenz. Reduktionstheorie des Entscheidungsproblems im Prädikatenkalkül der ersten Stufe nach der Methode von Kahr-Moore-Wang. Diplomarbeit Universität Münster (1965)
Uses reductions of domino problems in the spirit of [288] to strengthen the reduction classes $[\forall^3\exists, (\omega, 1)], [\forall^3\exists^*, (0, 1)], [\forall^*\exists, (0, 1)]$ by restricting the interpretations of the binary predicate to a reflexive relation.

180. J. Genenz. *Untersuchungen zum Entscheidungsproblem im Prädikatenkalkül der ersten Stufe.* PhD thesis, Institut f. math. Logik und Grundlagenforschung, Universität Münster, 1965.
Proof for the conservative reduction class $[\forall\exists^*\forall, (0, 1)]$, found independently from [317]. Proves also that the reduction classes $\forall\exists\forall, \forall^3\exists$ can be restricted to formulae with only one predicate, of arbitrarily high arity.

181. G. Germano. An arithmetical reconstruction of the liar's antinomy using addition and multiplication. *Notre Dame Journal of Formal Logic*, XVII:457–461, 1976.

182. M. Gladstone. Finite models for inequalities. *Journal of Symbolic Logic*, 31:581–592, 1966.
Proves that each finite consistent set of inequalities between terms of a first-order language has a finite model. This implies a new decision procedure for the class HERBRAND which in addition shows that this class has the finite model property [254].

183. Y. Glebskiĭ, D. Kogan, M. Liogonky, and V. Talanov. The volume and fraction of the satisfiability of first-order formulas. *Kibernetika (Kiev)*, 2:17–28, 1969. English translation (titled "range and degree of realizablility of formulas in the restricted predicate calculus") in Cybernetics 5 (1969), 142–154.
The 0-1 law for first-order logic is proved. The same result was obtained by Fagin [156] with a different proof. See [88] for a survey on 0-1 laws.

184. K. Gödel. Die Vollständigkeit der Axiome des logischen Funktionenkalküls. *Monatshefte f. Mathematik u. Physik*, 37:349–360, 1930. Reprinted and translated in [188, 102–123].

185. K. Gödel. Über formal unentscheidbare Sätze der *principia mathematica* und verwandter Systeme. *Monatshefte f. Mathematik u. Physik*, 38:173–198, 1931. Reprinted and translated in [188, 144–195].
186. K. Gödel. Ein Spezialfall des Entscheidungsproblems der theoretischen Logik. *Ergebnisse eines mathematischen Kolloquiums*, 2:27–28, 1932. Reprinted and translated in [188, pp. 230–233]
Proves the decidability of $[\exists^*\forall^2\exists^*, all]$. The same result was proved independently by Kalmár [293] and Schütte [457]. See also [187, 239] and Sect. 6.2.3 of this book.
187. K. Gödel. Zum Entscheidungsproblem des logischen Funktionenkalküls. *Monatshefte f. Mathematik u. Physik*, 40:433–443, 1933. Reprinted in [188, pp. 306–326]
Proves the finite model property for $[\exists^*\forall^2\exists^*, all]$ thus strengthening the results of [186, 293]. See [239] and Sect. 6.2.3 of this book for a simplified proof via a probabilistic argument. Improves Skolem's reduction class $[\forall^*\exists^*, all]$ (see [477]) to $[\forall^3\exists^*, (0,\omega)]$. A simpler proof for (a stronger form of) this result appears in [479], see [302] for the improvement to $[\forall^3\exists^*, (0,1)]$.
188. K. Gödel. *Collected Works, Vol. I: Publications 1929–1936.* Oxford University Press, 1986. Edited by S. Feferman, J. Dawson jr., S. Kleene, G. Moore, R. Solovay and J. van Heijenoort.
189. R. Goldberg. On the solvability of a subclass of the Surányi reduction class. *Journal of Symbolic Logic*, 28:237–244, 1963.
Simplifies the proof given in [132] for the decidability of the class of formulae $\forall y_1\forall y_2\exists x\forall y_3\alpha \in [\forall^2\exists\forall, (\omega,\omega)]$ in which in addition to monadic predicates only dyadic atoms from the following set occur: Py_iy_{i+1}, $Py_{i+1}y_i$, Py_ix, Pxy_i for $i \leq i \leq 2$, diagonal atoms Fzz for $z = \{y_1, y_2, x, y_3\}$. Note that if the atom Py_3y_1 is added then the resulting class is undecidable. Note that [132] shows also the finite model property of this class.
190. W. Goldfarb. *On Decision Problems for Quantification Theory.* PhD thesis, Harvard University, 1979.
Source of the book [133].
191. W. Goldfarb. On the Gödel class with identity. *Journal of Symbolic Logic*, 46:354–364, 1981.
Shows that there is no primitive recursive decision procedure for the Gödel-Kalmár-Schütte class with equality. In [192] the undecidability of this class is established.
192. W. Goldfarb. The unsolvability of the Gödel class with identity. *Journal of Symbolic Logic*, 49:1237–1252, 1984.
Proves that the class of relational $\forall^2\exists$-sentences with equality is a conservative reduction class, thus refuting a claim by Gödel [187]. See Sect. 4.3 of this book for a proof and Sect. 4.4 for historical remarks on this result.
193. W. Goldfarb. Random models and solvable Skolem classes. *Journal of Symbolic Logic*, 58:908–914, 1993.
Proposes random models for a general scheme for proofs of the finite model property for classes of formulae in Skolem normal form.
194. W. Goldfarb, Y. Gurevich, and S. Shelah. A decidable subclass of the minimal Gödel class with identity. *Journal of Symbolic Logic*, 49:1253–1261, 1984.
Consider statements of predicate logic with equality of the form "For every unordered pair $\{x, y\}$ of elements there exists an element z such that a quantifier-free formula $\varphi(x, y, z)$ holds". Here $\varphi(x, y, z)$ is symmetric with respect to x and y: $\varphi(y, x, z) = \varphi(x, y, z)$. These statements, written as first-order sentences, form a subclass X of $[\forall^2\exists, (\omega, 1)]_=$. The authors prove that X has the finite model property. Contrast this with the fact that $[\forall^2\exists, (\omega, 1)]_=$

is a conservative reduction class (see [192] and Chap. 4 of this book). Furthermore, the finite model property can be extended to conjunctions $\varphi \wedge \psi$ where $\varphi \in X$ and $\varphi \in [\forall\exists^*, all]_=$.

195. W. Goldfarb and H. Lewis. Skolem reduction classes. *Journal of Symbolic Logic*, 40:62–68, 1975.
It is shown that for each $n \geq 3$ and any distinct sets $Y_1, Y_2 \subseteq \{y_1, y_2, \ldots, y_n\}$ of variables of cardinality at least 2, the class $C(n, Y_1, Y_2)$ is a reduction class where $C(n, Y_1, Y_2)$ consists of all formulae $\forall y_1 \ldots \forall y_n \exists x \alpha \in [\forall^n \exists, all]$ such that for each atomic subformula β not containing x the set of variables occuring in β is either Y_1 or Y_2. The proof uses an encoding of the (origin-constrained) domino problem.

196. E. Grädel. On solvable cases of Hilbert's 'Entscheidungsproblem'. Habilitationsschrift, Universität Basel (1990), 73 pages.
A number of complexity results for decidable prefix-vocabulary classes are presented. Many results of Chapter 6 are taken from there.

197. E. Grädel. *The Complexity of Subclasses of Logical Theories.* PhD thesis, Universität Basel, 1987. A summary appeared in Bulletin of the EATCS vol. 34 (1988), 289–291.
See comments to [199, 201].

198. E. Grädel. Domino-games, with an application to the complexity of Boolean algebras with bounded quantifier alternation. In *Proceedings of the 5th Annual Symposium on Theoretical Aspects of Computer Science STACS 88, Bordeaux 1988*, Lecture Notes in Computer Science No. 294, pages 98–107. Springer, 1988.
See comment to [203].

199. E. Grädel. Subclasses of Presburger Arithmetic and the Polynomial-Time Hierarchy. *Theor. Computer Science*, 56:289–301, 1988.
It is shown that the bounded prefix classes in Presburger arithmetic, with $m+1$ finite alternating blocks of quantifiers, are complete for Σ_m^p or Π_m^p, i.e. the classes on the m-th level of the polynomial-time hierarchy. In particular it is shown that the $\exists\forall\forall$-prefix class is NP-complete. It was shown by Scarpellini [451] that the $\exists^t$ and $\forall^t$-classes are solvable in polynomial time (this is based on a result by H. Lenstra on integer linear programming with a fixed number of variables). See also comments to [176, 201].

200. E. Grädel. Complexity of formula classes in first order logic with functions. In *Fundamentals of Computation Theory (FCT '89)*, Lecture Notes in Computer Science No. 380, pages 224–233. Springer, 1989.
The complexity of decidable prefix vocabulary classes with function symbols is investigated. The results complement those of Lewis [352] and Fürer [175] on the classical solvable case. See Chapter 6 of this book.

201. E. Grädel. Dominoes and the complexity of subclasses of logical theories. *Annals of Pure and Applied Logic*, 43:1–30, 1989.
The complexity of prefix classes in logical theories (mainly Presburger and Skolem arithmetic) is studied. In particular it is shown that the $\exists\forall^*$-class in Presburger arithmetic has nondeterministic exponential complexity. It is a minimal prefix class with this property since the $\exists^*$-sentences form an NP-complete class and the finite prefixes define classes in the polynomial-time hierarchy [199].
Skolem arithmetic (the natural numbers with multiplication but without addition) is the weak direct power of Presburger arithmetic. In general, the complexities of most prefix classes in this theory are one exponential step higher than in the case of Presburger arithmetic. An exception is the class of existential sentences which is NP-complete. See also [89, 166, 168, 176, 199, 451].

To prove lower complexity bounds, finite versions of domino problems are introduced, which are used also in Chapter 6 of this book.

202. E. Grädel. Size of models versus length of computations. On inseparability by nondeterministic time complexity classes. In *Proceedings of the Second Workshop on Computer Science Logic CSL 88, Duisburg 1988*, Lecture Notes in Computer Science No. 385, pages 118–137. Springer, 1989.
For a number of (undecidable) prefix vocaculary classes X, results of the following kind are proved: There is no set in the complexity class $\text{NTIME}(T(cn))$ (for some $c > 0$) that separates the sentences $\psi \in X$ with a model of cardinality $\leq T(|\psi|$ from the unsatisfiable sentences in X. Such results depend on the size of models that are necessary to encode Turing machine computaions of a given length $T(n)$ via formulae of the class X. Also an application to a complexity results for Presburger arithmetic is presented.

203. E. Grädel. Domino games and complexity. *SIAM Journal of Computing*, 19:787–804, 1990.
Domino games are two-person games, where the players bouild domino tilings of a square of prescribed size. Domino games are used to describe computations of alternating Turing machines in a similar way as usual tiling problems encode computations of deterministic and nondeterminstic machines. It is shown that any problem in a complexity class $\text{ATIME}(T(n), m)$ (the sets accepted by alternating machines in time $T(n)$ with m alternations) can be reduced to a strategy problem for some domino game with m moves on a squere of size $T(n)$. Similar generalizations are presented for domino thread games. In particular games are found whose strategy problems are P-complete. For a similar approach, see [78], for applications see [206].

204. E. Grädel. Satisfiability of formulae with one $\forall$ is decidable in exponential time. *Archiv math. Logik u. Grundlagenforschung*, 29:265–276, 1990.
It is shown that the Gurevich-Orevkov-Maslov class $[\exists^*\forall\exists^*, \mathit{allall}]$, proved decidable in [226, 386] can actually be decided in deterministic exponential time. See Sect. 6.3. The result is established by a structural analysis of a particular infinite subset of the Herbrand universe and by an alternating, polynomially space bounded satisfiability test.

205. E. Grädel. Simple interpretations among complicated theories. *Information Processing Letters*, 35:235–238, 1990.
It is shown that several decidable, but non-elementary mathematical theories can be embedded into quantified propositional temporal logic (QPTL) by interpretations that do not increase the number of quantifier alternations. Using known complexity results for QPTL $k-1$-fold exponential space complexity bounds are derived for the prefix classes with k quantifier alternations in these theories.

206. E. Grädel. Simple sentences that are hard to decide. *Information and Computation*, 94:62–82, 1991.
Using the domino games developed in [198, 203] it is shown that in a number of first-order theories the prefix classes with bounded quantifier alternations have essentially the same complexity as the entire theory. Theories which such behaviour are the theory of Boolean algebras, the theory of polynomial rings over finite fields, the theory of idempotent rings, the theory of finite sets with inclusion, the theory of semilattices, the theory of natural numbers with divisibility or coprimeness and so on. The decision complexity of these theories is alternating exponential time with linear number of alternations. As a consequence one obtains simple prefix classes (with two or three alternations) with nondeterministic exponential lower complexity bounds.

207. E. Grädel. Capturing complexity classes by fragments of second-order logic. *Theor. Computer Science*, 101:35–57, 1992.
The expressive power on finite structures of certain fragments of second order logic is investigated and related to computational complexity. The fragments studied are second order Horn logic, second order Krom logic and certain variants of the latter. It is shown that all these logics collapse to their existential subfragments. In the presence of a successor relation they capture polynomial time and logspace complexity classes. See Sect. 2.2.3 of this book. Without a successor relation these logics still can express certain problems that are complete in the corresponding complexity classes, but on the other hand they are strictly weaker than previously known logics for these classes and fail to express some very simple properties. The paper also contains a proof that the decision problem for quantified propositional logic is in NLOGSPACE (see [25] for a linear-time algorithm). See [141] for background on descriptive complexity.

208. E. Grädel, P. Kolaitis, and M. Vardi. On the decision problem for two-variable first-order logic. In preparation (1996).
Proves that every satisfiable sentence of L_2 (relational first-order logic with only two variables) has a model whose cardinality is exponentially bounded with respect to the length of the sentence. It follows that the satisfiability problem for L_2 is decidable in nondeterministic exponential time. By a result in [177] (see also Sect. 6.2) this bound is essentially optimal.

209. E. Grädel and M. Otto. Inductive definability with counting on finite structures. In E. Börger, G. Jäger, H. Kleine Büning, S. Martini, and M.M. Richter, editors, *Computer Science Logic, 6th Workshop, CSL '92, San Miniato 1992, Selected Papers*, Lecture Notes in Computer Science No. 702, pages 231–247. Springer, 1993.

210. E. Grädel, M. Otto, and E. Rosen. Two-variable logic with counting is decidable. Preprint (1996).
It is shown that C_2, the extension of two-variable first-order logic L_2 by counting quantifiers $\exists^{\geq n}$ and $\exists^{\leq n}$, is decidable for satisfiability and finite satisfiability. See the comment to [210] and Sect. 5.3 and 8.1 of this book.

211. E. Grädel, M. Otto, and E. Rosen. Undecidability results on two-variable logics. Submitted for publication (1996).
Mortimer's Theorem that L_2, first-order logic with two variables, is decidable and has the finite model property (see [208, 396] and Sect. 8.1) motivates the study of more powerful two-variable logics. In this paper it is shown that going beyond L_2 by adding any one of the following leads to an undecidable logic: (1) very weak forms of recursion, such as transitive closure operations, or (restricted) monadic fixed-point operations; (2) weak access to cardinalities, through the Härtig (or equicardinality) quantifier, or (3) a choice construct known as Hilbert's ε-operator. In fact all these extensions of L_2 prove to be undecidable both for satisfiability, and for satisfiability in finite models. Moreover most of them are hard for Σ_1^1, the first level of the analytical hierachy, and thus have a much higher degree of undecidability than first-order logic. The proof use domino problems (including the recurring domino problems of [245, 246]) and projective characterization of local grids. See Sect. 5.3 for some of these results.

212. R. Graham, B. Rothschild, and J. Spencer. *Ramsey theory*. John Wiley & Sons, New York, 1990.

213. E. Grandjean. Universal quantifiers and time complexity of random access machines. *Math. Systems Theory*, 18:171–187, 1985.
Refines Fagin's Theorem by proving an exact correspondence between the

problems decidable in time $O(n^d)$ on a RAM and generalized spectra of first-order sentences with d variables, for all $d \geq 2$. The result is extended to cover also the (more difficult) case where $d = 1$ in [214].

214. E. Grandjean. First-order spectra with one variable. *Journal of Computer and System Sciences*, 40:136–153, 1990.
See comment to [213].

215. R. Greenlaw, J. Hoover, and W. Ruzzo. *Limits to Parallel Computation. P-Completeness Theory.* Oxford University Press, 1995.

216. S. Grigorieff. Decidabilité et Complexité des Théories Logiques. In *Logique et informatique: une introduction*, INRIA, Collection didactique, pages 7–97, 1991.
Contains a rich and well documented survey of decidability and complexity results for decidable theories.

217. Y. Gurevich. Existential interpretation. *Algebra and Logic*, 4:71–84, 1965. In Russian. A German translation is available at TIB Universität Hannover, Germany.
The method of existential interpretation is developed and used to prove the undecidability of $\exists^r\forall^*$ fragments of various first-order theories. [229] is a modified English version. See Section 3.2 of this book.

218. Y. Gurevich. The decision problem for pure predicate logic. *Dokl. Akad. Nauk SSSR*, 168:510–511, 1966. In Russian, English translation: Soviet Mathematics – Dokl. 7 (1966), 669–670.
The result of [220] is strengthened to $k = 0$. Superseded by [219].

219. Y. Gurevich. On the algorithmic decision of the satisfiability of predicate logic formulas. *Algebra i Logika*, 5:25–55, 1966. In Russian.
The main technical result is that class $[\forall\exists\forall\exists^*, (0,1)]$ is a conservative reduction class; see Section 3.3 of this book. This had settled the last open question about the prefix-vocabulary classification for pure predicate logic; see Sect. 3.3 of this book in this connection. In addition, the author settles the satisfiability and finite satisfiability problems for classes $[\Pi_1 \wedge \cdots \wedge \Pi_k, p] := \{\varphi_1 \wedge \cdots \wedge \varphi_k : \varphi_i \in [\Pi_i, p]\}$ in the case of $\sum_n p_n \geq 1 + k$ for some relatively small k; see Sect. 5.4 of this book. He shows in particular that the classes $[\forall\exists \wedge \forall^3, (\omega, 1)]$, $[\forall\exists\forall \wedge \exists^*, (k, 1)]$ and $[\forall\exists \wedge \forall^3 \wedge \exists^*, (k, 1)]$ are conservative reduction classes.

220. Y. Gurevich. On the decision problem for pure predicate logic. *Dokl. Akad. Nauk SSSR*, 166:10, 1966. In Russian, English translation: Soviet Mathematics – Dokl. 7 (1966), 217–219.
For a particular k, The fragment $[\forall\exists\forall\exists^*, (k, 1)]$ is proven to be a conservative reduction class. Superseded by [219].

221. Y. Gurevich. *The decision problem for some algebraic theories.* PhD thesis, The Ural University, 1968. In Russian.
In fact, the Russian "second doctor degree" (Doctor of Physico-Mathematical Sciences) thesis. Contrary to the name, the major part of the thesis is devoted to the classical decision problem.

222. Y. Gurevich. The decision problem for decision problems. *Algebra i Logika*, 8:640–642, 1969. In Russian, English translation: Algebra and Logic 8 (1969), 362–363.
Let D be the collection of first-order formulae α such that the first-order theory with the only axiom α is decidable. The author notes that D is neither r.e. nor co-r.e. The second (and therefore the undecidability of D) has been earlier noted in [506]. The precise complexity of D is Σ^0_3, see [58].

223. Y. Gurevich. The decision problem for the logic of predicates and operations. *Algebra i Logika*, 8:284–308, 1969. In Russian, English translation: Algebra and Logic 8 (1969), 160–174.

First, the classifiability theorem for prefix-vocabulary classes is proved; see Section 2.3 of this book. Second, the decision problem for (the prefix-vocabulary fragments of) pure logic of predicates and functions is completed, though the treatment of the most difficult decidable class is deferred to [226]. In particular, the classes $[\forall^2, (0,1), (1)]$ and $[\forall^2, (1), (0,1)]$ are proved to be conservative reduction classes; see Chapter 3 of this book.

224. Y. Gurevich. Minsky machines and the $\forall\exists\forall$ & $\exists^*$ case of the decision problem. *Mathematical Notes of the Ural University*, 7:77–83, 1970. In Russian. A German translation is available at TIB Universität Hannover, Germany.
The author remarks that Minsky machines maybe more convenient than Turing machines for reduction purposes and illustrates the point by simplifying the proof from [219] that some $[\forall\exists\forall\wedge\exists^*, (k,1)]$ is a reduction class. A different formalization of Minsky machines appears in [2, 39, 411].

225. Y. Gurevich. The decision problem for first order logic. Manuscript, Tbilisi, USSR, 1972. In Russian, 124 pages. A German translation is available at TIB Universität Hannover, Germany.
This is a survey on the classical decision problem which was withdrawn from press in 1973 when the author left USSR.

226. Y. Gurevich. Formulas with one $\forall$. In *Selected Questions in Algebra and Logic; in memory of A. Malćev*, pages 97–110. Nauka, Moscow, 1973. In Russian. A German translation is available at TIB Universität Hannover, Germany.
The author proves that the class $X = [\exists^*\forall\exists^*, all, all]$ has the finite model property (and therefore is decidable for satisfiability and finite satisfiability); this result had completed the classical decision problem for pure logic of predicate and functions. The decidability of the satisfiability problem for X was announced by Orevkov in [407] and proved by Maslov and Orevkov in [386] who cite Gurevich's proof (accepted for publication in 1968). Their method is proof theoretical and quite different from that of Gurevich. Gurevich's proof was refined to an exponential-time decision procedure in [204], see Sect. 6.3 of this book.

227. Y. Gurevich. The decision problem for standard classes. *Journal of Symbolic Logic*, 41:460–464, 1976.
The classification (decidable/undecidable) of prefix-vocabulary fragments of pure first-order logic with function symbols is extended to classes with equality and at least one function symbol of positive arity. The author settles all new minimal undecidable classes (see Chapter 4 of this book) but refers the reader to [430] and [468] in connection to two new difficult decidable classes.

228. Y. Gurevich. Semi-conservative reduction. *Archiv math. Logik u. Grundlagenforschung*, 18:23–25, 1976.
Gurevich's theorem on semi-conservative reductions; see Chapt 2 of this book.

229. Y. Gurevich. Existential interpretation II. *Archiv math. Logik u. Grundlagenforschung*, 22:103–120, 1982.
An improved exposition of (most of) [217]; see Section 3.2 of this book.

230. Y. Gurevich. Toward logic tailored for computational complexity. In *Computation and Proof Theory. Proceedings of the Logic Colloquium 83, Aachen, Part II*, Lecture Notes in Mathematics No. 1104, pages 175–216. Springer, 1984.

231. Y. Gurevich. Monadic second-order theories. In J. Barwise and S. Feferman, editors, *Model-Theoretical Logics*, pages 479–506. Springer-Verlag, 1985.

232. Y. Gurevich. Logic and the challenge of computer science. In E. Börger, editor, *Trends in Theoretical Computer Science*, pages 1–57. Computer Science Press, 1988.

233. Y. Gurevich. Average case completeness. *Journal of Computer and System Sciences*, 42:346–398, 1991.
Presents the theory of average case completeness in full details. Several problems are proved complete concerning tilings, the Post correspondence, etc.
234. Y. Gurevich. On the classical decision problem. In G. Rozenberg and A. Salomaa, editors, *Current Trends in Theoretical Computer Science*, pages 254–265. World Scientific, 1993.
A popular introduction into the classical decision problem published originally in the Bull. of European Association for Theoretical Computer Science, Oct. 1990, 140–150.
235. Y. Gurevich. Zero-one laws. In G. Rozenberg and A. Salomaa, editors, *Current Trends in Theoretical Computer Science*, pages 293–309. World Scientific, 1993.
A popular introduction into zero-one laws (mentioning the connection with the classical decision problem) published originally in the Bull. of European Association for Theoretical Computer Science, Feb. 1991, 90–106.
236. Y. Gurevich and L. Harrington. Trees, automata, and games. In *14th Annual ACM Symposium on Theory of Computing*, pages 60–65, 1982.
The authors formulate and prove the Forgetful Determinacy Theorem, explained in Sect. 7.1, which allows them to give a simple proof of Rabin's Complementation Lemma [430, 67].
237. Y. Gurevich and I. Koryakov. Remarks on Berger's paper on the domino problem. *Siberian Math. Journal*, 13:319–321, 1972. In Russian. A German translation is available at TIB Universität Hannover, Germany.
The author strengthen Berger's undecidability result for the unconstrained domino problem [33] by showing that the instances with no solution and instances with periodic solution are recursively inseparable; see the appendix of this book.
238. Y. Gurevich, M. Magidor, and S. Shelah. The monadic theory of ω_2. *Journal of Symbolic Logic*, 48:387–398, 1983. Assuming the consistency of a weakly compact cardinal, it is shown that, in different models of set theory, the theory in question may be arbitrary difficult (or easy) within obvious limits.
239. Y. Gurevich and S. Shelah. Random models and the Gödel case of the decision problem. *Journal of Symbolic Logic*, 48:1120–1124, 1983.
The first case of using probabilistic arguments in the classical decision problem. Gödel's ingenious construction of a finite model for a satisfiable formula in $[\exists^*\forall\forall\exists^*, all]$ is replaced by a simple probabilistic argument (but not as good a bound on the cardinality of the finite model). See Sect. 6.2.3 of this book.
240. Y. Gurevich and T. Turashvili. Strengthening a result of Suranyi. *Bulletin of the Academy of Sciences of the Georgian Soviet republic*, 70:290–292, 1973.
241. Th. Hailperin. A complete set of axioms for logical formulas invalid in some finite domain. *Zeitschr. f. math. Logik u. Grundlagen d. Math.*, 7:84–96, 1961.
242. J. Halpern and Y. Moses. A guide to completeness and complexity for modal logics of knowledges and belief. *Artificial Intelligence*, 54:319–379, 1992.
Reviews and examines possible-world-semantics for propositional logics of knowledge and belief. A number of complexity results for such logics are presented.
243. W. Hanf. Model-theoretic methods in the study of elementary logic. In *Proceedings of the 1963 International Symposium on Theory of Models*, pages 132–145. North-Holland, 1965.

244. P. Hanschke and J. Würtz. Satisfiability of the smallest binary program. *Information Processing Letters*, 45:237–241, 1993.
See the comment to [129].
245. D. Harel. Recurring dominoes: Making the highly undecidable highly understandable. *Annals of Discrete Mathematics*, 24:51–72, 1985.
The paper presents recurrence problems for solutions of domino problems, for example the Σ_1^1-complete problems whether a particular domino appears infinitely often (in the first column) or in every column at least once, or the Σ_2^0-complete problem whether for two given colours c_0, c_1 the colour pattern on the bottom of the first row is of the form c_0^n, c_1^* for some natural number n. These problems are reduced to decision problems of various temporal and dynamic logics.
246. D. Harel. Effective transformations on infinite trees, with applications to high undecidability, dominoes and fairness. *Journal of the ACM*, 33:224–248, 1986.
247. K. Härtig. Über einen Quantifikator mit zwei Wirkungsbereichen. In L. Kalmár, editor, *Colloquium on the foundations of mathematics, mathematical machines and their applications*, pages 31–36. Akadémiai Kiadó, Budapest, 1962.
248. K. Hauschild and W. Rautenberg. Interpretierbarkeit und Entscheidbarkeit in der Graphentheorie. *Zeitschr. f. math. Logik u. Grundlagen d. Math.*, 1971:47–55, 17.
249. K. Hauschild and W Rautenberg. Entscheidungsprobleme der Theorie zweier Äquivalenzrelationen mit beschränkter Zahl von Elementen in den Klassen. *Fundamenta Mathematicae*, 1973:35–41, 81.
250. L. Hay. Spectra and halting problems. *Zeitschr. f. math. Logik u. Grundlagen d. Math.*, 21:167–176, 1975.
The degree theoretic investigation of spectra in [462] is related to classical methods of degree theory.
251. K. Heidler. *Untersuchungen zur Reduktionstheorie des Entscheidungsproblems in der Prädikaten- und Termlogik*. PhD thesis, Universität Freiburg, 1973.
Solves the decision problem for ε-logic by proving the conservative reduction class property for first-order sentences containing no other function symbol than Hilbert's ε-operator and no other relation than equality (see Corollary 5.3.5) and by extending the decidability of Löwenheim's class to the class of monadic formulae of ε-logic without functions and equality (see Exercise 6.2.6). This result is extended to Hermes' term logic [257] for which it is shown that a vocabulary class $[\Omega, p]$ is undecidable (and then a conservative reduction class) if and only if either $\{=, \varepsilon\} \subseteq \Omega$ or $\{\varepsilon, \forall, \exists\} \cap \Omega \neq \emptyset$ and p contains at least one at least binary relation, see Exercise 5.3.7. The undecidability result for ε-logic was strengthened by Grädel, Otto and Rosen [211] who showed that already the two-variable fragment of ε-logic is undecidable (see Theorem 5.3.4).
252. L. Henschen and L. Wos. Unit refutations and Horn sets. *Journal of the ACM*, 21:590–605, 1974.
253. J. Herbrand. Recherches sur la théorie de la démonstration. *Travaux de la Société des Sciences et des Lettres de Varsovie*, page 128, 1930. Classe III, Warszawa. Engl. translation in: Herbrand, Logical Writings, Harvard Univ. Press, Cambridge, Mass., 1971, 46–188, 272–276
In this paper appears what is nowadays known as Herbrand's theorem (proved by Herbrand in terms of his formulation of first-order logic including functions). Herbrand states (and proves in more detail in [254]) the following applications and consequences of his theorem for the Entscheidungsproblem.

1. Reductions of the Entscheidungsproblem (Ch. 2 of [254], pp. 232–243 of the english translation): New proofs are given for Löwenheim's reduction to binary predicates [365] and for the elimination of functions; the classes [*all*, (0, 0, 1)] and [*all*, (0, 3)] are proved to be reduction classes. Being based upon Herbrand expansions the proofs are in terms of provability and use a finitistic notion of satisfiability instead of the set-theoretic one.
2. Solutions for "Special cases of the Entscheidungsproblem" (Ch .3 of [254], pp. 243–250 of the english translation): New proofs are given for the finite controllability of (a) monadic predicate logic ([365]), (b) the Bernays-Schönfinkel class ([35]), (c) the Ackermann class ([16]). Also the class HERBRAND of what are nowadays called Herbrand formulae is shown to be decidable for formulae without functions or equality (see the editor's note L, pp. 262–263 op. cit. and [182] for the finite controllability, see [137] for the completeness for P and [439] for resolution as decision procedure for this class). Note that when the equality is included, the class $\text{HERBRAND}_=$ is still decidable; for predicate logic with functions and equality the class HERBRAND is undecidable (see [505, 389] and Chap. 4 of this book). For a systematic treatment of decision problems for predicate logic without functions or equality based upon the theory of Herbrand expansions see [133, 351].

254. J. Herbrand. *Sur le problème fondamental de la logique mathématique*, volume 24, pages 12–56. Sprawozdania z posiedzen Towarzysta Naukowego Warszawskiego, Wydzial III, 1931. English translation in: Harvard Univ. J. Herbrand, Logical Writings, Harvard Univ. Press, Cambridge/Mass. (1971) pp.215-259.
See comment to [253].
255. H. Hermes. Ideen von Leibniz zur Grundlagenforschung: Die ars inveniendi und die ars iudicandi. *Studia Leibnitiana. Vierteljahresschrift für Philosophie und Geschichte der Wissenschaften. Sonderheft1: Systemprinzip und Vielheit der Wissenschaften, Wiesbaden. Akten des Internationalen Leibniz-Kongresses, Hannover*, pages 78–88, 1966.
256. H. Hermes. Entscheidungsprobleme und Dominospiele. In K. Jakobs, editor, *Selecta Mathematica II*, pages 114–140. Springer-Verlag, 1970.
Very well written introduction into the Entscheidungsproblem, focussing on its relation to domino problems. In particular the reduction class $[\exists\forall\exists\forall, (0, \omega)]$ is treated. For a further simplification to obtain also $[\forall\exists\forall, (0, \omega)]$ as a reduction class, see [258].
257. H. Hermes. *Term logic with choice operator*. Springer Lecture Notes in Mathematics No. 6 (2nd ed.), 1970. The first edition appeared in 1965 under the title 'Eine Termlogik mit Auswahloperator' (in German).
258. H. Hermes. A simplified proof for the unsolvability of the decision problem in the case $\bigwedge\bigvee\bigwedge$. In R.O Gandy and C.M.E. Yates, editors, *Logic Colloquium '69*, pages 307–310. North-Holland, Amsterdam, 1971.
Simplifies the dominoes used in the reduction of diagonal-constrained domino problems to $\forall\exists\forall$-formulae by [288] (see also [256]).
259. H. Herre, M. Krynicki, A. Pinus, and J. Väänänen. The Härtig quantifier: a survey. *Journal of Symbolic Logic*, 56:1153–1183, 1991.
260. G. Higman. Ordering by divisibility in abstract algebras. *Proc. of the London Math. Soc.*, 3:326–336, 1952.
261. D. Hilbert. Über die Grundlagen der Logik und Arithmetik. In *Verhandlungen des 3. Internationalen Mathematiker-Kongresses in Heidelberg*, pages 174–185. Leipzig, 1905.
262. D. Hilbert. Axiomatisches Denken. *Math. Annalen*, 78:405–415, 1918.

263. D. Hilbert. Die Neubegründung der Mathematik. *Abhandlungen aus dem Math. Seminar der Hamburger Universität*, 1:157–177, 1922.
264. D. Hilbert. Die logischen Grundlagen der Mathematik. *Math. Annalen*, 88:151–156, 1923.
265. D. Hilbert. Probleme der Grundlegung der Mathematik. In K. Reidemeister, editor, *Hilbert Gedenkband*, pages 9–19. Springer, Berlin, 1971.
Vortrag, gehalten am Internationalen Mathematiker-Kongress in Bologna, September 1928.
266. D. Hilbert and W. Ackermann. *Grundzüge der theoretischen Logik*. Springer, 1928/1938. English translation of the 2nd edition: "Principles of Mathematical Logic", Chelsea Publishing Company, New York (1950).
Contains in Chapter 12 the first systematic text book treatment of the Entscheidungsproblem, covering the results known at that time. In Chapters 11, 12 it is explained why the Entscheidungsproblem is considered to be "the main problem of mathematical logic".
267. D. Hilbert and P. Bernays. *Grundlagen der Mathematik*. Springer-Verlag, Berlin, Band I (1934, 1968), Band II (1939, 1970).
268. R. Hill. Lush resolution and its completeness. Dcl memo 78, University of Edinburgh, 1974.
269. K. Hintikka. Distributive normal forms in the calculus of predicates. *Acta Philosophica Fennica*, 6:71, 1953.
The normal form developed here establishes the refutability of sentences for certain cases of the Entscheidungsproblem. See [404].
270. W. Hodges. *Model theory*. Cambridge University Press, 1993.
271. J. Hopcroft and J. Ullman. *Introduction to Automata Theory, Languages and Computation*. Addison-Wesley, 1979.
272. A. Horn. On sentences which are true of direct unions of algebras. *Journal of Symbolic Logic*, 1951:14–21, 16.
273. V. Huber-Dyson. On the decision problem for theories of finite models. *Israel Journal of Math.*, 2(1):55–70, 1964. See comment to [274].
274. V. Huber-Dyson. On the decision problem for extensions of a decidable theory. *Fundamenta Mathematicae*, LXIV:7–40, 1969.
Various extensions T' of a given theory T by the theory of its finite, infinite etc. models are considered and it is shown that all combinations of decidability/undecidability of T and T' are possible.
275. C. Hughes. A reduction class containing formulas with one monadic predicate and one binary function symbol. *Journal of Symbolic Logic*, 41:45–49, 1976.
As variation of two results in [42] it is shown that $[\forall^2, (1), (0,1)] \cap$ HORN is a reduction class when restricted to formuale which are in HERBRAND except for one ternary disjunct. The proof is by reduction of partial implicational propositional calculi with only two variables. See 5.1.21.
276. N. Immerman. Relational queries computable in polynomial time. *Information and Computation*, 68:86–104, 1986.
Proves that least fixed-point logic captures polynomial time on ordered structures. The same result was also proved by Vardi [517]. See Sect. 2.2.3. For more results on descriptive complexity and finite model theory, see [141].
277. N. Immerman. Languages that capture complexity classes. *SIAM Journal on Computing*, 16:760–778, 1987.
Formulates explicitly the programme of descriptive complexity theory: to relate computational complexity with logical definability and to design logics whose expressive power directly corresponds to important complexity classes. Proves that on ordered structures, transitive closure logics capture logarithmic space complexity classes. See Sect. 2.2.3.

278. N. Immerman. Descriptive and computational complexity. In J. Hartmanis, editor, *Computational Complexity Theory, Proc. Symp. Applied Math., Vol. 38*, pages 75–91. American Mathematical Society, 1989.
A survey on descriptive complexity theory.
279. A. Itai and J. Makowsky. Unification as a complexity measure for logic programming. *Journal of Logic Programming*, 4:105–117, 1987.
280. A. Janiczak. Undecidability of some simple formalized theories. *Fundamenta Mathematicae*, 40:131–139, 1953.
281. R. Jensen. Ein neuer Beweis für die Entscheidbarkeit des einstelligen Prädikatenkalküls mit Identität. *Archiv math. Logik u. Grundlagenforschung*, 7:128–138, 1962.
Contains a new proof for Löwenheim's decidability result [365] for the monadic predicate logic.
282. J. Jones. Classification of quantifier prefixes over Diophantine equations. *Zeitschr. f. math. Logik u. Grundlagen d. Math.*, 1981:403–410, 27.
283. N. Jones and W. Laaser. Complete problems for deterministic polynomial time. *Theor. Computer Science*, 3:105–117, 1977.
Among various other problems also the unsatisfiability of propositional formulae by unit resolution is shown to be complete for P. See also [215].
284. N. Jones, E. Lien, and W. Laaser. New problems complete for nondeterministic log space. *Math. Systems Theory*, 10:1–17, 1976.
Among various other problems also the (provability by unit resolution of the) unsatisfiability of propositional Krom formulae is shown to be complete for nondeterministic logarithmic space. The proof adapts the Cook-Levin reduction from [91].
285. N. Jones and A. Selman. Turing machines and the spectra of first-order formulas. *Journal of Symbolic Logic*, 39:139–150, 1974.
The authors show that for unary notation of integers, the first-order spectra are precisely the NP-sets of positive integers. The proof introduces special "spectrum–automata" which are formalized by first-order formulae and by which nondeterministic 2^{cx}-time bounded Turing machine computations are simulated. A simpler proof by direct reduction of bounded nondeterministic Turing machine computations appears in [151, 152, 79] (where simultaneously the result is generalized to spectra of arbitrary order, see the Spectrum Theorem in Sect. 2.2.2 of this book).
286. W. Joyner jr. Resolution strategies as decision procedures. *Journal of the ACM*, 23:398–417, 1976.
This paper starts the investigation of resolution (see [439]) as a decision procedure for classes of first-order formulae. Various restrictions of the general resolution are developed which turn them into decision procedures for monadic predicate logic [365], the Ackermann class [16], the Gödel-Kalmar-Schütte class [186], the extended Skolem class (consisting of all $\exists z_1 \ldots \exists z_p \forall y_1 \ldots \forall y_n \exists x_1 \ldots \exists x_m$-formulae in which each atomic subformula has among its arguments either (a) at least one of the x_i, or (b) at most one of the y_j, or (c) each universally quantified variable), Maslov's class in [378].
287. A. Kahr. Improved reductions of the Entscheidungsproblem to subclasses of $\forall\exists\forall$ formulas. In *Proc. Symp. on Mathematical Theory of Automata*, pages 57–70. Brooklyn Polytechnic Institute, New York, 1962.
Strengthens the Kahr-Moore-Wang reduction class $[\forall\exists\forall, (0, \omega)]$ (see [288]) to $[\forall\exists\forall, (\omega, 1)]$ by inventing more sophisticated dominoes for asymmetric diagonal-constrained domino problems. A direct encoding of Turing machines into $[\forall\exists\forall, (\omega, 1)]$-formulae has been given by [441], building upon Kahr's ideas.

For another direct encoding of Turing machines see [319]. See Sect. 3.1 of this book.
288. A. Kahr, E. Moore, and H. Wang. Entscheidungsproblem reduced to the ∀∃∀ case. *Proc. Nat. Acad. Sci. U.S.A.*, 48:365–377, 1962.
Diagonal-constrained domino problems are proved to be recursively unsolvable and then used to establish $[\forall\exists\forall, (0, \omega)]$ as a reduction class. The formalization technique is taken from [64] and adapted to represent Turing machine configurations periodically on a diagonal of the Gaussian quadrant (see Sect. 3.1 of this book). For simplifications see [258, 441]. See also [134].
289. B. Kallick. A decision procedure based on the resolution method. In *Proc. IFIP Congr. 1968*, pages 269–275. North-Holland, Amsterdam, 1968.
Contains a decision procedure for the subclass $[\forall^2\exists, all]$ of the Gödel-Kalmár-Schütte class using Friedman's decision tables (see [173]).
290. L. Kalmár. Eine Bemerkung zur Entscheidungstheorie. *Acta Scientiarum Mathematicarum Universitatis Szegediensis*, 4:248–252, 1929.
Eliminates from Löwenheim's reduction formulae in [365] the use of the equality relation.
291. L. Kalmár. Ein Beitrag zum Entscheidungsproblem. *Acta Scientiarum Mathematicarum Universitatis Szegediensis*, 5:222–236, 1932.
292. L. Kalmár. Zum Entscheidungsproblem der mathematischen Logik. In *Verhandlungen des Internationalen Mathematischen Kongresses*, volume II, pages 337–338, Zürich, 1932.
Contains a proof for the reduction class $[\exists^*\forall^2\exists\forall^*, (0, 0, 1)]$.
293. L. Kalmár. Über die Erfüllbarkeit derjenigen Zählausdrücke, welche in der Normalform zwei benachbarte Allzeichen enthalten. *Math. Annalen*, 108:466–484, 1933.
Proves the decidability of the Gödel-Kalmár-Schütte class $[\exists^*\forall^2\exists^*, all]$. See also [186, 187, 239, 457, 456] and Sect. 6.2.3 of this book.
294. L. Kalmár. Über einen Löwenheimschen Satz. *Acta Scientiarum Mathematicarum Universitatis Szegediensis*, 7:112–121, 1934.
Simplifies the proof for Löwenheim's reduction class in [365].
295. L. Kalmár. Zurückführung des Entscheidungsproblems auf den Fall von Formeln mit einer einzigen, binären Funktionsvariablen. *Compositio Mathematica*, 4:137–144, 1936.
Improves Löwenheims reduction class $[all, (0, \omega)]$ (see [365]) to $[all, (0, 1)]$.
296. L. Kalmár. Zur Reduktion des Entscheidungsproblems. *Norsk Mat. Tidsskrift*, pages 1–10, 1937.
Strengthens Skolem's reduction class in [482].
297. L. Kalmár. On the reduction of the decision problem. *Journal of Symbolic Logic*, 4:1–9, 1939.
Improves Ackermann's reduction class $[\exists\forall\exists\forall^*]$ to $[\exists\forall\exists\forall^*, (0, 1)]$.
298. L. Kalmár. Contributions to the reduction theory of the decision problem. *Acta Mathematica Academiae Scientiarunm Hunagricae*, 1:64–73, 1950. First paper: Prefix $(x_1)(x_2)(Ex_3)\cdots(Ex_{n-1})(x_n)$, a single binary predicate.
299. L. Kalmár. Contributions to the reduction theory of the decision problem. *Acta Mathematica Academiae Scientiarunm Hunagricae*, 2:125–142, 1951. Fourth paper: Reduction to the case of a finite set of individuals
Containes a reduction of the decision problem for validity to that for finite satisfiability. This paper motivated the investigation in [520],(pg. 41, footnote 5). See also [509].
300. L. Kalmár. Contributions to the reduction theory of the decision problem. *Acta Mathematica Academiae Scientiarunm Hunagricae*, 2:19–38, 1951. Third paper: Prefix $(x_1)(Ey_1)\cdots(Ey_n)(x_2)(x_3)$, a single binary predicate.

301. L. Kalmár. Ein direkter Beweis für die allgemein-rekursive Unlösbarkeit des Entscheidungsproblems. *Zeitschr. f. math. Logik u. Grundlagen d. Math.*, 2:1–14, 1956.
The paper uses a diagonal argument to show the unsolvability of the Entscheidungsproblem by a direct formalisation of recursive functions. Systems of equations defining recursive functions are formalised; then the assumption that there is a recursive function which computes the satisfiability of arbitrary formulae yields a formula which is satisfiable if and only if it is not satisfiable. This proof is also used in [498, Chapter 8].
302. L. Kalmár and J. Surányi. On the reduction of the decision problem, second paper: Gödel prefix, a single binary predicate. *Journal of Symbolic Logic*, 12:65–73, 1947.
Improves Gödel's reduction class $[\forall^3\exists^*, (0,\omega)]$ to $[\forall^3\exists^*, (0,1)]$.
303. L. Kalmár and J. Surányi. On the reduction of the decision problem, third paper: Pepis prefix, a single binary predicate. *Journal of Symbolic Logic*, 15:161–173, 1950.
Improvement of the Pepis reduction class $[\forall^2\exists\forall^*]$ (see [420]) to $[\forall^2\exists\forall^*, (0,1)]$ and $[\forall^*\exists, (0,1)]$.
304. J. Kari. Reversibility and surjectivity problems of cellular automata. *Journal of Computer and System Sciences*, 48:149–182, 1994.
These problems are proved undecidable using tiling techniques developped by Robinson [440] and Berger [33] (see also [135]).
305. J. Ketonen and R. Weyhrauch. A decidable fragment of predicate calculus. *Theor. Computer Science*, 32:297–307, 1984.
Describes a decision procedure for the class of formulae which can be proved in the Gentzen sequent calculus without using the contraction rule.
306. D. Klaua. Systematische Behandlung der lösbaren Fälle des Entscheidungsproblems für den Prädikatenkalkül erster Stufe. *Zeitschr. f. math. Logik u. Grundlagen d. Math.*, 1:264–270, 1955.
Presents a short and transparent proof for Schütte's decision procedure for the Gödel-Kalmár-Schütte class $[\exists^*\forall^2\exists^*, all]$. The method is similar to the one used by [81] for other classes which at that time were known to be decidable.
307. S. Kleene. *Introduction to Metamathematics.* Elsevier, North-Holland, 1952.
308. H. Kleine Büning. Some undecidable theories with monadic predicates and without equality. *Archiv math. Logik u. Grundlagenforschung*, 21:137–148, 1981.
309. H. Kleine Büning and T. Lettmann. *Aussagenlogik: Deduktion und Algorithmen.* Teubner, 1994.
310. J. Klop. Term rewriting systems. In T. Maibaum S. Abramsky, D. Gabbay, editor, *Handbook of Logic in Computer Science, vol. II*, pages 1–116. Oxford University Press, 1992.
311. S. Kogalovskiĭ. Some reduction theorems for higher–order logic. *Soviet Mathematics - Dokl.*, 11(1):138–142, 1970.
312. A. Kokorin and A. Pinus. Decidability problems of extended theories. *Russian Math. Surveys*, 33:53–96, 1978.
Survey paper.
313. P. Kolaitis and M. Vardi. The decision problem for the probabilities of higher-order properties. In *19th Annual ACM Symposium on Theory of Computing*, pages 425–435., 1987.
The authors prove the 0-1 law for second-order predicate formulas obtained from existential first-order formulas by existential quantification over some of the predicate symbols. They also identify the complexity of the problem whether a given formula of that kind is true on almost all finite structures.

314. P. Kolaitis and M. Vardi. 0-1 laws and decision problems for fragments of second-order logic. *Information and Computation*, 87:302–338, 1990.
The authors prove the 0-1 law for second-order predicate formulas obtained from Ackerman first-order formulas by existential quantification over some of the predicate symbols. They also identify the complexity of the problem whether a given formula of that kind is true on almost all finite structures. In addition, they identify the complexity of the satisfiability problem for the Ackermann class with equality.

315. P. Kolaitis and M. Vardi. 0-1 laws for fragments of second-order logic: an overview. In Y. N. Moschovakis, editor, *Logic from computer science (Proc. of Workshop, 1989)*, pages 265–286, 1992.
A survey of the 0-1 laws for fragments of second-order logic. It also contains the observation that proving the 0-1 law for a fragment via the so-called transfer theorem implies the finite finite model property for the associated first-order fragment. (A transfer theorem asserts that a formula is almost true if and only if it is true in the random infinite model.).

316. V. Kostyrko. On a case of the decision problem in the restricted predicate calculus. *Uspekhi. Mat.Nauk*, 17:213, 1962.
Provides a sufficient condition for the existence of constructive models for certain axiomatizable theories, including Krom classes. This result generalizes the result proved in [11].

317. V. Kostyrko. Reduction class $\forall\exists^n\forall$. *Algebra i Logika*, 3:45–65, 1964. in Russian. A German translation is available at TIB Universität Hannover, Germany.
The author proves that $[\forall\exists^*\forall,(0,1)]$ is a conservative reduction class $[\forall\exists^*\forall,(0,1)]$; see Chap. 3.3 of this book. The result has been independently proved in [180].

318. V. Kostyrko. On the decision problem for the Ackermann case. *Siberian Math. Journal*, 6(2):342–364, 1965. in Russian. A German translation is available at TIB Universität Hannover, Germany.
This is about formulas of the form $\forall x\exists y F(x,y) \wedge \varphi$ where F is a binary predicate. In other words, consider the theory T with the axiom $\forall x\exists y F(x,y)$ and study prefix-vocabulary fragments of T. According to [17], the fragment $[\exists\forall^*,(all)]$ of T is a reduction class for satisfiability. A straightforward Skolemization of the formulas in the Kahr class $[\forall\exists\forall,(\omega,1)]$ gives that the fragment $[\forall^3,(\omega,2)]$ of T is a conservative reduction class. The author gives a decision algorithm for the finite satisfiability problem for the fragment $[\exists^*\forall^*,(\omega,1)]_=$ of T.

319. V. Kostyrko. The $\forall\exists\forall$ reduction class. *Kibernetika*, 2:17–22, 1966. in Russian. English translation in: Cybernetics 2, pp. 15–19.
The author gives an alternative proof (that uses a form of Turing machine rather than domino) of the following theorem of Kahr in [287]. $[\forall\exists\forall,(\omega,1)]$ is a conservative reduction class, and moreover the following restriction can be imposed: the formulae $\forall x\exists y\forall z\varphi(x,y,z)$ use only three binary atoms which can be any three out of the following four: $\{Pxz, Pzx, Pyz, Pzy\}$.

320. V. Kostyrko. The reduction class $\forall x\forall y\exists z F(x,y,z) \wedge \forall^n \mathcal{A}(F)$. *Cybernetics*, 5:1–3, 1971.
Consider sentences described in the title where F is ternary predicate symbol, $\mathcal{A}$ is quantifier-free and does not have any other predicate or function symbols. These sentences form a reduction class both for satisfiability and finite satisfiability. Compare this with [318].

321. D. Kozen. Complexity of Boolean algebras. *Theor. Computer Science*, 10:221–247, 1980.

The computational complexity for the elementary theory of various classes of Boolean algebras is investigated. In particular it is shown that the elementary theory of Boolean algebras is complete for alternating exponential time with n alternations. See also [206].

322. D. Kozen. Positive first-order logic is NP-complete. *IBM Journal for Research and Development*, 25:327–332, 1981.
Proves that the validity problem for first-order formulae without negations is NP-complete. The same holds for the following more general entailment problem: Given a finite set Σ of atomic sentences and a positive sentence ψ, determine whether $\Sigma \models \psi$. See Sect. 8.2.3.
323. D. Kozen. A finite model theorem for the propositional μ-calculus. *Studia Logica*, 47:233–241, 1983.
324. D. Kozen and R. Parikh. A decision procedure for the propositional μ-calculus. In E. Clarke and D. Kozen, editors, *Logics of Programs*, Lecture Notes in Computer Science No. 164, pages 313–325. Springer, 1982.
325. G. Kreisel. Note on arithmetic models for consistent formulae of predicate calculus. *Fundamenta Mathematicae*, 37:265–285, 1950.
326. G. Kreisel. Note on arithmetic models for consistent formulae of the predicate calculus II. In *Proc. XIth Int. Cong. Philos.*, pages 39–49. North-Holland, Amsterdam, 1953. vol.14.
327. G. Kreisel and J. Krivine. *Eléments de logique mathématique.* Dunod, Paris, 1967.
328. S. Kripke. The undecidability of monadic modal quantification theory. *Zeitschr. f. math. Logik u. Grundlagen d. Math.*, 8:113–116, 1962.
The paper proves that monadic predicate logic — even with only two monadic predicate symbols — becomes undecidable if it is extended to a system of modal quantification theory by adding the -operator, the rule of substitution (as primitive or derived rule) and the axioms and rules of $S5^*$. The proof is by a reduction of [*all*, (0, 1)] interpreting Rxy by $(Px \wedge Qy)$. Another proof is given in [402]. For a strengthening of this result, see [484].
329. M. Krom. A decision procedure for a class of formulas of first order predicate calculus. *Pacific Journal of Math.*, 14:1305–1319, 1964.
Extends the decision procedure for Herbrand's class. See [332].
330. M. Krom. A property of sentences that define quasi-order. *Notre Dame Journal of Formal Logic*, 7:349–352, 1966.
Shows that any sentence in a first-order language without equality and in prenex conjunctive normal form which states that a binary predicate is (reflexive and) transitive has a disjunction with more than two terms.
331. M. Krom. The decision problem for a class of first-order formulas in which all disjunctions are binary. *Zeitschr. f. math. Logik u. Grundlagen d. Math.*, 13:15–20, 1967.
Shows the decidability of KROM $\cap$ $[\forall^* \wedge \exists^*\forall^2\exists^*]$.
332. M. Krom. The decision problem for segregated formulas in first-order logic. *Mathematica Scandinavica*, 21:233–240, 1967.
Let S_k^+ be the class of formulae in prenex conjunctive normal form in which each disjunction has only negated or only unnegated, in the latter case exactly k, atomic formulae; similarly define S_k^-. It is shown that if predicates of arbitrary arity are allowed, then $S_2^+ \cap S_3^-$ and $S_3^+ \cap S_2^-$ are reduction classes. The proof provides a reduction of the Kalmar-Suranyi class $[\forall^3\exists^*, (0,1)]$. The class $S_1^+ \cup S_1^-$ is shown to be decidable; the proof uses and strengthens the decidability of Herbrand's class $S_1^+ \cap S_1^-$ in [254].
333. M. Krom. Some interpolation theorems for first-order formulas in which all disjunctions are binary. *Logique et Analyse*, 43:403–412, 1968.

334. M. Krom. The decision problem for formulas in prenex conjunctive normal form with binary disjunctions. *Journal of Symbolic Logic*, 35:210–216, 1970. Proves $[\forall\exists^*\forall, (0,\omega)] \cap \text{KROM} \cap \text{HORN}$ to be a reduction class by reducing to it Post's Tag problem. Improved in [442].
335. B. Kruskal. The theory of well–quasi–ordering: A frequently discovered concept. *Journal of Combinatiorial Theory*, 3:297–305, 1972.
336. D. Kuehner. A note on the relation between resolution and Maslov's inverse method. *Machine Intelligence*, 6:73–76, 1971.
337. R. Ladner. The computational complexity of provability in systems of modal propositional logic. *SIAM Journal of Computing*, 6:467–480, 1977.
338. I. Lavrov. Effective inseparability of the sets of identically true and finitely refutable formulae for certain elementary theories. *Algebra i Logika*, 2:5–18, 1963.
339. A. Leisenring. *Mathematical logic and Hilbert's ε-symbol.* Gordon & Breach, New York, 1969.
340. A. Leitsch. *The Resolution Calculus.* In preparation.
341. A. Leitsch. Deciding Horn classes by hyperresolution. In E. Börger, H. Kleine Büning, and M.M. Richter, editors, *CSL'89*, Lecture Notes in Computer Science No. 440, pages 225 – 241. Springer, 1989.
This is – together with [158] and [501] – one of the first papers investigating clausal decision classes with full functional structure (the function symbols need not be Skolem symbols). Two classes KI and KII are defined on which hyperresolution terminates. KI is a functional generalization of DATALOG. Hyperresolution is also used as basis for a decision algorithm for a subclass of the Horn clause implication problem HI; HI was later shown undecidable [377].
342. A. Leitsch. Deciding clause classes by semantic clash resolution. *Fundamenta informaticae*, 18:163 – 182, 1993.
The classes KI and KII in [341] are generalized to a nonHorn class PVD. A further generalization is achieved in replacing term-depth by arbitrary *atom complexity measures.* A set of clauses $\mathcal{C}$ is called *positive T-dominated* for an atom complexity measure T if there exists a number d s.t. for all $C \in \mathcal{C}$ and substitutions λ either $T(C_+\lambda) \leq d$ or $T(C_+\lambda) \leq T(C_-\lambda)$ holds (C_+ is the set of positive literals in C, C_- is defined accordingly). It is shown that hyperresolution terminates on all positive T-dominated classes. The paper thus defines a general method to obtain decidable classes and their corresponding decision procedures.
343. L. Levin. Universal search problems. *Problems of Information Transmission*, 9(3):265–266, 1973.
344. L. Levin. Average case complete problems. *SIAM J. Comput*, 15(1):285–286, February 1986.
This is the first (but very short) paper on average case completeness. A randomized version of the bounded tiling problem is proved complete. The theory is explained in [235] and studied for search problems in [37]. See also [136] for the use of tiling problems in this field.
345. H. Lewis. The decision problem for formulas with a bounded number of atomic subformulae. *Notices of the AMS*, 20:A–23, 1973.
States that any class of prenex formulae which is determined by restricting both the number of atomic formulae and the number of universal quantifiers is reducible to a finite class of formulae.
346. H. Lewis. *Herbrand expansions and reductions of the decision problem.* PhD thesis, Center for Research in Computing Technology, Harvard University, 1974.

Encodings of the Herbrand expansion of equality and function free first-order formulae of a given class in the Herbrand expansion of formulae of another class are studied and proposed as a unifying method for reductions among unsolvable decision problems. The method is extended in [351].

347. H. Lewis. Descriptions of restricted automata by first-order formulae. *Math. Systems Theory*, 9:97–104, 1975.
Formalizes context-free grammars (in Chomsky normal form), finite (tree) automata and pushdown automata by monadic formulae with restricted prefix of form $\exists^*\forall^*\exists^*$. A similar formalization of program schemes allows to derive the decidability of the totality and the total-equivalence of full schemes from the decidability of the class of the corresponding reduction formulae.

348. H. Lewis. Krom formulas with one dyadic predicate letter. *Journal of Symbolic Logic*, 41:341–362, 1976.
The class of Krom formulae with a single predicate, a binary one, is shown to be a reduction class. It is proved that the class of prenex formulae having disjunctive normal form with only two disjuncts and having just two predicate letters, both pentadic, is a reduction class for validity (see [409, 411]). The proof uses a reduction of the Post Correspondence Problem [422] and encoding techniques introduced by Shannon in [465] for his construction of a universal Turing machine with two internal states.

349. H. Lewis. Complexity of solvable cases of the decision problem for the predicate calculus. In *Proc. 19th Symp. on the Foundations of Computer Science*, pages 35–47, 1978.
Preliminary account of [352].

350. H. Lewis. Satisfiability problems for propositional calculi. *Math. Systems Theory*, 13:45–53, 1979.
It is shown that a condition sufficient for NP-completeness is that the function $x \wedge \neg y$ be representable, and that any set of connectives not capable of representing this function has a polynomial-time satisfiability problem.

351. H. Lewis. *Unsolvable Classes of Quantificational Formulas*. Addison-Wesley, 1979.
Book on unsolvable classes of first-order formulae without functions and equality classified by vocabulary, prefix and truth-functional form of the quantifier-free part. The treatment is mostly based on tiling problems and uses the machinery of Herbrand expansions. Also a simplified proof for the undecidablity of Aanderaa's linear sampling problem (see [1, 13]) is given.

352. H. Lewis. Complexity results for classes of quantificational formulas. *Journal of Computer and System Sciences*, 21:317–353, 1980.
This paper and [175] initiated the study of the complexity of decidable cases of the decision problem. Complexity results are given for the four classical solvable cases, see Sect. 6.2 of this book. In addition, for the class T defined below an upper bound $\mathrm{DTIME}(c^{c^{n/\log n}})$ is established for some $c > 0$ and a double exponential time lower bound is shown to hold by a reduction of the non acceptance problem of alternating stack automata. T is the class of $\exists z_1 \ldots \exists z_k \forall y_1 \exists x_1 \ldots \exists x_m \forall y_2$-formulae containing binary predicates only, and not having any atomic subformulae of the form Py_2y_1 or $Py_2x_j (j = 1, \ldots, m)$. See [421].

353. H. Lewis and L. Denenberg. A hard problem for $\mathrm{NTIME}(n^d)$. In *Proc. of 19th Allerton Conference on Control, Communication, and Computing*, 1981.

354. H. Lewis and W. Goldfarb. The decision problem for formulas with a small number of atomic subformulas. *Journal of Symbolic Logic*, 38:471–480, 1973.
Shows the undecidability of the class of all equality and function free $\forall\exists\forall^*$-formulae containing five atomic subformulae and with quantifier-free part of

the form $(\neg\pi_1 \wedge \pi_2 \wedge \pi_3) \vee (\neg\pi_4 \wedge \pi_5)$ (or of the form $(\pi_1 \vee \pi_2) \wedge (\neg\pi_3 \vee \neg\pi_4) \vee (\neg\pi_3 \vee \neg\pi_5)$ or of the form $(\pi_1 \wedge (\pi_2 \rightarrow \pi_3) \wedge (\pi_2 \rightarrow \pi_4) \wedge \neg\pi_5)$. This sharpens the result in [409]. As a corollary one obtains the undecidability of equality and function free $\forall^*\exists$-formulae containing six atomic subformulae whose quantifier free part is in conjunctive normal form with three conjuncts. See Theorem 5.2.2.

355. H. Lewis and C. Papadimitriou. *Elements of the Theory of Computation.* Prentice-Hall, 1981.

356. H. Lewis and R. Statman. Unifiability is complete for co-NLOGSPACE. *Information Processing Letters*, 15,5:220–222, 1983.
Shows that the problem of unifiability of two first-order terms is complete for nondeterministic logarithmic space with respect to logspace reductions. The hardness claim is established by reducing the acyclicity problem for directed graphs.

357. V. Lifshitz. Deductive validity and reduction classes. In A. Slisenko, editor, *Sem. in Math, vol. 4*, pages 26–28. Steklov Mathematikal Institute, Leningrad, 1969. Russian Original 1967.
A method is proposed to obtain new reduction classes for classical or intuitionist predicate calculus.

358. V. Lifshitz. Problem of decidability for some intuitionistic theories of equality. In A. Slisenko, editor, *Sem. in Math, vol. 4*, pages 29–31. Steklov Mathematikal Institute, Leningrad, 1969. Russian Original 1967.
The author proves that the pure intuitionistic theory of equality is undecidable, the intuitionistic theory of normal equality is undecidable, but the intuitionistic theory of decidable equality is decidable.

359. V. Lifshitz. Some reduction classes and undecidable theories. In A. Slisenko, editor, *Studies in Constructive Mathematics and Mathematical Logic I*, Seminars in Math. 4, pages 24–25. Steklov Mathematikal Institute, Leningrad, 1969. Russian Original 1967.
The author announces several theorems. In particular, $[\exists, (0), (2)]_=$ is a reduction class for validity even if the quantifier free part is a disjunction of atomic and negated atomic formulas. It follows that $[\forall, (0), (2)]_=$ is a reduction class for satisfiability; see Chapter 4 in this book.

360. V. Lifshitz. What is the inverse method. *Journal of Automated Reasoning*, 5:1–23, 1989.

361. P. Lincoln. Deciding provability of linear logic formulas. In *Advances in Linear Logic*, volume 222 of *London mathematical society lecture notes.* Cambridge University Press, 1995.
This paper surveys a series of results about the decidability of linear logic, including the undecidability of propositional linear logic (by reduction from counter machines). Other results include the PSPACE-completeness of propositional and constant-only multiplicative-additive linear logic (by reduction from classical QBF) and the NEXPTIME-completeness of that fragment with first-order quantifiers, the NP-completeness of multiplicative linear logic (3-Partition) at the first-order, propositional, and constant-only levels, and the undecidability of some second-order fragments. Some preliminary results are mentioned regarding the only significant pure fragment for which decidability is not known: multiplicative exponential linear logic. MELL is Petri-net reachability-hard, but not known to be decidable.

362. P. Lincoln. *Computational Aspects of Linear Logic.* MIT Press, 1996.
Investigates the complexity of the decision problems for fragments and variants of linear logic, as well as developing some issues in linear proof theory. A computational interpretation of linear logic is used to study the computational

complexity of various decision problems, yielding proofs of tight complexity bounds for many of them, including the undecidability of propositional linear logic (by reduction from counter machines). Includes the results of [361].

363. M. Löb. Decidability of the monadic predicate calculus with unary function symbols. *Journal of Symbolic Logic*, 32:pg.563, 1967.
This abstract defines a procedure for the successive elimination of monadic function symbols preserving validity. Thereby monadic predicate logic with monadic functions is reduced to Löwenheim's class [365] and thus proved decidable. See also [223] and Sect. 6.2 of this book.
364. L. Lovász and P. Gács. Some remarks on generalized spectra. *Zeitschr. f. math. Logik u. Grundlagen d. Math.*, 23:547–554, 1977.
365. L. Löwenheim. Über Möglichkeiten im Relativkalkül. *Math. Annalen*, 76:447–470, 1915. English translation in [515].
Besides a proof of Löwenheim's Theorem this pioneering paper contains the following results about the Entscheidungsproblem, in terms of validity, with equality but without functions: A proof that satisfiability of formulae with only monadic predicates is decidable and finitely controllable; recognition of the existence of infinity axioms (formulae which are valid in every finite domain but not valid in any infinite domain); a reduction of the Entscheidungsproblem to formulae with only binary predicates. For a sharpening to $[all, (0,3)]$ and $[all, (0,0,1)]$ see [254], to $[all, (0,1)]$ see [295]. See [290] (for the elimination of the equality symbol in the reduction formulae) and [294] (for a simplified reduction). Other proofs and extensions of Löwenheim's decidability result appear in [31, 35, 82, 145, 253, 254, 426, 476, 526, 527, 528, 281, 363, 488]. Löwenheim's proof principle for the decidability of monadic predicate logic is used again by [387] to solve the decision problem for various first-order algebraic theories. See also [434]. See [311] for the reducibility of arbitrary formulae of order n to monadic formulae of order $n+1$ for $n \geq 2$.
366. J. Lynch. Complexity classes and theories of finite models. *Math. Systems Theory*, 15:127–144, 1982.
The paper shows how to encode m moves of a Turing machine into a structure of size $O(m)$ by using the graph of the addition. It follows that each language which is accepted in nondetermistic time n^d (for some natural number d) is spectrum of a sentence whose predicates are at most d-ary (see [152, 153, 367]).It also follows that the k-th level in the Meyer-Stockmeyer hierachy (see [491]) is characterized by sentences with $k-1$ second-order quantifier alternations; similarly the time complexity classes correspondig to the Ritchie functions [438] are characterized by higher-order sentences whose order reflects the Ritchie hierachie level.
367. J. Lynch. On sets of relations definable by addition. *Journal of Symbolic Logic*, 47, 659–679 1982.
368. M. Machtey and P. Young. *An Introduction to General Theory of Algorithms.* North Holland, New York, 1978.
369. S. MacLane and D. Siefkes. *The Collected Works of J. Richard Büchi.* Springer-Verlag, 1990.
370. A. Malcev. The effective inseparability of the set of valid sentences from the set of finitely refutable sentences in several elementary theories. *Dokl. Akad. Nauk SSSR*, 139:802–805, 1961.
English translation in Russian Math. Surveys 2, 1961, 1005-1008.
371. A. Malcev. *Algorithmen und rekursive Funktionen.* Vieweg, Braunschweig, 1974. German translation.

372. S. Marchenkov. Undecidability of the $\forall\exists$-positive theory of a free semigroup. *Siberian Math. Journal*, 23:196–198, 1982.
In Russian.
373. J. Marcinkowski. Undecidability of the Horn clause finite implication problem.
Improves the result of [377] to hold also for finite implication.
374. J. Marcinkowski. Undecidability of uniform boundedness for single rule Datalog programs.
Proves the undecidability of the problem to decide whether, given a Bernays-Schönfinkel Horn formula with only one disjunction, recursion can be eliminated from this formula. The proof uses Conway's functions [90] to show that an appropriately defined class of machines, the so-called Achilles-turtle machines, is computation universal. See also [376] where formulae are considered which have only one occurrence of a recursive predicate, a ternary one, in the body.
375. J. Marcinkowski. A Horn clause that implies an undecidable set of Horn clauses. In *CSL'93. Proceedings of the 1993 Annual Conference of the European Association of Computer Science Logic.*, Lecture Notes in Computer Science No. 832, pages 223–237. Springer, 1994.
Sharpens the result of [377] to a fixed ternary disjunction.
376. J. Marcinkowski. The 3 Frenchmen method proves undecidability of the uniform boundedness for single recursive rule ternary Datalog programs. In *STACS 96, 13th Annual Symposium on Theoretical Aspects of Computer Science*, Lecture Notes in Computer Science No. 1046, pages 427–438. Springer, 1996.
See the comment to [374].
377. J. Marcinkowski and L. Pacholski. Undecidability of the Horn-clause implication problem. In *Annual IEEE Symposium on Foundations of Computer Science*, pages 354–362, 1992.
It is shown that the class of Horn formulae with only one, a ternary, disjunction and with arbitrary many variable-free units is undecidable. For a sharpening of the result, see [375],[373].
378. S. Maslov. An inverse method for establishing deducibilities in the classical predicate calculus. *Soviet Mathematics – Dokl.*, 5:1420–1424, 1964.
Proposes a general method for deciding solvable cases of first-order decision problems. It is proved that the so called Maslov class (the prenex class $\exists^*\forall^*\exists^*$ with Krom matrix) is decidable.
379. S. Maslov. Application of the inverse method for establishing deducibility to the theory of decidable fragments in the classical predicate calculus. *Soviet Mathematics – Dokl.*, 7:1653–1657, 1966.
Also in: Dokl. Akad. Nauk SSSR 171 (1966) 1282–1285.
380. S. Maslov. An inverse method for establishing deducibility of nonprenex formulas of the predicate calculus. *Soviet Mathematics – Dokl.*, 8:16–19, 1967.
Also in: Dokl. Akad. Nauk SSSR 172 (1967) 22–25.
381. S. Maslov. The inverse method for establishing deducibility for logical calculi. *Trudy Math. Inst. Steklov*, XCVIII:25–95, 1968. Engl. Transl. AMS 1971.
This is one of the first papers designing a particular logic calculus to be used as a decision procedure: every proof search terminates for the target class of formulae; (although some features of the method already appear in [378]). The calculus is of Gentzen type and is based on rules closely related to hyperresolution. It had been applied to decide a new large class K of first-order formulae without equality including e.g. the Gödel-Kalmár-Schütte class. The exact definition of K is involved. See also [544, 160, 163].

382. S. Maslov. The conection between tactics of inverse method and the resolution method. *Zapiski Nauchnykh Seminarov LOMI*, 16, 1969. English translation in: J.Siekmann, G.Wrightson, eds. Automated Reasoning, Springer-Verlag, Berlin, 1983, v2., 264–272.

383. S. Maslov. Proof search strategies for methods of the resolution type. *Machine Intelligence*, 6:77–90, 1971.

384. S. Maslov. The inverse method and tactics for establishing deducibility for a calculus with functional symbols. *Trudy Math. Inst. Steklov*, 121:14–56, 1972.
Also in: Proc. Steklov Inst. Math. 121, 11–60.

385. S. Maslov, G. Mints, and V. Orevkov. Unsolvability in the intuitionistic predicate calculus of certain classes of formulas containing only monadic predicate variables. *Soviet Mathematics - Dokl.*, 6:918–920, 1965. also in: Dokl. Akad. Nauk SSSR 163 (1967).
The principal aim of the paper is a proof, by constructive means, of the unsolvability of the deducibility problem for $[all, (\omega)]$ in the intuitionistic, minimal and positive predicate calculi.

386. S. Maslov and V. Orevkov. Decidable classes reducing to one-quantifier class. *Proc. Steklov Inst. Steklov Math*, 121:61–72, 1972. Russian original in Trudy Math. Inst. Steklov.
See the comments to [226].

387. J. McKinsey. The decision problem for some classes of sentences without quantifiers. *Journal of Symbolic Logic*, 8:61–76, 1943.
Provides decision procedures for various classes of formulae in the theory of lattices, based upon a general reduction procedure and using Löwenheim's principle for the decision of the monadic predicate logic in [365].

388. R. McNaughton. Testing and generating infinite sequences by a finite automaton. *Information and Control*, 9:521–530, 1966.

389. G. McNulty. Undecidable properties of finite sets of equations. *Journal of Symbolic Logic*, 41:589–604, 1976.

390. A. Meyer. The inherent computational complexity of theories of ordered sets. In *Proc. 1974 Int. Cong. of Mathematicians*, pages 477–482. Vancouver, 1974.

391. A. Meyer. Weak monadic second order theory of successor is not elementary-recursive. In *Proc. of Boston University Logic Colloquium, Boston 1972*, Lecture Notes in Mathematics No. 453, pages 132–154. Springer-Verlag, 1975.

392. A. Meyer and J. Stockmeyer. The equivalence problem for regular expressions with squaring requires exponential space. In *Conf. Rec. of 13th Annual Symp. on Switching and Automata Theory*, pages 125–129, 1972.
The paper introcuces the idea that Gödel's proof method for showing undecidability can be used to show that sets which are difficult to decide can be transformed into interesting mathematical and logical decision problems. For a development of this idea see [168, 490, 491, 492, 493, 538].

393. M. Minsky. Recursive unsolvability of Post's problem of 'tag' and other topics in the theory of Turing machines. *Annals of Math.*, 74:437–455, 1961.
See comment to [469].

394. G. Mints. Solvability of the problem of deducibility in LJ for a class of formulas not containing negative occurrences of quantifiers. *Trudy Math. Inst. Steklov*, 98:134–146, 1968.
English translation AMS 1971.

395. G. Mints. Decidability of the class ∃ by Maslov's inverse method. In Chang and Lee, editors, *Symbolic Logic and Mechanical Theorem Proving*, pages 304–314. Nauka, Moscow, 1983.
Supplement A, section 4,5 to the Russian translation of the book. The paper

contains a proof for Orevkov's result [407] that the derivability problem for the class $[\forall^*\exists\forall^*, all, all]$ is decidable. For a proof of the finite model property for the corresponding class $[\exists^*\forall\exists^*, all, all]$ see [226].

396. M. Mortimer. On languages with two variables. *Zeitschr. f. math. Logik u. Grundlagen d. Math.*, 21:135–140, 1975.
Proves the finite model property for relational first-order sentences with only two variables. See [208] and Sect. 8.1 for a simpler proof with a better bound on the model size.

397. A. Mostowski. On a system of axioms which has no recursively enumerable model. *Fundamenta Mathematicae*, 40:56–61, 1953.
See Chap. 2.1.1 of this book.

398. A. Mostowski. A formula with no recursively enumerable model. *Fundamenta Mathematicae*, 42:125–140, 1955.
See Chap. 2.1.1 of this book.

399. A. Mostowski. Concerning a problem of H. Scholz. *Zeitschr. f. math. Logik u. Grundlagen d. Math.*, 2:210–214, 1956.
Complementing the results of [26] this paper shows that for each function f in a class of functions defined from simple initial functions by composition and a bounded primitive recursion, the set of solutions of $f(n) = 0$ yields a spectrum, i.e. $\{n+1 : f(n) = 0\}$ is a first-order spectrum. (Since the proof works also for simultaneous bounded primitive recursion this shows that every Grzegorczyk-E_2-set is a first-order spectrum.) Examples are the set of primes, of powers of a given integer, of all numbers $n!$, and the set $\{n : n^2+1 \text{ is prime}\}$. It is posed as an open question whether also the set of Fermat-primes and its complement are spectra. See [542] for the (positive) answer.

400. A. Muchnik. Games on infinite trees and automata with dead-ends: A new proof for the decidability of the monadic second-order theory of two successors. *Semiotics and Information*, 24:17–40, 1984. (In Russian). An English translation appeared in the Bulletin of the EATCS, 48 (1992), 219–267.

401. D. Muller and P. Schupp. Simulating alternating tree automata by nondeterministic automata: New results and new proofs of the theorems of Rabin, McNaughton and Safra. *Theor. Computer Science*, 141:69–107, 1995.

402. A. Nakamura. On the undecidability of monadic modal predicate logic. *Zeitschr. f. math. Logik u. Grundlagen d. Math.*, 16(3):257–260, 1970. See [328].

403. S. Nash-Williams. On well-quasi-ordering finite trees. *Proc. Cambridge Phil. Soc.*, 59:833–835, 1963.

404. F. Oglesby. An examination of a decision procedure. *Memoirs of the AMS*, 44:148 pages, 1962.
See abstract in Bulletin AMS 67, 1961, 300–304. Detailed investigation of the proof procedure defined in [488]. Stanley's result that his proof procedure is a decision procedure for monadic predicate logic is extended to (a subclass of) the Bernays-Schönfinkel class. Stanley's proof procedure is also shown to be a decision procedure for the class of all closed second degree distributive normal forms defined by Hintikka in [269]. A peculiar extension of Stanley's procedure is defined which is shown to yield a decision "procedure" for the class of what is called there second degree sentences.

405. D. Oppen. An upper bound on the complexity of Presburger arithmetic. *Journal of Computer and System Sciences*, 16:323–332, 1978.

406. V. Orevkov. Certain reduction classes and solvable classes of sequents for the constructive predicate calculus. *Soviet Mathematics – Dokl.*, 6:888–891, 1965. Russian original in Dokl. Akad. Nauk SSSR163 (1965) 30–32.

407. V. Orevkov. One solvable class of formulas of the classical predicate calculus with functional symbols. In *II. Symp. on Cybernetics, Tbilisi*, page 176, 1965. In Russian.
It is announced that the derivability problem for the class $[\forall^*\exists\forall^*, all, all]$ is decidable. See [395] for a proof and [226] for a proof of the finite model property for the corresponding class $[\exists^*\forall\exists^*, all, all]$.
408. V. Orevkov. Unsolvability of the class of formulas of the type $\neg\neg\forall\exists$ in the intuitionistic predicate calculus. *Soviet Mathematics – Dokl.*, 6:977–980, 1965. Russian original in Dokl. Akad. Nauk SSSR163 (1965) 581–583.
The deducibility problem for prenex formulas is solvable in the intuitionistic predicate calculus; see e.g. [130]. The deducibility problem for a class of negated prenex formulas is solvable in the intuitionistic predicate calculus if and only it is solvable in the classical predicate calculus; see e.g. [82]. The author proves that the collection of sentences of the form $\neg\neg\forall x_1 \ldots \forall x_n \exists y_1 \ldots \exists y_m M$, where the quantifier-free part M contains only one predicate symbol, is a reduction class for the intuitionistic predicate calculus as well as for the minimal predicate calculus.
409. V. Orevkov. Two undecidable classes of formulas in classical predicate calculus. In *Seminars in Math*, volume 8, pages 98–102. V.A. Steklov Inst. Leningrad, 1969. Russian original in Zapiski Nauchnykh Seminarov LOMI, 8 (1968).
The two theorems, as stated by the author, are these. (1) The class $[\exists\exists, all, all]$ remains a reduction class for validity if the quantifier part is restricted to be a conjunction of two simple disjunctions (disjunctions of atomic formulas and their negations). (2) The same applies to the class $[\forall^*\exists\forall\exists^4, all]$.
410. V. Orevkov. Unsolvability in the modal predicate calculus of the class of formulae containing only one monadic predicate variable. In *Seminars in Math*, volume 4. V.A. Steklov Inst. Leningrad, 1969. Russian original in Zapiski Nauchnykh Seminarov LOMI, 4 (1967), 168-173.
Proves unsolvability of the modal logic S5 with just one monadic predicate. See also [484].
411. V. Orevkov. On biconjunctive reduction classes. *Journal of Soviet Math.*, 1:106–109, 1973. Russian original in Zapiski Nauchnykh Seminarov LOMI, 20 (1971).
Using the same logical description of 2-register machines as the one found independently by Aanderaa and Börger in [2, 39] it is shown that the following classes are conservative reduction classes for some fixed k: $[\forall\exists\exists\forall, (\omega, k)] \cap$ KROM, $[\exists\forall\exists\forall, (\omega, k)] \cap$ KROM, $[\exists^*\forall\exists\forall, (0, k)] \cap$ KROM, $[\forall\exists^*\forall, (0, k)] \cap$ KROM. The first three classes appeared independently in [2, 39], the fourth class strengthens Krom's reduction class $[\forall\exists^*\forall, (0, \omega)] \cap$KROM in [334]. Note that a footnote in this paper states that "the principle results of this article were reported at the Leningrad Seminar on Math. Logic on May 21, 1970".
412. V. Orevkov. Decidable classes of pseudo-prenex formulae. In *Seminars in Math*, volume 60, pages 109–170. V.A. Steklov Inst. Leningrad, 1976. English translation in Journal of Soviet Math..
413. M. Otto. Bounded Variable Logics and Counting — a Study in Finite Models. Habilitationsschrift RWTH Aachen (1995). A revised version will appear in the Springer Lecture Notes in Logic series.
414. L. Pacholski and W. Szwast. The 0-1 law fails for the class of existential second-order Gödel sentences with equality. In *30th Annual IEEE Symposium on Foundations of Computer Science*, pages 280–285, 1989. The title states the result (see also [415] where the authors show that the 0-1 law fails even for the minimal Gödel class). Together with the results of Kolaitis and Vardi

[313, 314, 315] these papers complete the classification of 0-1 laws for prefix fragments of Σ_1^1 (with equality).

415. L. Pacholski and W. Szwast. On the 0-1 law for the existential second-order minimal Gödel sentences with equality. In *6th Annual IEEE Symposium on Logic in Computer Science*, pages 280–285, 1991.
416. C. Papadimitriou. *Computational Complexity*. Addison-Wesley, 1994.
417. M. Paterson and M. Wegman. Linear unification. *Journal of Computer and System Sciences*, 16:158–167, 1978.
418. J. Pepis. Beiträge zur Reduktionstheorie des logischen Entscheidungsproblems. *Acta Scientiarum Mathematicarum Universitatis Szegediensis*, 8:7–41, 1936.
Predecessor paper to [420], see the review by P. Bernays in Journal of Symbolic Logic 2 (1937) 84–85.
419. J. Pepis. Ein Verfahren der mathematischen Logik. *Journal of Symbolic Logic*, 3:61–76, 1938.
See comment to [420] and P. Bernays' review in Journal of Symbolic Logic 3 (1938) 161–162.
420. J. Pepis. Untersuchungen über das Entscheidungsproblem der mathematischen Logik. *Fundamenta mathematicae*, 30:257–348, 1938.
Improved the results of [418]. Contains various prefix-vocabulary reduction classes which have been improved shortly later by other authors, in particular the class $[\forall^4\exists^*, (0,0,1)]$ and the class $[\forall^2\exists\forall^*, (1,0,1)]$ improved to the vocabulary $(0,2)$ or $(0,0,1)$ in [419] and to $[\forall^2\exists\forall^*, (0,1)]$ by [303]. Pepis' reduction formulae are of the form $\forall x\forall y\exists u Sxyu \wedge \forall y_1 \ldots \forall y_n \alpha$ where S is a ternary predicate and α contains besides S only one other predicate symbol, a monadic one. Therefore they establish also the reduction class $[\forall^*\exists]$. See also [494] and P. Bernays' review in Journal of Symbolic Logic 3 (1938), 160–161.
421. D. Plaisted. Complete problems in the first-order predicate calculus. *Journal of Computer and System Sciences*, 29:8–35, 1984.
Continues the investigations of the computational complexity of decidable classes started in [352]. By encoding of appropriate Turing machine computations the following formulae classes F are shown to be complete for the following complexity classes: Bernays-Schönfinkel with ternary disjunctions for NEXPTIME, restricted to Horn formulae for DEXPTIME, for PSPACE if restricted to Horn formulae with unique matches or to Krom formulae or to Krom and Horn formulae or to Krom formulae with unique matches. For propositional formulae the corresponding complexity classes are shown to be NP, P, NLOGSPACE, LOGSPACE respectively. By encoding of resolution and natural deduction proof systems similar completeness results are established for provability by natural deduction or (hyper-) resolution refutation of various restricted depths from propositional or Bernays-Schönfinkel or arbitrary formulae. For an extensive treatment of such results see the book [309].
422. E. Post. A variant of a recursively unsolvable problem. *Bulletin of the American Math. Soc.*, 52:264–268, 1946.
The unsolvability of the Post Correspondence Problem is shown: Given a finite set of pairs $(v_i, w_i)(1 \leq i \leq n)$ of words v_i and w_i, determine whether there is a sequence $1 \leq i_1, \ldots, i_k \leq n$ such that $v_{i_1} \cdots v_{i_k} \equiv w_{i_1} \cdots w_{i_k}$. For a simplified proof see [57, pp 73–74].
423. M. Presburger. Über die Vollständigkeit eines gewissen Systems der Arithmetik ganzer Zahlen, in welchem die Addition als einzige Operation hervortritt. In *Comptes Rendus, I. Congrès des Math. des Pays Slaves*, pages 192–201, 395, Warsaw, 1929.
Contains one of the historically first decidability results on first-order theories,

a decision procedure for the first-order theory of the integers with addition, today called Presburger arithmetic. Fischer and Rabin [168] initiated the study of the complexity of Presburger arithmetic (and other theories) and established a double exponential lower bound. See also [34, 176, 199, 201, 405, 436, 452] for complexity results on Presburger arithmetic. For a comprehensive treatment of the complexity of decidable logical theories see [89, 166]. See also [543] where it is shown that each subtheory of Presburger arithmetic has an NTIME($2^{2^{cn}}$) lower bound (in analogy to the hereditary version of Gödel's theorem stating that all subtheories of Peano arithmetic are undecidable). See the line of research reported in [543].

424. P. Pudlak. Bounds for the Hodes–Specker Theorem. In E. Börger, G. Hasenjäger, and D. Rödding, editors, *Logic and Machines: Desision Problems and Complexity*, Lecture Notes in Computer Science No. 171, pages 421–445. Springer, 1984.
Shows the asymptotic bound $n \log\log n$ for the Hodes-Specker theorem which implies that certain Boolean functions have nonlinear formula size complexity.

425. H. Putnam. Decidability and essential undecidability. *Journal of Symbolic Logic*, 22:39–54, 1957.

426. W. Quine. On the logic of quantification. *Journal of Symbolic Logic*, 10:1–12, 1945.
Contains a new decidability proof for monadic predicate logic.

427. W. Quine. *Methods of Logic.* Holt, Reinhard & Winston, 1963.

428. M. Rabin. On recursively enumerable and arithmetic models of set theory. *Journal of Symbolic Logic*, 23:408–416, 1958.

429. M. Rabin. A simple method for undecidability proofs and some applications. In *1964 Int. Congr. Logic, Methodology and Philosophy of Science*, pages 58–68. North-Holland, Amsterdam, 1964.
Presents a new principle for interpreting an undecidable theory T in a theory T' to prove the undecidability of T'. The method is illustrated by giving new proofs for the undecidability of some theories and by showing the theory of finite commutative rings to be undecidable.

430. M. Rabin. Decidability of second-order theories and automata on infinite trees. *Trans. Amer. Math. Soc.*, 141:1–35, 1969.
The monadic second-order theory of two successors (S2S) is proven to be decidable. This is one of the most important results on the decidability of formalized mathematical theories; it has been used to prove the decidability of various other formalized mathematical theories; see an example in Sect.7.2. The decidability proof is conceptually transparent with the notable exception of the so-called Complementation Lemma; see Sect. 7.4 in this connection. Additional results related to S2S are found in [431].

431. M. Rabin. Automata on infinite objects and church's problem. *Regional conference series in math. Amer. Math. Soc.*, 13, 1972.

432. M. Rabin. Decidable theories. In J. Barwise, editor, *Handbook of Mathematical Logic*, pages 595–629. North Holland, Amsterdam, 1977.
A good, but narrow-focused survey paper.

433. C. Rackoff. The emptiness and complementation problems for automata on infinite trees. Technical Report Project MAC Technical Report TR-106, MIT, 1973.
The author gives a simple algorithm (with a simple correctness proof) for the emptiness problem for the tree automata. In addition, the author gives a simplified complementation construction whose correctness proof "however, is difficult and very similar in complexity to Rabin's proof in [430].

434. C. Rackoff. Complexity of some logical theories. Technical Report Project MAC Technical Report TR-144, MIT, 1975.
435. F. Ramsey. On a problem of formal logic. *Proc. of the London Math. Soc. 2nd series*, 30:264–286, 1930.
Reprinted in: F.R. Ramsey, The Foundations of Mathematics. Routledge and Kegan Paul, London 1931, pp.82–111.
This paper contains a proof for the decidability of the class of universal prenex sentences with equality and without functions, i.e. $[\forall^*, all]_=$ and shows that the spectrum of every formula in $\exists^*\forall^*$ is finite or co-finite. The proof is remarkable because Ramsey develops for this proof a combinatorial theorem which marks the beginning of a still very active subfield of combinatorial analysis (see [212]).
436. C. Reddy and D. Loveland. Presburger arithmetic with bounded quantifier alternation. In *10th Annual ACM Symposium on Theory of Computing*, pages 320–325, 1978.
437. J. Reynolds. Transformational systems and the algebraic structure of atomic formulas. *Machine Intelligence*, 5:135–151, 1969.
It is shown that there is no decision procedure for transformational systems, i.e. finite sets of clauses which contain only units or clauses with exactly one positive and one negative literal. This implies the unsolvability of the decision problem for Krom formulae, proved independently and with different proofs also by [409] and [334].
438. R. Ritchie. Classes of predictably computable functions. *Trans. Amer. Math. Soc.*, 106:139–173, 1963.
439. J. Robinson. A machine-oriented logic based on the resolution principle. *Journal of the ACM*, 12(1):23–41, 1965.
This seminal paper introduces the resolution method. For a use of resolution as decision procedure for classes of formulae see [286].
440. R. Robinson. Undecidability and non-periodicity for tilings of the plane. *Invent. Math.*, 12:177–209, 1971.
Simplifies Berger's proof (see [33]) of the undecidability of the unconstrained domino problem introduced by [531]. See Appendix A of this book.
441. D. Rödding. Reduktionstypen der Prädikatenlogik. Lecture notes, Universität Münster i.W., 1969/70.
See review in: Zentralblatt für Mathematik vol. 207, No.02034.
Lectures on reduction classes containing in particular a simplified new proof for the class $[\forall\exists\forall, (1, \omega)]$ avoiding domino problems and using a geometrical interpretation of 2-register machines in the Gaussian quadrant. See Sect. 3.1 of this book.
442. D. Rödding and E. Börger. The undecidability of $\wedge \vee \wedge(0,4)$ - formulae with binary disjunctions. *Journal of Symbolic Logic*, 39:412–413, 1974.
Improves Krom's reduction class $[\forall\exists^*\forall, (0, \omega)] \cap$ KROM $\cap$ HORN [334] to $[\forall\exists^*\forall, (0, 4)] \cap$ KROM $\cap$ HORN, see Theorem 5.1.10.
443. D. Rödding and H. Schwichtenberg. Bemerkungen zum Spektralproblem. *Zeitschr. f. math. Logik u. Grundlagen d. Math.*, 18:1–12, 1972.
444. H. Rogers. Certain logical reductions and decision problems. *Annals of Math.*, 64:264–284, 1956.
Continues the investigation of [84, 93] into the decision problem for classes of first-order formulae with restrictions on the interpretation of the occurring predicates. Reductions of the decision problem (in terms of validity) are given to the case of one binary predicate which represents a partial ordering, or a single disjoint (or transitive or transitive-reflexive or symmetric or symmetric-reflexive) relation, or a lattice ordering; similarly for the case of

two predicates representing equivalence relations. For similar reductions with a bounded prefix length of the reduction formulae see [229, 499].
445. H. Rogers. *Theory of recursive functions and effective computability.* McGrw Hill, 1967.
446. B. Russel and A. Whitehead. *Principia Mathematica I–III.* Cambridge University Press, 1910–1913.
447. S. Safra and M. Vardi. On ω-automata and temporal logic. In *Proceedings of 21st Annual ACM Symposium on Theory of Computing*, pages 127–137., 1989.
448. M. Savelsbergh and P. van Emde-Boas. Bounded tiling, an alternative to satisfiability? In G. Wechsung, editor, *Proc. of 2nd Frege Conference*, pages 354–363. Akademie Verlag, Schwerin, 1984. Mathematische Forschung Nr.20
The paper proposes to replace the propositional logic satisfiability problem as a foundation of the NP-completeness theory by bounded domino problems. See [514, 355].
449. W.J. Savitch. Relationships between nondeterministic and deterministic tape complexities. *Journal of Computer and System Sciences*, 4:177–192, 1970.
450. B. Scarpellini. Complete second-order spectra. *Zeitschr. f. math. Logik u. Grundlagen d. Math.*, 30:509–524, 1984.
It is shown that in second-order predicate logic each standard prefix class has a spectrum of deterministic exponential time complexity which is accepted by a prefix-length-alternating multitape machine in linear time and to which every other second-order spectrum within this prefix class is reducible by a polynomial injective function. The proof uses alternating Turing machines [73] and the relation between second-order spectra and the polynomial-time hierachy discussed in [491]. See [452] for a strengthening of the result.
451. B. Scarpellini. Complexity of subcases of Presburger arithmetic. *Trans. Amer. Math. Soc.*, 284:203–218, 1984. See comments to [199, 201].
452. B. Scarpellini. Second order spectra. In E. Börger, G. Hasenjäger, and D. Rödding, editors, *Logic and Machines: Desision Problems and Complexity*, Lecture Notes in Computer Science No. 171, pages 380–389. Springer, 1984.
Proves that the theorem on complete second-order spectra proved in [450] can be proved with formulae containing only monadic and binary predicates. The proof uses an appropriate description of alternating Turing machines by second-order formulae.
453. B. Scarpellini. Lower bound results on lengths of second-order formulas. *Annals of Pure and Applied Logic*, 29:29–58, 1985.
454. M. Schmidt-Schauss. Implication of clauses is undecidable. *Theor. Computer Science*, 59:287–296, 1988.
The result of Lewis and Goldfarb [354] that the class of equality and function free Krom and Horn formulae with only two disjunctions (and only two units) is a reduction class is considered for Horn formulae with functions; for the sharpening to such formulae with only one disjunction see the comment to [129]. It is shown that the class is decidable if only one disjunction and arbitrary many, but variable-free, units are allowed. It is shown that the class is undecidable with only one disjunction, but with four disjuncts, and with only variable-free units; for the sharpening of this result to such formulae with only one, a ternary, disjunction see [377]. The decidability result is sharpened in [125] to units in which each variable occurs at most once.
455. H. Scholz. Ein ungelöstes Problem in der symbolischen Logik. *Journal of Symbolic Logic*, 17:160, 1952.
Motivated by Trakhtenbrot's theorem on spectral representation of all recur-

sively enumerable predicates (see [509]) the notion of a spectrum and the spectrum problem are formulated. See Sect. 2.2.2 of this book.

456. K. Schütte. Über die Erfüllbarkeit einer Klasse von logischen Formeln. *Math. Annalen*, 110(2):161–194, 1934. See the comment to [457].

457. K. Schütte. Untersuchungen zum Entscheidungsproblem der mathematischen Logik. *Math. Annalen*, 109(4):572–603, 1934.
In this paper and [456], Schütte proves the decidability and the finite model property of the Gödel-Kalmár-Schütte class $[\exists^*\forall^2\exists^*, all]$. See also [186, 187, 239, 293] and Sect. 6.2.3 of this book.

458. T. Schwartz. A simple treatment of Church's Theorem on the decision problem. *Logique et Analyse*, 46:153–156, 1969.

459. D. Scott. A decision method for validity of sentences in two variables. *Journal of Symbolic Logic*, 27:477, 1962.
Reduces $Sat(L_2)$, i.e. the satisfiability problem for relational first-order sentences with only two variables, to satisfiability of relational $\forall^2\exists^*$-sentences. By the results of Gödel, Kalmár and Schütte this implies that satisfiability of L_2-sentences without equality is decidable. Scott claims the latter result also for L_2-sentences with equality, since at that time, the error in Gödel's statement concerning the $\forall^2\exists^*$-class with equality had not been detected yet. The first proof that the full class $Sat(L_2)$ is decidable was given by Mortimer [396]. See Sect. 8.1.

460. J. Sebelik and P. Stepanek. Horn clause programs for recursive functions. In K. Clark and S.-A. Tärnlund, editors, *Logic Programming*, pages 325–341. Academic Press, 1982.
See the comment to Example 2.1.21.

461. J. Seiferas, M. Fischer, and A. Meyer. Separating nondeterministic time complexity classes. *Journal of the ACM*, 25:146–167, 1978.

462. A. Selman. Sets of formulas valid in finite structures. *Trans. Amer. Math. Soc.*, 177:491–504, 1973.
The paper investigates Mostowski's problem about the relation between the degree complexity of sets A of natural numbers and the degree complexity of the set of first-order formulae which are valid in all finite structures of cardinality in A. The paper makes use of the original proof of Trakhtenbrot's Theorem [509]. See [250] where the results of this investigation are related to classical results of degree theory.

463. A. Semenov. Decidability of monadic theories. In *Lecture Notes in Computer Science No.* , pages 162–175. Springer, 1984.

464. V. Sevjakov. Formulas of the restricted predicate calculus which distinguish certain classes of models with simply computable predicates. *Soviet Mathematics – Dokl.*, 14:743–745, 1973.
Three satifiable formulae with certain lower bounds for their model complexity are provided, namely: (a) a satisfiable formula whose satisfying predicates cannot be computet by n-tape finite automata, (b) a formula with a model whose predicates can be computed by an n-tape finite machine but without (generalized) Presburger models, (c) a formula with a (generalized) Presburger model but without ordering model.

465. C. Shannon. A universal Turing machine with two internal states. In C. Shannon and J. McCarthy, editors, *Automata Studies*, pages 157–165. Princeton Univ. Press, Princeton, New Jersey, 1956.

466. E. Shapiro. Alternation and the computational complexity of logic programs. In *Proc. of 1st Int. Logic Programming Conf.*, pages 154–163, 1982.

467. S. Shelah. The monadic theory of order. *Ann. of Math.*, 102:379–419, 1975.
A seminal paper with new powerful methods. Some outstanding problems of

the monadic theory of order has been proved, notably the monadic theory of (the order of) the real line has been proved undecidable. Also, the known results have been presented in a uniform way. See [231].

468. S. Shelah. Decidability of a portion of the predicate calculus. *Israel Journal of Math.*, 28:32–44, 1977.
Shelah proves (confirming a conjecture of Gurevich) that the satisfiability and finite satisfiability problems for $[\exists^*\forall\exists^*, (all), (1)]_=$ are decidable; see Chap. 7 of this book.

469. J. Shepherdson and H. Sturgis. Computability of recursive functions. *Journal of the ACM*, 10:217–255, 1963.
In this paper (and in [393]) register machines are defined. Among other things it is shown that each partial recursive function is computable by a register machine program using only two registers; therefore the halting problem of these machines is undecidable.

470. O. Shmueli. A single recursive predicate is sufficient for pure Datalog. *Information and Computation*, 117:91–97, 1995.

471. J. Shoenfield. *Mathematical Logic.* Addison-Wesley, 1967.

472. J. Shoenfield. *Recursion Theory.* Lecture Notes in Logic No. 1, Springer, 1993.

473. D. Siefkes. Decidable theories I:Büchi's monadic second order successor arithmetic. In *Lecture Notes in Mathematics No. 120.* Springer, 1970.

474. W. Sieg. Mechanical procedures and mathematical experience. In A. George, editor, *Mathematics and Mind*, pages 71–117. Oxford Univesity Press, 1994.

475. A. Sistla, M. Vardi, and P. Wolper. The complementation problem for Büchi automata with applications to temporal logic. *Theor. Computer Science*, 49:217–237, 1987.

476. T. Skolem. Untersuchungen über die Axiome des Klassenkalküls und über Produktations- und Summationsprobleme, welche gewisse Klassen von Aussagen betreffen. *Skrifter utgit av Vidensk. i Kristiania I, Math.-nat. Klasse*, 3:pp. 37, 1919.
Extends the decidability of the monadic predicate logic from the first to the second order. Includes an improved form of the decidability of Löwenheim's class [365]. See also [31].

477. T. Skolem. Logisch-kombinatorische Untersuchungen über die Erfüllbarkeit oder Beweisbarkeit mathematischer Sätze nebst einem Theorem über dichte Mengen. *Norsk Vid-Akad. Oslo Mat.-Natur Kl. Skr.*, 4, 1920.
In this paper the Skolem normal form for satisfiability is introduced (and is used to obtain a simpler proof and a generalization of Löwenheim's Theorem). The paper adopts the point of view of satisfiability rather than of validity (as does [365]). As a corollary of Skolem's normal form the class $[\forall^*\exists^*, all]$ is established as a conservative reduction class. Gödel [187] has improved this prefix class to $[\forall^3\exists^*, all]$. Skolem's reduction class has been the starting point for numerous other stronger reduction classes, culminating in Surányi's reduction class $[\forall^3\exists, (\omega, 1)]$ (see [498]).

478. T. Skolem. Über die mathematische Logik. *Norsk Mat. Tidsskrift*, 106:125–142, 1928.
This paper contains among other results another proof for the decidability of the Ackermann class (see [16]). It establishes the decidability of the class of $[\exists^*\forall^*\exists^*, all]$-formulae in which every atomic formula that contains any universally quantified variable contains either all of them or at least one of the following existentially quantified variables. It also establishes the decidability of the class of prefix sentences with a prefix terminating with universal

quantifiers such that every atomic subformula contains at least one variable bound by these universal quantifiers. See [480] for an extension.

479. T. Skolem. Ein Satz über Zählausdrücke. *ACTA Scientiarum Mathematicarum*, 7:193–199, 1935.
The author simplifies Gödel's proof [187] that $[\forall^3\exists^*]$ is a reduction class. In fact, he proves that $[\forall\exists \wedge \forall^2\exists^* \wedge \forall^3, (\omega, \omega)]$ is a reduction class; moreover, the $\forall^2\exists^*$ sentence can be written as a conjunction of $\forall^2\exists$ sentences. See [480, 481] about decidable subclasses of the reduction class in question.

480. T. Skolem. Über die Erfüllbarkeit gewisser Zählausdrücke. *Norsk Vid-Akad. Oslo Mat.-Natur Kl. Skr.*, 6, 1935.
This paper gives a more complete treatment of the decision procedure outlined in [478] and shows the finite model property for the class of $[\forall\exists^* \wedge \forall^3, (0, \omega)]$-formulae in which the second conjunct is the axiom stating that all the binary relations occurring in the first conjunct are symmetric and transitive. In [481] Skolem shows the decidability of an extension of the class of $[\forall\exists^* \wedge \forall^3, (0, \omega)]$-formulae where more general equivalences are allowed in the second conjunct and where the first conjunct is allowed to contain also other not necessarily binary predicate symbols.

481. T. Skolem. Ein Satz über die Erfüllbarkeit von einigen Zählausdrücken der form $(x)(ey_1, \ldots, y_n)k_1(x, y_1, \ldots, y_n) \wedge (x_1, x_2, x_3)k_2(x_1, x_2, x_3)$. *Norsk Vid-Akad. Oslo Mat.-Natur Kl. Skr.*, 8:3–10, 1936.
See comment to [480].

482. T. Skolem. Einige Reduktionen des Entscheidungsproblems. *Norsk Vid-Akad. Oslo Mat.-Natur Kl. Skr.*, 6:3–17, 1936.
In reaction to Church's undecidability proof for Hilbert's *Entscheidungsproblem* a reduction method is proposed to be used for a systematic investigation of the classification problem. In particular it is shown that some reduction classes due to Kalmar, Gödel, Ackermann can be strenghthened by the restriction to a single (binary) predicate symbol. See [296] for a strngthening of the reduction.

483. A. Slomson. The monadic fragment of predicate calculus with the Chang quantifier and equality. In M. Löb, editor, *Proc. Summer School in Logic*, Lecture Notes in Mathematics No. 70, pages 279–301. Springer, 1968.

484. A. Slomson. An undecidable two-sorted predicate calculus. *Journal of Symbolic Logic*, 34:21–23, 1969.
Proves the undecidability of two-sorted first-order logic with a single dyadic predicate over $A \times B$ where A and B are the universes over the first and the second sort, respectively. As a consequence, the modal logic S5 with just one monadic predicate is undecidable; this strengthens a result of Kripke [328]. See also [410].

485. R. Smullyan. *Theory of formal systems*. Number 147 in Annals of Math. Studies. Princeton Univesity Pree, 1961.

486. E. Specker and V. Strassen. *Komplexität von Entscheidungsproblemen*. Lecture Notes in Computer Science No. 43. Springer, 1976.
A very valuable and carefully written introduction into the field of computational complexity of logical and combinatorial decision problems. It emphasizes the analogies between classical concepts and methods in logic and those employed in computational complexity theory and contains many elegant new proofs and extensions of known results. The contributions in this book cover polynomial reductions and NP-complete problems, the spectrum problem, the relation between polynomial transformations and equivalence proofs for weak forms of the axiom of choice, the double exponentional complexity of Presburger arithmetic, network complexity for Boolean functions and the length

of formulae to represent Boolean functions (including the Hodes-Specker theorem for which the asymptotic bound $n \log \log n$ is shown in [424]).

487. G. Stalmark. A note on the computational complexity of the pure classical implicational calculus. *Information Processing Letters*, 6:277–278, 1989.

488. R. Stanley. An extended procedure in quantificational logic. *Journal of Symbolic Logic*, 18:97–104, 1953.
Develops a proof procedure for predicate logic which shows validity by reductio ad absurdum. For monadic predicate logic this procedure is shown to be a decision procedure. See the extensions in [404], see also the review by W. Ackermann in Journal of Symbolic Logic 21 (1956) pg. 197.

489. R. Statman. Intuitionistic propositional logic is polynomial-space complete. *Theor. Computer Science*, 9:67–72, 1979.

490. L. Stockmeyer. *The complexity of decision problems in automata theory and logic.* PhD thesis, MIT, Project MAC, Cambridge/Mass., 1974. Report TR-133
See comment to [392].

491. L. Stockmeyer. The polynomial-time hierarchy. *Theor. Computer Science*, 3:1–22, 1977. The paper introduces and studies basic properties of the subrecursive analogue of the Kleene-Mostowski arithmetical hierachy in which deterministic (nondeterministic) polynomial time plays the role of recursive (recursively enumerable) time. The Cook-Lewin Theorm on the NP-completeness of the satisfiability problem for propositional logic is extended to the Σ-levels of this hierachy. The inequivalence problem for integer expresions — regular expresions defining nonnegative integers — is shown to be complete in Σ_2^p with respect to logspace-reducibility. The set of sentences which are valid in the first-order theory of equality is shown to be PSPACE-complete with respect to logspace-reducibility. A connection between polynomial-time hierachy and second-order spectra is discussed. See also [538].

492. L. Stockmeyer. Classifying the computational complexity problem. *Journal of Symbolic Logic*, 52:1–43, 1987.
A very valuable survey paper.

493. L. Stockmeyer and A. Meyer. Word problems requiring exponential time: preliminary report. In *Proc. 5th. ACM Symp. on Theory of Computing*, pages 1–9, 1973.
Contains in particular the theorem that the decision problem of quantified propositional logic is PSPACE-complete (see Chap. 2 of this book). This result is generalized in [321] to the elementary theory of any finite collection of finite Boolean algebras. For a proof using domino games see [78].

494. J. Surányi. A logikai függvńykalkulus eldöntésproblémájának redukciójŕól. *Mathematikai és Fizikai Lapok*, 1943.
Contains the improvement of a reduction class of Pepis (see [420]) to $[\forall^*\exists, (0,1)]$. Contains also the reduction class of all formulae of form $\forall x \forall y \forall z \alpha \wedge \forall u \forall v \exists w \beta$ with quantifier-free α, β containing only binary predicates and the reduction class $[\forall\exists\forall\exists, (0,\omega)]$. The results are also published in [496].

495. J. Surányi. Reduction of the decision problem to formulas containing a bounded number of quantifiers only. In *Proc. of the the Xth International Congress of Philosophy*, volume I, pages 759–762, Amsterdam, 1948.
See [497].

496. J. Surányi. Contributions to the reduction theory of the decision problem, second paper: Three universal, one existential quantifier. *Acta Mathematica Academiae Scientiarunm Hunagricae*, 1:261–270, 1950.
See [494].

497. J. Surányi. Contributions to the reduction theory of the decision problem. Fifth paper: Ackermann prefix with three universal quantifiers. *Acta Mathematica Academiae Scientiarunm Hunagricae*, 2:325–335, 1951.
Contains the following reduction classes: $[\exists\forall\exists\forall\forall, (\omega, 7)]$, $[\forall\exists^2\forall^2, (\omega, 7)]$.

498. J. Surányi. *Reduktionstheorie des Entscheidungsproblems im Prädikatenkalkül der ersten Stufe.* Verlag der Ungarischen Akademie der Wissenschaften, Budapest, 1959.
A comprehensive treatment of the reduction classes known in 1959, complementing Ackermann's book [18] on solvable cases. Many proofs from the literature are simplified and numerous known results are strengthened. The outstanding single new result is the proof that $[\forall^3 \wedge \forall^2\exists, (\omega, 1)]$ is a reduction class. For finite prefixes the classes $[\exists\forall\exists\forall^2, (\omega, 4)]$, and $[\exists\forall\exists^3\forall, (\omega, 4)]_=$ are shown to be reduction classes. The same holds when in the prefix the first two quantifiers are interchanged: $\forall\exists^2\forall^2$, $\forall\exists^4\forall$. For infinite prefixes the classes $[\Pi, (0, 1)]$ are shown to be reduction classes for the following Π: $\exists^*\forall^3\exists$, $\exists^*\forall^2\exists\forall$, $\forall^2\exists^*\forall$, $\forall^3\exists^*$, $\forall^*\exists$, $\forall^2\exists\forall^*$. For a not furthermore specified $k \leq 50$, $\exists^*\forall\exists^k\forall^2$, $\exists\forall\exists\forall^*$ and similar when the first two quantifiers are interchanged. With equality the classes $[\exists^*\forall\exists^k\forall, (0, 1)]_=$ and $[\forall\exists^*\forall, (0, 1)]_=$ are shown to be reduction classes. This complex and incomplete classification has been simplified and completed only through [64, 219, 287]. See Chap. 3 of this book.

499. J. Surányi. Reduction of the decision problem of the first order predicate calculus to reflexive and symmetrical binary predicates. *Periodica Mathematica Hungarica*, 1:97–106, 1971.
Shows that the Suranyi reduction class $[\forall^3 \wedge \forall^\exists, (\omega, 1)]$ can be strengthened by restricting the interpretation of the binary predicate to a reflexive relation; similarly for an irreflexive or two reflexive and transitive relation etc. For related results see [229].

500. M. Taitslin. Some further examples of undecidable theories. *Algebra i Logika*, 6 and 7:105–111 and 94–97, 1967 and 1968.
English translation in Algebra and Logic7, 1968, pages 127-129.

501. T. Tammet. Using resolution for deciding solvable classes and building finite models. In *Baltic Computer Science*, Lecture Notes in Computer Science No. 502, pages 33 – 64. Springer, 1991.
Some classes of clause sets are proved to be decidable via special ordering resolution refinements. For a class containing the Ackermann and the monadic class, a backtracking free algorithm is presented for the extraction of descriptions of finite models via ground term equations from inference-stable clause sets.

502. T. Tammet. *Resolution Methods for Decision Problems and Finite-Model Building.* PhD thesis, Chalmers University of Technology, Göteborg, 1992.
Contains a general and unifying treatment of the completeness of ordering refinements. A resolution decision procedure for the class $E+$ (corresponding to an extension of the Ackermann class) is developed (see also [163]). Two finite-model building methods for classes AM and AMS (corresponding to extensions of the union of the Ackermann- and the monadic class) are defined. These methods work on inference-stable sets of clauses, use paramodulation and are backtracking-free. The two last chapters illustrate the applicability of resolution decision procedures to ordinary theorem proving (by clever selections of refinements) and to decide the consistency problem of KL-ONE type languages.

503. S.-A. Tärnlund. Horn clause computability. *BIT*, 17:215–226, 1977.
See the comment to Example 2.1.21 to the Aanderaa-Börger Theorem in Chap. 2 of this book.

504. A. Tarski. *A decision method for elementary algebra and geometry.* University of California Press, Berkley and Los Angeles, 1951.

505. A Tarski. Equational logic. In K. Schütte, editor, *Contributions to mathematical logic*, pages 275–288. North-Holland, 1968.

506. A. Tarski, A. Mostowski, and R. Robinson. *Undecidable Theories.* North-Holland, Amsterdam, 1953.

507. W. Thomas. Automata on infinite objects. In J. van Leeuwen, editor, *Handbook of Theoretical Computer Science, vol. B: Formal Models and Semantics*, pages 133–191. Elsevier, 1990.

508. A. Thue. Probleme über Veränderungen von Zeichenreihen nach gegebenen Regeln, 1914. in: Skr.Vidensk. Selsk.I,10,1914, pp.34.

509. B. Trakhtenbrot. The impossibility of an algorithm for the decision problem for finite models. *Dokl. Akad. Nauk SSSR*, 70:596–572, 1950. English translation in: AMS Transl. Ser. 2, vol.23 (1963),1-6
The paper contains a proof for what became known as Trakhtenbrot's Theorem, namely that the class of finitely satisfiable formulae of predicate logic with equality and without functions is undecidable (see also [92]). Simpler proofs with smaller classes of reduction formulae α_f have been found in [64] and [48], see Sect. 2.1.1, of this book. See also [300] where the decision problem for validity is reduced to the decision problem for finite satisfiability. See also [520] where the result is shown using besides monadic predicates only one binary predicate (see [521]). Actually Trakhtenbrot's proof shows more, namely that precisely the graphs of partial recursive functions f have a "spectral representation", i.e. admit a first-order formula α_f containing besides binary predicates distinguished monadic predicates $P_1 \ldots, P_{n+1}$, equality, but no function such that for all $m_1, \ldots, m_{n+1}$ the following is true:
$f(m_1, \ldots, m_n) = m_{n+1}$ if and only if there is a finite model satisfying α_f in which the cardinality of the interpretation of P_i is m_i for $1 \leq i \leq n+1$. This result has been strenghtened in [110] to prenex sentences $\alpha_f \in [\forall\exists^*\forall, (\omega, 1)]_=$ in which the equality occurs only once. Trakhtenbrot's spectral representation of graphs of partial recursive functions has motivated the definition of the notion of spectrum and the formulation of the spectrum problem [455] which has played a major role in complexity theory and finite model theory. See Sect. 2.1.2 and 2.2.2 of this book.

510. B. Trakhtenbrot. On recursive separability. *Dokl. Akad. Nauk SSSR*, 88:953–955, 1953. in Russian. A German translation is available at TIB Universität Hannover, Germany.
Trakhtenbrot's undecidability result from [509] is improved to the theorem that there is no recursive set that separates the set of all valid sentences from the set of all negations of finitely satisfiable formulae. See Sect. 2.1.1 of this book.

511. B. Trakhtenbrot. Finite automata and the logic of monadic predicates. *Dokl. Akad. Nauk SSSR*, 140:326–329, 1961. in Russian.
Proves that a language is regular if and only if it is definable in monadic second-order logic. The same result was obtained by Büchi [62].

512. B. Trakhtenbrot and Y. Barzdin. *Finite Automata.* North-Holland, Amsterdam, 1973.

513. A. Turing. On computable numbers, with an application to the 'Entscheidungsproblem'. *Proc. of the London Math. Soc.*2nd *series*, 42:230–265, 1937. Correction ibid Vol.43 pp 544–546
In the "application to the Entscheidungsproblem" Turing shows the unsolvability of the latter, independently from [80]. Turing's seminal proof introduces a reduction method which not only is more direct than Church's ap-

proach but also later turned out to be crucial for the solution of sophisticated classification problems for logical decision problems. Turing formalized Turing machines by logical formulae and reduces an unsolvable class of particular word problems for Turing machines to the validity problem for the reduction formulae. Turing observes that his reduction shows the undecidability of he class of sentences of form $\exists\alpha \wedge \beta$ with $\beta \in [\exists\forall^6, (0, \omega)]$. Turing's idea to describe machine computations by logical formulae was taken up only much later by [64] who simplified if dramatically by using canonical interpretations over Herbrand domains to provide the possibly complicated descriptions of the data structures of the machine for free. See Sect. 2.1.1 of this book.

514. P. van Emde-Boas. Dominoes are forever. In *1st GTI-Workshop*, pages 75–95. University of Paderborn, 1983.

515. J. van Heijenoort. *From Frege to Gödel: a source book in mathematical logic, 1879–1931.* Harvard Univ. Press, Cambridge, Mass., 1967.

516. S. van Westrhenen. A probabilistic machine for the estimation of provability in the first order predicate calculus. *Zeitschr. f. math. Logik u. Grundlagen d. Math.*, 15:291–297, 1969.
Uses Herbrand's Theorem to establish the existence of a machine for checking the provability of logical formulae using probabilities. It is discussed how such machine can be used as decision procedures for first-order logic.

517. M. Vardi. The complexity of relational query languages. In *Procedings of 14th Annual ACM Symposium on Theory of Computing*, pages 137–146, 1982.
Proves that least fixed-point logic captures polynomial time on ordered structures. The same result was also proved by Immerman [276]. See Sect. 2.2.3. For more results on descriptive complexity and finite model theory, see [141].

518. M. Vardi and P. Wolper. Reasoning about infinite computation paths. *Information and Computation*, 115:1–37, 1994.

519. R. Vaught. Applications of the Löwenheim-Skolem-Tarski theorem to problems of completeness and decidability. *Indagationes Mathematicae*, 16:467–472, 1954.

520. R. Vaught. Sentences true in all constructive models. *Journal of Symbolic Logic*, 25:39–58, 1960.
Three inseparability results are shown:
(a) the set of sentences which are valid in recursively enumerable structures and the set of sentences which are valid in primitive recursive (finite) structures cannot be separated by an arithmetical (recursively enumerable) set;
(b) Trakhtenbrot's Theorem. Also a new proof is obtained for the theorem of [326] and [397] stating that there exists a satisfiable sentence having no recursively enumerable model (see Sect. 2.1.1 of this book). In all cases it is shown that besides monadic predicates only one binary predicate is needed. The proof uses ideas from [509], [510] and [398].

521. R. Vaught. On a theorem of Cobham concerning undecidable theories. In *Proc. 1960 Int. Congr. Logic, Phil. and Methodology of Sci.*, Stanford, CA, 1962. Stanford Univ. Press.
Trakhtenbrot's inseparability theorem (see [510]) and the result from [520] are used to prove Cobham's theorem that every theory which is consistent with a weak fragment of number theory (see [506]) is undecidable.

522. M. Veanes. Uniform representation of recursively enumerable sets with simultaneous rigid *e*-unification. UPMAIL Technical Report 126, Uppsala University, Computing Science Department, 1996.
The author gives a description of recursively enumerable sets in terms of solutions to simultaneous rigid E-unification. This description proves the following result: the $\exists\exists$-fragment of intuitionistic logic with equality in the signature

with one constant and one function symbol of arity ≥ 2 is undecidable. This result improves the undecidability result of [103].

523. K. Venkataraman. Decidability of the purely existential fragment of the theory of term algebras. *Journal of the ACM*, 34:492–510, 1987.
524. M. Venturini-Zilli. Complexity of the unification algorithm for first-order expressions. *Calcolo*, 12:361–371, 1975.
The first published polynomial-time unification algorithm.
525. J. von Neumann. Zur Hilbertschen Beweistheorie. *Math. Zeitschrift*, 26:1–46, 1927.
526. G. von Wright. On the idea of logical truth I. *Societas scientiarum fennica commentationes physico-matematicae*, 14, 1948.
See comment to [365].
527. G. von Wright. On the idea of logical truth II. *Societas scientiarum fennica commentationes physico-matematicae*, 15, 1950.
See comment to [365].
528. G. von Wright. On double quantification. *Societas scientiarum fennica commentationes physico-matematicae*, 16:1–14, 1952.
See comment to [365].
529. A. Voronkov. Proof search in intuitionistic logic based on constraint satisfaction. In *Theorem Proving with Analytic Tableaux and Related Methods. 5th International Workshop, TABLEAUX '96, Terrasini, Italy*, Lecture Notes in Artificial Intelligence No. 1071, pages 312–329, 1996.
A description of provability in first-order intuitionistic logic is given in terms of derivation skeletons and constraints. As a consequence, it is proved that the prefix fragment of intuitionistic logic with function symbols is PSPACE-complete.
530. K. Wagner and G. Wechsung. *Computational Complexity.* Deutscher Verlag der Wissenschaften, 1986.
531. H. Wang. Proving theorems by pattern recognition II. *Bell System Technical Journal*, 40:1–41, 1961.
Domino problems are introduced, motivated by the work of Büchi [64]. The origin constrained domino problem is shown to be undecidable and the unconstrained domino problem is formulated. The latter has been proved to be undecidable in [33]. For reductions of domino problems to logical decision problems see [201, 206, 227, 288, 532]; for reductions to various combinatorial decision problems see [351, 355, 246, 514] where also bounded versions of domino problems appear. See [245] for further applications of domino problems to establish lower bounds for logical decision problems, in particular for dynamic and temporal logics.
532. H. Wang. Dominoes and the $\forall\exists\forall$-case of the decision problem. In *Proc. Symp. on Mathematical Theory of Automata*, pages 23–55. Brooklyn Polytechnic Institute, New York, 1962.
533. M. Wirsing. *Das Entscheidungsproblem der Prädikatenlogik 1. Stufe mit Identität und Funktionszeichen in Herbrandformeln.* PhD thesis, Universität München, 1976.
Proves the results published in [534, 535] and the decidability of the class $[all, all, (1)]_= \cap$ HERBRAND; the decidability is shown by a reduction to formulae of monadic logic with only one, a monadic, function variable to which Rabin's algorithm in [430] is applied.
534. M. Wirsing. Das Entscheidungsproblem der Klasse von Formeln, die höchstens zwei Primformeln enthalten. *Manuscripta Mathematica*, 22:13–25, 1977.
It is shown that in first order logic with functions and equality, the class

of Herbrand formulae with only two subformulae, one equality and one inequality, is a conservative reduction class even when restricted to formulae in $[\forall^6, (0), (0,1)]_=$. The proof is by encoding of 2-register machines. It is also observed that for formulae in prenex *disjunctive* normal form, the class of formulae with only two subformulae is decidable. See Exercise5.2.1.

535. M. Wirsing. Kleine unentscheidbare Klassen der Prädikatenlogik mit Identität und Funktionszeichen. *Archiv math. Logik u. Grundlagenforschung*, 19:97–109, 1978.
It is shown that in first order logic with functions and equality, the class of Herbrand formulae with only three subformulae, only one of which is an inequality, is a conservative reduction class when restricted to formulae in $[\forall^3, (0), (0,1)]_=$. The same is shown for formulae in $[\forall, (0), (0,1)]_=$ or $[\forall, (0), (2)]_=$ with four atomic subformulae. The proofs are by encoding of 2-register machines.

536. M. Wirsing. A proof by Turing machines of the undecidability of the class of first-order formulas with only one quantifier. Manuscript, Techn. Universität München, 1978.
Shows by encoding of Turing machines that the class $[\forall, (0), (\omega)]_= \cap$ HERBRAND is a conservative reduction class when restricted to formulae which contain only equations and one inequality. Wirsing's proof is given for the proof of Corollary 4.1.3.

537. M. Wirsing. "Small" universal Post systems. *Zeitschr. f. math. Logik u. Grundlagen d. Math.*, 25:559–564, 1979.

538. C. Wrathall. Complete sets and the polynomial-time hierarchy. *Theor. Computer Science*, 3:23–33, 1977.
It is shown that Stockmeyer's definition of the polynomial-time hierachy which uses a restricted Turing reducibility is equivalent to a definition by means of polynomially bounded quantifiers. This characterisation is used to give a simpler proof of the theorem of Stockmeyer-Meyer in [493] that each Σ-level of the polynomial-time hierachy contains a set of propositional formulae whose decision problem is complete for this level with respect to LOGSPACE-reductions.

539. J. Würtz. Unifying cycles. In *Proc. European Conference on Artificial Intelligence*, pages 60–64. John Wiley & Sons, New York, 1992.

540. A. Yakhnis and V. Yakhnis. Extension of Gurevich-Harrington's restricted memory determinacy theorem: a criterion for the winning player and an explicit class of winning strategies. *Annals of Pure and Applied Logic*, 48:277–297, 1990.
The authors strengthen Gurevich-Harrington's Forgetful Determinacy Theorem [236] in a substantial way, providing in particular explicit winning strategies for the players.

541. A. Yakhnis and V. Yakhnis. Gurevich-Harrington's games defined by finite automata. *Annals of Pure and Applied Logic*, 62:265–294, 1993.

542. M. Yasuhara. On a problem of Mostowski on finite spectra. *Zeitschr. f. math. Logik u. Grundlagen d. Math.*, 17:17–20, 1971.
Mostowski's question (see [399]) whether the set of FERMAT primes and its complement are spectra is answered positively. The same result is shown for MERSENNE primes. The resultsa are obtained as corollaries to the following theorem: let f be any function obtained from successor and $\max(x_1, \dots, x_k)$ by composition and primitive recursion restricted by max (i.e. with $f(\boldsymbol{x}, y+1) = \max(k(\boldsymbol{x}, y, f(\boldsymbol{x}, y)), y+1)$); then the range of f and its complement, provided this is nonempty, are both finite spectra.

543. P. Young. Gödel theorems, exponential difficulty and undecidability of arithmetic theories: an exposition. In A. Nerode and R. Shore, editors, *Recursion Theory. Proc. Symp. Pure Math.*, volume 42, pages 503–522, New York, 1985. AMS.
The paper shows each subtheory of Presburger arithmetic has an $\text{NTIME}(2^{2^{cn}})$ lower bound. The proof technique is an adaptation of classical techniques to show the hereditary version of Gödels theorem stating that each subtheory of Peano arithmetic is undecidable.

544. N. Zamov. Maslov's inverse method and decidable classes. *Annals of Pure and Applied Logic*, 42:165–194, 1989.
An account of Maslov's inverse method [378] is given as well as a complete and self-contained proof for the decidability of Maslov's class K [381]. The class K contains a number of the known decidable classes, such as e.g. the Löwenheim class and the Gödel-Kalmár-Schütte class.

545. S. Zeitman. *The composition method.* PhD thesis, Wayne State University, Detroit, Michigan, 1994.

546. S. Zeitman. Unforgettable forgetful determinacy. *Journal of Logic and Computation*, 4:273–283, 1994.
A good exposition of the Forgetful Determinacy Theorem [236, 540] generalized to graph games.

547. I. Zhegalkine. Sur l'Entscheidungsproblem. *Mat. SB. Akad. Nauk. SSSR*, 6:185–198, 1939.
In Russian (French summary).

548. W. Zielonka. Infinite games on finitely coloured graphs with applications to automata on infinite trees. pp. 48, University of Bordeaux I, 1995.

549. A. Zykov. The spectrum problem in the extended predicate calculus. *Dokl. Akad. Nauk SSSR*, 17:63–76, 1953. English translation in: AMS Transl. (2) 3, 1956, 1–14
Contains various reductions of the spectrum problem for second-order formulae.

Index

GPSR Compliance

The European Union's (EU) General Product Safety Regulation (GPSR) is a set of rules that requires consumer products to be safe and our obligations to ensure this.

If you have any concerns about our products, you can contact us on ProductSafety@springernature.com

In case Publisher is established outside the EU, the EU authorized representative is:

Springer Nature Customer Service Center GmbH
Europaplatz 3
69115 Heidelberg, Germany

Batch number: 10761599

Printed by Printforce, the Netherlands